Hamiltonian Dynamical Systems and Applications

NATO Science for Peace and Security Series

This Series presents the results of scientific meetings supported under the NATO Programme: Science for Peace and Security (SPS).

The NATO SPS Programme supports meetings in the following Key Priority areas: (1) Defence Against Terrorism; (2) Countering other Threats to Security and (3) NATO, Partner and Mediterranean Dialogue Country Priorities. The types of meeting supported are generally "Advanced Study Institutes" and "Advanced Research Workshops". The NATO SPS Series collects together the results of these meetings. The meetings are co-organized by scientists from NATO countries and scientists from NATO's "Partner" or "Mediterranean Dialogue" countries. The observations and recommendations made at the meetings, as well as the contents of the volumes in the Series, reflect those of participants and contributors only; they should not necessarily be regarded as reflecting NATO views or policy.

Advanced Study Institutes (ASI) are high-level tutorial courses intended to convey the latest developments in a subject to an advanced-level audience

Advanced Research Workshops (ARW) are expert meetings where an intense but informal exchange of views at the frontiers of a subject aims at identifying directions for future action

Following a transformation of the programme in 2006 the Series has been re-named and re-organised. Recent volumes on topics not related to security, which result from meetings supported under the programme earlier, may be found in the NATO Science Series.

The Series is published by IOS Press, Amsterdam, and Springer, Dordrecht, in conjunction with the NATO Public Diplomacy Division.

Sub-Series

A.	Chemistry and Biology	Springer
B.	Physics and Biophysics	Springer
C.	Environmental Security	Springer
D.	Information and Communication Security	IOS Press
E.	Human and Societal Dynamics	IOS Press

http://www.nato.int/science
http://www.springer.com
http://www.iospress.nl

Series B: Physics and Biophysics

Hamiltonian Dynamical Systems and Applications

edited by

Walter Craig
McMaster University, Hamilton, ON, Canada

 Springer

Published in cooperation with NATO Public Diplomacy Division

Proceedings of the NATO Advanced Study Institute on
Hamiltonian Dynamical Systems and Applications
Montreal, Canada
18–29 June 2007

Library of Congress Control Number: 2008920287

ISBN 978-1-4020-6963-5 (PB)
ISBN 978-1-4020-6962-8 (HB)
ISBN 978-1-4020-6964-2 (e-book)

Published by Springer,
P.O. Box 17, 3300 AA Dordrecht, The Netherlands.

www.springer.com

Printed on acid-free paper

Preface

This volume is a collection of lecture notes from the courses that were given during the 2007 Séminaire de Mathématiques Supérieure in Montréal (SMS), which was conceived and supported as a NATO Advanced Study Institute. The courses took place during the two-week period from June 18 to June 29, 2007, at the Centre de Recherches Mathématiques (CRM), and they were funded by a grant from NATO and from the ISM, which is the combined graduate mathematics program of the Montréal area. The organising committee for this event was D. Bambusi (Milan), W. Craig (McMaster), S. Kuksin (Edinburgh and Paris), and A. Neishtadt (Moscow). There were more than 80 participants, coming from around the world, and in particular there were a good number of students from France, from Italy, from Spain, from the United States and from Canada. The program of lectures occupied two complete weeks, with five or six one-hour lectures each day, so that in total 57 h of courses were presented.

The topic of the 2007 NATO-ASI was Hamiltonian dynamical systems and their applications, which concerns mathematical problems coming from physical and mechanical systems of evolution equations. Many aspects of the modern theory of the subject were covered; topics of the principal lectures included low dimensional problems as well as the theory of Hamiltonian systems in infinite dimensional phase space, and and their applications to problems in classical mechanics, continuum mechanics, and partial differential equations. Applications were also presented to several important areas of research, including to celestial mechanics, control theory, the partial differential equations of fluid dynamics, and the theory of adiabatic invariants.

It is a good thing to do to articulate the relevance of the subject matter of these SMS lectures to the physical sciences. Physical laws are for the most part expressed in terms of differential equations, and the most natural classes of these are in the form of conservation laws or of problems of the calculus of variations for an *action functional*. These problems can often be posed as Hamiltonian systems, whether dynamical systems on finite dimensional phase space as in classical mechanics, or partial differential equations (PDE) which are naturally of infinitely many degrees of freedom. For instance, the well known N-body problem of celestial mechanics is still of great relevance to modern mathematics and more broadly to science; indeed in applications the mission design of interplanetary exploration regularly uses the gravitational boost of close encounters to manoeuvre their spacecraft (first used in the Mariner–10 mission, 1974). This is also true on the level of theoretical results, which can be traced to the work of Laplace, Lagrange and Poincaré, but whose modern successes date to the celebrated theory of Kolmogorov, Arnold and Moser (KAM) (1954/1961/1963). Recent mathematical progress includes the discoveries of new choreographies of many body orbits (Chenciner & Montgomery, 2000), and the constructions of Poincaré's second species orbits (Bolotin & MacKay, 2001). Furthermore, the development of rigorous averaging methods (Nekhoroshev 1979) gives hope for realistic long time stability results (Neishtadt 1981, Treschev 1996, Pöschel 1999). Additionally, the last several years has seen major progress in the long outstanding problem of Arnold diffusion, with the advent of Mather's variational techniques (2003) related to a generalised Morse–Hedlund theory, including Cheng's subsequent work on variational methods, and the geometrical approach to the 'gap problem' due to de la Llave, Delshams & Seara (2006).

Over the last decade the field of Hamiltonian systems has taken on completely new directions in the extension of the analytical methods of Hamiltonian mechanics to partial differential equations. The results of Kuksin, Wayne, Pöschel, Craig, Bambusi and Bourgain have introduced a new paradigm to the study of partial differential equations of evolution, where research focuses on the fundamental structures invariant under the dynamics of the PDE in an appropriate phase space of functions. Two basic examples of this direction of enquiry include (i) the development of several approaches to a KAM theory, with very recent contributions by Yuan (2006) and Eliasson & Kuksin (2007), and (ii) Nekhoroshev stability results for systems with infinitely many degrees of freedom (Bambusi 1999). These considerations show an exciting and extremely promising connection between Hamiltonian dynamical systems and harmonic analysis techniques in PDE. A case in point is the relationship between upper bounds on the growth of higher Sobolev norms of solutions of nonlinear evolution equations, and the bounds on orbits given by Nekhoroshev theory; similarly there is a possibly surprising connection between lower bounds on such growth and the existence of solution of PDE which exhibit phenomena related to Arnold diffusion. This research area of evolution equations and Hamiltonian systems is one of the most active and exciting fields of PDE in the last several years.

The subjects in question involve by necessity some of the most technical aspects of analysis coming from a number of diverse fields, and before our event there has not been one venue nor one course of study in which advanced students or otherwise interested researchers can obtain an overview and sufficient background to enter the field. What we have done with the Montréal Advanced Studies Institute 2007 is to offer a series of lectures encompassing this wide spectrum of topics in PDE and dynamical systems. Most of the major developers in this field were speakers at this ASI, including the top international leaders in the subject. This has made it a unique opportunity for junior mathematicians to hear a focused set of lectures given by major researchers and contributors to the field. The organizers are grateful for the time and energy that the speakers devoted to the thoughtful preparation of their lectures, and to the subsequent written and complete versions that appear in this volume. And in addition the students at this ASI, who were for the most part advanced graduate students and postdoctoral fellows, included many very promising and active young mathematicians in the field, with their own well-developed research programs. The participants' enthusiasm for the ASI, their help in writing lecture notes for the courses, and their general cheerfulness and good attitude during the course of the two weeks of lectures, made the event an experience not to be forgotten.

Last but not least, the organizers of the SMS 2007 would like to acknowledge the generous and timely support of the Public Diplomacy Division of NATO, without which the two weeks of this Advanced Study Institute would not have taken place, the additional financial support of the Montréal Centre de Recherches Mathématiques (CRM), the ISM and the Université de Montréal, and for the dependable guidance and initiative of Sakina Benhima, our Directrice de Programme at the CRM in Montréal.

The series of lectures in this volume includes the following topics: Hamiltonian systems and optimal control (A. Agrachev, SISSA, Trieste), Birkhoff normal form for some semilinear PDEs (D. Bambusi, Universita degli Studi di Milano), Variational methods for Hamiltonian PDEs (M. Berti, Università degli Studi di Napoli), The N-body problem (A. Chenciner, Observatoire de Paris), Variational methods for the problem of Arnold diffusion (C.-Q. Cheng, Nanjing University), The transformation theory of Hamiltonian PDE and the problem of water waves (W. Craig, McMaster University), Geometric approaches to diffusion and instability (R. de la Llave, University of Texas at Austin), KAM for the nonlinear Schrödinger equation (H. Eliasson, Université de Paris 7), Groups and topology in Euler hydrodynamics and the KdV (B. Khesin, University of Toronto), Three theorems on perturbed KdV (S. Kuksin, Heriot-Watt University), Averaging methods and adiabatic invariants (A. I. Neishtadt, Space Research Institute, Russian Academy of Science), Periodic KdV equation in weighted Sobolev spaces (J. Pöschel, Universität Stuttgart), The forced pendulum as a model for dynamical behavior (P. Rabinowitz, University of Wisconsin), Normal forms of holomorphic dynamical systems (L. Stolovitch, Université Paul Sabatier), Some aspects of finite dimensional Hamiltonian systems (D. Treschev, Moscow State University), Infinite dimensional dynamical systems and the Navier–Stokes equations (C. E. Wayne, Boston University), and KAM theory with applications to nonlinear wave equations. (X. Yuan, Fudan University).

Hamilton and Montréal, Canada

Walter Craig
July 2007

Contents

Some aspects of finite-dimensional Hamiltonian dynamics 1
D.V. Treschev
1 Symplectic structure. Invariant form of the Hamiltonian equations 1
  1.1 Hamiltonian equations 1
  1.2 The Poisson bracket 3
  1.3 Liouville theorem on completely integrable systems 4
2 A pendulum with rapidly oscillating suspention point 5
3 Anti-integrable limit 8
  3.1 The standard map 8
  3.2 Anti-integrable limit 9
  3.3 Proof of the Aubry theorem 11
  3.4 Some remarks 12
4 Separatrix splitting 13
  4.1 Poincaré's observation 13
  4.2 The Poincaré integral 14
  4.3 Proof of Theorem 5 15
  4.4 Standard example 18
  References 19

Four lectures on the N-body problem 21
Alain Chenciner
1 The Poincaré–Birkhoff–Conley twist map of the annulus
  for the planar circular restricted three-body problem 21
  1.1 The Kepler problem as an oscillator 21
  1.2 The restricted problem in the lunar case 22
  1.3 Hill's solutions................................... 24
  1.4 The annulus twist map 26
2 The Arnold–Herman stability theorem for the spatial
  $(1+n)$-body problem 29
  2.1 The secular Hamiltonian........................... 30
  2.2 Herman's normal form theorem and how to use it 33

2.3   A stability theorem 35
2.4   Herman's degeneracy 36
3   Minimal action and Marchal's theorem 37
3.1   Central configurations and their homographic motions... 37
3.2   Variational characterizations of Lagrange's equilateral
        solutions 38
3.3   Marchal's theorem 40
3.4   Minimization under symmetry constraints 42
4   Global continuation via minimization 43
4.1   Bifurcations from the Lagrange equilateral relative
        equilibrium 44
4.2   From the equilateral triangle to the Eight 46
4.3   From the square to the Hip-Hop 48
4.4   The avatars of the regular n-gon relative equilibrium:
        eights, chains and generalized Hip-Hops 50
References 50

Averaging method and adiabatic invariants 53
Anatoly Neishtadt
1   Introduction 53
2   Adiabatic invariance in one-frequency systems 54
3   On adiabatic invariance in multi-frequency systems 63
References 65

**Transformation theory of Hamiltonian PDE and the problem
of water waves** 67
Walter Craig
1   Hamiltonian systems 67
2   Partial differential equations as Hamiltonian systems........... 68
3   The problem of water waves 71
4   The Dirichlet–Neumann operator 73
5   Perturbation theory 74
6   The calculus of transformations 75
References 82

Three theorems on perturbed KdV 85
Sergei B. Kuksin
1   KdV equation 85
1.1   Integrability of (KdV) 86
1.2   Normal forms for (KdV)........................... 87
2   KAM-theory 88
3   Averaging: Hamiltonian perturbations 89
4   Averaging: case of non-Hamiltonian perturbations 90
4.1   Deterministic perturbations 90
4.2   Random perturbations............................ 90
References 91

Groups and topology in the Euler hydrodynamics and KdV 93
Boris Khesin
    1   Euler equations and geodesics 93
           1.1   The Euler hydrodynamics equation 93
           1.2   Geodesics on Lie groups 94
           1.3   Geodesic description for various equations 95
    2   Topology of steady flows 95
           2.1   Arnold's classification of steady fluid flows 95
           2.2   Variational principles for steady flows 97
    3   Euler equations and integrable systems 98
           3.1   Hamiltonian reformulation of the Euler equations 98
           3.2   The Virasoro algebra and the KdV equation 99
           3.3   Equations–relatives and conservation laws 100
    References 101

Infinite dimensional dynamical systems and the Navier–Stokes equation . 103
C. Eugene Wayne
    1   First lecture: infinite dimensional dynamical systems 103
    2   Second lecture: invariant manifolds for partial differential
           equations on unbounded domains 111
    3   Third lecture: an introduction to the Navier–Stokes equations 122
    4   Fourth lecture: the long-time asymptotics of solutions
           of the two-dimensional Navier–Stokes equation 129
    5   Conclusions 139
    References 140

Hamiltonian systems and optimal control 143
Andrei Agrachev
    1   Introduction 143
    2   First lecture 144
           2.1   First order conditions 145
    3   Second lecture 146
           3.1   Second variation 147
    4   Third lecture 149
           4.1   Curves in the Lagrange Grassmannians 150
           4.2   Curvature-type invariants 151
    5   Fourth lecture 153
    References 156

**KAM theory with applications to Hamiltonian partial
differential equations** 157
Xiaoping Yuan
    1   Brief history and basic ideas of KAM theory 157
    2   Derivation of the linearized equations 163
    3   Solutions of the linearized equations 167
    4   Applications to partial differential equations
           in higher dimensions 175

    Appendix 176
    References 176

Four lectures on KAM for the non-linear Schrödinger equation 179
L.H. Eliasson and S.B. Kuksin
  1   The non-linear Schrödinger equation 179
       1.1   The non-linear Schrödinger equation 179
       1.2   An ∞-dimensional Hamiltonian system 180
       1.3   The topology 182
       1.4   Action-angle variables 183
       1.5   Statement of the result 184
       1.6   KAM-tori 185
       1.7   Consequences of Theorem 1 186
       1.8   References 188
       1.9   Notation 189
  2   The homological equation 190
       2.1   Normal form Hamiltonians 190
       2.2   The KAM-iteration 191
       2.3   The components of the homological equation 192
       2.4   Small divisors and the second Melnikov condition 194
  3   Normal form Hamiltonians 195
       3.1   Blocks 195
       3.2   Lipschitz domains 197
       3.3   Töplitz at $\infty (d = 2)$ 199
       3.4   Töplitz–Lipschitz matrices $(d = 2)$ 200
       3.5   Normal form Hamiltonians 201
  4   Estimates of small divisors 202
       4.1   A basic estimate 202
       4.2   The second Melnikov condition $(d = 2)$ 203
  5   Functions with the Töplitz–Lipshitz property $(d = 2)$ 208
       5.1   Töplitz structure of the Hessian 208
       5.2   Töplitz–Lipschitz matrices $\mathscr{L} \times \mathscr{L} \to gl(2, \mathbb{R})$ 209
       5.3   Functions with the Töplitz–Lipschitz property 211
       5.4   A short remark on the proof of Theorem 1.1 211
    References 212

A Birkhoff normal form theorem for some semilinear PDEs 213
D. Bambusi
  1   Introduction 213
  2   Birkhoff's theorem in finite dimensions 214
       2.1   Statement 214
       2.2   Proof 216
  3   The case of PDEs 221
       3.1   Hamiltonian formulation of the wave equation 221
       3.2   Extension of Birkhoff's theorem to PDEs:
                     heuristic ideas 223

4   A Birkhoff normal form theorem for semilinear PDEs 224
    4.1   Maps with localized coefficients and their properties 224
    4.2   Statement of the normal form theorem
         and its consequences 228
    4.3   Application to the nonlinear wave equation 229
5   Discussion 231
6   Proofs 231
    6.1   Proof of the properties of functions with localized
         coefficients 231
    6.2   Proof of the Birkhoff normal form Theorem 4.3
         and of its dynamical consequences 236
    6.3   Proof of Proposition 4.2 on the verification
         of the property of localization of coefficients 240
    6.4   Proof of Theorem 4.4 on the nonresonance condition 244
References 246

Normal form of holomorphic dynamical systems 249
Laurent Stolovitch
1   Definitions and examples 249
    1.1   Vector fields and differential equations 250
    1.2   Normal forms of vector fields 253
    1.3   Examples about linearization 256
    1.4   Examples about nonlinearizable vector fields 258
2   Holomorphic normalization 259
    2.1   Theorem of A.D. Brjuno 259
    2.2   Theorems of J. Vey 261
    2.3   Singular complete integrability–Main result 261
    2.4   How to recover Brjuno's and Vey's theorems
         from Theorem 2.3.6 268
    2.5   Sketch of the proof 269
3   Proof of main Theorem 2.3.6 273
    3.1   Bounds for the cohomological equations 273
    3.2   Iteration scheme 277
    3.3   Proof of the theorem 278
4   Miscellaneous results 280
    4.1   Normal forms again 280
    4.2   KAM theory 281
    4.3   Poisson structures 282
References 283

**Geometric approaches to the problem of instability in Hamiltonian
systems. An informal presentation** 285
Amadeu Delshams, Marian Gidea, Rafael de la Llave, and Tere M. Seara
1   Introduction 285
    1.1   Two types of geometric programs 287

2   Exposition of the Arnold example 289
        2.1   The obstruction property 291
        2.2   Some final remarks on the example in [Arn64] 293
3   Return to a normally hyperbolic manifold. The two dynamics
        approach 294
        3.1   The basics of the mechanism of return to a normally
                   hyperbolic invariant manifold 294
        3.2   The scattering map 296
        3.3   The scattering map and homoclinic intersections
                   of submanifolds 298
        3.4   Monodromy of the scattering map....................299
        3.5   Smoothness and smooth dependence on parameters 299
        3.6   Geometric properties of the scattering map 300
        3.7   Calculation of the scattering map 301
4   The large gap model......................................303
        4.1   Generation of intersections. Melnikov theory
                   for normally hyperbolic manifolds 304
        4.2   Computation of the scattering map 307
        4.3   The averaging method. Resonant averaging 308
        4.4   Repeated averaging 312
        4.5   Invariant objects generated by resonances: secondary
                   tori, lower dimensional tori 312
        4.6   Heteroclinic intersections between the invariant objects
                   generated by resonances 315
5   The method of correctly aligned windows 316
6   The large gap model: the method of correctly aligned windows ...318
7   The large gap model in higher dimensions.....................319
8   Instability caused by normally hyperbolic laminations...........320
        8.1   Models with two time scales: geodesic flows, billiards
                   with moving boundaries, Littlewood problems 321
References 324
Appendix 330

Variational methods for the problem of Arnold diffusion 337
Chong-Qing Cheng
1   Introduction to Mather theory 337
2   Existence of Homoclinic Orbits 341
3   Pseudo-connecting orbit set $\tilde{\mathscr{C}}_{\eta,\mu,\psi}$ 342
4   Existence of heteroclinic orbits 345
5   Construction of global connecting orbits 353
6   Application to a priori unstable systems.....................364
References 365

Contentsxv

The calculus of variations and the forced pendulum 367
Paul H. Rabinowitz
    1   Introduction 367
    2   Periodic solutions 369
    3   Heteroclinic solutions 374
    4   Multitransition solutions: the simplest case 380
    5   Multitransition solutions: general case 387
    6   The tip of the iceberg 389
    References 390

Variational methods for Hamiltonian PDEs 391
Massimiliano Berti
    1   Finite dimensions: resonant center theorems 391
          1.1   The variational Lyapunov–Schmidt reduction 393
    2   Infinite dimensions 396
    3   The variational Lyapunov–Schmidt reduction 397
          3.1   The bifurcation equation 399
    4   The small divisor problem 402
    5   The $(Q1)$-equation 406
    6   A variational principle on a Cantor set 408
    7   Forced vibrations 412
          7.1   The variational Lyapunov–Schmidt reduction 413
          7.2   The bifurcation equation 414
    References 419

Spectral gaps of potentials in weighted Sobolev spaces 421
Jürgen Pöschel
    1   Results 421
    2   Reduction 425
    3   Gap Estimates 426
    4   Coefficient Estimates 426
    5   Modified Weights 428
    References 429

**On the well-posedness of the periodic KdV equation
in high regularity classes** 431
Thomas Kappeler and Jürgen Pöschel
    1   Results 431
    2   Birkhoff Coordinates 435
    3   Regularity 438
    References 440

Some aspects of finite-dimensional Hamiltonian dynamics

D.V. Treschev[*]

Abstract These lectures touch upon two aspects of Hamiltonian mechanics. The first one (geometric) establishes fundamental role of symplectic geometry as the language of Hamiltonian mechanics. The second aspect (dynamical) exhibits the main problem in the domain, which is the interplay between regular and chaotic motion.

1 Symplectic structure. Invariant form of the Hamiltonian equations

1.1 Hamiltonian equations

Hamiltonian system[1] is an ODE-system which in certain coordinates $q=(q_1,\ldots,q_n)$, $p=(p_1,\ldots,p_n)$ can be presented in the form

$$\dot{q}=\frac{\partial H}{\partial p}, \quad \dot{p}=-\frac{\partial H}{\partial q}, \tag{1}$$

the function $H(q,p)$ is called the *Hamiltonian function*. Frequently, non-autonomous systems are considered, where $H = H(q,p,t)$.

This definition looks very non-geometrical. Although all calculations are anyway presented in coordinates (partially we will see this below), it would be good to present an equivalent invariant (coordinate independent) definition.

Recall that a symplectic structure on a smooth manifold M is a closed non-degenerate differential two-form ω. The pair (M,ω) is a *symplectic manifold*.

---

[*] Steklov Mathematical Institute
e-mail: treschev@mi.ras.ru.

[1] We will consider only the case of Hamiltonian ODE's.

W. Craig (ed.), Hamiltonian Dynamical Systems and Applications, 1–19.

Theorem 1 *(Darboux) In a neighborhood of any point of M there are local coordinates $(q, p) = (q_1, \ldots, q_n, p_1, \ldots, p_n)$, in which the symplectic structure has the form $\omega = dp \wedge dq$.*

Corollary 1 *Any symplectic manifold is even dimensional.*

Such coordinates (q, p) are called symplectic, canonical, or Darboux coordinates. Note that ω associates to any vector field v on M the differential 1-form f:

$$f(\cdot) = \omega(\cdot, v),$$

where on the empty place $\cdot$ an arbitrary vector field can be posed. Let J be the inverse operator. It exists because ω is non-degenerate, and the dimensions of the vector spaces $T_x M$ and $T_x^* M$ $(x \in M)$ coincide. Then

$$f(\cdot) = \omega(\cdot, Jf).$$

Let $H : M \to \mathbb{R}$ be a smooth function. It determines the 1-form dH.

Definition 1 *The vector field $v_H = JdH$ on M is called the Hamiltonian vector field with Hamiltonian H.*

Hence $dH(\cdot) = \omega(\cdot, v_H)$.

Problem 1 *Check that in canonical coordinates the Hamiltonian vector field takes the traditional form $v_H = (H_p, -H_q)$.*

Any map $T : M \to M$ preserving the symplectic structure is called *symplectic*. Symplectic maps can be regarded as discrete analogs of Hamiltonian systems.

Problem 2 *Let (q, p) be canonical local coordinates on M and let $T : M \to M$ be symplectic. Prove that the functions $(P, Q) = (q \circ T, p \circ T)$ are also canonical local coordinates on M.*

Problem 3 *Let (q, p) and (P, Q) be local coordinates on M such that for some smooth function $W = W(q, P)$*

$$p = \frac{\partial W}{\partial q}, \quad Q = \frac{\partial W}{\partial P}. \tag{2}$$

Suppose also that the coordinates (q, p) are canonical. Prove that (P, Q) are also canonical.

Hence in variables P, Q equations (1) remain the same:

$$\dot{Q} = \frac{\partial \mathscr{H}}{\partial P}, \quad \dot{P} = -\frac{\partial \mathscr{H}}{\partial Q},$$

where the Hamiltonian is the same: $\mathscr{H}(P, Q) = H(p, q)$.

The function W can depend on t. It is called a generating function of the canonical transformation $(q,p) \mapsto (P,Q)$. In the non-autonomous case the new Hamiltonian equals

$$\mathscr{H}(P,Q,t) = \partial W(q,P,t)/\partial t + H(p,q,t).$$

1.2 The Poisson bracket

Let (M,ω) be a symplectic manifold. For any two functions H,F on M we define the Poisson bracket

$$\{H,F\} := \partial_{v_H} F = dF(v_H).$$

Here ∂_{v_H} is the operator of differentiation w.r.t. the vector field v_H. The first equality is a definition, while the second one is just an identity.

We have the following simple properties of the Poisson bracket.

1. A smooth function F is a first integral of the Hamiltonian equations with Hamiltonian $H \iff \{H,F\} = 0$.
2. $\{H,F\} = \omega(v_H, v_F)$.
3. The operation $\{\cdot,\cdot\}$ is bilinear and skew-symmetric.
   According to 1 and 3 in any (autonomous) Hamiltonian system the Hamiltonian is a first integral.
4. In canonical coordinates $\{H,F\} = \sum_{j=1}^{n} \left(\dfrac{\partial H}{\partial p_j} \dfrac{\partial F}{\partial q_j} - \dfrac{\partial H}{\partial q_j} \dfrac{\partial F}{\partial p_j} \right)$.
   A direct calculation in canonical coordinates gives
5. The Leibnitz identity:

$$\{FG,H\} = F\{G,H\} + \{F,H\}G.$$

6. The Jacobi identity:

$$\{F,\{G,H\}\} + \{G,\{H,F\}\} + \{H,\{F,G\}\} = 0$$

for any three functions $F,G,H : M \to \mathbb{R}$.

This Poisson bracket is non-degenerate, i.e., for any $z \in M$ and any function F such that $dF \neq 0$ at z there exists G such that $\{F,G\} \neq 0$. In some physical problems degenerate Poisson brackets appear,[2] but we will not deal with these cases below.

For any two vector fields u,v on M let $[u,v]$ be their commutator:

$$\partial_{[u,v]} = \partial_u \partial_v - \partial_v \partial_u.$$

Theorem 2 *For any two functions F,G on M*

$$[v_F, v_G] = v_{\{F,G\}}.$$

---

[2] Such Poisson brackets are not generated by symplectic structures.

Proof. For an arbitrary function φ on M we have:

$$\partial_{v_{\{F,G\}}}\varphi = \{\{F,G\},\varphi\} = -\{\{G,\varphi\},F\} - \{\{\varphi,F\},G\}$$
$$= \{F,\{G,\varphi\}\} - \{G\{F,\varphi\}\} = (\partial_{v_F}\partial_{v_G} - \partial_{v_G}\partial_{v_F})\varphi. \qquad \square$$

Proposition 1 *(Poisson). Let F and G be first integrals of the Hamiltonian system (M,ω,H). Then $\{F,G\}$ is also a first integral.*

Indeed, if $\{H,F\} = \{H,G\} = 0$ then by the Jacobi identity $\{H,\{F,G\}\} = 0$. $\quad\square$

Unfortunately, this statement is not of much use in the problem of the search for new integrals of motion. Usually the Poisson bracket of two integrals is an already known integral or zero.

We say that two functions F,G are in involution or commute if $\{F,G\} = 0$.

1.3 Liouville theorem on completely integrable systems

Suppose that the system (M,ω,H) ($\dim M = 2m$) has m first integrals $F_1,\ldots,F_m$ in involution: $\{F_j,F_k\} = 0$. Consider the joint integral level

$$M_f = \{z \in M : F_j(z) = f_j = \text{CONST}, \ j = 1,\ldots,m\}. \tag{3}$$

Theorem 3 *(Liouville–Arnold) Suppose that on M_f the functions F_j are independent. Then*

1. *M_f is a smooth manifold, invariant with respect to the Hamiltonian system $\dot{z} = v_H$.*
2. *Each compact connected component of M_f is diffeomorphic to an m-dimensional torus [3] $\mathbb{T}^m$.*
3. *In some coordinates $(\varphi_1,\ldots,\varphi_m)$ mod 2π on $\mathbb{T}^m$ the Hamiltonian equations have the form $\dot{\varphi} = \lambda$, where $\lambda = \lambda(f) \in \mathbb{R}^m$ is a constant vector.*

Proof. Assertion (1) follows from the implicit function theorem. To check (2) and (3), we note that the vector fields $v_j = v_{F_j}$ are tangent to M_f. (Indeed, $\partial_{v_j}F_k = \{F_j,F_k\} = 0$.) Since the functions F_j are independent on M_f, the vector fields v_j are also independent on M_f. Moreover,

$$[v_j,v_k] = v_{\{F_j,F_k\}} = 0.$$

It remains to use the following geometric fact (see for example, [2]):

Lemma 2 *Any compact connected m-dimensional manifold on which there are m everywhere independent commuting vector fields is diffeomorphic to the torus $\mathbb{T}^m$. Moreover there are angular coordinates $(\varphi_1,\ldots,\varphi_m)$ mod 2π on it such that all the m vector fields become constant ($v_j = const \in \mathbb{R}^m$).* $\qquad\square$

---

[3] In the non-compact case M_f turns out to be $\mathbb{T}^k \times \mathbb{R}^{m-k}$, $0 \leqslant k < m$ (see [2]).

Problem 4 *Check that the tori $\mathbb{T}_f^m$ from Theorem 3 are Lagrangian, i.e., $\dim \mathbb{T}_f^m = m$ and restriction of the symplectic structure to $\mathbb{T}_f^m$ vanishes.*

Hamiltonian systems having a complete set (i.e., m) of almost everywhere independent first integrals in involution are said to be *completely*, or *Liouville integrable*.

In Liouville integrable systems there are convenient, so-called, action-angle coordinates (φ, I) (I are the actions and φ are the angles) such that

- $\omega = dI \wedge d\varphi$ (symplecticity),
- $H = H(I)$ (i.e., I are first integrals),
- $\varphi = \varphi \bmod 2\pi$ (i.e. φ are angular coordinates on the tori M_h).

2 A pendulum with rapidly oscillating suspention point

Mathematical pendulum is a (classical) mechanical system that consists of the rigid weightless rod AB with fixed end A. A point with mass m is attached to the end B. The motion is assumed to take place in a fixed vertical plane in the constant gravity force field. This system is well-known and Liouville integrable.

Consider a more complicated problem. Let the point A vertically periodically oscillate. Period and amplitude of the oscillations is assumed to be small (of order ε). We are interested in the action of the oscillations of the suspension point on the dynamics.

Consider in the plane of motion a fixed coordinate system such that the x-axis is horizontal, the y-axis is vertical, and A lies on the y-axis. We assume that in this coordinate system

$$A(t) = \left(0, a\varepsilon \cos \frac{\omega t}{\varepsilon}\right), \qquad \omega = \sqrt{\frac{g}{l}}.$$

Here g is the gravity acceleration, $l = |AB|$ is the length of the pendulum, and ε is small. The frequency ω is introduced so that ε is dimensionless. The dimension of a is length.

The system is non-autonomous and has one degree of freedom. It is convenient to take the angle φ between the pendulum and the vertical, directed downward, as a variable, that determines position of the system.

Problem 5 *Obtain the Lagrangian of the system.*

Hint. $L = T(\varphi, \dot{\varphi}, t) - V(\varphi, t)$, where T and V are kinetic and potential energy of the pendulum.

Answer.

$$L = \frac{m}{2}\left(l^2\dot{\varphi}^2 - 2al\omega\dot{\varphi}\sin\varphi\sin\frac{\omega t}{\varepsilon} + a^2\omega^2\sin^2\frac{\omega t}{\varepsilon}\right) - mg\left(a\varepsilon\cos\frac{\omega t}{\varepsilon} - l\cos\varphi\right).$$

It is convenient to remove from L all terms which depend only on time and to divide L by ml^2. Let $\hat{L}$ be the Lagrangian, obtained in this way:

$$\hat{L} = \frac{\dot{\varphi}^2}{2} - \frac{a\omega}{l}\dot{\varphi}\sin\varphi\sin\frac{\omega t}{\varepsilon} + \omega^2\cos\varphi.$$

Problem 6 *Prove that Lagrangian systems with Lagrangians L and $\hat{L}$ are the same.*

Problem 7 *Obtain the Hamiltonian of the system.*

Hint. H and $\hat{L}$ are related by the Legendre transform: $H(\varphi,p,t) = p\dot{q} - \hat{L}(\varphi,\dot{\varphi},t)$, where $\dot{\varphi}$ in the right-hand side should be expressed in terms of (φ,p,t) from the equation $p = \partial\hat{L}/\partial\dot{\varphi}$.

Answer.

$$p = \dot{\varphi} - \frac{a\omega}{l}\sin\varphi\sin\frac{\omega t}{\varepsilon},$$

$$H = \frac{p^2}{2} + p\frac{a\omega}{l}\sin\varphi\sin\frac{\omega t}{\varepsilon} + \frac{a^2\omega^2}{2l^2}\sin^2\varphi\sin^2\frac{\omega t}{\varepsilon} - \omega^2\cos\varphi.$$

We will construct a canonical change of variables which removes dependence of H on t in the main (zero) approximation in ε. We look for a change $(\varphi,p) \mapsto (\Phi,P)$ in the form

$$p = \frac{\partial W}{\partial\varphi}, \qquad \Phi = \frac{\partial W}{\partial P}, \qquad W = P\varphi + \varepsilon f\left(\varphi,P,\frac{\omega t}{\varepsilon}\right),$$

where f is 2π-periodic in the last argument.[4] We have:

$$p = P + \varepsilon f_\varphi, \quad \Phi = \varphi + \varepsilon f_P.$$

The new Hamiltonian reads

$$\mathscr{H}\left(\Phi,P,\frac{\omega t}{\varepsilon}\right) = \varepsilon f_t + H\left(\varphi,p,\frac{\omega t}{\varepsilon}\right) = \omega D_3 f + H\left(\Phi - \varepsilon f_P, P + \varepsilon f_\varphi, \frac{\omega t}{\varepsilon}\right),$$

where D_3 is the derivative in the third argument. We obtain:

$$\mathscr{H} = \omega D_3 f(\Phi,P,\tau) + H(\Phi,P,\tau) + O(\varepsilon), \qquad \tau = \frac{\omega t}{\varepsilon}.$$

Therefore $\mathscr{H}$ does not depend on t in zero approximation in ε provided the function

$$F = \omega D_3 f(\Phi,P,\tau) + P\frac{a\omega}{l}\sin\Phi\sin\tau + \frac{a^2\omega^2}{2l^2}\sin^2\Phi\sin^2\tau$$

does not depend on τ. We choose

$$f(\Phi,P,\tau) = P\frac{a}{l}\sin\Phi\cos\tau + \frac{a^2\omega}{8l^2}\sin^2\Phi\sin 2\tau,$$

---

[4] This periodicity condition is necessary to have a change uniformly close to the identity for all t.

and get[5]: $F = \frac{a^2\omega^2}{4l^2}\sin^2\Phi$. Hence, in the new variables

$$\mathscr{H} = \frac{P^2}{2} - \omega^2\cos\Phi + \frac{a^2\omega^2}{4l^2}\sin^2\Phi + O(\varepsilon),$$

where the part of $\mathscr{H}$, contained in $O(\varepsilon)$, is 2π-periodic in τ.

Remark 1 *In fact it is possible to move the dependence on time to order $O(\varepsilon^N)$ for an arbitrary $N > 0$, and even to $O(e^{-c/|\varepsilon|})$ for some positive constant c. However it is impossible to reach more: for any 2π-periodic in τ canonical near-identity change of variables the dependence of $\mathscr{H}$ on t will be greater than of order $e^{-C/|\varepsilon|}$ for a certain constant $C > 0$.*

Now let us study the system we have just obtained, neglecting the terms $O(\varepsilon)$. The system can be interpreted as the one describing the motion of a particle on a line (or on the circle Φ mod 2π) in the force field with potential

$$V = \omega^2\left(-\cos\Phi + \frac{a^2}{4l^2}\sin^2\Phi\right).$$

The phase portrait of the system is (by definition) the set of level lines of the energy integral $\frac{P^2}{2} + V(\Phi) = \text{const}$. As usual, it is convenient to draw it under the graph of the potential energy. There are two cases, see Fig. 1.

The left-hand side of the figure contains the case of "small" amplitude $a^2 < 2l^2$. In this situation there are no qualitative differences with the case of the ordinary pendulum ($a = 0$).

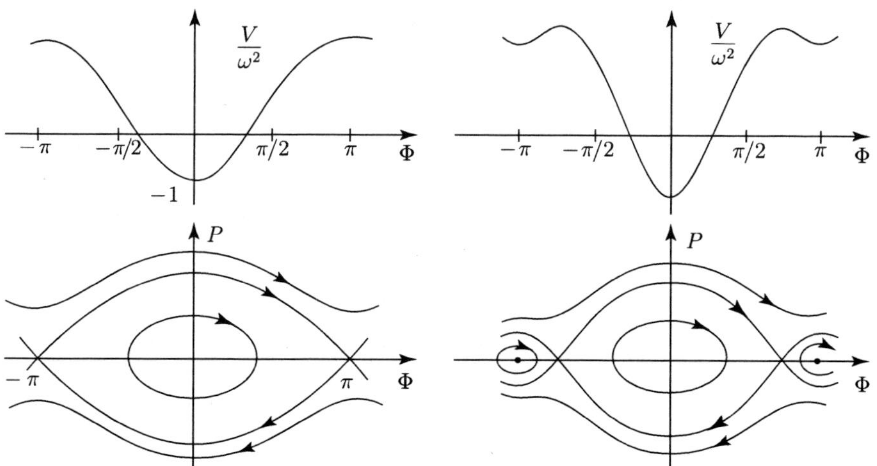

Fig. 1 Phase portraits. Left: $a^2 < 2l^2$, and right: $a^2 > 2l^2$

---

[5] Recall once more that f should be periodic in τ.

The situation changes drastically, when $a^2 > 2l^2$ (the right-hand side of the figure). In this case a bifurcation occurs and the equilibrium $\Phi = \pm\pi$ becomes stable. Moreover, the terms $O(\varepsilon)$ in the Hamiltonian do not destroy this effect, but we will not go into the detail.

Problem 8 *Draw the phase portrait in the case* $a^2 = 2l^2$.

3 Anti-integrable limit

3.1 The standard map

The standard map is, probably, the basic conceptual model for Hamiltonian dynamics in two degrees of freedom. Consider the cylinder

$$\mathscr{L} = \{(x,y) : x \bmod 2\pi\}$$

and its self-map $T_\varepsilon : \mathscr{L} \to \mathscr{L}$, $(x,y) \mapsto T_\varepsilon(x,y) = (X,Y)$, where

$$X = x + y + \varepsilon\sin x, \quad Y = y + \varepsilon\sin x. \tag{1}$$

Here ε is a real parameter which controls the type of the dynamics (regular or chaotic). The cylinder $\mathscr{L}$ is said to be the phase space of the system. The dynamics should be understood as properties of the trajectories, i.e., sequences of points $(x_k, y_k) \in \mathscr{L}$ such that for any integer k

$$(x_{k+1}, y_{k+1}) = T_\varepsilon(x_k, y_k).$$

The cylinder $\mathscr{L}$ is a two-dimensional symplectic manifold with symplectic structure $\omega = dy \wedge dx$.

Problem 9 *Check that the map T_ε is symplectic, i.e., $T_\varepsilon^*\omega = \omega$.*

Any of you can easily look at trajectories of T_ε by using a computer. To this end we remark that the variable y can be also regarded as angular. Indeed, T_ε "respects" not only the shift of x by 2π, but also the analogous shift of y in the sense that for any integer k and n

$$T_\varepsilon(x + 2\pi k, y + 2\pi n) = (X + 2\pi k, Y + 2\pi k + 2\pi n)$$

(shifts of X and Y also have the form $2\pi\cdot$ (integer number)). Hence we can ask the computer to draw on the screen the square

$$\mathscr{S} = \{(x,y) : 0 \leqslant x \leqslant 2\pi, 0 \leqslant y \leqslant 2\pi\},$$

to take an initial point $(x_0, y_0) \in \mathscr{S}$ and to put it on the screen, to compute the point $(x_1, y_1) = T_\varepsilon(x_0, y_0)$ and to put it on the screen, etc. If a point (x_n, y_n) leaves the square, it should be returned[6] to $\mathscr{S}$ by the shift of x and/or y by $2\pi k$ with a proper integer k. I recommend you to do this and to look at the trajectories for various values of ε.

Consider the case $\varepsilon = 0$. The system becomes a discrete analog of a Liouville integrable Hamiltonian system. The variables x, y play the role of action-angle variables. In particular, the action y is a first integral. Any trajectory lies on the curve (on the one-dimensional torus)

$$l_c = \{(x, y) \in \mathscr{Z} : y = c = \text{const}\}.$$

The curve l_c rotates by the angle c. If c/π is rational, the trajectory is periodic. If c/π is irrational, the trajectory fills l_c densely. Such curves l_c are said to be non-resonant.

In the case $\varepsilon \neq 0$ the situation gets much more complicated. One should not hope that any regular first integral exists, because trajectories (at least, some of them) stop to lie on smooth curves (like the circles l_c) and begin to demonstrate a chaotic behavior.

However, the chaos appears gradually. According to the KAM-theory for small values of ε many of nonresonant curves l_c, slightly deformed, exist as invariant curves for T_ε. These curves can be easily seen on pictures, produced by numerical simulations. Trajectories, lying on these circles, are regarded as regular.

Chaotic trajectories are presented on a computer screen as clouds, more or less densely filled with points. If ε is small and initial conditions are taken randomly, regular trajectories are more probable. When ε increases, the curves $l_{c,\varepsilon}$ destroy and chaos becomes more noticeable. For large ε numerical simulations show that a "typical" trajectory fills $\mathscr{S}$ almost without holes.

3.2 Anti-integrable limit

Chaotic trajectories can be constructed analytically. We show how to do this in the anti-integrable limit, i.e., for large ε.

First, we rewrite the dynamical equations (1) in the "Lagrangian form". Let (x_k, y_k), $k \in \mathbb{Z}$ be a trajectory of the standard map. Then for all integer k

$$x_{k+1} = x_k + y_k + \varepsilon \sin x_k, \quad y_{k+1} = y_k + \varepsilon \sin x_k. \tag{2}$$

Eliminating the momenta y_k, we get:

$$x_{k+1} - 2x_k + x_{k-1} = \varepsilon \sin x_k. \tag{3}$$

---

[6] In fact, we have replaced the (non-compact) phase space $\mathscr{Z}$ by (compact) $\mathbb{T}^2$, where $\mathbb{T}^2 = \{(x, y) \bmod 2\pi\}$.

The map takes the form $(x_{k-1}, x_k) \mapsto (x_k, x_{k+1})$, and the phase cylinder becomes: $\{(x_-, x) \in \mathbb{R}^2\}/\sim$, where the equivalence relation $\sim$ identifies any two points (x'_-, x') and (x''_-, x'') such that

$$x'_- - x''_- = x' - x'' = 2\pi l, \qquad l \in \mathbb{Z}.$$

Now trajectories of the map are the sequences $\{x_k\}_{k \in \mathbb{Z}}$, satisfying (3). In case of necessity y_k can be calculated by using the first equation (2).

Consider the case $\varepsilon = \infty$. Formally speaking, for $\varepsilon = \infty$ there is no dynamics: x_{k+1} can not be expressed in terms of x_{k-1} and x_k. However still there are some "trajectories". Indeed, dividing by ε, we obtain:

$$\sin x_k = \frac{1}{\varepsilon}(x_{k+1} - 2x_k + x_{k-1}) = 0.$$

Hence, for $\varepsilon = \infty$ trajectories are sequences of the form

$$x_k = \pi l_k, \qquad l_k \in \mathbb{Z}. \tag{4}$$

It turns out that for large ε the standard map has many trajectories similar to (4).

Take a large positive number Λ and define the space of codes C_Λ which consists of sequences

$$a = \{a_k\}_{k \in \mathbb{Z}}, \qquad a_k = \pi l_k, \quad l_k \in \mathbb{Z}, \quad |a_{k+1} - a_k| \leqslant \Lambda.$$

Hence C_Λ is the space of sequences (4) such that the distances between the points a_{k+1} and a_k are bounded from above by Λ.

For any code $a \in C_\Lambda$ we define the metric space of sequences Π_a:

$$x = \{x_k\}_{k \in \mathbb{Z}}, \qquad \sup_{k \in \mathbb{Z}} |x_k - a_k| < \infty.$$

Metric on Π_a has the form

$$\rho(x', x'') = \sup_{k \in \mathbb{Z}} |x'_k - x''_k|, \qquad x', x'' \in \Pi_a.$$

Theorem 4 *Given $\Lambda > 0$ and $\sigma > 0$ there exists $\varepsilon_0 = \varepsilon_0(\Lambda, \sigma) > 0$ such that for any code $a \in C_\Lambda$ and any $\varepsilon > \varepsilon_0$ the standard map has a trajectory $\hat{x} \in \Pi_a$ with $\rho(\hat{x}, a) < \sigma$.*

The trajectory x from Theorem 4 follows the code a in the sense that any point x_k differs from a_k not more than by σ. Hence, we have constructed a set of trajectories of the standard map which are in one-to-one correspondence with C_Λ.

Problem 10 *What is the cardinality of C_Λ?*

It is natural to regard the trajectories $\hat{x}$ as chaotic because according to our order they jump along σ-neighborhoods of the set $\pi\mathbb{Z}$. In fact, there is a more serious motivation to say about chaos in this situation.[7]

3.3 Proof of the Aubry theorem

The proof is based on the contraction principle in the metric space (Π_a, ρ).

Equations (3) can be presented in the form

$$x_k = \arcsin_k \left(\frac{x_{k+1} - 2x_k + x_{k-1}}{\varepsilon} \right), \tag{5}$$

where $\arcsin_k$ is the branch of arcsinus such that $\arcsin_k(0) = a_k \in \pi\mathbb{Z}$. Hence $\arcsin_k$ maps the interval $(-1,1)$ onto the interval $(a_k - \frac{\pi}{2}, a_k + \frac{\pi}{2})$, and the trajectory $x = a$ satisfies (5) for $\varepsilon = \infty$.

For big ε it is natural to construct $\hat{x}$, satisfying (5), as follows. Consider the map $x \mapsto \tilde{x} = W(x)$ such that

$$\tilde{x}_k = \arcsin_k \left(\frac{x_{k+1} - 2x_k + x_{k-1}}{\varepsilon} \right).$$

Any fixed point of W is obviously a trajectory of the standard map.

Lemma 3 *Let $\varepsilon > \varepsilon_0$, where $\varepsilon_0 = \varepsilon_0(\Lambda, \sigma)$ is sufficiently large. Then*

1. *W is defined on the ball $B_{a,\sigma} \subset \Pi_a$ with center a and radius σ;*
2. *$W(B_{a,\sigma}) \subset B_{a,\sigma}$;*
3. *W is a contracting map on $B_{a,\sigma}$, i.e.,*

$$\rho(W(x'), W(x'')) < \frac{1}{2}\rho(x', x'') \quad \text{FOR ANY } x', x'' \in B_{a,\sigma}. \tag{6}$$

Theorem 4 follows from Lemma 3. Now we will prove the lemma. Below without loss of generality we assume that $\sigma < \pi/2$.

(1)+(2). To check that $W(B_{a,\sigma}) \subset B_{a,\sigma}$ it is sufficient to show that for any $x \in B_{a,\sigma}$

$$\left| \frac{x_{k+1} - 2x_k + x_{k-1}}{\varepsilon} \right| < \sin \sigma. \tag{7}$$

Since $\rho(x,a) < \sigma$ and $a \in C_\Lambda$, we have:

$$|x_{k+1} - 2x_k + x_{k-1}| \leqslant |x_{k+1} - x_k| + |x_k - x_{k-1}| \leqslant 2(\Lambda + 2\sigma).$$

---

[7] It is easy to show that the trajectories $\hat{x}$ form a hyperbolic set in the standard map.

Hence, inequality (7) holds if

$$\varepsilon_0 > \frac{2(\Lambda + 2\sigma)}{\sin \sigma}.$$

(3). Note that for any pair of real numbers $u', u'' \in (-\sin \sigma, \sin \sigma)$

$$\left| \arcsin_k u' - \arcsin_k u'' \right| \leqslant \frac{1}{\cos \sigma} |u' - u''|.$$

Here the multiplier $\frac{1}{\cos \sigma} = \sup\limits_{u \in (-\sin \sigma, \sin \sigma)} \left| \frac{d}{du} \arcsin_k u \right|$.

We put $\tilde{x}' = W(x')$, $\tilde{x}'' = W(x'')$. Then for any $k \in \mathbb{Z}$

$$
\begin{aligned}
|\tilde{x}'_k - \tilde{x}''_k| &= \left| \arcsin_k \left(\frac{x'_{k+1} - 2x'_k + x'_{k-1}}{\varepsilon} \right) - \arcsin_k \left(\frac{x''_{k+1} - 2x''_k + x''_{k-1}}{\varepsilon} \right) \right| \\
&\leqslant \frac{1}{\cos \sigma} \left| \frac{x'_{k+1} - 2x'_k + x'_{k-1}}{\varepsilon} - \frac{x''_{k+1} - 2x''_k + x''_{k-1}}{\varepsilon} \right| \\
&\leqslant \frac{|x'_{k+1} - x''_{k+1}| + 2|x'_k - x''_k| + |x'_{k+1} - x''_{k+1}|}{\varepsilon \cos \sigma} \\
&\leqslant \frac{4}{\varepsilon \cos \sigma} \rho(x', x'').
\end{aligned}
$$

Hence, inequality (6) holds if

$$\varepsilon_0 > \frac{8}{\cos \sigma}.$$

□

3.4 Some remarks

I would like to mention one unpleasant fact, which is that all methods that are known to date give a metrically negligible chaotic set in T_ε and analogous systems. I mean the following. Given an arbitrary ε consider a set of chaotic trajectories that can be constructed by all methods, known by now. This subset of the cylinder $\mathscr{Z}$ has zero measure.

This contradicts to our physical intuition, for large ε chaos should dominate. The results of computer simulations also show that this should be the case. But maybe we should not believe these computer pictures, as the precision of computations is necessarily finite. Nevertheless, most of specialists believe that the following conjecture is true.

Conjecture. For $\varepsilon \neq 0$ in the standard map, chaos lives on sets of positive measure.

4 Separatrix splitting

4.1 Poincaré's observation

Consider a Hamiltonian system, obtained as a non-autonomous perturbation of a system with one degree of freedom:

$$\dot{x} = \frac{\partial H}{\partial y}, \quad \dot{y} = -\frac{\partial H}{\partial x}, \qquad (x,y) \in D \subset \mathbb{R}^2. \tag{1}$$

Here D is a domain and

$$H(x,y,t,\varepsilon) = H_0(x,y) + \varepsilon H_1(x,y,t) + O(\varepsilon^2). \tag{2}$$

We assume that H is 2π-periodic in t and ε is a small parameter.

Let $z_0 = (x_0, y_0) \in D$ be an equilibrium in the unperturbed ($\varepsilon = 0$) system: $\operatorname{grad} H_0(z_0) = 0$. In the extended phase space $D \times \mathbb{T}$ instead of the equilibrium we have the 2π-periodic solution $z_0 \times \mathbf{T}$.

Suppose that the equilibrium position (and therefore, the corresponding periodic solution) is hyperbolic. This means the following. Let

$$A = \begin{pmatrix} \dfrac{\partial^2 H_0}{\partial x \partial y} & \dfrac{\partial^2 H_0}{\partial y^2} \\ -\dfrac{\partial^2 H_0}{\partial x^2} & -\dfrac{\partial^2 H_0}{\partial y \partial x} \end{pmatrix}(z_0)$$

be the matrix determined by the linearization of $(1)|_{\varepsilon=0}$ at z_0. Then $\operatorname{tr} A = 0$. Hyperbolicity means that eigenvalues of A are outside the imaginary axis, i.e., $\det A < 0$. Hyperbolic equilibria of Hamiltonian systems are exponentially unstable.

On the critical energy level $H_0(x,y) = H_0(z_0)$ asymptotic curves (separatrices) $\Lambda^{s,u}$ are situated.[8] We assume that the separatrices are doubled: $\Lambda^s = \Lambda^u = \Lambda$. In the extended phase space we have 2-dimensional asymptotic surfaces $\Lambda^s \times \mathbb{T} = \Lambda^u \times \mathbb{T} = \Lambda \times \mathbb{T}$.

Problem 11 *Prove that for small values of ε the perturbed system has a 2π-periodic solution $(\sigma_\varepsilon(t), t)$, $\sigma_\varepsilon(t) = z_0 + O(\varepsilon) \in D$.*

The periodic solution $(\sigma_\varepsilon(t), t)$ is hyperbolic. Hence by the Hadamard-Perron theorem [9] there are surfaces $W_\varepsilon^{s,u} \subset D \times \mathbb{T}$, asymptotic to $(\sigma_\varepsilon(t), t)$. They are small deformations of the unperturbed surfaces $W_0^{s,u} = \Lambda^{s,u} \times \mathbb{T}$.

Poincaré discovered that W_ε^s and W_ε^u are generically distinct for $\varepsilon \neq 0$. Let us draw these surfaces. We will present a picture on the Poincaré section $D \times \{0\}$. Hence, the periodic solution $(\sigma_\varepsilon(t), t)$ is presented by the point $z_\varepsilon = \sigma_\varepsilon(0)$, and instead of the surfaces $W_\varepsilon^{s,u}$ we have the curves $\Lambda_\varepsilon^{s,u} = W_\varepsilon^{s,u} \cap \{t = 0\}$.

---

[8] s from "stable" and u from "unstable": not very good, but traditional notation.

[9] Poincaré could prove this theorem, for analytic systems.

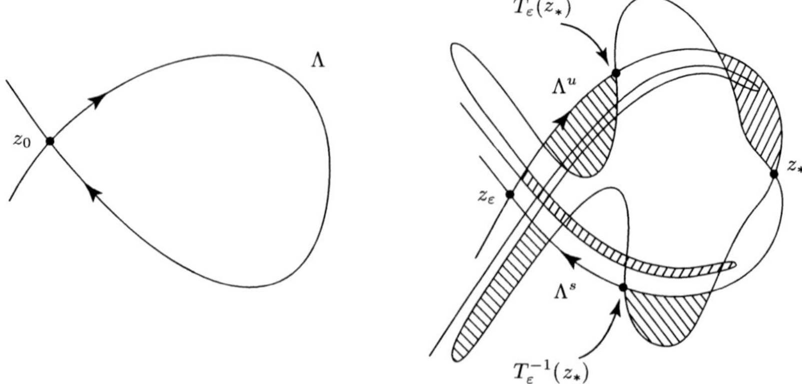

Fig. 2 A complicated behavior of the separatrices for $\varepsilon \neq 0$ (right) unlike the unperturbed case (left) on the Poincaré section $\{(x,y,t) : t = 0 \bmod 2\pi\}$. The dashed domains are mapped by T_ε to each other. Hence, their areas are the same

To obtain the right-hand part of Fig. 2, one should keep in mind the following:

(a) For small ε the curves Λ^u and Λ_ε^u (and also Λ^s and Λ_ε^s) differ just a little, at least, till $\Lambda_\varepsilon^{s,u}$ are not far away from z_ε.
(b) $\Lambda_\varepsilon^{s,u}$ are invariant w.r.t. the Poincaré map T_ε.
(c) $\Lambda_\varepsilon^{s,u}$ have no self-intersections, but can intersect each other.
(d) Any intersection point $z_* \neq z_\varepsilon$ of the curves Λ_ε^s and Λ_ε^u (a *homoclinic point*) is mapped by T_ε (and by T_ε^{-1}) to a homoclinic point.
(e) Near the fixed point z_ε T_ε is approximately determined by its linear approximation: it extends along Λ_ε^u and contracts along Λ_ε^s.
(f) T_ε and T_ε^{-1} preserve area.

Now it remains to assume that the curves Λ_ε^s and Λ_ε^u intersect transversally at some point z_*, and the right-hand side of Fig. 2 readily appears. The complicated entangled net formed by the curves $\Lambda_\varepsilon^{s,u}$ is an evidence of the complicated dynamics in the perturbed system.

4.2 The Poincaré integral

To measure the separatrix splitting, we calculate the area of a lobe, presented in Fig. 2. The main tool for this and similar calculations is the Poincaré integral.

Let $\gamma(t)$ be the natural parametrization of Λ, i.e.,

$$\gamma(t) = (\hat{x}(t), \hat{y}(t)) \tag{3}$$

is a solution of (1). Since addition to the Hamiltonian of a function, depending only on t and ε, does not influence on the dynamics, we will assume that $H_1(z_0, t) \equiv 0$.

Then the *Poincaré integral*

$$\mathscr{P}(\tau) = \int_{-\infty}^{+\infty} H_1(\gamma(t+\tau),t)\,dt$$

converges.

Problem 12 *Prove that $\mathscr{P}(\tau)$ is 2π-periodic.*

Problem 13 *Prove the identity*

$$\frac{d\mathscr{P}(\tau)}{d\tau} = \int_{-\infty}^{+\infty} \{H_0,H_1\}(\gamma(t+\tau),t)\,dt,$$

The function $\mathscr{P}$ contains all information on the separatrix splitting in the first approximation in ε.

Theorem 5 *Let τ_1 and τ_2 be two neighboring non-degenerate critical points of $\mathscr{P}$. Then there are two associated to them homoclinic points such that the area $\mathscr{A}(\varepsilon)$ of the corresponding lobe equals*

$$\mathscr{A}(\varepsilon) = |\varepsilon\mathscr{P}(\tau_1) - \varepsilon\mathscr{P}(\tau_2)| + O(\varepsilon^2). \tag{4}$$

4.3 Proof of Theorem 5

4.3.1 Hamilton–Jacobi equation

Following Poincaré, consider the case when Λ projects one-to-one to the axis. In the general case the proof is based on the same ideas.

The curve Λ (see Fig. 3) can be determined by the equation $y = \frac{\partial\varphi}{\partial x}(x)$ for some function $\varphi(x)$. We have an analogous equation in the extended phase space, i.e., the surface $W_0^s = W_0^u$ has the form

$$\left\{ (x,y,t) : y = \frac{\partial\varphi}{\partial x}(x) \right\}.$$

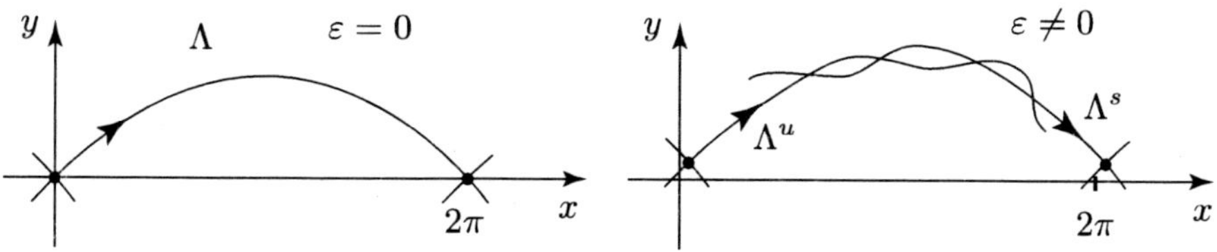

Fig. 3 The case, considered in the proof of Theorem 5, appears when x is an angular variable. For example, for non-autonomous perturbation of a pendulum. The corresponding separatrices look as in the figure

The perturbed asymptotic surfaces are as follows:

$$\left\{(x,y,t) : y = \frac{\partial S^{s,u}}{\partial x}(x,t,\varepsilon)\right\}, \qquad S^{s,u}(x,t,0) = \varphi(x).$$

Remark 2 *The functions $S^{s,u}$ are defined non-uniquely: up to an addition of arbitrary functions $f^{s,u}(t,\varepsilon)$.*

Proposition 4 *One can assume that $S^{s,u}$ satisfy the Hamilton-Jacobi equation*

$$\frac{\partial S^{s,u}}{\partial t}(x,t,\varepsilon) + H\left(x,\frac{\partial S^{s,u}}{\partial x}(x,t,\varepsilon),t,\varepsilon\right) = 0. \tag{5}$$

Remark 3 *Equation (5) for $\varepsilon = 0$ shows that if we want the equations $S^{s,u}(x,t,0) = \varphi(x)$ to hold exactly (not up to an addition of a function of t), we should put $H_0|_\Lambda = 0$.*

Proof of Proposition 4 is based on a direct calculation. Let $(x,y,t) = (x,\frac{\partial S}{\partial x}(x,t,\varepsilon),t)$ be a point, lying on W_ε (for brevity we do not write the indices s,u), and $()^{\cdot} = \frac{d}{dt}$, denotes the time derivative w.r.t. equations (1). Then

$$\begin{aligned}
\dot{y} &= \frac{\partial^2 S}{\partial x \partial t}(x,t,\varepsilon) + \frac{\partial^2 S}{\partial x^2}(x,t,\varepsilon)\dot{x} \\
&= -\frac{\partial H}{\partial x}(x,y,t,\varepsilon) \\
&= -\frac{\partial}{\partial x}H\left(x,\frac{\partial S}{\partial x}(x,t,\varepsilon),t,\varepsilon\right) + \frac{\partial H}{\partial y}(x,y,t,\varepsilon)\frac{\partial^2 S}{\partial x^2}(x,t,\varepsilon).
\end{aligned}$$

Since $\frac{\partial^2 S}{\partial x^2}\dot{x} = \frac{\partial H}{\partial y}\frac{\partial^2 S}{\partial x^2}$, we get:

$$\frac{\partial}{\partial x}\left(\frac{\partial S}{\partial t}(x,t,\varepsilon) + H\left(x,\frac{\partial S}{\partial x}(x,t,\varepsilon),t\right)\right) = 0.$$

Hence for some function $\alpha(t,\varepsilon)$

$$\frac{\partial S}{\partial t}(x,t,\varepsilon) + H\left(x,\frac{\partial S}{\partial x}(x,t,\varepsilon),t\right) = \alpha(t,\varepsilon).$$

By Remark 2 α can be taken equal to zero. $\square$

4.3.2 The function $S_1^{s,u}$ and the Poincaré integral

Expand equations (5) in power series in ε. Let $S = \varphi(x) + \varepsilon S_1(x,t) + O(\varepsilon^2)$. In zero approximation we have:

$$\frac{\partial \varphi}{\partial t}(x) + H_0\left(x,\frac{\partial \varphi}{\partial x}\right) = 0.$$

(compare with Remark 3).

The first approximation is as follows:

$$\frac{\partial S_1^{s,u}}{\partial t}(x,t) + H_1\left(x,\frac{\partial \varphi}{\partial x},t\right) + \frac{\partial H_0}{\partial y}\left(x,\frac{\partial \varphi}{\partial x}\right)\frac{\partial^2 S_1^{s,u}}{\partial x \partial t}(x,t) = 0. \tag{6}$$

Since $\partial H_0/\partial y = \dot{x}$, equation (6) can be rewritten in the form

$$\frac{d}{dt}S_1^{s,u}(x,t) + H_1\left(x,\frac{\partial \varphi}{\partial x},t\right) = 0. \tag{7}$$

Plugging in (7) instead of x its parametrization $\hat{x}(t+\tau)$ (see (3)), we get:

$$\frac{d}{dt}S_1^{s,u}(\hat{x}(t+\tau),t) = H_1(\gamma(t+\tau),t).$$

Now we integrate in t:

$$S_1^s(\hat{x}(t+\tau),t) - S_1^s(\hat{x}(+\infty),t) = \int_t^{+\infty} H_1(\gamma(s+\tau),s)\,ds,$$

$$S_1^u(\hat{x}(t+\tau),t) - S_1^u(\hat{x}(-\infty),t) = -\int_{-\infty}^t H_1(\gamma(s+\tau),s)\,ds.$$

(Recall that $\hat{x}(-\infty) = \hat{x}(+\infty) = x_0$.) Hence

$$S_1^s(\hat{x}(t+\tau),t) - S_1^u(\hat{x}(t+\tau),t) = \mathscr{P}(\tau) + \beta(t),$$

where $\beta(t) = S_1^s(x_0,t) - S_1^u(x_0,t)$. Differentiating in τ, we get:

$$\dot{x}(t+\tau)\frac{\partial}{\partial x}\left(S_1^s(\hat{x}(t+\tau),t) - S_1^u(\hat{x}(t+\tau),t)\right) = \mathscr{P}'(\tau). \tag{8}$$

4.3.3 Homoclinic points and lobes

Homoclinic points are determined by the equations $(x,\frac{\partial S^s}{\partial x}) = (x,\frac{\partial S^u}{\partial x})$ i.e.,

$$\frac{\partial \varphi}{\partial x}(\hat{x}(t+\tau)) + \varepsilon\frac{\partial S_1^s}{\partial x}(\hat{x}(t+\tau),t) - \frac{\partial \varphi}{\partial x}(\hat{x}(t+\tau)) - \varepsilon\frac{\partial S_1^u}{\partial x}(\hat{x}(t+\tau),t) + O(\varepsilon^2) = 0,$$

where we take again $\hat{x}(t+\tau)$ instead of x. According to (8) and the relation $\dot{x}(t+\tau) \neq 0$ we get:

$$\mathscr{P}'(\tau) + O(\varepsilon) = 0.$$

Hence non-degenerate critical points of $\mathscr{P}(\tau)$ generate homoclinic points.

Question. *Why we need non-degeneracy and in what sense we use the word "generate"?*

Let τ_1 and τ_2 be two consecutive non-degenerate critical points of $\mathscr{P}(\tau)$. The corresponding homoclinic points $z_1 = (x_1, y_1)$, $z_2 = (x_2, y_2)$ on the Poincaré section $\{t = 0 \bmod 2\pi\}$ are "angles" of a lobe. Let $\mathscr{A}(\varepsilon)$ be its area. Then

$$\mathscr{A}(\varepsilon) = \left| \int_{x_1}^{x_2} \left(\frac{\partial S^s}{\partial x}(x, 0, \varepsilon) - \frac{\partial S^u}{\partial x}(x, 0, \varepsilon) \right) dx \right|$$

$$= \left| \int_{x_1}^{x_2} \left(\varepsilon \frac{\partial S_1^s}{\partial x}(x, 0) - \varepsilon \frac{\partial S_1^u}{\partial x}(x, 0) \right) dx \right| + O(\varepsilon^2).$$

We change variables $x = \hat{x}(\tau)$ in the integral and use (8):

$$\mathscr{A}(\varepsilon) = \left| \int_{\tau_1}^{\tau_2} \varepsilon \frac{\partial}{\partial x} \left(S_1^s(\hat{x}(\tau), 0) - S_1^u(\hat{x}(\tau), 0) \right) \dot{\hat{x}}(\tau) \, d\tau \right| + O(\varepsilon^2)$$

$$= \left| \int_{\tau_1}^{\tau_2} \varepsilon \mathscr{P}'(\tau) \, d\tau \right| + O(\varepsilon^2).$$

This implies (4). $\square$

4.4 Standard example

Consider a pendulum with a vertically oscillating suspension point, i.e., the system with Hamiltonian

$$H(x, y, t, \varepsilon) = \frac{1}{2}y^2 + \Omega^2 \cos x + \varepsilon \theta(t) \cos x. \qquad (9)$$

Performing in case of necessity the change $t \mapsto \lambda t$, we can assume that θ is 2π-periodic.

A natural parametrization on the unperturbed separatrix $\gamma(t)$ can be computed explicitly.

Problem 14 *Check that* $\cos(\hat{x}(t)) = 1 - 2\cosh^{-2}(2\Omega t)$.

Hence $\mathscr{P}(\tau) = \int_{-\infty}^{+\infty} \theta(t)(\cos(\hat{x}(t + \tau)) - 1) \, dt$.

Problem 15 *Check that for* $\theta(t) = \cos t$

$$\mathscr{P}(\tau) = -\frac{\pi \cos \tau}{2\Omega^2 \sinh(\frac{\pi}{2\Omega})}.$$

If $\theta(t) = \cos t$ lobes have the areas

$$\mathscr{A}(\varepsilon) = \frac{\varepsilon \pi}{\Omega^2 \sinh(\frac{\pi}{2\Omega})} + O(\varepsilon^2).$$

References

1. Abraham, R, Marsden, J. Foundations of mechanics. Benjamin/Cummings Publishing Co., Inc., Advanced Book Program, Reading, MA, 1978.
2. Arnold, V. I. Mathematical methods of classical mechanics. Graduate Texts in Mathematics, 60. Springer, New York, 1989.
3. Arnold, V. I., Kozlov, V. V., Neishtadt, A. I. Mathematical aspects of classical and celestial mechanics. Encyclopaedia Math. Sci., 3, Springer, Berlin, 1993. Springer, Berlin, 1997.

Four lectures on the N-body problem

Alain Chenciner[1]

Abstract In the first two lectures, Hamiltonian techniques are applied to avatars of the N-body problem of interest to astronomers: the first one introduces one of the simplest non integrable equations, the planar circular restricted problem in the lunar case, where most degeneracies of the general (non-restricted) problem are not present; the second one is a quick introduction to Arnold's theorem on the stability of the planetary problem where degeneracies are dealt with thanks to Herman's normal form theorem. The last two lectures address the general (non-perturbative) N-body problem: in the third one, a sketch of proof is given of Marchal's theorem on the absence of collisions in paths of N-body configurations with given endpoints which are local action minimizers; in the last one, this theorem is used to prove the existence of various families of periodic and quasi-periodic solutions with pre-scribed symmetries and in particular to extend globally Lyapunov families bifurcating from polygonal relative equilibria. Celestial mechanics is famous for demanding extensive computations which hardly appear here: these notes only describe the skeleton on which these computations live.

1 The Poincaré–Birkhoff–Conley twist map of the annulus for the planar circular restricted three-body problem

1.1 The Kepler problem as an oscillator

The (normalized) motions in a plane of a particle submitted to the Newtonian attraction of a fixed center – the so-called *Kepler problem* – are the solutions of the equation

$$\ddot{x} = -x/|x|^3,$$

---

[1] University Paris 7 and IMCCE (Paris Observatory)
e-mail: chenciner@imcce.fr

W. Craig (ed.), Hamiltonian Dynamical Systems and Applications, 21–52.

where $x \in \mathbb{R}^2 = \mathbb{C}$ is identified with a complex number and the dot denotes the time derivative. These equations are the Hamilton equations

$$\dot{x} = \frac{\partial H}{\partial \bar{y}}, \; \dot{y} = -\frac{\partial H}{\partial \bar{x}}$$

associated to the Hamiltonian $H : (\mathbb{C} \setminus \{0\}) \times \mathbb{C} \to \mathbb{R}$ and the symplectic form ω respectively defined by

$$H(x,y) = |y|^2 - 2/|x|, \quad \omega = dx \wedge d\bar{y} + d\bar{x} \wedge dy.$$

The Levi-Civita mapping $(z, w) \mapsto (x = 2z^2, \, y = w/\varepsilon\bar{z})$ defines a two-fold covering

$$(L.C.) \qquad\qquad K^{-1}(0) \setminus \{z = 0\} \to \Sigma_\varepsilon = H^{-1}(-1/\varepsilon^2)$$

from the complement of the plane $z = 0$ in the 0-energy three-sphere $K^{-1}(0)$ of the harmonic oscillator

$$K(z, w) = |z|^2 + |w|^2 - \varepsilon^2 = \varepsilon^2 |z|^2 \left[H \left(2z^2, w/\varepsilon\bar{z}\right) + 1/\varepsilon^2 \right],$$

to the energy hypersurface $\Sigma_\varepsilon = H^{-1}(-1/\varepsilon^2)$ of the Kepler problem (both diffeomorphic to $S^1 \times \mathbb{R}^2$). It is conformally symplectic and sends integral curves of the harmonic oscillator with energy ε^2 to those of the Kepler problem with energy $-1/\varepsilon^2$ after the change of time $dt = 2\varepsilon |x| dt'$ which prevents the velocity to become infinite at collision. In the coordinates $u_1 = w + iz$, $u_2 = \bar{w} + i\bar{z}$ these integral curves are $u_1(t) = c_1 e^{it}$, $u_2(t) = c_2 e^{it}$, $|c_1|^2 + |c_2|^2 = 2\varepsilon^2$, that is the intersections of the three-sphere with the complex lines $u_1/u_2 = cste$, or in other words the fibers of the *Hopf fibration* $(u_1, u_2) \mapsto u_1/u_2 : S^3 \to P_1(\mathbb{C})$. The closest approximation to a section of the Hopf map, the annulus

$$\arg u_1 + \arg u_2 = 0 \quad (\mathrm{mod}\; 2\pi)$$

is a global surface of section of the flow of the Harmonic oscillator in a sphere of constant energy: with the exception of the two fibers which form its boundary, all the fibers cut this annulus transversally in two points; hence, the second return map is the identity. Thus perturbations of the Kepler problem with negative energy are essentially perturbations of the identity map. This is one of the main sources of degeneracies in celestial mechanics.

1.2 The restricted problem in the lunar case

The equations of the N-body problem

$$\ddot{\mathbf{r}}_i = g \sum_{j \neq i} \frac{m_j(\mathbf{r}_j - \mathbf{r}_i)}{||\mathbf{r}_i - \mathbf{r}_j||^3}$$

make sense even if some of the masses vanish. Such masses are influenced by the non-zero masses but do not influence them. We shall consider two primaries, say the Sun (mass μ) and the Earth (mass ν) which have a uniform circular motion around their center of mass and a zero-mass third body, say the Moon, which stays close to the Earth. We shall use the normalization $g = 1$ and $\mu + \nu = 1$. We identify the inertial plane with $\mathbb{C}$ (coordinate $X = X_1 + iX_2$ centered on the center of mass of the couple Sun-Earth) and introduce a rotating complex coordinate $x = x_1 + ix_2 = Xe^{-i\omega t} - \mu$ centered on the Earth. Setting $y = \dot{x} + i\omega x$ (up to a translation, this is the velocity in the inertial frame), the equations of motion of the Moon take the Hamiltonian form

$$\dot{x} = \frac{\partial H}{\partial \bar{y}}, \ \dot{y} = -\frac{\partial H}{\partial \bar{x}},$$

where H is the *Jacobi integral* (the constant 2μ is added for convenience)

$$H(x,y) = |y|^2 + i\omega(\bar{x}y - x\bar{y}) - \frac{2\nu}{|x|} - \frac{2\mu}{|x+1|} - \mu(x+\bar{x}) + 2\mu.$$

More precisely, the vector field is the symplectic gradient of the symplectic form

$$\omega = dx \wedge d\bar{y} + d\bar{x} \wedge dy = 2(dx_1 \wedge dy_1 + dx_2 \wedge dy_2).$$

As in the first section, we consider the energy hypersurface $H^{-1}(1/\varepsilon^2)$, with ε a small parameter. Its projection on the x plane is made of three connected components: a neighborhood of the Sun, a neighborhood of the Earth and a neighborhood of infinity (the so-called Hill's regions, which imply Hill's stability result, praised by Poincaré). We shall be interested in the connected component of $H^{-1}(1/\varepsilon^2)$ where $|x|$ stays small. Then

$$H(x,y) = |y|^2 + i\omega(\bar{x}y - x\bar{y}) - \frac{2\nu}{|x|} - 2y'\mu\left[\frac{1}{4}|x|^2 + \frac{3}{8}(x^2 + \bar{x}^2) + O_3(x)\right].$$

We see that the influence of the Sun on the Moon becomes negligible with respect to the one of the Earth and that at the collision limit, it disappears and one is left with a Kepler problem. To make this apparent, we again apply the Levi–Civita transformation. We get

$$K(z,w) = \varepsilon^2|z|^2\left[H\left(2z^2, \frac{w}{\varepsilon\bar{z}}\right) + \frac{1}{\varepsilon^2}\right] = f^2(z,w)|z|^2 + |w|^2 - \nu\varepsilon^2 - \varepsilon^2\mu\frac{1}{2}g(z),$$

where

$$f(z,w) = \sqrt{1 + 2i\varepsilon(\bar{z}w - z\bar{w})}, \quad g(z) = 2|z|^2\left(\frac{1}{|2z^2 + 1|} - 1 + z^2 + \bar{z}^2\right).$$

As in the Kepler case, the direct image of the restriction to $K^{-1}(0) \setminus \{z = 0\}$ of the Hamiltonian flow $\dot{z} = \frac{\partial K}{\partial \bar{w}}$, $\dot{w} = -\frac{\partial K}{\partial \bar{z}}$ becomes the flow of the restricted problem with Jacobi constant $-1/\varepsilon^2$ after the change of time $dt = 2\varepsilon|x|dt'$.

Each truncation of the Taylor expansion of $K(z,w)$ at the origin,

$$K(z,w) = -\nu\varepsilon^2 + |z|^2 + |w|^2 + 2i\varepsilon|z|^2(\bar{z}w - \bar{w}z) - \varepsilon^2\mu(2|z|^6 + 3|z|^2(z^4 + \bar{z}^4)) + 0_8(z)),$$

makes sense dynamically when restricted to $K^{-1}(0)$: we get

At order 2, the harmonic oscillator, which regularizes the Kepler problem
At order 4, the regularization of the Kepler problem in a rotating frame
At order 6, *Hill's problem*. This is the highest order of interest to us

1.3 Hill's solutions

The truncation $\hat{K}(z,w) = -\nu\varepsilon^2 + f^2(z,w)|z|^2 + w^2$ of K at fourth order is a completely integrable Hamiltonian, a first integral being the angular momentum or, what is equivalent, the function $f^2(z,w)$. This is not surprising as we already knew that the restriction to $K^{-1}(0)$ corresponds to the completely integrable Kepler problem in a rotating frame. The intersection of level hypersurfaces of K and f^2 defines in general a two-dimensional torus, except when the two hypersurfaces are tangent, that is when $w = \pm if(z,w)z$. In this case the intersection degenerates to a circle; in $K^{-1}(0)$, this defines two solutions which project (by a 2-1 map) onto the two circular solutions (one direct, one retrograde) of the rotating Kepler problem with the given value $-1/\varepsilon^2$ of the Jacobi constant.

From now on, two roads may be followed: one can, along with Kummer [Ku], stick to symplectic coordinates or one can, as did Conley, use the simpler but not symplectic coordinates

$$\xi_1 = w + if(z,w)z, \quad \xi_2 = \bar{w} + if(z,w)\bar{z}.$$

We shall follow Conley. The equations $\dot{z} = \frac{\partial K}{\partial \bar{w}}$, $\dot{w} = -\frac{\partial K}{\partial \bar{z}}$ take the form

$$\dot{\xi}_1 = i\xi_1\left(1 - \frac{\varepsilon}{2}|\xi_1 - \bar{\xi}_2|^2\right) + \varepsilon^2 O_5(\xi_1, \xi_2),$$

$$\dot{\xi}_2 = i\xi_2\left(1 + \frac{\varepsilon}{2}|\xi_1 - \bar{\xi}_2|^2\right) + \varepsilon^2 O_5(\xi_1, \xi_2).$$

For this section, we do not need the exact expression of the terms of order 5.

We shall show that the energy hypersurface $K^{-1}(0)$ contains two periodic solutions of minimal periods close to 2π, corresponding to the so-called *Hill's lunar orbits*, direct and retrograde, which are almost circular periodic motions of the Moon around the Earth in the rotating frame. The value 0 of the energy does not play a special role and it is in fact possible to prove the existence of two "Lyapunov" families of periodic solutions stemming from the origin and foliating two smooth (even analytical) germs of invariant surfaces in the (z,w) four-dimensional phase space. This is a degenerate version of Lyapunov' theorem, the degeneracy being the double eigenvalues $\pm i$ of the linearization $\dot{\xi}_1 = i\xi_1$, $\dot{\xi}_2 = i\xi_2$, of the vector-field at

$\xi_1 = \xi_2 = 0$. Recall that this degeneracy comes from the fact that all solutions of the Kepler problem with a given energy are periodic with the same period. Here are the main steps of the proof of the existence of Hill's orbits.

1. *Putting the vector-field into normal form at order 3:* the idea, which goes back to Poincaré's thesis and was much developed by Birkhoff, is to simplify as much as possible a finite part of the vector-field's Taylor expansion at the origin by means of local change of variables tangent to Identity. It relies on the fact that replacing $X = (x_1, \cdots, x_n)$ by $Y = X + h(X)$, where the components of $h(X)$ start with terms homogeneous in X of degree r, transforms the equation $\dot{X} = AX + F(X)$ into the equation $\dot{Y} = AY + [A, h](Y) + O_{r+1}$, where $[,]$ is the Lie bracket of the two vector-fields. If $A = diag(\lambda_1, \cdots, \lambda_n)$ and $h = (h_1, \ldots, h_n)$ with $h_s(Y) = y_1^{i_1} \cdots y_n^{i_n}$ and $h_j = 0$ if $j \neq s$, one checks that $[A, h] = k$ with $k_s(Y) = (i_1 \lambda_1 + \cdots + i_n \lambda_n - \lambda_s) y_1^{i_1} \cdots y_n^{i_n}$ and $k_j = 0$ if $j \neq s$. It follows that one can suppress only *non-resonant* terms, i.e. those for which no *resonance relation* $i_1 \lambda_1 + \cdots + i_n \lambda_n - \lambda_s =$ is satisfied.

   In our case, this allows to replace the equations by the following (we kept the same name for the variables):

   $$\dot{\xi}_1 = i\xi_1 \left(1 + \alpha|\xi_1|^2 + \beta|\xi_2|^2\right) + \varepsilon^2 \varphi_1(\xi_1, \xi_2),$$

   $$\dot{\xi}_2 = i\xi_2 \left(1 + a|\xi_1|^2 + b|\xi_2|^2)\right) + \varepsilon^2 \varphi_2(\xi_1, \xi_2),$$

   with $\alpha = \beta = -\frac{\varepsilon}{2}$, $a = b = +\frac{\varepsilon}{2}$, φ_1 and φ_2 of order 5 in $\xi_1, \xi_2, \bar{\xi}_1, \bar{\xi}_2$. In the neighborhood of the origin, the flow $\Phi_t(\xi_1, \xi_2) = (\xi_1(t), \xi_2(t))$ can be written

   $$\xi_1(t) = e^{it} \left[\xi_1(1 + i(\alpha|\xi_1|^2 + \beta|\xi_2|^2)t) + \varepsilon^2 \alpha_1(\xi_1, \xi_2, t)\right],$$

   $$\xi_2(t) = e^{it} \left[\xi_2(1 + i(a|\xi_1|^2 + b|\xi_2|^2)t) + \varepsilon^2 \alpha_2(\xi_1, \xi_2, t)\right],$$

   with α_1, α_2 of order 5 in $\xi_1, \xi_2, \bar{\xi}_1, \bar{\xi}_2$ uniformly in t belonging to a compact.

2. *Regularizing the equations for a periodic solution by means of a blow-up:* We look for a periodic solution whose period T is close to the period 2π of the solution $\xi_2 = 0$ of the rotating Kepler problem approximation (an analogous reasoning can be made for a solution close to $\xi_1 = 0$). Because of the existence of the energy first integral, the equations which define a periodic solution of period T, that is $\xi_1(T) = \xi_1$, $\xi_2(T) = \xi_2$, are consequence of the equations

   $$Arg\,\xi_1(T) - Arg\,\xi_1 = 2\pi, \quad \xi_2(T) - \xi_2 = 0.$$

Writing down directly these equations would lead to possibly non differentiable terms like $\alpha_1(\xi_1, \xi_2)/\xi_1$. Indeed, they read

$$2\pi = T + \arg\left[1 + i(\alpha|\xi_1|^2 + \beta|\xi_2|^2)T + \varepsilon^2 \frac{\alpha_1(\xi_1, \xi_2, T)}{\xi_1}\right],$$

$$\left[e^{iT} \left(1 + i(a|\xi_1|^2 + b|\xi_2|^2)T\right) - 1\right] \xi_2 + \varepsilon^2 e^{iT} \alpha_2(\xi_1, \xi_2, T) = 0.$$

We solve this problem by a further localization in a domain of the form $|\xi_2| \leqslant |\xi_1|$ by means of a complex blow-up

$$\xi_1 = z_1, \quad \xi_2 = z_1 z_2$$

which replaces such a term by $\alpha_1(z_1, z_1 z_2)/z_1$ which is now differentiable. The first equation determines T as a C^3 function of $z_1, \bar{z}_1, z_2, \bar{z}_2$,

$$T = 2\pi - 2\pi |z_1|^2 (\alpha + \beta |z_2|^2) + o_3,$$

where o_3 vanishes at order 3 along $z_1 = 0$. The second one becomes

$$2\pi i |z_1|^2 (a - \alpha + (b - \beta)|z_2|^2) z_2 + o_3 = 0_3.$$

As $a - \alpha = \varepsilon \neq 0$, solving this equation leads to a C^1 surface tangent to the plane $z_2 = 0$, that is in the (ξ_1, ξ_2) space to a C^2 surface N_1 tangent at order 2 to the plane $\xi_2 = 0$. Intersecting with the energy hypersurface $K = 0$ gives the seeked for periodic solution. In the same way, one proves the existence of N_2 tangent to $\xi_1 = 0$.

3. *Proving the analyticity of N_1 and N_2:* This is done in Conley's thesis by closely following the proof given in the non-resonant case by Siegel and Moser. To understand the formulas, one suppresses the resonant terms of any order by means of a formal (not convergent !) transformation. One gets new (formal coordinates) ζ_1, ζ_2 such that $\dot{\zeta}_1$ and $\dot{\zeta}_2$ become formal series in the resonant terms $\zeta_i |\zeta_j|^2$ and $\zeta_i(\zeta_j \bar{\zeta}_k)$. Rewriting the computation of periodic solutions as above leads to formal surfaces N_1 and N_2 where, for example, N_1 is defined by a (formal) equation of the form $\zeta_2 = \gamma(|\zeta_1|^2)\zeta_1$, the restriction of the vector-field being of the form $\dot{\zeta}_1 = \alpha(|\zeta_1|^2)\zeta_1$ where α has purely imaginary values (this corresponds to the fact that N_1 is foliated by periodic solutions surrounding the origin). One proves the convergence of γ and α by writing down majorant series.

1.4 The annulus twist map

Replacing the boundaries $\xi_1 = 0$ and $\xi_2 = 0$ of the Kepler annulus by the two Hill orbits, one can now construct a global annulus of section of the flow in the three-sphere $K^{-1}(0)$ and analyze the first return map. Such an annulus is of course not unique and it will be convenient to chose it so as to contain the "collision circle" of equation $z = 0$.

In order to get precise enough information on the first return map, one must analyze the equations up to the 5th order where the influence of the Sun comes into play. Writing down a normal form up to this order implies first computing the effect on terms of order five of the change of variables leading to a normal form at order 3. In fact, one can dispense with this: it is enough to suppress only the non resonant terms of order 5, keeping the terms of order 3 as they stood initially. Moreover, the above analysis of the submanifolds N_1 and N_2 whose intersection with $K = 0$ defines Hill's orbits, shows that there exists an analytic change of variables which transforms them into coordinate planes. A finer analysis shows that such a straightening change of variables differs from Id only by terms $\varepsilon A + \varepsilon^2 B$, where A is resonant of order 5 and B is of order 7. One deduces that such a straightening of N_1 and N_2 does not bring any new change to the differential equation up to order 5. Finally, we get new coordinates (ζ_1, ζ_2) such that N_1 and N_2 are respectively defined by $\zeta_1 = 0$ and $\zeta_2 = 0$, and the energy hypersurface $K^{-1}(0)$ and the collision circle $z = 0$ by

$$\frac{1}{2}(|\zeta_1|^2 + |\zeta_2|^2) - v\varepsilon^2 + \varepsilon O_6(\zeta) = 0, \quad \text{and} \quad \zeta_1 - \bar{\zeta}_2 + \varepsilon O_5(\zeta) = 0.$$

It follows that an annulus of section in $K^{-1}(0)$ containing the collision circle and bounded by the Hill orbits can be defined by the equation

$$Arg\, \zeta_1 + Arg\, \zeta_2 + \varepsilon O_4(\zeta) = 0 \quad (\mathrm{mod}\ 2\pi).$$

Computing a little more, one can find coordinates (φ, ρ) on this annulus, such that the two boundaries are close to $\rho = \pm 1$ and the first return map takes the form

$$P_\varepsilon(\varphi, \rho) = \left(\varphi + \frac{1}{2} - \frac{v}{2}\varepsilon^3 - \frac{3v^2}{2}(1 - \frac{\mu}{4})\varepsilon^6 \rho + 0(\varepsilon^7),\ \rho + O(\varepsilon^7) \right).$$

Coming back to the definition of this annulus, one checks that the return map corresponds essentially to the passages of the orbit of the Moon through aphelium in the rotating frame. Originating from a Hamiltonian system, this map necessarily preserves a measure defined by a smooth density. Moreover, it is a $O(\varepsilon^7)$ perturbation of an integrable twist map whose twist is of size ε^6. This is a perfect ground for applying the main results of the general theory of conservative twist maps, a particular case of the theory of Hamiltonian systems with two degrees of freedom:

1. Applied to the iterates of the return map, the *Birkhoff fixed point theorem* yields an infinite number of periodic orbits of higher and higher periods to which correspond periodic orbits of long period of the Moon around the Earth in the rotating frame.
2. The *Moser invariant curve theorem* implies the existence of a positive measure Cantor set of invariant curves on which the map is conjugated to a diophantine irrational rotation and to which correspond quasi periodic orbits of the Moon.
3. To the Liouville rotation numbers, the *Aubry–Mather theory* associates invariant Cantor sets to which correspond orbits of the Moon with a Cantor caustic.
4. Finally, it is possible to prove that the image of the collision circle intersects itself transversally at eight points [CL]; in particular, it is not contained in an invariant curve. Varying the value of ε moves the invariant curve of a given rotation number across the annulus which forces intersection with the collision curve. This proves the existence of invariant "punctured" tori which correspond

to orbits of the Moon which persistently change their direction of rotation around the Earth in the rotating frame (generalization of the punctured tori to the full planar three-body problem were given by Féjoz in his thesis [Fe1]).

Remark. For writing down formulas, working in the two-fold covering $K^{-1}(0)$ of the energy hypersurface diffeomorphic to S^3 is convenient but one can prefer to state the results downstairs in the compactification (regularization), diffeomorphic to $SO(3)$ (that is to the real projective space of dimension 3), of the original energy hypersurface $H^{-1}(-\frac{1}{\varepsilon^2})$. The first return map then becomes a perturbation of the Identity (the Kepler case) of the form

$$\mathscr{P}_\varepsilon(\tilde{\varphi},\rho) = \left(\tilde{\varphi} - v\varepsilon^3 - 3v^2(1 - \frac{\mu}{4})\varepsilon^6\rho + 0(\varepsilon^7), \rho + O(\varepsilon^7) \right).$$

and the collision curve intersects its image only four times.

A problem. When the collision curve intersects the set of invariant curves, the closure of the union of its iterates, containing the set of intersected curves, is of positive measure. What if the collision curve is contained in a Birkhoff region of instability?

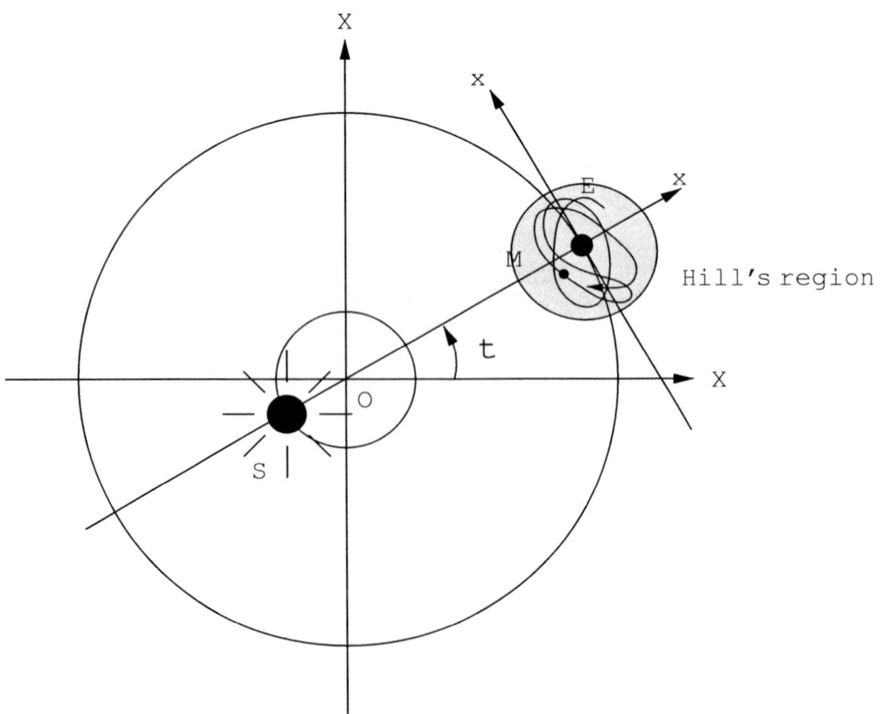

Fig. 1 Hill's regions

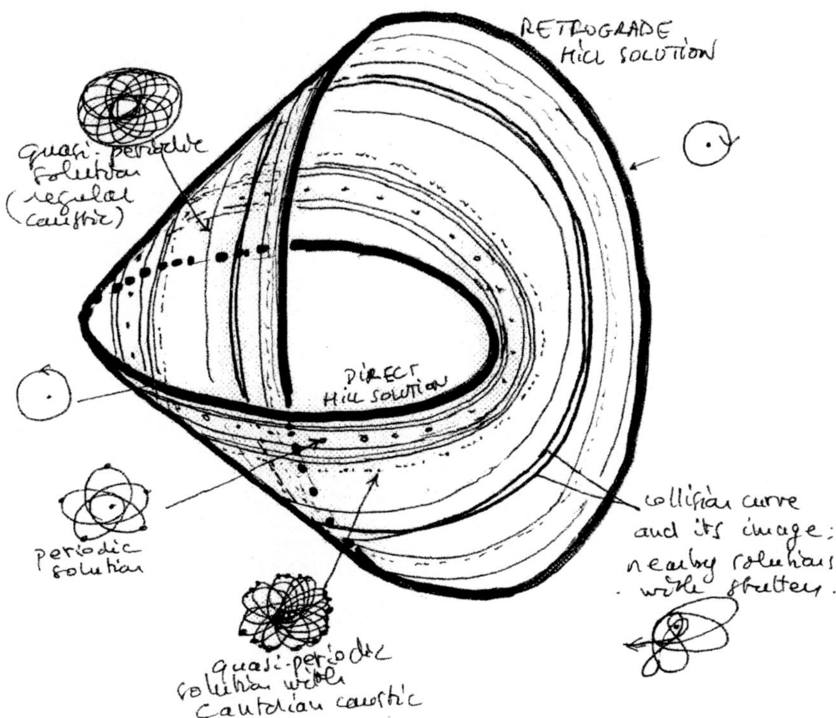

Fig. 2 The annulus of section

2 The Arnold–Herman stability theorem for the spatial (1 + n)-body problem

In the so-called *planetary problem*, one mass m_0 is dominant (the Sun) and the others, the planets are of the form $\varepsilon m_1, \ldots, \varepsilon m_n$, where ε is small (around 10^{-3} for the "real" solar system). If $x_0 = (x_0^1, x_0^2, x_0^3), x_1, \ldots, x_n \in \mathbb{R}^3$ are the positions and $||.||$ the euclidean norm, Newton's equations read

$$\ddot{x}_j = m_0 \frac{x_0 - x_j}{||x_0 - x_j||^3} + \varepsilon \sum_{k \neq j} m_k \frac{x_k - x_j}{||x_k - x_j||^3}, \quad j = 1, \ldots, n.$$

The solutions are the projections on the configuration space of the integral curves of the Hamiltonian vector field defined in the phase space, whose coordinates are denoted by $(x_0, \ldots, x_n, y_0, \varepsilon y_1, \ldots, \varepsilon y_n)$ and symplectic form is $\sum_{1 \leq k \leq 3} dx_0^k \wedge dy_0^k + \varepsilon \sum_{1 \leq j \leq n} \sum_{1 \leq k \leq 3} dx_j^k \wedge dy_j^k$, by the Hamiltonian

$$\frac{1}{2} \frac{||y_0||^2}{m_0} + \varepsilon \left(\frac{1}{2} \sum_{1 \leq j \leq n} \frac{||y_j||^2}{m_j} - \sum_{1 \leq j \leq n} \frac{m_0 m_j}{||x_j - x_0||} \right) - \varepsilon^2 \sum_{1 \leq j < k \leq n} \frac{m_j m_k}{||x_j - x_k||}.$$

One reduces the translation symmetry by restricting to the value $Y_0 = 0$ the total linear momentum and going to the quotient by translations in the so-called Poincaré heliocentric canonical coordinates

$$X_0 = x_0, Y_0 = y_0 + \varepsilon y_1 + \cdots + \varepsilon y_n, \quad X_j = x_j - x_0, \quad Y_j = y_j, \quad j = 1, \ldots, n.$$

After dividing the new Hamiltonian and symplectic form by ε one obtains a Hamiltonian defined on $T^* \mathbb{R}^{3n}$ (coordinates $(X_1, \ldots, X_n, Y_1, \ldots, Y_n)$) deprived of the collision set ($X_j = 0$ or $X_j = X_k$) with its canonical symplectic structure :

$$F_\varepsilon = \sum_{1 \leqslant j \leqslant n} \left(\frac{\|Y_j\|^2}{2\mu_j} - \frac{\mu_j M_j}{\|X_j\|} \right) + \varepsilon \sum_{1 \leqslant j < k \leqslant n} \left(-\frac{m_j m_k}{\|X_j - X_k\|} + \frac{Y_j \cdot Y_k}{m_0} \right).$$

It describes an ε perturbation of n uncoupled *Kepler problems* with fictitious masses defined by $M_j = m_0 + \varepsilon m_j$ and $\mu_j M_j = m_0 m_j$. Whe shall be interested in solutions which stay close to solutions of F_0 where the planets describe circular coplanar motions with the same orientation around the sun.

Theorem 2.1. *Given $m_0, \ldots, m_n, a_1, \ldots, a_n$, there exists $\varepsilon_0 > 0$ with the following property: if $\varepsilon < \varepsilon_0$, in the phase space of the spatial $(1+n)$-body problem, in the neighborhood of the circular coplanar positively oriented Keplerian motions with semi major axes $a_1, \ldots, a_n$, there exists a set of positive Lebesgue measure of initial conditions which lead to quasi-periodic motions with $3n - 1$ frequencies (resp. $2n$ frequencies for the planar problem)*

These solutions are slow (*secular*) modulations of the quasi-periodic motions with n frequencies corresponding to n independent elliptic motions (case $\varepsilon = 0$), the new secular frequencies being associated to a slow precession of the perihelia and the nodes. A complete proof of this theorem for the $(1+2)$-planar problem was given by Arnold in 1963. What follows s a guide to Herman's proof of the general case as written by Féjoz in [Fe2].

2.1 The secular Hamiltonian

We make again a symplectic change of coordinates, using the so-called *Poincaré coordinates* $(\lambda_j, \Lambda_j, \xi_j, \eta_j, p_j, q_j)_{i=1,\ldots,n} \in (\mathbb{T}^1 \times \mathbb{R}_+ \times \mathbb{R}^2 \times \mathbb{R}^2)^n$, analytic in the neighborhood of the circular and horizontal Keplerian motions. These coordinates are defined by the following formulas where the unnamed letters are defined on the figure: $\lambda_j = l_j + g_j + \theta_j$ is the mean longitude, $\Lambda_j = \mu_j \sqrt{M_j a_j}$ is its conjugate variable, while $r_j = \xi_j + i\eta_j = \sqrt{2\Lambda_j \left(1 - \sqrt{1 - \varepsilon_j^2}\right)} e^{-i(g_j + \theta_j)}$ and

$z_j = p_j + iq_j = \sqrt{2\Lambda_j \sqrt{1 - \varepsilon_j^2}(1 - \cos \iota_j)} e^{-i\theta_j}$ describe each a symplectic plane. The modules $|r_j| = \sqrt{\Lambda_j/2} \varepsilon_j (1 + O(\varepsilon_j^2))$ and $|z_j| = \sqrt{\Lambda_j/2} \iota_j (1 + O(\varepsilon_j^2) + O(\iota_j^2))$

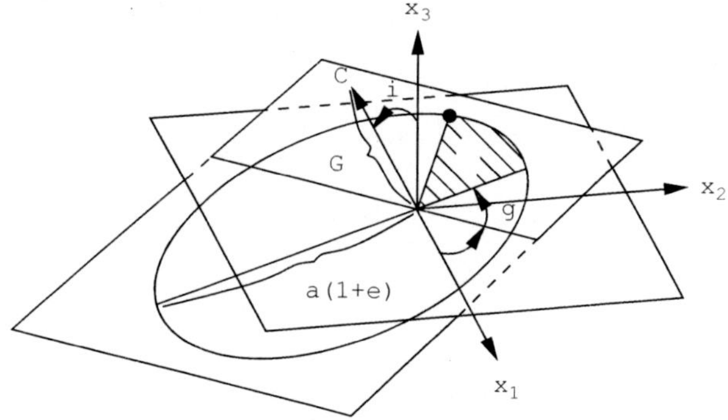

Fig. 3 Coordinates for Keplerian motion

describe respectively the eccentricity and inclination of a Keplerian ellipse; the horizontal circular motions we are interested in correspond to $|r_j| = |z_j| = 0$ for all j. We shall abbreviate the Poincaré coordinates by $(\lambda, \Lambda, Z) \in \mathbb{T}^n \times (\mathbb{R}_+)^n \times \mathbb{C}^{2n}$, with $Z = (r_1, \ldots, r_n, z_1, \ldots, z_n)$.

In these coordinates, the Hamiltonian H becomes an ε-perturbation of a sum of n uncoupled Keplerian Hamiltonians

$$H^0(\Lambda) = \sum_{1 \leqslant j \leqslant n} -\frac{\mu_j^3 M_j^2}{2\Lambda_j^2}.$$

This is a very degenerate situation indeed, as H^0 depends only on n action variables instead of $3n$. The averaging method tells us to write down H in the form

$$H(\lambda, \Lambda, Z) = H^0(\Lambda) + \varepsilon H_\varepsilon^1(\Lambda, Z) + \varepsilon H_\varepsilon^2(\lambda, \Lambda, Z),$$

where $\varepsilon H_\varepsilon^1(\Lambda, Z)$ is the average of the perturbation $H - H^0$ over the so-called *fast* angles $\lambda = (\lambda_1, \ldots, \lambda_n) \in \mathbb{T}^n$ (the only ones which move if $\varepsilon = 0$) and H_ε^2 has zero average over these angles. The hamiltonian H_ε^1 defines the *first order secular system*. As it does not depend on the mean longitudes λ_j, the conjugate variables Λ_j remain constant under its flow (they are supposed to be such that the (not too excentric) ellipses remain far enough from each other so that the perturbation function deserves its name). Hence, for given values of the Λ_j, i.e. of the semi major axes a_j, H_ε^1 defines a flow

$$\frac{dZ_k}{dt} = i\frac{\partial H_\varepsilon^1}{\partial \bar{Z}_k}, \; k = 1, \ldots, 2n,$$

on an open set, diffeomorphic to $\mathbb{R}^{4n} = \mathbb{C}^{2n}$ of the space of n-tuples of normalized ellipses in $\mathbb{R}^3$, which is diffeomorphic to $(S^2 \times S^2)^n$. The detailed study of the secular hamiltonian is a sequence of long computations, started by Laplace and Lagrange in the 18th century, of which we only summarize the results:

1. Each of the terms $Y_j \cdot Y_k$ is readily seen to have zero average, which implies

$$H_\varepsilon^1(\Lambda, Z) = - \sum_{1 \leqslant j < k \leqslant n} \int_{\mathbb{T}^n} \frac{m_j m_k}{||X_j - X_k||} d\lambda_1 \ldots d\lambda_n.$$

This is the Newtonian potential of a set of elliptic rings whose mass repartition would follow Kepler's area law.

2. Being only interested in the neighborhood of the origin, one writes down the expansion up to second order (actually third because of parity) of H_ε^1. This depends on computations, using the so-called Laplace coefficients, of the Fourier expansion of the inverse distance function of two planets considered as a periodic function of their mean longitudes.
   One gets $H_\varepsilon^1(\Lambda, Z) = h^0(\Lambda) + Q_\Lambda(Z) + O(|Z|^4)$, with

$$Q_\Lambda(Z) = Q'_\Lambda(\xi_1, \ldots, \xi_n) + Q'_\Lambda(\eta_1, \ldots, \eta_n) - Q''_\Lambda(p_1, \ldots, p_n) - Q''_\Lambda(q_1, \ldots, q_n),$$

$$Q'_\Lambda(\xi_1, \ldots, \xi_n) = \sum_{1 \leqslant j < k \leqslant n} m_j m_k \left(C_1(a_j, a_k) \left(\frac{\xi_j^2}{\Lambda_j} + \frac{\xi_k^2}{\Lambda_k} \right) + 2 C_2(a_j, a_k) \frac{\xi_j \xi_k}{\sqrt{\Lambda_j \Lambda_k}} \right),$$

$$Q''_\Lambda(p_1, \ldots, p_n) = \sum_{1 \leqslant j < k \leqslant n} m_j m_k C_1(a_j, a_k) \left(\frac{p_j}{\sqrt{\Lambda_j}} - \frac{p_k}{\sqrt{\Lambda_k}} \right)^2.$$

The value $h^0(\Lambda)$ of Q_Λ at $Z = 0$ (which is a critical point corresponding to circular horizontal motions) depends on the masses and the semi-major axes while the coefficients $C_1(a_j, a_k)$ and $C_2(a_j, a_k)$ are independent of the masses. All of them have simple expressions in terms of Laplace coefficients. As a good exercise, the reader will show for two planets that this form of the quadratic terms is essentially dictated by the symmetries of the problem.

3. If $\rho' \in SO(n)$ and $\rho'' \in SO(n)$ respectively diagonalize Q' and Q'', the linear transformation $\rho = diag(\rho', \rho', \rho'', \rho'') \in SO(4n)$ is symplectic and transforms Q_Λ into a hamiltonian of the form

$$Q_\Lambda \circ \rho(\Lambda, Z) = h^0(\Lambda) + \sum_{1 \leqslant j < k \leqslant n} \sigma_j(\xi_j + \eta_j^2) + \sum_{1 \leqslant j < k \leqslant n} \zeta_j(p_j^2 + q_j^2) + O(|Z|^4).$$

Applying the above coordinate changes to the full Hamiltonian leads to a Hamiltonian which we shall still write H, defined in a neighborhood of $\mathbb{T}^n \times \mathbb{R}_+^n \times \{0\}$ in $\mathbb{T}^n \times \mathbb{R}_+^n \times \mathbb{C}^{2n}$ (symplectic form $d\lambda \wedge d\Lambda + \sum_{1 \leqslant j \leqslant 2n} \frac{1}{2i} d\bar{Z}_j \wedge dZ_j$), of the form

$$H_\varepsilon(\lambda, \Lambda, Z) = H^0(\Lambda) + \varepsilon \left[h^0(\Lambda) + \sum_{1 \leqslant j \leqslant 2n} \tau_j(\Lambda) |Z_j|^2 + O(|Z|^4) + H_\varepsilon^2(\lambda, \Lambda, Z) \right],$$

where $\tau_j = \sigma_j$ if $1 \leqslant j \leqslant n$, $\tau_j = \zeta_j$ if $n + 1 \leqslant j \leqslant 2n$, the term $O(|Z|^4)$ does not depend of λ and H_ε^2 has zero average with respect to $\lambda \in \mathbb{T}^n$.

The degeneracy of the integrable approximation $H_\varepsilon - \varepsilon(O(|Z|^4) + H_\varepsilon^2)$ appears clearly: for $\varepsilon = 0$ or for $Z = 0$, the dimension of the invariant tori drops down to n. We shall later encounter other degeneracies which affect the spatial problem but we first turn to Herman's way of proving an appropriate KAM theorem.

2.2 Herman's normal form theorem and how to use it

Herman's powerful idea is to separate a normal form theorem for Hamiltonians close to what could be called a *Kolmogorov Hamiltonian* – one such that $\mathbb{T}^m \times \{0\}$ is a diophantine invariant torus – from the actual verification of a non-degeneracy hypothesis which allows a tuning of the available parameters turning such a normal form into a conjugacy to some Kolmogorov Hamiltonian. For a hint of the complicated history of KAM theorems with weak non degeneracy conditions, see [Se] and the references therein.

The following theorem is a far reaching generalization of the Arnold-Moser theorem on vector fields on the torus which states that, *among all C^∞ vector-fields on $\mathbb{T}^2$ close enough to a constant vector-field (noted $\omega = (\omega_1, \omega_2)$) whose frequencies ω satisfy a diophantine condition $HD_{\gamma,\tau}$ (defined below), the ones which are C^∞-conjugated to it form a submanifold of codimension 2; more precisely, that the mapping*

$$\Phi_\omega : \mathrm{Diff}^\infty(\mathbb{T}^2, 0) \times \mathbb{R}^2 \to \mathscr{X}^\infty(\mathbb{T}^2)$$

defined by $\Phi_\omega(h, \lambda) = h_* \omega + \lambda$ (where $h_* \omega$ is the direct image by h of the constant vector-field ω) is a C^∞ (more precisely *tame* in the sense of Hamilton) diffeomorphism of a neighborhood of $(Id, 0)$ onto a neighborhood of ω in $\mathscr{X}^\infty(\mathbb{T}^2)$.

We study hamiltonians $H(r, \theta)$ on $T^*\mathbb{T}^m \equiv \mathbb{T}^m \times \mathbb{R}^m$ (in our case, $m = 3n, r = (\Lambda - \Lambda_0, |Z| - |Z|_0), \theta = (\lambda, ArgZ)$). The role of the constant vector field of frequencies ω on the torus is now held by the set $\mathscr{N}_\omega$ of *Kolmogorov Hamiltonians* $N(r, \theta) = N_\omega(r) + O(r^2)$, where $N_\omega(r) = \omega \cdot r$. This is the set of Hamiltonians whose Hamiltonian vector-field leaves invariant the torus $r = 0$ and induces on it the constant vector-field with frequency vector ω. Let also $\mathscr{G}$ be a space of Hamiltonian diffeomorphisms close to Identity, defined on $\mathbb{T}^m \times B^m$, where B^m is the unit ball in $\mathbb{R}^m$ as follows: the elements of $\mathscr{G}$ are defined as truncations (described in [Fe2]) of diffeomorphisms g of $\mathbb{T}^m \times \mathbb{R}^m$ of the form $g(\theta, r) = (\varphi(\theta), {}^t d\varphi(\theta)^{-1}(r + \rho))$, where φ is a diffeomorphism of $\mathbb{T}^m$ and $\rho = df : \mathbb{T}^m \to \mathbb{R}^{m*} \equiv \mathbb{R}^m$ is an exact one form. Let $C_+^\infty(\mathbb{T}^m \times \mathbb{R}^m)$ be the quotient of the space of Hamiltonians by the real constants. We denote

$$HD_{\gamma,\tau} = \left\{ \omega \in \mathbb{R}^m, \forall k \in \mathbb{Z}^m \setminus 0, |l \cdot \omega| \geqslant \gamma \|k\|^{-\tau} \right\}.$$

Theorem 2.2 (Herman's normal form). *For every $\omega \in HD_{\gamma,\tau}$ and for every $N^o \in \mathscr{N}_\omega$, the map*

$$\Phi_\omega : \; \mathscr{N}_\omega \times \mathscr{G} \times \mathbb{R}^m \;\to\; C_+^\infty(\mathbb{T}^m \times \mathbb{R}^m)$$
$$(N, G, \Delta\omega) \quad \mapsto\; H = N \circ G + N_{\Delta\omega},$$

is a local C^∞-diffeomorphism in a neighborhood of $(N^o, id, 0)$. Moreover, the inverse map Φ_ω^{-1} depends smoothly in the sense of Whitney on $\omega \in HD_{\gamma,\tau}$.

As in the Arnold–Moser theorem, this theorem asserts that the set of Hamiltonians which are conjugated to a normal form with a diophantine frequency vector (i.e. those of the form $H = N \circ G$ with $N = N_\omega + O(r^2)$) form a submanifold of codimension m of the set of Hamiltonians modulo constants. Herman's theorem is in fact more general (see [Fe2]) in that it works also with normal forms which leave invariant tori of dimension lower than n. Following Herman, the proof given in [Fe2] uses a "hard" implicit function theorem, that is one valid in a scale of Fréchet spaces. The key feature of such theorems is the necessity of inverting (or inverting approximately) the differential of the mapping Φ_ω on a whole neighborhood of $(N^0, Id, 0)$ (invertibility is not an open property in Fréchet spaces).

Of course, it is only when the frequency correction $\Delta\omega$ vanishes that Herman's normal form implies the existence of an invariant torus. The beautiful idea of Herman was to use the Whitney extension theorem and the usual implicit function theorem to draw the following corollary (I use the name given by Féjoz): let $\mathscr{N} = \cup_{\omega \in \mathbb{R}^m} N_\omega = \{\omega \cdot r + O(r^2)\}_{\omega \in \mathbb{R}^m}$ be the set of all normal forms.

Corollary 2.1 (hypothetical conjugacy). *For every $N^0 \in \mathscr{N}$, there is a (non unique) germ of C^∞-diffeomorphism*

$$C_+^\infty(\mathbb{T}^m \times \mathbb{R}^m) \ni H \mapsto \Theta(H) = (N_H = \omega_H \cdot r + O(r^2), G_H) \in \mathscr{N} \times \mathscr{G}$$

at $N^0 \mapsto (N^0, Id)$ such that $H = N_H \circ G_H$ for each H verifying $\omega_H \in HD_{\gamma,\tau}$.

The proof is in two steps: first, the Whitney extension theorem allows to extend (non uniquely) from $C_+^\infty(\mathbb{T}^m \times \mathbb{R}^m) \times HD_{\gamma,\tau}$ to $C_+^\infty(\mathbb{T}^m \times \mathbb{R}^m) \times \mathbb{R}^m$ the map $(H, \omega) \mapsto \Phi_\omega^{-1}(H) = (N, G, \Delta\omega)$; then, from the identity $N^0 = (N^0 + N_{\omega - \omega^0}) \circ Id + N_{\omega^0 - \omega}$, one deduces that, at (N^0, Id), one has $\frac{\partial \Delta\omega}{\partial \omega} = -Id$. Hence, from the usual implicit function theorem, it is possible to define a function $\omega \mapsto \omega_H$ by locally solving the equation $\Delta\omega(\omega) = 0$.

We are now left with a serious problem: how to check that ω_H which we do not know satisfies a diophantine condition? The magic word here is "parameters".

If we were in the non-degenerate case of Kolmogorov where the *frequency map* from the actions to the frequencies of the corresponding invariant torus is a local diffeomorphism the existence of a positive measure set of "good" values of the actions would follow immediately from the fact that $HD_{\gamma,\tau}$ has positive measure. But in our case, the frequency map $H \mapsto \omega_H$ is of the form

$$(\Lambda, \rho) \mapsto \left[\nu(\Lambda) + O(\varepsilon), \varepsilon\left(\tau(\Lambda) + O(\rho^2)\right) \right].$$

Going back to Arnold and later used by Parasyuk, Bakhtin and Rüssmann, the key idea is that in the analytic case, the non-degeneracy hypothesis implying a positive measure set of good actions can be much weakened; thanks to the following result, it is enough that the image of the mapping $s \mapsto \omega_s^0$ lies in no proper vector subspace of $\mathbb{R}^m$:

Theorem 2.3 (Arnold, Margulis, Pyartli). *If some real-analytic map $s \mapsto \omega_s^o$ from a domain of $\mathbb{R}^p$ to $\mathbb{R}^m$ is non-planar in the sense that its image is nowhere locally contained in some proper vector space of $\mathbb{R}^m$, the Lebesgue measure of $\{s,\ \omega_s^o \in HD_{\gamma,\tau}\}$ is positive provided that γ is small enough and τ large enough.*

2.3 A stability theorem

We come back to Hamiltonians on $\mathbb{T}^n \times (\mathbb{R}_+)^n \times \mathbb{R}^{2p}$ of the form obtained at the end of section 2.1 (for the spatial (resp. planar) secular system $p = 2n$ (resp. $p = n$)).

$$H_\varepsilon(\lambda,\Lambda,Z) = H^0(\Lambda) + \varepsilon H_\varepsilon^1(\Lambda,Z) + \varepsilon H_\varepsilon^2(\lambda,\Lambda,Z),$$

with $H_\varepsilon^1(\Lambda,Z) = h^0(\Lambda) + \sum_{1 \leqslant j \leqslant 2n} \tau_j(\Lambda)|Z_j|^2 + 0(|Z|^4)$, and H_ε^2 has zero average with respect to $\lambda \in \mathbb{T}^n$. We denote as before $v_i = \frac{\partial H^0}{\partial \Lambda_i}(\Lambda)$.

Theorem 2.4 (Herman's stability theorem). *If, for Λ near Λ_0, the frequency map $\alpha : \Lambda \mapsto (v_1,\ldots,v_n,\tau_1,\ldots,\tau_{2p})$ is non planar, there is a positive measure set of Lagrangian invariant tori close to $\mathbb{T}^n \times \{\Lambda_0\} \times \{0\} \in \mathbb{T}^n \times (\mathbb{R}_+)^n \times \mathbb{R}^{2p}$.*

One starts by changing coordinates so that H_ε appears as a close enough approximation of an integrable Hamiltonian in the neighborhood of a Lagrangian invariant torus. There are standard ways of simplifying such a Hamiltonian by symplectic transformations defined by polynomial generating functions; the non-planarity hypothesis implies that the set A_2 of Λ's on which this is possible has positive measure and moreover that it intersects any neighborhood of Λ_0. In the case of the $(1+n)$-body problem, the assertion on the bigger set A_1 defined below is directly ensured by the non degeneracy of the map $\Lambda \mapsto v(\Lambda) = (v_1(\Lambda),\cdots v_n(\Lambda))$.

1. *Elimination "à la Lindstedt" of the dependence on the fast angles λ_j at a sufficiently high order N_1.* This is possible if Λ belongs to the set A_1 on which $v(\Lambda) \in HD_{\gamma,\tau}$. Moreover, Whitney regularity allows to extend this to a (non unique) symplectic transformation L such that $H_\varepsilon \circ L$ keeps the same form with $H_\varepsilon^2(\lambda,\Lambda,Z))$ replaced by $R_1(\varepsilon,\lambda,\Lambda,Z) + O(\varepsilon^{N_1})$, where R_1 vanishes at infinite order along $\{(\varepsilon,\lambda,\Lambda,Z)|\Lambda \in A_1\}$.

2. *Transformation to Birkhoff normal form up to order N_2.* This is possible if Λ belongs to the subset A_2 of A_1 defined by diophantine conditions on the set $(v_1,\ldots,v_n,\tau_1,\ldots,\tau_p)$ of all frequencies. As above, one can get a symplectic transformation B such that

$$H_\varepsilon \circ L \circ B(\lambda,\Lambda,Z) = H^0(\Lambda) + \varepsilon \tilde{H}^1(\varepsilon,\Lambda,Z) + \varepsilon R_2(\varepsilon,\lambda,\Lambda,Z) + 0(\varepsilon^{N_1}),$$

$$\tilde{H}^1(\varepsilon,\Lambda,Z) = h^0(\Lambda) + \sum_{1\leqslant j\leqslant p} \tau_j(\Lambda)|Z_j|^2 + K(\Lambda,|Z^2|) + O(|Z|^{2N_2}),$$

where K is a polynomial in the $|Z_j|^2$ with terms of degree between 2 and $N_2 - 1$ and R_2 vanishes at infinite order along $\{(\varepsilon,\lambda,\Lambda,Z)|\Lambda \in A_2\}$. On this subset, H_ε appears now as a $O(\varepsilon^{N_1},|Z|^{N_2})$-perturbation of the completely integrable system with Hamiltonian $H^0(\Lambda) + \varepsilon\left[h^0(\Lambda) + \sum_{1\leqslant j\leqslant p}\tau_j(\Lambda)|Z_j|^2 + K(\Lambda,|Z_1|^2,\ldots,|Z_p|^2)\right]$. To focus the attention on the Lagrangian invariant tori $\Lambda = \Lambda_0, |Z| = |Z|_0$ of this integrable approximation, one moves to symplectic polar coordinates $Z_k = \sqrt{\rho_k}e^{i\theta_k}$, which leads to

$$\mathscr{H}_\varepsilon = H^0(\Lambda) + \varepsilon\left[h^0(\Lambda) + \mathscr{K}(\Lambda,\rho)\right] + \varepsilon R_3 + O(\varepsilon^{N^1},\rho^{N_2}),$$

where R_3 vanishes at infinite order along $\{(\varepsilon,\lambda,\Lambda,Z)|\Lambda \in A_2\}$. In order to show that enough of these tori do survive the perturbation, one considers the $(m = n + p)$-parameter family $H_{(\Lambda,\rho)}$ of Hamiltonians H obtained by translating the origin of the actions at (Λ,ρ). If $\Lambda^0 \in A_2$, $\rho^0 > 0$ and if (Λ,ρ) is close to (Λ^0,ρ^0), the flow of $H_{(\Lambda,\rho)}$ is close to the flow of $H^0(\Lambda) + \varepsilon\left[h^0(\Lambda) + \mathscr{K}(\Lambda,\rho)\right]$ in the neighborhood of the Lagrangian torus $\mathscr{T}_{(\Lambda,\rho)} = \mathbb{T}^n \times \{\Lambda\} \times \{|Z|^2 = \rho\}$. The non-planarity being an open condition, it will be verified at Λ and the conclusion follows from the hypothetical conjugacy theorem.

2.4 Herman's degeneracy

For the planar $1 + n$-body problem, a thorough study of the Laplace coefficients after complexification of the semi major axes, allows proving by induction on the number of planets (letting one semi major axis go to zero) that the frequency map is non planar. For the spatial problem, this map presents an expected degeneracy, say $\zeta_n = 0$, due to the invariance under rotation of the problem, as well as an unexpected one: the trace $\sum_{1\leqslant j\leqslant n}\sigma_j + \sum_{1\leqslant j\leqslant n}\zeta_j$ of Q_Λ is always zero. In the study of the motion of the Moon, this resonance is responsible for the well-known fact that "at the first order of the theory of perturbations" the retrograde motion of the node is exactly opposite to the mean motion of the apogee. Nevertheless, it is only Herman who noticed it in its generality. An induction similar to the one done in the planar case shows that these are the only degeneracies. The first resonance is well known to disappear when the direction of the (non-zero) angular momentum is fixed (here, vertically), which corresponds to restricting the system to a codimension-2 symplectic submanifold $\mathscr{V}$; the second one disappears when completing the reduction by fixing the angular momentum and quotienting by the rotations around its axis. This comes from the fact that in the Poincaré coordinates, the vertical component of the angular momentum becomes the quadratic form $\mathscr{C}_z = \sum_{1\leqslant j\leqslant n}\left(\Lambda_j - \frac{1}{2}(|r_j|^2 + |z_j|^2)\right)$ whose trace, when restricted to $T_0\mathscr{V}$ is different from zero. Hence, after reduction, the frequency map becomes non planar and the stability theorem yields diophantine Lagrangian invariant tori of dimension $3n - 2$. To these tori correspond, for the non-reduced system, invariant tori of dimension $3n - 1$ whose number of independent frequencies is $3n - 2$ or $3n - 1$.

3 Minimal action and Marchal's theorem

3.1 Central configurations and their homographic motions

The equations of the n-body problem in an euclidean space E can be given the particularly simple form

$$\ddot{x} = \nabla U(x), \qquad\qquad (*)$$

where $x = (\mathbf{r}_1, \dots, \mathbf{r}_n) \in E^n$ and $U(x) = \sum_{i<j} m_i m_j \|\mathbf{r}_i - \mathbf{r}_j\|^{-1} \in \mathbb{R}$ are respectively an n-body configuration and its *potential function*, and where the gradient is relative to the *mass scalar product* (or *kinetic energy scalar product*), defined by

$$x' \cdot x'' = (\mathbf{r}_1', \dots, \mathbf{r}_n') \cdot (\mathbf{r}_1'', \dots, \mathbf{r}_n'') = \sum_{i=1}^{n} m_i \langle \mathbf{r}_i' - \mathbf{r}_G', \mathbf{r}_i'' - \mathbf{r}_G'' \rangle_E .$$

The presence of the centers of mass $\mathbf{r}_G = \frac{1}{\sum m_i} m_i \mathbf{r}_i$ makes the formula translation invariant; one may as well consider only configurations x such that $\mathbf{r}_G = 0$.

In addition to being invariant under translation, equation $(*)$ is invariant under isometries of E and it inherits from the homogeneity of U the following scaling property : if $x(t)$ is a solution, so is $\lambda^{-\frac{2}{3}} x(\lambda t)$ for any positive real number λ. When $n = 2$, any change in the configuration is necessarily a similarity (a segment has no shape !); when n is at least 3, the simplest motions (called *homographic*) are such that the similarity class of their configuration does not change. If $\dim E \leq 3$, such motions are necessarily of Keplerian type: if for example, the total energy $\frac{1}{2} \|\dot{x}\|^2 - U(x)$ is negative, the solution is periodic, each body following an ellipse of the same excentricity according to Kepler law. Such solutions were first discovered for $n = 3$ by Euler and Lagrange at the end of 18th century. The configurations x which admit homographic motions are called *central configurations* and their determination for $n \geq 4$ is a very difficult problem. They are characterized by the existence of a negative energy Keplerian motion with excentricity 1, which means that they collapse on their center of mass when released with 0 initial velocity. In other words, $\nabla U(x)$ is proportional to x. But $x = \frac{1}{2} \nabla I(x)$, where $I(x) = \|x\|^2$ is the *moment of inertia* of the configuration with respect to its center of mass. Hence central configurations are the critical points of the restrictions of the potential function U to the spheres $I = $ constant. As an exercise, the reader will use (squared) mutual distances as coordinates on the space of "triangles mod isometries" and prove Lagrange's result that, whatever be the masses, the only non-colinear central configuration of three masses is the equilateral triangle.

Another important fact, already proved by Lagrange for $n = 3$, is that a homographic solution with excentricity $e < 1$ is necessarily planar. Note that only the case of a *relative equilibrium* (that is $e = 0$) is "physically" obvious.

3.2 Variational characterizations of Lagrange's equilateral solutions

Equations of the type $\ddot{x} = \nabla U(x)$ are known, since Lagrange, to be the so-called Euler–Lagrange equations of an action functional, the *Lagrangian action*

$$\int L(x(t),\dot{x}(t))dt, \quad L(x,\dot{x}) = \frac{1}{2}||\dot{x}||^2 + U(x),$$

where the Lagrangian $L(x,\dot{x})$ is the difference between the kinetic energy $\frac{1}{2}||\dot{x}||^2$ and the potential energy $-U(x)$. This means that the solutions of $(*)$ are exactly the set of "extremal" curves of the action functional. It is the mathematical formulation of the so-called *principle of least action*. Poincaré was the first to try to obtain new solutions of an n-body problem using minimization. In a short note written in 1896, he looked for quasi-periodic (periodic in a rotating frame) solutions of the three-body problem in $\mathbb{R}^2$ as functions $x(t)$ defined on $[0,T]$ and with values in three-body configurations, which minimize the Lagrangian action $\int_0^T L(x(t),\dot{x}(t))$ among those with the following property: after the "period" T, the new triangle $x(T)$ is the image of the initial one $x(0)$ by a rigid rotation and the three sides have respectively turned by the real (not mod 2π) angles $\alpha, \alpha + k_1, \alpha + k_2$ where k_1 and k_2 are fixed integers. This amounts to fixing a one-dimensional homology class in the space of triangles up to rotation (this space has the topology of $\mathbb{R}^3$ deprived of three half-lines from the origin). Assuming existence (this is a consequence of Tonelli's theorem, proved around 1930, because $k_1 \neq 0$ and $k_2 \neq 0$ garantee *coercivity*, that is the impossibility that a minimizer be at infinity), he was blocked by the collision problem caused by the weakness of the Newtonian attraction. Indeed, around 1913 Sundman proved that in any solution of the n-body problem which ends in a collision (partial or total) at time t_0, two bodies i, j involved in the collision satisfy the estimates

$$||\mathbf{r}_i(t) - \mathbf{r}_j(t)|| = O(|t - t_0|^{\frac{2}{3}}), \quad ||\dot{\mathbf{r}}_i(t) - \dot{\mathbf{r}}_j(t)|| = O(|t - t_0|^{-\frac{1}{3}}).$$

For the two-body problem, these estimates are an easy exercise which was enough to convince Poincaré that the action of a solution ending in collision might (in fact always does) converge, hence that a minimizer could a priori be the mere concatenation through collisions of segments of solutions. He eliminated the problem by assuming a "strong force" potential (proportional to the inverse squared distance).

Poincaré's retreat was in a sense wise because very often such homology constraints indeed lead to minimizers with collisions. The simplest example is given by the Kepler problem of attraction by a fixed center in the plane (the two-body problem can be reduced to this). Let us look for periodic solutions of the equation $\ddot{x} = -\frac{x}{|x|^3}$ in $\mathbb{R}^2 \setminus 0$. The action is $\int_0^T (|\dot{x}(t)|^2 - \frac{1}{|x(t)|})dt$ and one seeks for minimizers in the space of loops $x(t)$ of period T going k times around the origin (i.e. loops belonging to a fixed homology class). Coercivity is insured as soon as the integer k is different from 0. It was proved by Gordon that for $k = \pm 1$, minimizers are exactly the elliptic solutions of the given period T, with any excentricity (along a curve of critical points, a function stays constant !) while, if $k \neq 0, \pm 1$, minimizers are only collision-ejection solutions (ellipses with excentricity 1). The main point was to notice that, by convexity of the action, a sequence of ejection collisions in a given time T has a higher action than a single ejection collision solution during the same time.

A partial generalization of this result exists for the three body problem (Venturelli, Zhang-Zhou): action minimizers among loops of configurations $x(t)$ of a given period T such that, during time T the three sides of the triangle make respectively k_1, k_2, k_3 complete turns, where the k_i are fixed integers, are the equilateral elliptic homographic solutions of the given period and any excentricity if $(k_1, k_2, k_3) = \pm(1, 1, 1)$, a collision ejection of the given period if this is not the case and all k_i are different from 0, unknown if one of the k_i is 0. Let us give a sketch of proof of the case $(1, 1, 1)$. In a frame fixing the center of mass, a classical identity going back to Leibniz allows to write the action as the sum of three Keplerian actions:

$$\sum_{i<j} \frac{m_i m_j}{M} \int_0^T \left[\frac{||\dot{\mathbf{r}}_{ij}(t)||^2}{2} + \frac{M}{||\mathbf{r}_{ij}(t)||} \right] dt,$$

where $M = \sum m_i$ and $\mathbf{r}_{ij}(t) = \mathbf{r}_j(t) - \mathbf{r}_i(t)$. By the result of Gordon, an a priori lower bound of the action is obtained by replacing each term by its minimum, obtained if each $\mathbf{r}_{ij}(t)$ is a Kepler elliptic solution of period T. The end of the proof consists in showing that the Lagrange equilateral solution is the only one which achieves this lower bound: from $\sum \mathbf{r}_{ij}(t) \equiv 0$ it follows that $\sum \ddot{\mathbf{r}}_{ij}(t) \equiv 0$ that is $\sum \frac{\mathbf{r}_{ij}(t)}{||\mathbf{r}_{ij}(t)||^3} \equiv 0$ from which it follows that the $\mathbf{r}_{ij}(t)$ cannot be colinear and the three mutual distances $|\mathbf{r}_{ij}(t)|$ must be equal at each instant of time.

Notice that in all the cases considered above, collision solutions exist among minimizers. This will not be the case anymore if we minimize the action among loops $x(t)$ of configurations of period T satisfying the *italian symmetry*

$$x(t - T/2) = -x(t).$$

This symmetry selects the relative equilibria (excentricity 0) among all Keplerian motions and indeed, minimizers for the two-body and three-body problem are exactly the circular solutions (with equilateral configuration in the latter case). The proof (Chenciner-Desolneux, Long-Zhang) is even simpler than above, the reason for the selection of the equilateral triangle among central configurations being more clearly seen to originate from the fact that it is the unique configuration which realizes the minimum of the restriction of U to $I = $ constant or, what amounts to the same, the minimum U_0 of the normalized potential function $\tilde{U}(x) = I^{\frac{1}{2}}(x)U(x)$.

On the other hand, the Fourier series of a symmetric loop has no constant term and this implies the inequality

$$\int_0^T ||\dot{x}(t)||^2 dt \geq \frac{4\pi^2}{T^2} \int_0^T ||x(t)||^2 dt.$$

Hence, the action A of a symmetric loop satisfies

$$A \geq A_0 = \int_0^T \left[\frac{2\pi^2}{T^2} I(x(t)) + U_0 I^{-\frac{1}{2}}(x(t)) \right] dt \geq T \inf_I \left(\frac{2\pi^2}{T^2} I + U_0 I^{-\frac{1}{2}} \right),$$

with equality if and only if there exist two configurations α and β such that $x(t) = \alpha \cos \frac{2\pi}{T} + \beta \sin \frac{2\pi}{T}$ (no harmonics of order higher than 1), and the function $\frac{2\pi^2}{T^2} I(x(t)) + U_0 I^{-\frac{1}{2}}(x(t))$ is constant and equals its absolute minimum. Hence $I(x(t)$ is constant, from which it follows that the two configurations α and β are orthogonal and have the same norm. Finally, $x(t)$ is a rigid circle in the configuration space. One concludes that the motion is a relative equilibrium by using the fact that the similitude classes of 3-bodies central configurations are isolated.

The two proofs above are misleading. As soon as the constraints select more complicated (non a priori known) solutions, one has to prove the existence of collision-free minimizers. In the next paragraph, an idea is given of the proof of Marchal's theorem which is the basic tool explaining why action minimizers under symmetry constraints are very often collision-free.

3.3 Marchal's theorem

Theorem 3.1. *Let* $x' = (\mathbf{r}'_1, \mathbf{r}'_2, \cdots, \mathbf{r}'_n)$ *and* $x'' = (\mathbf{r}''_1, \mathbf{r}''_2, \cdots, \mathbf{r}''_n)$ *be two arbitrary configurations, possibly with collisions, of n material points with positive masses* $m_1, m_2, \cdots, m_n$ *in the plane or in space. For any* $T > 0$, *any local minimizer of the action among paths* $x(t) = (\mathbf{r}_1(t), \mathbf{r}_2(t), \cdots, \mathbf{r}_n(t))$ *in the configuration space which start at* $x(0) = x'$ *and end at* $x(T) = x''$ *is collision-free, and hence a true solution of Newton's equations, in the open interval* $]0, T[$.

Already in the case of two bodies, this theorem is non-trivial. Translated in terms of the Kepler problem, it asserts that given two points $x', x'' \in \mathbb{R}^2 \setminus 0$ and $T > 0$, a minimizing path $x(t) \in \mathbb{R}^2 \setminus 0$ $x(0) = x', x(T) = x''$, is a collision-free solution of the equation $\ddot{x}(t) = -x/||x(t)||^3$. Many proofs can be given of this special case but Marchal's one is still among the simplest.

In what follows, I give the main idea of the proof of Marchal's theorem (see [Ma3, C3, FT]) . Suppose that the minimum of the action is attained by a path $x(t)$ which has a collision at time t_0. In order to get a contradiction, we try to slightly modifiy the path in such a way as to decrease the action. The problem which was faced in the early attempts to prove that minimizers of some kind are collision-free is that, except in the case of three bodies, not much is known about the configuration taken by the bodies entering the collision. There is Sundman's theory, which says that the normalized configuration tends to the set of central configurations, but the latter ones are so poorly understood that it is of no use (for five bodies and more one even does not know if the number of similitude classes is finite!). Marchal proposes to chose any one of the bodies, say $\mathbf{r}_i$ involved in the collision and to shift slightly its position at time t_0, replacing $r_i(t)$ by $r_i(t) + \varepsilon \varphi(t)\mathbf{v}_i$, where $\mathbf{v}_i$ is a unit vector and $\varphi(t)$ is a smooth function of time such that $\varphi(t_0) = 1$, supported by a small interval $[t_0 - \eta, t_0 + \eta]$. Controling the modification brought to the action by this single modification is impossible but Marchal makes the striking observation that replacing the original action by the average of the modified action when $\mathbf{v}_i$ takes every possible direction amounts to replacing the perturbed body i by a uniform repartition of its mass over a sphere in the spatial case (resp. a circle in the planar case). But, in the spatial case, the potential generated by a homogeneous sphere is constant inside the ball bounded by the sphere and equal to the potential of a point mass at the center with the same total mass outside. This is a strong hint that the averaged action is strictly smaller than the original one.

Let us prove that it is indeed the case in the simplest possible situation, to which it is indeed possible to reduce the general case. We suppose that the minimizer $x(t)$ is a parabolic homothetic collision-ejection solution of the n-body problem in $\mathbb{R}^3$, that is:

$$x(t) = |t|^{\frac{2}{3}} x_0, \ t \in [-T, T]$$

where x_0 is some central configuration (x_0 could be different for $t < 0$ and $t > 0$). Thanks to the linearity of the mean, we may treat separately ejection and collision, hence we can restrict the attention to the time interval $[0, T]$. We study deformations of $x(t)$ of the form

$$x_{\mathbf{s}}^k(t) = (\mathbf{r}_1(t), \ldots, \mathbf{r}_k(t) + R(t)\mathbf{s}, \ldots, \mathbf{r}_n(t)),$$

where $1 \leq k \leq n$ and $R(t) = (1 - \frac{t}{T})\rho$ with ρ a small positive real number and $\mathbf{s}$ belongs to the unit sphere. Taking the mean of the actions over $\mathbf{s}$ and exchanging the order of integration amounts to truncating the potential of the (k, j)-interactions to $m_j m_k / R(t)$ for t belonging to the interval $[0, t_j]$, where t_j is the characteristic time after which this potential is the same as the one for the original path, that is

$$R(t_j) = r_{jk}(t_j) = r_{jk}^0 t_j^{\frac{2}{3}},$$

which implies

$$\rho = r_{jk}^0 t_j^{\frac{2}{3}} (1 + O(t_j)).$$

Hence

$$\mathscr{A}_m^k - \mathscr{A} \leq \frac{m_k}{2} \frac{\rho^2}{T} + \sum_{j \neq k} m_j m_k \int_0^{t_j} \left[\frac{1}{R(t)} - \frac{1}{r_{jk}(t)} \right] dt,$$

(the inequality sign comes from the fact that the deformations do not keep the center of mass fixed).

In other words, the last term is the integral over the whole interval $[0,T]$ of the function $\left[\frac{1}{R(t)} - \frac{1}{r_{jk}(t)}\right]^-$, where for any $f : [0,T] \to \mathbb{R}$, we have denoted by $f(t)^-$ the function which is equal to $f(t)$ when $f(t) \leq 0$ and to 0 otherwise.

Hence

$$\mathscr{A}_m^k - \mathscr{A} \leq \frac{m_k}{2T}\rho^2 - \sum_{j \neq k,\, j \leq p} m_j m_k \Delta_j,$$

where

$$\Delta_j = \frac{T}{\rho} \log\left(1 - \frac{t_j}{T}\right) + \int_0^{t_j} \frac{1}{r_{jk}(t)} dt.$$

Hence

$$\mathscr{A}_m^k - \mathscr{A} \leq \frac{m_k}{2T}\left(r_{jk}^0\right)^2 t_j^{\frac{4}{3}} + O\left(t_j^{\frac{7}{3}}\right) - \sum_{j \neq k,\, j \leq p} m_j m_k \left(\frac{1}{r_{jk}^0} t_j^{\frac{1}{3}} + o\left(t_j^{\frac{1}{3}}\right)\right),$$

and we conclude that $\mathscr{A}_m^k - \mathscr{A} < 0$.

The proof that one can reduce the general problem to this special case is given in [C3]. It uses the ideas of R. Montgomery, S. Terracini and A. Venturelli; the two main steps in this proof are (1) the existence of an *isolated* collision in any local minimizer $x(t)$ and (2) the reduction, via *blow-up*, of the case of an arbitrary isolated collision to the case of a parabolic homothetic collision-ejection solution. In [FT] an important generalization is given, with detailed proofs, to some equivariant cases, to other exponents of the potential and any space dimension greater than 1. The main remark is that in many cases (the ones possessing the *rotating circle property*), averaging over a well-chosen circle is sufficient.

3.4 Minimization under symmetry constraints

The simplest case where Marchal's theorem applies directly is the already mentioned *italian symmetry* $x(t - T/2) = -x(t)$, which corresponds to an action of the group $\mathbb{Z}/2\mathbb{Z}$ on the space of T-periodic loops in the configuration space of the n-body problem in $\mathbb{R}^p$. Indeed, let $[t_0, t_0 + T/2] \subset [0,T]$ be a fundamental domain of this action: the restriction of x to $[t_0, t_1]$ must be an unrestricted local minimizer of the action $\mathscr{A}$ among paths with the same endpoints, and as such collision-free in the open interval $]t_0, t_1[$. As the starting point t_0 may be chosen arbitrarily, we deduce that x cannot have a collision.

For the planar problem ($p = 2$), this result is somewhat disapointing as one can prove that a relative equilibrium whose configuration minimizes the scaled potential $U_0 = I^{\frac{1}{2}} U$ is always an absolute minimizer and that these are the sole minimizers provided certain technical conditions are satisfied (which are at least satisfied for $n = 3$ and $n = 4$). Hence, in order to get interesting minimizers, one must either look at the spatial problem ($p = 3$) or impose stronger symmetry constraints. These two routes lead to interesting new families of periodic solutions of the n-body problem, the *Hip-Hops* and the *choreographies*.

1. **The Hip-Hops** (see [CV, C4]) Combined with known results on central configurations [Mo3] and the above remark that a relative equilibrium solution whose configuration minimizes U_0 is a minimizer for the italian symmetry, a simple analysis of Hessian of the action along such a relative equilibrium solution shows that a minimizer for the spatial problem cannot be a planar solution as soon as the number n of bodies is at least 4. The simplest case is the one of four equal masses for which a minimizer should be (this is not proved) the original Hip-Hop with its $D_4 \times \mathbb{Z}_2$ symmetry. In this solution, to the relative equilibrium of the square is added a vertical oscillation of the two diagonals; twice per period, the shape is the one of a regular tetrahedron. It is a remarkable compromise between the relative equilibrium of the square and the relative equilibrium of the regular tetrahedron which should have been the minimizer if it existed (it does in $\mathbb{R}^4$). More generally, whatever be the masses, the corresponding minimizers are likely to be among the "simplest" non-planar solutions of the corresponding n-body problem.

2. **The choreographies** (see [CM, Si, CGMS]) In this case, one imposes equal masses and a symmetry constraint which implies that after time T/n, the bodies occupy the same positions save for a circular permutation (i.e. the symmetry group G contains as a subgroup a copy of $\mathbb{Z}/n\mathbb{Z}$ which acts in the indicated way). This implies the existence of a curve along which the bodies move, separated by equal time lags. It is likely that the equality of the masses is a necessary condition for such a solution to exist but up to now this is proved only when $n \leqslant 5$ [C6]. The simplest choreographies are the relative equilibria of n equal masses which are the vertices of a regular n-gon. Surprizingly we shall see in the next section that they are related through families of relatively periodic solutions to more complicated choreographies (in particular the figure eight solution when $n = 3$) and to Hip-Hops. An extensive search for choreographies was done by Carles Simó (see his website for animations).

4 Global continuation via minimization

We study the three-dimensional dynamics in the neighborhood of the equilateral relative equilibrium of the regular n-gon with equal masses ($\forall i, m_i = 1$).

The fact that, when perturbed in an orthogonal direction, the length of a straight-line segment stays constant at the first order of approximation, implies a splitting of the variational equation of the n-body problem along any planar solution into a part (HVE) describing the "horizontal variations" (along the plane of motion) and one (VVE) describing the vertical ones (orthogonal to the plane of motion). When the planar solution is a relative equilibrium, this last equation takes the particularly simple form

$$\ddot{z}_i = \sum_{j \neq i} \frac{m_j}{r_{ij}^3} (z_j - z_i), \qquad\qquad (VVE)$$

where the r_{ij} are the (constant) mutual distances of the bodies in the relative equilibrium and $(z_1, z_2, \cdots, z_n) \in \mathbb{R}^n$ are supposed to be such that $\sum_{i=1}^n z_i = 0$, which amounts to fixing the center of mass at the origin. In what follows, we suppose that all the m_i are equal.

After reducing the rotation symmetry by fixing the angular momentum and quotienting by the rotations around its axis, the relative equilibrium becomes an isolated equilibrium. One reads directly from the variational equation the spectrum of the linearized vector-field at this equilibrium: the corresponding $6n - 10$ dimensional matrix splits into a $4n - 6$ *horizontal* block and a $2n - 4$ *vertical* block whose eigenvalues are all purely imaginary because the Newton force is attractive.

In the next sections, we concentrate essentially on the case $n = 3$, giving only hints at the end for the cases $n = 4$ (partially understood) and $n > 4$ more conjectural.

4.1 Bifurcations from the Lagrange equilateral relative equilibrium

When $n = 3$, after reducing the rotation symmetry and restricting to a center manifold one gets into a situation very similar to the one in the lunar problem, with a 1–1 resonant spectrum and energy surfaces diffeomorphic to the three-sphere. Here also the local existence of two Lyapunov families of (relatively) periodic solutions can be proved: one is already known, it is the homographic family; the other one, when globally continued (see the next section) goes all the way to the reverse equilateral relative equilibrium through the planar figure eight solution. In an energy surface close to the relative equilibrium, the flow admits an annulus of section whose Poincaré return map is a twist map which, because of a resonance which persists all along the homographic family, is the identity on the corresponding boundary. I shall not reproduce the computations of [CF2] but be content with explaining the similarities and the differences with the first chapter.

For the relative equilibrium of an equilateral triangle whose edges have length 1 and vertices have masses m_i, (VVE) reads $\ddot{z}_i = \sum_{j \neq i} m_j (z_j - z_i)$, $i = 0, 1, 2$. As $\sum_{i=0}^2 m_i z_i = 0$, this becomes the following (with $M = \sum_{i=0}^2 m_i$):

$$\ddot{z}_i = -M z_i, \quad i = 0, 1, 2.$$

We shall choose the masses to be $1/3$ so that the period of the relative equilibrium solution is 2π and the $2n - 4 = 2$ "vertical" eigenvalues are $\pm i$.

On the other hand, the $4n - 6 = 6$ "horizontal" eigenvalues are $\pm i$ and a quadruple $\pm \frac{1}{\sqrt{2}} \pm i$ (see for instance [Mo2]), so that the spectrum is completely resonant. Using Maple, an analogue of the normal form described in the first chapter can be computed. This leads to complex coordinates (u, v, h, k) (I keep the notations of [CF2]) in the tangent space (identified to $\mathbb{C}^4$) of the eight-dimensional reduced phase space such that the linearized vector field becomes free of non-resonant terms up to order three. The normal form, which is not unique at a general order, can be chosen so that the vector field is invariant under $\mathcal{T}(u,v,h,k) = (u,-v,h,k)$. This corresponds to the symmetry with respect to the invariant horizontal plane, which is defined by the equation $v = 0$.

The result is of the following form:

$$\dot{u} = iu[1 + \alpha|u|^2 + \beta|v|^2 + \gamma hk + \bar{\gamma}\bar{h}\bar{k}] + O_5$$
$$\dot{v} = iv[1 + a|u|^2 + b|v|^2 + chk + \bar{c}\bar{h}\bar{k}] + A\bar{v}h\bar{k} + O_5$$
$$\dot{h} = \lambda h[1 + r|u|^2 + s|v|^2 + thk + t'\bar{h}\bar{k}] + Rv^2\bar{h} + O_5$$
$$\dot{k} = -\lambda k[1 + r|u|^2 + s|v|^2 + thk + t'\bar{h}\bar{k}] - R\bar{v}^2\bar{k} + O_5,$$

where the coefficients have the following non-zero values:

$$\alpha = -1, \beta = -1, \gamma = \frac{9}{2} + 6i\sqrt{2},$$

$$a = -1, b = -\frac{21}{19}, c = \frac{186}{19} + \frac{126\sqrt{2}}{19}i, A = -\frac{120}{19},$$

$$r = -\frac{11}{12} - \frac{\sqrt{2}}{12}i, s = -\frac{73}{57} + \frac{10\sqrt{2}}{57}i, t = \frac{275}{57} + \frac{334\sqrt{2}}{57}i,$$

$$t' = \frac{105}{19}(1 - i\sqrt{2}), R = \frac{5\sqrt{2}}{19}i,$$

and where O_5 stands for real analytic functions of order 5 in $u,\bar{u},v,\bar{v},h,\bar{h},k,\bar{k}$.

Even if the situation looks more complicated than in the restricted problem, it is not really so. This is because one can restrict the attention to a "center manifold" tangent to the invariant space associated to the purely imaginary part of the spectrum, and containing all the local recurrence near the equilibrium. A simple analysis shows that, when lifted up to the non-reduced phase space, such a four-dimensional center manifold at the equilibrium becomes a six-dimensional manifold tangent to the one obtained from the relative equilibrium solution by making the rotations act independently on positions and momenta. From this description of the tangent space one can deduce that the restriction of the reduced Hamiltonian to a center manifold has the equilibrium as a non-degenerate minimum, which implies that its levels close enough to the equilibrium are three spheres (and in fact, as noted by Moeckel, that the center manifold is unique). In restriction to the center manifold (coordinates $u,\bar{u},v,\bar{v}$), the normal form, still invariant under the mapping $\tau : (u,v) \mapsto (u,-v)$, is of the form

$$\dot{u} = iu[1 + \alpha|u|^2 + \beta|v|^2] + O_5$$
$$\dot{v} = iv[1 + a|u|^2 + b|v|^2] + O_5,$$

with $v = 0$ defining the Lyapunov family of equilateral homographic motions. Moreover, the energy becomes

$$H = -\frac{1}{2} + \frac{|u|^2}{36} + \frac{|v|^2}{6} + O_4.$$

The problem is now similar to the planar circular restricted problem in the Lunar case (see [Co, C0, 28] or [Du] in a more general situation), where the Lyapunov orbits are Hill's direct and retrograde orbits. The proof of existence and local uniqueness of the vertical Lyapunov family (the one tangent to $u = 0$) follows exactly as in the first chapter because $b \neq \beta$; moreover, if we knew that our center manifold is analytic, we would get also analyticity of the family. On the contrary, the higher order resonance $a = \alpha$ would prevent us from applying the same proof to the horizontal homographic family tangent to $v = 0$ if we did not know that it exists. A simple analysis of the vertical variational equation along the homographic family shows that this resonance must persist in normal forms of any order: the coefficients of the monomials $u|u|^{2k}$ in $\dot{u}$ and $v|u|^{2k}$ in $\dot{v}$ are necessarily equal. One can nevertheless prove that no other Lyapunov family bifurcates from the relative equilibrium by showing that the Poincaré return map in an annulus of section, whose one boundary belongs to the homographic family and the other one to the vertical family, is a monotone twist map.

4.2 From the equilateral triangle to the Eight

The vertical Lyapunov family is highly symmetric. Indeed, after choosing appropriately a phase, it is tangent to the "linear" family

$$\mathbf{r}_j(t) = \left(\frac{1}{\sqrt{3}} \zeta^j e^{i2\pi t}, A\mathrm{Re}(\bar{\zeta}^j e^{i2\pi t}) \right) \in \mathbb{R}^2 \times \mathbb{R} = \mathbb{R}^3, \quad j \in \mathbb{Z}/3\mathbb{Z}, \qquad (S_1)$$

where $\zeta = e^{\frac{2\pi}{3}}$ and the amplitude A is a real parameter. The discrete symmetry group of (S_1) is seeked as a subgroup of

$$G_0 = O(\mathbb{R}/\mathbb{Z}) \times \Sigma(3) \times O(\mathbb{R}^3),$$

where $g = (\tau, \sigma, \rho) \in G_0$ acts naturally on the space of 1-periodic loops:

$$\begin{array}{ccc} x: \mathbb{R}/\mathbb{Z} \times \{1,2,3\} & \to & \mathbb{R}^3 \\ \tau\downarrow \quad\quad \sigma\downarrow & & \rho\downarrow \\ gx: \mathbb{R}/\mathbb{Z} \times \{1,2,3\} & \to & \mathbb{R}^3. \end{array}$$

If $x = (\mathbf{r}_1, \mathbf{r}_2, \mathbf{r}_3)$ is a loop in the configuration space, the transformed loop by the (left) action of $g = (\tau, \sigma, \rho)$ is

$$g\mathbf{r}_j(t) = \rho\mathbf{r}_{\sigma^{-1}(j)}(\tau^{-1}(t)).$$

Lemma 4.1. *The stabilizer $G_1 \subset G_0$ of (S_1) is isomorphic to the dihedral group D_6 with 12 elements.*

The proof is an easy exercise. One finds that the elements (τ, σ, ρ) of G_1 act as follows (vectors in $\mathbb{R}^3$ are decomposed into a horizontal part h and a vertical part v):

$$\tau^{-1}(t) = \xi(t-\theta), \ \sigma^{-1}(j) = \xi(j+\delta), \ \rho(h,v) = (e^{i2\pi\alpha}\bar{h}^{\xi}, e^{i\pi\beta}v),$$

with $\xi = \pm 1$ (and $\bar{h}^{\xi} = h$ or $\bar{h}$ according to whether $\xi = +1$ or $\xi = -1$) and $\alpha \in \mathbb{R}/\mathbb{Z}, \beta \in \mathbb{Z}/2\mathbb{Z}, \delta \in \mathbb{Z}/3\mathbb{Z}, \theta \in \mathbb{R}/\mathbb{Z}$ satisfying

$$\alpha = \theta - \frac{\delta}{3} \quad (\text{mod } 1), \quad \theta = \frac{\beta}{2} - \frac{\delta}{3} \quad (\text{mod } 1).$$

The choices of $(\xi = 1, \beta = 1, \delta = 1)$ and $(\xi = -1, \beta = 0, \delta = 0)$ define generators g_1 and g_2 of G_1 which satisfy the relations $g_1^6 = g_2^2 = 1, g_1 g_2 = g_2 g_1^{-1}$, which is a presentation of D_6.

In a frame which rotates uniformly in the opposite direction with the same frequency as the relative equilibrium, (S_1) becomes

$$\hat{\mathbf{r}}_j(t) = \left(\zeta^j e^{i4\pi t}, \text{Re}(\bar{\zeta}^j e^{i2\pi t})\right) \in \mathbb{R}^2 \times \mathbb{R} = \mathbb{R}^3, \quad j \in \mathbb{Z}/3\mathbb{Z}, \tag{S_1}$$

The symmetry group does not change but its action does: the formula defining α is changed to $\alpha = 2\theta - \frac{\delta}{3} = \beta - \delta \ (\text{mod } 1) = 0 \ (\text{mod } 1)$. The resulting curve in rotating frame is now a *choreography*. Indeed, the group element defined by $\xi = 1, \beta = 0, \delta = 1$, transforms $(h_j(t), v_j(t))$ into $(h_{j+1}(t-\frac{1}{3}), v_{j+1}(t-\frac{1}{3}))$: all bodies lie on one and the same spatial curve. Now, it follows from unicity that

Lemma 4.2. *In a family of rotating frames parametrized by ϖ close to -2π, an appropriate lift of the local vertical Lyapunov family becomes a family of D_6 invariant choreographies (called the P_{12} family).*

Global continuation of the family is based on the following remark [Ma2, CF2]: we consider the following family (parametrized by ϖ) of paths in the configuration space:

$$\mathbf{r}_j^{\varpi}(t) = \left(\frac{1}{\sqrt{3}}\left(\frac{4\pi + \varpi}{2\pi}\right)^{-\frac{2}{3}} \zeta^j e^{i(4\pi + \varpi)t}, 0\right), \quad j \in \mathbb{Z}/3\mathbb{Z}. \tag{L}$$

In a frame which rotates uniformly with frequency ϖ, each member of the family becomes a loop with the G_1 symmetry, the equilateral triangle formed by the bodies making two complete rotations during the period 1. Its action during the period 1 is readily computed to be proportional to $\left(\frac{4\pi + \varpi}{2\pi}\right)^{\frac{2}{3}}$. In particular, it tends to its absolute minimum zero as ϖ tends to -4π, the limit situation corresponding in the inertial frame to bodies at rest at infinity. When ϖ varies from -4π to 0, the action increases. It can stop being a relative minimum among paths which, in the rotating frame become loops with the G_1 symmetry, only when appears a 1-periodic Jacobi field, that is a solution of the variational equation which, in the rotating frame, is 1-periodic and possesses the required G_1 symmetry. This is the case only when $\varpi = -2\pi$. For values of ϖ closer to 0, the minimum is no more the (L) family but an appropriate lift of the vertical Lyapunov family. The global continuation is obtained by looking, for each value of ϖ between -2π and 0, to such a minimizer among paths which are G_1-symmetric in the rotating frame. The end of the family is the figure Eight solution for which the D_6 symmetry can be interpreted as the symmetry of the space of similarity classes of plane oriented triangles (the so-called *shape sphere* (see [CM, Mo1]). It is the maximal discrete symmetry that a solution of the three-body problem may possess in the case of equal masses (see [Ma1] 10-8-2).

Technically, one is faced with the problem of showing that, for each value of ϖ, a (local) minimizer has no collision. This is not a direct consequence of Marchal's theorem because of the time reversal symmetry which implies that the boundaries of a fundamental domain of the τ action on the time circle cannot be chosen arbitrarily. Nevertheless, this can be proved by a direct estimation of a lower bound of the action of paths with collision with the given symmetry: this lower bound happens to be exactly the value of the action of the member of (L) corresponding to $\varpi = 0$.

Remarks.

1. Using obvious symmetries, the P_{12} family can be continued into a loop of quasi-periodic solutions containing the horizontal equilateral relative equilibria rotating in both directions (the first line of Fig. 4 shows half of it in the rotating frame). Applying isometries and scaling, this defines in the 12-dimensional (after reduction of translations) phase space a compact invariant six-dimensional submanifold entirely foliated by relatively periodic solutions. Topologically, this manifold is a fibre space over the lens space $L(4,1)$.

2. It is interesting to recall a remark made by C. Marchal at page 257 of his book [Ma1]: after having determined the expansion ot the vertical Lyapunov family up to order 6 in a small parameter c_1 corresponding to the vertical extension of the solution (opening of the mouth of the oyster described in the rotating frame), he asks for their continuation, mentioning as an example of surprising continuation the family of retrograde Hill solutions up to the colinear "Schubart" solution (see [He]).

4.3 *From the square to the Hip-Hop*

In the case of the square relative equilibrium of four equal masses, there are two Lyapunov families in addition to the homographic family; one of these leads by continuation to the Hip-Hop, which is the simplest non-planar solution of the four-body problem (line 2 of Fig. 4). The possibility of obtaining this family by minimization of the action is related to the fact that, in $\mathbb{R}^3$, a relative equilibrium must be planar (I recalled in Sect. 3.1 that this is true of any homographic solution of the n-body problem).

The case of the Hip-Hop corresponds to a frequency which is not in resonance with the frequency of the relative equilibrium; the local study is done in [Ba] (compare also with [MS] for the case of an additional central mass). The global continuation of the Hip-Hop family is done in [TV]. Here also, the proof that there are no collisions for minimizers in this family cannot appeal to Marchal's theorem

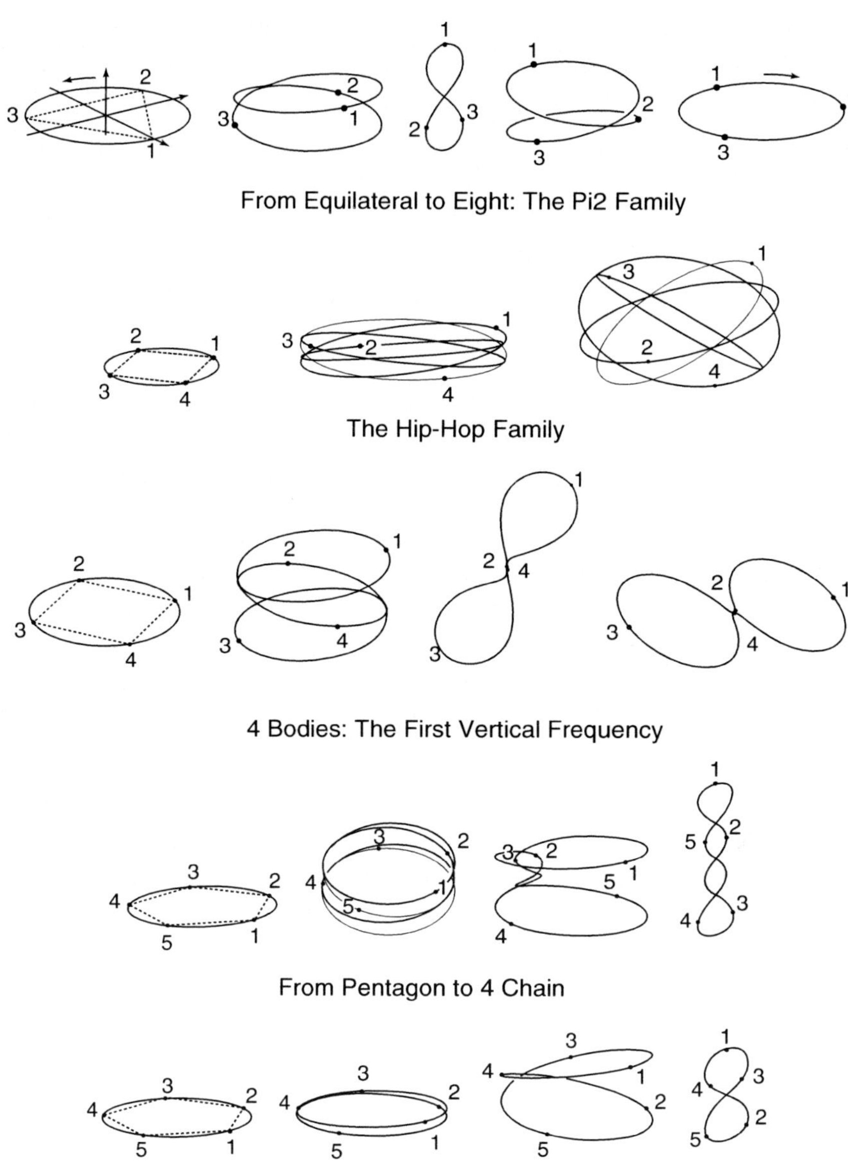

From Equilateral to Eight: The Pi2 Family

The Hip-Hop Family

4 Bodies: The First Vertical Frequency

From Pentagon to 4 Chain

From Pentagon to Eight

Fig. 4 Lyapunov families seen in the rotating frame (horizontal parameter ϖ)

or to its equivariant strengthening given in [FT]). The problem is the topological constraint attached to the rotating frame: one has to minimize among paths such that, in the inertial frame, the starting point and end point of each body make a fixed real (not mod 2π) angle $\alpha = -\varpi T$ between 0 and 2π, and this is a topological condition as soon as $\pi < \alpha < 2\pi$ (think of the same problem for the planar Kepler problem). The proof that no collision occurs in a minimizer is by contradiction via the introduction of an obstacle. The end of the family should be a simultaneous double collision but this is not proved.

Remark. The fate of the first vertical Lyapunov family, associated to the frequency of the relative equilibrium is complicated (line 3 of Fig. 4), probably leading through a secondary bifurcation to a planar solution proved to exist at first numerically by J. Gerver and then with a computer assisted proof by Kapela and Zglyczinski (this solution lies in the horizontal plane and not the vertical one because its angular momentum, in contrast with the figure eight solution, is not zero).

4.4 The avatars of the regular n-gon relative equilibrium: eights, chains and generalized Hip-Hops

Symmetries of the solutions of VVE along the regular n-gon relative equilibrium are easily analyzed [CF3] and may lead to Lyapunov families with interesting continuation [CF1] (lines 4 and 5 of Fig. 4). Possible problems connected to minimization under the corresponding symmetry constraints could appear for $n \geqslant 6$ because of the appearance of new imaginary eigenvalues of the Horizontal Variational Equation [Mo2] which could lead to different types of bifurcations with the given symmetries.

Remark. It is easy to prove that when, observed in the inertial frame, the members of the vertical Lyapunov families attached to the regular n-gon relative equilibrium are choreographies for a dense set of values of the parameter ϖ.

Thanks to Jacques Féjoz, Laurent Niederman and David Sauzin for various comments about these notes. Special thanks to Jacques Féjoz for his help with the figures.

References

[A]   V.I. Arnold *Méthodes mathématiques pour la mécanique classique*, Editions Mir 1976

[Ba]   E. Barrabes, J.M. Cors, C. Pinyol, J. Soler *Hip-hop solutions of the 2N-body problem*, Celestial Mechanics **95**, 55–66 (2006)

[B]   J.B. Bost *Tores invariants des systèmes dynamiques hamiltoniens (d'après Kolmogorov, Arnol'd, Moser, Rüssmann, Zehnder, Herman, Pöschel,...)*, (séminaire Bourbaki 639) Astérisque 133–134, p. 113–157, Société Mathématique de France (1986)

[C0]   A. Chenciner *Le problème de la lune et la théorie des systèmes dynamiques*, unpublished notes, Université Paris 7 (1985)

[C1]   A. Chenciner Ravello (1997) Notes available at `http://www.imcce.fr/Equipes/ASD/person/chenciner/chen_preprint.html`

[C2]   A. Chenciner *Conservative Dynamics and the Calculus of Variations*, ELAM (December 2005). Notes available at http://www.imcce.fr/Equipes/ASD/person/chenciner/chen_preprint.html

[C3]   A. Chenciner *Action minimizing solutions of the n-body problem: from homology to symmetry*, Proceedings ICM, Beijing 2002, vol. III, PP. 279–294

[C4]   A. Chenciner *Simple non-planar periodic solutions of the n-body problem*, proceedings NDDS, Kyoto (2002)

[C5]   A. Chenciner *Symmetries and "simple" solutions of the n-body problem*, Proceedings ICMP03 (Lisbonne 2003), 4–20 World Scientific (2005)

[C6]   A. Chenciner *Are there perverse choreographies?*, pages 63–76 of "New advances in Celestial Mechanics and Hamiltonian Systems" (HAMSYS 2001), J. Delgado, E.A. Lacomba, J. Llibre & E. Pérez-Chavela editors, Kluwer (2004)

[CF1]   A. Chenciner, J. Féjoz, *L'équation aux variations verticales d'un équilibre relatif comme source de nouvelles solutions périodiques du problème des N corps*, C.R. Acad. Sci. Paris Ser I 1340, 593–598 (2005)

[CF2]   A. Chenciner, J. Féjoz, *The flow of the equal-mass spatial 3-body problem in the neighborhood of the equilateral relative equilibrium*, to appear in Discrete and Continuous Dynamical Systems (2007)

[CF3]   A. Chenciner, J. Féjoz *Unchained polygons and the n-body problem*, in preparation

[CGMS]   A. Chenciner, J. Gerver, R. Montgomery, C. Simó *Simple Choreographic Motions of N bodies: A Preliminary Study*, in Geometry, Mechanics, and Dynamics, pp. 287–308, Springer (2002)

[CL]   A. Chenciner, J. Llibre *A note on the existence of invariant punctured tori in the planar circular restricted three-body problem*, Ergodic theory and Dynamical Systems, 8^*, 63–72 (1988)

[CM]   A. Chenciner, R. Montgomery *A remarkable periodic solution of the three-body problem in the case of equal masses*, Annals of Mathematics **152**, 881–901 (2000)

[CV]   A. Chenciner, A. Venturelli, *Minima de l'intégrale d'action du Problème newtonien de 4 corps de masses égales dans $\mathbb{R}^3$: orbites "hip-hop"*, Celestial Mechanics **77**, pp. 139–152 (2000)

[Co]   C. Conley *On some new long periodic solutions of the plane restricted three-body problem*, Communications on Pure and Applied Mathematics, **XVI**, 449–467 (1963)

[Du]   Duistermaat *Bifurcations of periodic solutions near equilibrium points of Hamiltonian systems*, CIME lectures Montecatini, LN in Mathematics **1057**, Springer (1984)

[Fe1]   J. Féjoz *Quasiperiodic motions in the planar three-body problem*, Journal of Differential Equations **183**, 303–341 (2002)

[Fe2]   J. Féjoz *Démonstration du 'théorème d'Arnold' sur la stabilité du système planétaire (d'après Herman)*, Michael Herman Memorial Issue, *Ergodic Theory and Dynamical Systems* **24, 5** (2004) 1521–1582. Updated version at http://www.institut.math.jussieu.fr/~fejoz/

[FT]   D. Ferrario, S. Terracini *On the Existence of Collisionless Equivariant Minimizers for the Classical n-body Problem* Invent. Math. **155**, no. 2, 305–362 (2004)

[He]   M. Hénon *A family of periodic solutions of the planar three-body problem, and their stability*, Celestial Mechanics **13**, 267–285 (1976)

[Ku]   M. Kummer *On the stability of Hill's solutions of the plane restricted three-body problem*, American Journal of Maths **101**, 1333–1354 (1979)

[Ma1]   C. Marchal *The Three-Body Problem*, Elsevier (1990)

[Ma2]   C. Marchal *The P_{12} family ot the three-body problem - the simplest family of periodic orbits with twelve symmetries per period*, Celestial Mechanics **78**, 279–298 (2000)

[Ma3]   C. Marchal *How the method of minimization of action avoids singularities*, Celestial Mechanics **83**, 325–353 (2002)

[MS]   K. Meyer, D. Schmidt *Librations of central configurations and braided Saturn rings*, Celestial Mechanics **55**, 289–303 (1993)

[Mo1]   R. Moeckel *Some qualitative features of the three-body problem*, Contemporary Mathematics **81** (1988)

[Mo2]   R. Moeckel *Linear stability analysis of some symmetrical classes of relative equilibria*,

[Mo3]   R. Moeckel *On central configurations*, Mat. Zeit. **205**, 499–517 (1990)

[Se]   M. Sevryuk *The classical KAM theory at the dawn of the twenty-first century*, Moscow Math J. **3** (2003)

[Si]   C. Simó *New families of Solutions in N-Body Problems*, Proceedings third european congress of Mathematics, Progress in Mathematics **201**, 101–115 (2001)

[Te]   S. Terracini *On the variational approach to the periodic N-body problem*, Celestial Mechanics and Dynamical Astronomy **95**, 3–25 (2006)

[TV]   S. Terracini, A. Venturelli *Symmetric trajectories for the 2N-body problem with equal masses*, to appear in ARMA (2007)

Averaging method and adiabatic invariants

Anatoly Neishtadt[1]

Abstract There are many problems that lead to analysis of Hamiltonian dynamical systems in which one can distinguish motions of two types: slow motions and fast motions. Adiabatic perturbation theory is a mathematical tool for the asymptotic description of dynamics in such systems. This theory allows to construct adiabatic invariants, which are approximate first integrals of the systems. These quantities change by small amounts on large time intervals, over which the variation of slow variables is not small. Adiabatic invariants usually arise as first integrals of the system after having been averaged over the fast dynamics. Adiabatic invariants are important dynamical quantities. In particular, if a system has sufficiently many adiabatic invariants, then the motion over long time intervals is close to regular. On the other hand, the destruction of adiabatic invariance leads to chaotic dynamics.

1 Introduction

Adiabatic invariance is a remarkable phenomenon in dynamics of systems with slowly varying parameters. It can be described as follows. Consider a system which depends on a parameter. Suppose that the system has a first integral for every fixed value of this parameter. If this parameter changes in time, the system in general does not have any, even approximate, first integrals. However, if the parameter is changing slowly, such an approximate first integral exists. This approximate first integral is called an *adiabatic invariant*. It is a function of phase variables and the parameter such that its value along a trajectory remains approximately constant on long time intervals on which the parameter changes considerably. Here are several classical examples.

[1] Space Research Institute RAS, ul.Profsouznaya, 84/32, Moscow, Russia, GSP-7, 117997
e-mail: aneishta@iki.rssi.ru
Loughborough University, Loughborough LE11 3TU, UK
e-mail: A.Neishtadt@lboro.ac.uk

W. Craig (ed.), Hamiltonian Dynamical Systems and Applications, 53–66.

Example 1 (Rayleigh, or Lorentz-Einstein, pendulum). Consider a mathematical pendulum whose length l is varying slowly in time: $l = l(\varepsilon t)$, t is time and ε is a small positive parameter. On a long time interval of length $\sim 1/\varepsilon$, the energy E of the pendulum will change considerably, by a value ~ 1. However, there is a function $I = I(E, l)$ of energy and length of pendulum which is an adiabatic invariant; its value changes only by $O(\varepsilon)$ during time $1/\varepsilon$. This function I is the "action" of the pendulum. If the amplitude of oscillations is small then I is approximately equal to $E/(g/l)^{1/2}$, where g is the acceleration of gravity (we assume that E is normalized so that it is zero when the amplitude of oscillations vanishes).

Example 2 (Fermi–Ulam model). Consider the one-dimensional motion of a particle bouncing between slowly moving ideally reflecting walls, where the distance between the walls is $d = d(\varepsilon t)$. Then the product vd of the particle's velocity v and the distance between the walls d is an adiabatic invariant: its value changes only by $O(\varepsilon)$ during time $1/\varepsilon$.

Adiabatic invariants became widely known and used only after the work of P. Ehrenfest, who suggested the name "adiabatic invariant". He discovered the adiabatic invariance of the "action" variable in Hamiltonian systems with one degree of freedom (see Sect. 2) as a particular case of some thermodynamical relations that had been established by L. Boltzmann, R. Clausius and C. Szily [16]. Subsequently, adiabatic invariance was a very important concept in the early stage of development of quantum mechanics. Later, after the work of H. Alfvén, adiabatic invariance becomes important in problems of plasma physics. But in this case one should consider not systems with slowly varying in time parameters, but systems with slow dependence on some of the phase variables.

Example 3 (Magnetic moment). In a constant magnetic field a charged particle moves along a spiral around a force line of the field. This motion is a composition of rotation around the field line (along a circle which is called the Larmor circle) and a drift motion of this circle. In the case when there is a small relative change of the field over a distance of order of the Larmor radius and when the pitch of the spiral is small of order ε, the magnetic moment of the particle is an adiabatic invariant; its value changes only by $O(\varepsilon)$ during time $1/\varepsilon$. The magnetic moment of the particle is defined to be the value $E_\perp/B$, where $E_\perp$ is the energy of the Larmor motion, $E_\perp = v_\perp^2/2$, $v_\perp$ is the value of projection of velocity of the particle onto the plane perpendicular to the magnetic field, and B is the strength of the magnetic field.

Adiabatic invariants have also applications in celestial mechanics, hydrodynamics, optics, radio-physics, chemical kinetics, to name other areas in the physical sciences.

2 Adiabatic invariance in one-frequency systems

A somewhat general framework in which adiabatic invariants appear can be described as follows. Consider a Hamiltonian system with one degree of freedom, and suppose that the Hamiltonian E of this system depends on a parameter λ which is slowly varying in time;

$$E = E(p,q,\lambda), \ \lambda = \lambda(\tau), \ \tau = \varepsilon t, \ 0 < \varepsilon \ll 1. \tag{1}$$

Here (p,q) are canonical conjugate variables, $(p,q) \in R^2$. All functions are assumed to be smooth enough. Because the parameter λ is changing slowly, it is reasonable to first consider the problem at frozen values of λ. For every frozen value of λ in the plane p,q let there be a domain filled by closed phase curves of E (Fig. 1). In this domain one can introduce "action-angle" variables $I = I(p,q,\lambda), \varphi = \varphi(p,q,\lambda) \bmod 2\pi$ [4]. The "action" $I(p,q,\lambda)$ is the area surrounded by the phase curve passing through the point (p,q), divided by 2π. The "angle" $\varphi(p,q,\lambda)$ is a uniformly varying angular variable on this phase curve transcribed by the motion of the system with Hamiltonian E. Now let the parameter λ change in time. Denote by $(p(t),q(t))$ a solution of our Hamiltonian system.

Theorem 2.1. *The action variable $I(p,q,\lambda)$ is an adiabatic invariant:*

$$|I(p(t),q(t),\lambda(\varepsilon t)) - I_0| < C\varepsilon \ \text{for} \ 0 \leqslant t \leqslant 1/\varepsilon.$$

Here $I_0 = I(p(0),q(0),\lambda(0))$, $C = \text{const} > 0$.

Proof. For fixed λ the canonical transformation $p,q \mapsto I,\varphi$ is defined by means of a generating function $W = W(I,q,\lambda)$. The old and the new variables are related via the expressions

$$p = \partial W/\partial q, \ \varphi = \partial W/\partial I. \tag{2}$$

Denote $H_0 = H_0(I,\tau)$ the old Hamiltonian expressed in the new variables. Let us make a canonical transformation of variables by means of formulas (2) in the case when the parameter λ is changing in time. According to a standard recipe of

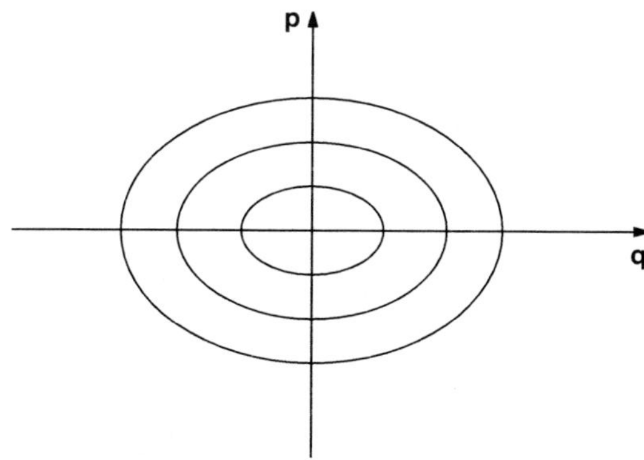

Fig. 1 Domain in the phase plane filled by closed phase curves

analytical dynamics, the behaviour of the variables I, φ is described by a Hamiltonian system with Hamiltonian $H(I, \varphi, \tau, \varepsilon) = E + \partial W / \partial t = H_0(I, \tau) + \varepsilon H_1(I, \varphi, \tau)$, where the function H_1 is 2π - periodic in φ. The equations of motion have the form

$$\dot{I} = -\varepsilon \frac{\partial H_1}{\partial \varphi}, \dot{\tau} = \varepsilon, \quad \dot{\varphi} = \frac{\partial H_0}{\partial I} + \varepsilon \frac{\partial H_1}{\partial I}. \tag{3}$$

This system contains slowly varying variables I, τ and angular variable (the phase) φ, which rotates with frequency approximately equal to $\omega = \partial H_0 / \partial I$. Systems of such form are standard objects for application of the averaging method (see, e.g. [4]). In order to describe approximately the behaviour of the slow variables, the averaging method prescribes to average the r.h.s of equations for these variables over the fast phase. The averaged equation for I has the form $\dot{I} = 0$, which implies that $I = \text{const}$ along the trajectory. The theory about accuracy of the averaging method (see, e.g. [4]) says that behavior of slow variables in the exact system is described by solutions of the averaged system with the same initial conditions, with accuracy $O(\varepsilon)$, over a time interval of length $1/\varepsilon$. This implies the assertion of the theorem. $\square$

Example 4 (Quadratic Hamiltonians). Consider the quadratic Hamiltonian

$$E = \frac{1}{2}(ap^2 + 2bpq + cq^2),$$

and assume that $\omega^2 = ac - b^2 > 0$. For a, b, c constant, the Hamiltonian E describes linear oscillations with frequency ω, and phase trajectories (level lines of E) are ellipses. It is easy to check that the "action" is $I = E/\omega$. If a, b, c are changing slowly in time, then according to Theorem 1 the action I is an adiabatic invariant. This explains the result of Example 1 for the case of small oscillations of a pendulum.

For Example 2 we have $I = vd/\pi$. However in this case adiabatic invariance of I does not follow from Theorem 1 because of the lack of smoothness in the system. "The proof of adiabatic invariance of vd in this system is an instructive elementary problem" [4] (however, see [13] for general consideration of adiabatic invariance in systems with impacts).

A more general framework for adiabatic invariance, which is needed for problems similar to that in Example 3, is a framework of slow-fast Hamiltonian systems. Consider Hamiltonian system with Hamiltonian of the form

$$E = E(p, q, y, x) \tag{4}$$

and with the symplectic structure

$$dp \wedge dq + \varepsilon^{-1} dy \wedge dx.$$

The equations of motion have the form

$$\dot{p} = -\frac{\partial E}{\partial q}, \ \dot{q} = \frac{\partial E}{\partial p}, \ \dot{y} = -\varepsilon \frac{\partial E}{\partial x}, \ \dot{x} = \varepsilon \frac{\partial E}{\partial y}, \tag{5}$$

and the variables p, q are called fast while variables y, x are called slow. System (5) is called a Hamiltonian system with fast and slow variables or a slow-fast Hamiltonian system. (In particular when the Hamiltonian has the form $E = y + E_0(p,q,x)$, we get a system depending on the slow time $x = \tau$.)

Consider the case when the fast variables correspond to one degree of freedom (i.e. p, q are scalar variables). The dimensions of y, x are not important for now. Because variables y, x are changing slowly, it is reasonable first to consider the problem at frozen values of y, x. For every frozen value of y, x in the plane p, q suppose that there is a domain filled by closed phase curves of Hamiltonian E (Fig. 1). In this domain one can introduce "action-angle" variables $I = I(p,q,y,x)$, $\varphi = \varphi(p,q,y,x) \bmod 2\pi$. Denote by $H_0(I,y,x)$ the Hamiltonian E expressed via variables I, y, x. Denote as $(p(t), q(t), y(t), x(t))$ a solution of our Hamiltonian system (5).

Theorem 2.2. *The action variable $I(p,q,y,x)$ is an adiabatic invariant:*

$$|I(p(t),q(t),y(t),x(t)) - I_0| < C\varepsilon \text{ for } 0 \leqslant t \leqslant 1/\varepsilon.$$

Moreover

$$|y(t) - Y(t)| + |x(t) - X(t)| < C\varepsilon \text{ for } 0 \leqslant t \leqslant 1/\varepsilon.$$

Here $(Y(t), X(t))$ is the solution of Hamiltonian system with Hamiltonian $H_0(I_0, Y, X)$ and with initial data $Y(0) = y(0)$, $X(0) = x(0)$, $I_0 = I(p(0),q(0),y(0),x(0))$.

Proof. At frozen values of the slow variables y, x a canonical transformation of variables $(p,q) \mapsto (I,\varphi)$ is determined by a generating function $W(q,I,y,x)$ containing y, x as parameters. In the system with Hamiltonian (4) perform a canonical transformation of variables $(p,q,y,x) \mapsto (\hat{I}, \hat{\varphi}, \hat{y}, \hat{x})$ with generating function $\varepsilon^{-1}\hat{y}x + W(q,\hat{I},\hat{y},x)$. This transformation of variables takes the form

$$\hat{\varphi} = \frac{\partial W}{\partial \hat{I}}, \ p = \frac{\partial W}{\partial q}, \ \hat{x} = x + \varepsilon \frac{\partial W}{\partial \hat{y}}, \ y = \hat{y} + \varepsilon \frac{\partial W}{\partial x}. \tag{6}$$

In the new variables the Hamiltonian (4) has the form

$$H = H_0(I,y,x) + \varepsilon H_1(I,\varphi,y,x,\varepsilon), \ H_1 = \frac{\partial E}{\partial y}\frac{\partial W}{\partial x} - \frac{\partial H_0}{\partial x}\frac{\partial W}{\partial y} + O(\varepsilon),$$

where the function H_1 is 2π - periodic in φ. The equations of motion have the form

$$\dot{\hat{I}} = -\varepsilon \frac{\partial H_1}{\partial \hat{\varphi}}, \ \dot{\hat{y}} = -\varepsilon \frac{\partial H_0}{\partial \hat{x}} - \varepsilon^2 \frac{\partial H_1}{\partial \hat{x}}, \ \dot{\hat{x}} = \varepsilon \frac{\partial H_0}{\partial \hat{y}} + \varepsilon^2 \frac{\partial H_1}{\partial \hat{y}}, \tag{7}$$

$$\dot{\hat{\varphi}} = \frac{\partial H_0}{\partial \hat{I}} + \varepsilon \frac{\partial H_1}{\partial \hat{I}}.$$

This system contains slowly varying variables $\hat{I}, \hat{y}, \hat{x}$, and an angular variable (the phase) $\hat{\varphi}$ which rotates with frequency approximately equal to $\omega = \partial H_0(\hat{I}, \hat{y}, \hat{x})/\partial \hat{I}$. In order to describe the approximate behaviour of the slow variables, the averaging method prescribes to average the r.h.s. of the equations for these variables over the fast phase, and to neglect terms of $O(\varepsilon^2)$. The averaged system is a Hamiltonian system with Hamiltonian $H_0(J, Y, X)$:

$$
\dot{J} = 0, \ \dot{Y} = -\varepsilon \frac{\partial H_0}{\partial X}, \ \dot{X} = \varepsilon \frac{\partial H_0}{\partial Y}. \tag{8}
$$

The theorem about the accuracy of the averaging method (see, e.g. [4]) says that the behaviour of the slow variables in the exact system (7) is described by the solution of the averaged system with the same initial conditions, up to accuracy $O(\varepsilon)$ over the time interval $1/\varepsilon$. This, together with the fact that the variables with "hat" differ from the variables without "hat" by $O(\varepsilon)$ as well, implies the assertion of the theorem. $\qquad\square$

In the problem under consideration, a description of the motion by means of equations (8) is called the adiabatic approximation. Trajectories of system (8) are called the adiabatic trajectories.

Example 5 (Magnetic traps). In Example 3, the dynamics of particles is described by a slow-fast Hamiltonian system with three degrees of freedom; one degree of freedom corresponds to fast variables, while two degrees of freedom correspond to slow variables. The kinetic energy E of a particle is a first integral of the motion. For E we have an expression $E = mv_{\parallel}^2/2 + mv_{\perp}^2/2$, where $v_{\parallel}$ and $v_{\perp}$ are the values of the projections of the velocity of a particle onto the direction of the magnetic field and the plane perpendicular to the magnetic field respectively, and m is the particle's mass. Denote μ the magnetic moment of the particle; $\mu = v_{\perp}^2/(2B)$. Thus, along the trajectory $E = m(v_{\parallel}^2/2 + \mu B)$ is constant. At frozen values of the slow variables, the motion of the fast variables is Larmor motion. The frequency of this motion is proportional to the strength of the magnetic field B. The Hamiltonian for this motion in principal approximation is a quadratic function of the fast variables, and it coincides with $E_{\perp} = mv_{\perp}^2/2$. So, the action of the fast motion, which is equal in the principal approximation to the ratio of $E_{\perp}$ to the Larmor frequency (see Example 3), is an adiabatic invariant. Therefore, the magnetic moment μ is an adiabatic invariant. Consider the problem in the adiabatic approximation $\mu = \text{const}$, for which the motion takes place in the domain where $B \leqslant E/(m\mu)$. The surface where $B = E/(m\mu)$ is called a "magnetic mirror". This surface "reflects" particles with energy E and magnetic moment μ. If for a given particle, its magnetic line crosses two magnetic mirrors, then the particle will bounce between these mirrors. The construction of traps for a plasma which are called "adiabatic traps" or "traps with magnetic mirrors" is based on this phenomenon. A gigantic natural adiabatic trap is the Earth's magnetosphere. In plasma physics the adiabatic approximation is called "guiding centre approximation".

Theorems 1 and 2 guarantee the conservation of adiabatic invariance on time intervals of the length of order $1/\varepsilon$. For a magnetic trap in Example 5 this is just the time for several oscillations between magnetic mirrors. In fact, if the system is smooth enough, the time of conservation of adiabatic invariance is much longer. We consider this matter for Hamiltonian systems with a parameter which is slowly varying in time (a Hamiltonian of the form (1)). For slow-fast Hamiltonian systems (a Hamiltonian of the form (4)) the results are similar (see [6]). Let I, φ be action-angle variables for a system with Hamiltonian (1) with "frozen" values of λ. If the parameter λ changes in time, then dynamics of variables I, φ is described by a Hamiltonian system with Hamiltonian of the form

$$H(I, \varphi, \tau, \varepsilon) = H_0(I, \tau) + \varepsilon H_1(I, \varphi, \tau) \tag{9}$$

(see the proof of Theorem 1). The function H_1 is 2π - periodic in φ, and both functions H_0, H_1 are defined in the domain $G = \{I, \varphi, \tau : I \in D \subset R^1,\ \varphi \in T^1,\ \tau \in R^1\}$. Assume that these functions can be continued analytically into a complex δ-neighborhood $G + \delta$ of the set G, and in this neighborhood the following estimates are satisfied:

$$|H_0| < M,\ |H_1| < M,\ |\partial H_0/\partial I| > c^{-1},$$

where δ, M and c are positive constants.

Theorem 2.3. [19] *For $0 < \varepsilon < \varepsilon_0$ and (I, ϕ, τ) in a complex $\delta/2$-neighborhood of the domain G, there exists a canonical transformation of variables $I, \varphi \mapsto J, \psi$ such that*

$$|J - I| + |\varphi - \psi| < c_1 \varepsilon$$

and the Hamiltonian expressed in the new variables has the form

$$\mathscr{H}(J, \psi, \tau, \varepsilon) = \mathscr{H}_\Sigma(J, \tau, \varepsilon) + \varepsilon \alpha(J, \psi, \tau, \varepsilon),$$
$$|\mathscr{H}_\Sigma - H_0| < c_1 \varepsilon,\ |\alpha| < \exp(-c_2^{-1}/\varepsilon). \tag{10}$$

Here ε_0, c_1 and c_2 are positive constants.

Corollary 2.1. *Along a trajectory the value of the variable I undergoes only oscillations of order ε over time intervals of length $\exp(\frac{1}{2}c_2^{-1}/\varepsilon)$. Thus, in analytic one-frequency systems, adiabatic invariance is conserved over exponentially long time intervals.*

To prove this Corollary we note that the dynamics of the variable J as introduced in the Theorem 3 is described by the equation

$$\dot{J} = -\varepsilon \frac{\partial \alpha}{\partial \psi} = O(\varepsilon \exp(-c_2^{-1}/\varepsilon)). \tag{11}$$

Hence over exponentially long time $\exp(\frac{1}{2}c_2^{-1}/\varepsilon)$ the change of J along trajectory is exponentially small. On the other hand, the difference between I and J is $O(\varepsilon)$. This completes the proof.

The origin of the exponentially small term in (10) can be explained heuristically as follows. Let us try by means of a canonical transformation which is close to the identity $I, \varphi \mapsto J, \psi$ to eliminate the dependence of the Hamiltonian on the fast angular variable. Let us try to find a new Hamiltonian $\mathscr{H}(J, \tau, \varepsilon)$ and a generating function of this transformation $J\varphi + \varepsilon S(J, \varphi, \tau, \varepsilon)$ in the form of a series

$$\mathscr{H}(J, \tau, \varepsilon) = H_0(J, \tau) + \varepsilon \mathscr{H}_1(J, \tau) + \varepsilon^2 \mathscr{H}_2(J, \tau) + \dots, \tag{12}$$

$$S(J, \varphi, \tau, \varepsilon) = S_1(J, \varphi, \tau) + \varepsilon S_2(J, \varphi, \tau) + \varepsilon^2 S_3(J, \varphi, \tau) + \dots.$$

The functions S_l are 2π-periodic in φ, and the new and old variables are related by the expressions

$$I = J + \varepsilon \frac{\partial S}{\partial \varphi}, \quad \psi = \varphi + \varepsilon \frac{\partial S}{\partial J}. \tag{13}$$

The new and the old Hamiltonian are related via the formula

$$\mathscr{H}(J, \tau, \varepsilon) = H(I, \varphi, \tau, \varepsilon) + \varepsilon^2 \frac{\partial S}{\partial \tau}. \tag{14}$$

Plugging (13) into (14), taking into account the expansions (12) and equating terms of the same order in ε we get a sequence of equations

$$\mathscr{H}_1 = \frac{\partial H_0}{\partial J} \frac{\partial S_1}{\partial \varphi} + H_1, \tag{15}$$

$$\mathscr{H}_2 = \frac{\partial H_0}{\partial J} \frac{\partial S_2}{\partial \varphi} + \frac{1}{2} \frac{\partial^2 H_0}{\partial J^2} (\frac{\partial S_1}{\partial \varphi})^2 + \frac{\partial S_1}{\partial \tau}, \tag{16}$$

$$\mathscr{H}_k = \frac{\partial H_0}{\partial J} \frac{\partial S_k}{\partial \varphi} + X_k(J, \varphi, \tau) + \frac{\partial S_{k-1}}{\partial \tau}, \quad k \geqslant 3. \tag{17}$$

The function X_k is well defined if functions $\mathscr{H}_l$, S_l, $l = 1, 2, \dots, k-1$ are defined and the expression for X_k does not include $\partial S_{k-1}/\partial \tau$. This sequence of equations allows to define step by step all the functions $\mathscr{H}_k$, S_k (one should remember that functions S_k are 2π-periodic in φ). In particular, $\mathscr{H}_1$ is equal to the average of H_1 over φ. After defining $\mathscr{H}_1$ one can find S_1 via quadrature, and so on. However, the series (12) for $\mathscr{H}$, S as a rule should diverge. Indeed, on the k-th step of our procedure one should differentiate function S_{k-1} with respect to τ. So, basically in order to define $\mathscr{H}_k$, S_k one should differentiate H_1 $k-1$ times with respect to τ. The n-th derivative of an analytic function can be estimated from above as $a^n n!$, $a = $ const, and this estimate can not be improved. This informal reasoning indicates the divergence of the series (12). It indicates also that these series should be of Gevrey type 1 [23]. If we truncate the series for S at terms of order ε^{n-3} and make transformation of variables (13) with this truncated generating function S, then the new Hamiltonian will have form (10) with $|\varepsilon\alpha| < \text{const}\,\varepsilon^n a^n n!$. According to Stirling's formula the right hand side of this inequality for large n grows approximately as $\exp(n(\log(a\varepsilon) + \log n - 1))$. This function of n has a minimum at $n = 1/(a\varepsilon)$. Now choosing n to be the integer part of $1/(a\varepsilon)$ implies the estimate (10).

The proof of Theorem 3 in [19] is based on a representation for the transformation $I, \varphi \mapsto J, \psi$ as a composition of many, $\sim 1/\varepsilon$, symplectic transformations which eliminate the dependence of the Hamiltonian on fast variables in subsequent orders in ε. The proof, based on establishing a Gevrey 1 type of formal series $\mathscr{H}$, S, is given in [23]. The method of continuous averaging [25] gives a sharp estimate of the constant c_2 in Theorem 3.

A certain modification of Theorem 3 explains a remarkable property known as the exponential accuracy of the conservation of adiabatic invariants [15]. Assume that value $\lambda(\tau)$ tends sufficiently fast to definite the limits $\lambda_\pm$ as $\tau \to \pm\infty$. Then the value of the action I along a trajectory also tends to some limits $I_\pm$ as $t \to \pm\infty$. The difference

$$\Delta I = I_+ - I_-$$

is called the accuracy of conservation of this adiabatic invariant [15]. Although for finite t the quantity I undergoes oscillations of order ε, the value ΔI is much smaller then ε. In particular, if the system is analytic, then ΔI is exponentially small; $\Delta I = O(\exp(-c_2^{-1}/\varepsilon))$. With the help of Theorem 3 this can be explained as follows. The function H_1 in (9) is proportional to $d\lambda/d\tau$ and so it tends to 0 fast enough as $\tau \to \pm\infty$. This implies that function α in Theorem 3 tends to 0 fast enough as $\tau \to \pm\infty$. Together with estimate (10) this implies that

$$\int_{-\infty}^{+\infty} |\varepsilon\alpha|\,dt = O(\exp(-c_2^{-1}/\varepsilon)).$$

Together with (11) this implies that the values of J along the trajectory tend to certain limits $J_\pm$ as $t \to \pm\infty$, and $J_+ - J_- = O(\exp(-c_2^{-1}/\varepsilon))$. But $J - I$ tends to 0 as $\tau \to \pm\infty$ (this is again because H_1 tends to 0 as $\tau \to \pm\infty$). Therefore, $I_\pm = J_\pm$. This implies our assertion about the exponential accuracy of conservation of adiabatic invariant in analytic systems. An analysis of the analytic continuation in the plane of complex time of solutions with complex data at infinity allows one to get sharp estimates for the constant c_2 [21,24] and, in some cases, to obtain asymptotic expressions for ΔI [12,24].

Now consider the case where the parameter λ varies periodically in time; λ is a 2π-periodic function of τ. Over infinite time the adiabatic invariants can undergo considerable evolution due to the accumulation of small perturbations.

Example 6 (Parametric resonance). Consider the linear oscillator

$$\ddot{x} = -\omega^2(1 + \kappa\cos\varepsilon t)x, \ \kappa = \text{const} < 1.$$

The equilibrium $x = 0$ can be unstable for arbitrarily small ε (the phenomenon of parametric resonance [4]). The adiabatic invariant changes unboundedly.

However, it turns out that for a periodic variation of the parameter, such non-conservation of the adiabatic invariant is due to the linearity of the system (more precisely, to the fact that the frequency of oscillations is independent of the amplitude). In a nonlinear system, as the amplitude increases, the frequency changes, and the oscillations do not have enough time to accrue before the resonance condition is violated.

Denote

$$\bar{\omega}(I) = \frac{1}{2\pi} \int_0^{2\pi} \frac{\partial H_0(I,\tau)}{\partial I} d\tau$$

Theorem 2.4. [2] *If the Hamiltonian function of a nonlinear oscillatory system with one degree of freedom depends on time slowly and periodically, the action variable I of the system is a perpetual adiabatic invariant:*

$$|I(p(t),q(t),\lambda(\varepsilon t)) - I_0| < C\varepsilon \quad \text{for} \quad -\infty < t < +\infty.$$

The required nonlinearity condition is $\partial\bar{\omega}(I)/\partial I \neq 0$.

The proof is based on a construction of invariant surfaces (tori), which fill the phase space $p, q, \tau \bmod 2\pi$ of the problem, up to a residue of exponentially small, $O(\exp(-\text{const}/\varepsilon))$, measure, and are $O(\varepsilon)$-deformations of tori $I = \text{const}$. The phase space is three-dimensional, and invariant tori are two-dimensional. Therefore, a phase point that started to move in a gap between the tori remains confined in this gap forever. For this phase point, the value of I remains $O(\varepsilon)$-close to its initial value. Evidently, the same estimate is valid if the phase point started to move on an invariant torus.

A perpetual adiabatic invariant also exists (under certain conditions) in slow-fast Hamiltonian systems with two degrees of freedom and with a Hamiltonian function of the form (4) [2]. According to Theorem 2, the motion in such a problem is approximately described by the Hamiltonian $H_0(I,y,x)$. Suppose that the phase curves of this Hamiltonian for fixed I are closed. Then in the approximation under consideration the motion in the phase space takes place on the two-dimensional tori defined by the conditions $I = \text{const}$, $H_0 = \text{const}$. This motion has two frequencies, and one of the frequencies is $1/\varepsilon$ times smaller than the other. If for a given $H_0 = \text{const}$ the frequency ratio changes as I varies, then in the exact system on each hyper-surface of constant Hamiltonian there are many invariant tori close to the invariant tori of the approximate system. This implies that the action variable I is perpetually close to its initial value. From this conclusion it follows, in particular, that if the magnetic trap of Example 5 is axially symmetric then it confines charged particles perpetually [2].

Previous analysis was based on the assumption that at frozen values of parameter λ for Hamiltonian (1) (or at frozen values of slow variables y, x for Hamiltonian (5)) the domain under consideration in the plane of fast variables p, q is filled with closed phase curves. However, it often happens that in this domain there is a saddle point with separatrices passing through it, as in Fig. 2.

As a result of the slow variation of the parameter (or of the slow variables) the phase points may cross this separatrix. This leads to an interesting phenomenon associated with jumps of adiabatic invariant at separatrices and destruction of adiabatic invariance; see details in [6] and in the references therein.

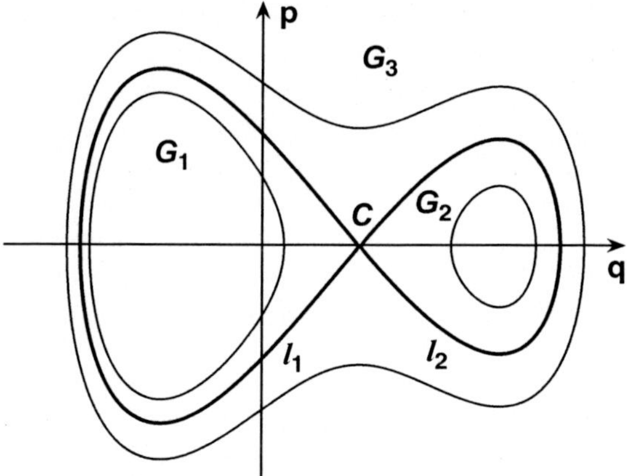

Fig. 2 Separatrices in the phase plane of fast variables

3 On adiabatic invariance in multi-frequency systems

Now consider the case when a Hamiltonian system with Hamiltonian (1) possesses $n \geqslant 2$ degrees of freedom; $(p,q) \in R^{2n}$. First, consider the Hamiltonian at frozen λ, and suppose the system with this Hamiltonian is completely integrable. This means that in the phase space of the Hamiltonian there exists a domain filled up by n-dimensional invariant tori and the "action-angle" variables $(I,\varphi), I \in R^n$, $\varphi \in T^n \bmod 2\pi$ are defined [4]; $I = (I_1, I_2, \ldots, I_n)$, $\varphi = (\varphi_1, \varphi_2, \ldots, \varphi_n)$. (The "angle" variables φ are called phases as well.) The transformation of the variables $(p,q) \mapsto (I,\varphi)$ is canonical (symplectic). This transformation can be defined with a generating function $W(I,q,\lambda)$. The old and the new variables are related via formulas (2). In terms of the new variables, the Hamiltonian has the form $E = H_0(I,\tau)$. The motion is a multidimensional rotation with a vector of frequencies $\omega(I,\tau) = (\omega_1(I,\tau), \omega_2(I,\tau), \ldots, \omega_n(I,\tau))$.

Now let λ change slowly in time: $\lambda = \lambda(\tau)$, $\tau = \varepsilon t$. Here on the level of formal computations everything is completely analogous to the one-frequency case. In the system with Hamiltonian (1) let us make the canonical transformation $(p,q) \mapsto (I,\varphi)$ by means of formulas (2). As in Section 2, the Hamiltonian for the new variables has the form $H(I,\varphi,\tau,\varepsilon) = H_0(I,\tau) + \varepsilon H_1(I,\varphi,\tau)$, where $H_1 = \partial W / \partial \tau$. The differential equations of the motion have the form (3). The Hamiltonian and the equations of the motion are in the standard form to which averaging method can be applied. In order to describe the approximately behaviour of the variables I, the averaging method prescribes to average the rate of changing of this variables over the fast phases φ. Averaging a certain function $f(\cdot)$ means calculating the following value;

$$\langle f \rangle = \frac{1}{(2\pi)^n} \int_0^{2\pi} \int_0^{2\pi} \cdots \int_0^{2\pi} f(\varphi) d\varphi_1 d\varphi_2 \ldots d\varphi_n . \tag{1}$$

The averaged equation for I has the form $\dot{I} = 0$, implying that $I = \text{const}$ along the trajectory in approximation of the averaging method. Thus, "actions" I are natural candidates to be adiabatic invariants. They would be adiabatic invariants provided the averaging method describes the dynamics on the time interval $1/\varepsilon$ with accuracy which goes to zero as $\varepsilon \to 0$. These calculations and reasoning are due to J.M. Burgers [8]. (He was a PhD student of P. Ehrenfest and his adviser assigned to him the problem of adiabatic invariance in multi-frequency systems [16].)

The justification of the averaging method in multi-frequency systems encounters problems with resonances. The resonance condition is a relation of the form

$$k_1 \omega_1(I, \tau) + k_2 \omega_2(I, \tau) + \ldots + k_n \omega_n(I, \tau) = 0, \tag{2}$$

where $k = (k_1, k_2, \ldots, k_n)$ is an integer non-zero vector. If for some I, τ a resonance condition (2) is satisfied, then for this (I, τ) the unperturbed ($\tau = \text{const}$) motion takes place on an $(n-1)$-dimensional torus. So independent averaging over all phases, like in (1), which is actually averaging over an n-dimensional torus, may not be a correct tool for approximate description of the direction of the evolution. In the process of the motion, even if I would be approximately constant, the value τ changes, and resonance conditions (2) with different vectors k are satisfied on a dense set of points in time. This problem was first addressed by P.A.M. Dirac [9] for two-frequency case ($n = 2$, $\omega = (\omega_1, \omega_2)$). In [9] the problem was considered under the following condition

$$\omega_2 \frac{\partial \omega_1}{\partial \tau} - \omega_1 \frac{\partial \omega_2}{\partial \tau} - \left(\omega_2 \frac{\partial \omega_1}{\partial I} - \omega_1 \frac{\partial \omega_2}{\partial I} \right) \frac{\partial H_1}{\partial \varphi} > c^{-1}, \quad c = \text{const} > 0. \tag{3}$$

In the two-frequency case, the resonance condition means that the ratio of the frequencies is a rational number. Condition (3) means that the ratio of the frequencies changes at a non-zero rate along trajectories of the system. So, the phase point cannot stay for a long time near any given resonance. If this condition is satisfied, then the "action" I is an adiabatic invariant; its variation along a trajectory on time interval of the length $1/\varepsilon$ tends to zero as $\varepsilon \to 0$ [9]. It follows from the general result of V.I. Arnold [3] about averaging in two-frequency systems, that this variation is $O(\sqrt{\varepsilon})$.

If the condition (3) is not satisfied, then along some trajectories the value of I during time $1/\varepsilon$ may change considerably, by a value of order 1, due to the phenomenon of capture into resonance, see example in [20]. However, results of D.V. Anosov [1] and T. Kasuga [14] imply that under very general conditions the measure of the initial data for such trajectories tends to zero as $\varepsilon \to 0$. The value I can be called an almost adiabatic invariant [5]. Some estimates of this measure are contained in [7, 10, 17, 18]. For a description of dynamics with capture into resonance on time intervals of the length of order $1/\varepsilon$ see, e.g., review [22]. On time intervals of the length $1/\varepsilon^{3/2}$ the adiabatic invariance of I may be completely destroyed; the value of I along a trajectory may change by a value which is bounded from below by a constant for a set of initial data of measure which is bounded from below by another constant. The corresponding example is constructed in [20], and general approach is developed in [11], see also discussion in [22].

Acknowledgements This work was supported in part by grants RFBR 06-01-00117 and NSh-1312.2006.1.

References

1. Anosov, D. V.: Averaging in systems of ordinary differential equations with rapidly oscillating solutions. Izv. Akad. Nauk SSSR, Ser. Mat. **24**, 721–742 (1960) (Russian)
2. Arnold, V. I.: Small denominators and problems of stability of motion in classical and celestial mechanics. Russ. Math. Surv. **18**, No. 6, 85–191 (1963)
3. Arnold, V. I.: Conditions for the applicability, and estimate of the error, of an averaging method for systems which pass through states of resonance in the course of their evolution. Sov. Math., Dokl. **6**, 331–334 (1965)
4. Arnold, V. I.: Mathematical Methods of Classical Mechanics. Springer, New York (1978)
5. Arnold, V. I.: Geometrical Methods in the Theory of Ordinary Differential Equations. Springer, New York (1988)
6. Arnold, V. I., Kozlov, V. V., Neishtadt, A. I.: Mathematical Aspects of Classical and Celestial Mechanics (Dynamical Systems III. Encyclopedia of Mathematical Sciences), 3rd edition. Springer, Berlin (2006)
7. Bakhtin, V. I.: Averaging in multifrequency systems. Funct. Anal. Appl. **20**, 83–88 (1986)
8. Burgers, J. M.: Adiabatic invariants of mechanical systems, I, II, III. Proc. Roy. Acad. Amsterdam **20**, 149–169 (1917/18)
9. Dirac, P. A. M.: The adiabatic invariance of the quantum integrals. Proc. R. Soc. Lond., Ser. A, Math. Phys. Eng. Sci. **107**, 725–734 (1925)
10. Dodson, M. M., Rynne, B. P., Vickers, J. A. G.: Averaging in multifrequency systems. Nonlinearity **2**, 137–148 (1989)
11. Dolgopyat D.: Repulsion from Resonances. Preprint Univ. of Maryland (2004)
12. Dykhne, A. M.: Quantum transitions in the adiabatic approximation. Sov. Phys., JETP **11**, 411–415 (1960)
13. Gorelyshev, I. V., Neishtadt, A. I.: On the adiabatic perturbation theory for systems with impacts. J. Appl. Math. Mech. **70**, No. 1, 4–17 (2006)
14. Kasuga, T.: On the adiabatic theorem for the Hamiltonian system of differential equations in the classical mechanics. I, II, III. Proc. Japan Acad. **37**, 366–371, 372–376, 377–382 (1961)
15. Landau, L. D., Lifshits, E. M.: Course of Theoretical Physics. Vol. 1: Mechanics. Addison-Wesley, Reading, MA (1958)
16. Navarro, L., Pérez, E.: Paul Ehrenfest: The genesis of the adiabatic hypothesis, 1911–1914. Arch. Hist. Exact Sci. **60**, 209–267 (2006)
17. Neishtadt, A. I.: Passage through a resonance in the two-frequency problem. Sov. Phys., Dokl. **20**, 189–191 (1975)
18. Neishtadt, A. I.: Averaging in multifrequency systems. II. Sov. Phys., Dokl. **21**, 80–82 (1976)
19. Neishtadt, A. I.: On the accuracy of conservation of the adiabatic invariant. J. Appl. Math. Mech. **45**, No. 1, 58–63 (1982)
20. Neishtadt, A. I.: On destruction of adiabatic invariants in multi-frequency systems. In: Perelló, C. (ed.) et al. Proc. Int. Conf. on Differential Equations EQUADIFF 91 (Barcelona, Spain, 1991). Vol. 1, 2. World Scientific, Singapore, 195–207 (1993)
21. Neishtadt, A. I.: On the accuracy of persistence of an adiabatic invariant in single-frequency systems. Regul. Chaotic Dyn. **5**, No. 2, 213–218 (2000)

22. Neishtadt, A. I.: Capture into resonance and scattering on resonances in two-frequency systems. Proc. Steklov Inst. Math. **250**, 183–203 (2005)
23. Ramis, J.-P., Schäfke, R.: Gevrey separation of fast and slow variables. Nonlinearity **9**, 353–384 (1996)
24. Slutskin, A. A.: On the motion of a one-dimensional non-linear oscillator in adiabatic conditions. Sov. Phys., JETP **18**, 676–682 (1963)
25. Treshchev, D.V.: The method of continuous averaging in the problem of separation of fast and slow motions. Regul. Chaotic. Dyn. **2**, No. 3, 9–20 (1997)

Transformation theory of Hamiltonian PDE and the problem of water waves

Walter Craig[1]

Abstract This set of lecture notes gives (i) a formal theory of Hamiltonian systems posed in infinite dimensions, (ii) a perturbation theory in the presence of a small parameter, adapted to reproduce some of the well-known formal computations of fluid mechanics, and (iii) a transformation theory of Hamiltonian systems and their symplectic structures. A series of examples is given, starting with a rather complete description of the problem of water waves, and, following a series of scaling and other simple transformations placed in the above context, a derivation of the well known equations of Boussinesq and Korteweg deVries.

1 Hamiltonian systems

A *Hamiltonian system* is given in terms of a Hamiltonian function $H : M \to \mathbb{R}$, where M is the phase space. We will restrict ourselves to phase spaces which are Hilbert spaces, denoting the inner product between two vectors $V_1, V_2 \in T(M)$ by $\langle V_1 | V_2 \rangle$. The symplectic structure is as usual given by a two-form ω on (M), which can be represented by the inner product, namely $\omega(V_1, V_2) = \langle V_1 | J^{-1} V_2 \rangle$, where, because of the antisymmetry of two-forms, the operator J satisfies $J^{-T} = -J^{-1}$. The Hamiltonian vector field X_H is defined through the relation $dH(V) = \omega(V, X_H)$ which is asked to hold for all $V \in T(M)$. The system of equations that we study, known as *Hamilton's canonical equations*, is given by

$$\dot{v} = X_H(v), \qquad v(0) = v_0. \tag{1}$$

The inner product enters into the definition of the gradient of functions on M, which is in particular that for all $V \in T(M)$, $dH(V) = \langle grad_v H | V \rangle$, therefore Hamiltonian vector fields are expressed by

[1] Department of Mathematics and Statistics, McMaster University, Hamilton, Ontario L8S 4K1, Canada
e-mail: craig@math.mcmaster.ca

W. Craig (ed.), Hamiltonian Dynamical Systems and Applications, 67–83.
© 2008 *Springer Science + Business Media B.V.*

$$X_H = J\,grad_{\,v}H(v). \tag{2}$$

We will denote the solution map, or the *flow*, for the initial value problem for system (1) by $v(t) = \varphi_t(v_0)$. From the usual theory of ordinary differential equations, whenever the Hamiltonian vector field $X_H(v)$ is $C^1(M, T(M))$ (usually meaning when the Hamiltonian $H(v)$ itself is $C^2(M, \mathbb{R})$) then the flow is defined and unique, at least locally in time. The disclaimer is that this regularity property holds very rarely the case when equation (1) describes a partial differential equation (the BBM equation is a notable exception), and much effort has gone into the study of the well posedness of the initial value problem and the properties of the solution map for numerous important examples of evolution equations. Furthermore, in this effort it is not clear that the property of being a Hamiltonian system is of particular importance in general. Nonetheless, because of its interest in various special cases, and because Hamiltonian partial differential equations (PDE) appear naturally in many areas of physics, it seems reasonable to take seriously the analogy between Hamiltonian dynamical systems and PDEs. This is one purpose of the presentation in this note.

2 Partial differential equations as Hamiltonian systems

It seems most useful to discuss Hamiltonian PDEs with a good set of examples. These are supplied by problems in physics, and in particular the ones I bear in mind most often come from the problems in wave propagation in fluid mechanics.

(i) The wave equation

Consider a scalar field $u(x,t)$ defined for $x \in \Omega \subseteq \mathbb{R}^d$ which satisfies the equation

$$\partial_t^2 u = \Delta_x u - g(u,x), \qquad u(x,t) = 0 \text{ when } x \in \partial\Omega. \tag{1}$$

This can be written in the form of equation (1); indeed define

$$H(u,p) := \int_{\Omega} \tfrac{1}{2}p^2 + \tfrac{1}{2}|\nabla u|^2 + G(u,x)\,dx, \tag{2}$$

where $\partial_u G = g$. Then the second order equation (1) can be equivalently written as a first order system of PDEs

$$\dot{u} = p = grad_{\,p}H \tag{3}$$

$$\dot{p} = \Delta u - \partial_u G = -grad_{\,u}H.$$

The gradient is taken with respect to the $L^2(\Omega)$ inner product, which dictates as well which Hilbert space we should propose for M. Actually, as operators such as Δ are unbounded, the initial value problem should normally be posed only on an appropriate subdomain of M. In any case, this problem is in the form of a Hamiltonian system with $v = (u, p)^T$ and

$$\dot{v} = J grad_v H, \qquad J := \begin{pmatrix} 0 & I \\ -I & 0 \end{pmatrix}. \tag{4}$$

We will say that a Hamiltonian system with J of this form is in *Darboux coordinates*.

(ii) Burger's equation

A famous example in the theory of shoch waves is *Burger's equation*, which can be written in Hamiltonian form as well.

$$\partial_t w = w \partial_x w \qquad x \in \mathbb{R}^1. \tag{5}$$

Define the Hamiltonian as

$$H := \int_{\mathbb{R}} \tfrac{1}{6} w^3 \, dx, \tag{6}$$

from which we compute the form of Hamiltonian's canonical equations

$$\dot{w} = \partial_x(\tfrac{1}{2} w^2) = J grad_w H, \qquad J := \partial_x. \tag{7}$$

Notice that the symplectic structure is given by an operator with no direct finite dimensional analog; it furthermore is not invertible, meaning that our formal discussion of the representation of the symplectic form in Section 1 has to be taken with a grain of salt. It is well known that every nonconstant solution of Burger's equation develops discontinities, or *shocks*. The standard law of conservation of the Hamiltonian function, $H(\varphi_t(w)) = H(w)$ holds for smooth solutions, however it does not hold in most cases for time t after the time T of formation of a shock.

(iii) The Korteweg deVries equation

The classical Korteweg deVries (KdV) equation was derived as a model equation for the propagation of waves in the surface of a fluid. The beautiful fact about the KdV is that it is an example of an infinite dimensional completely integrable system, with algebraic integrals viewed in the proper coordinates. This integrability is not the topic of the present discussion. Rather, we show that it can be posed as a Hamiltonian PDE, and furthermore we discuss its relationship to fluid dynamics. The KdV equation for a function $r(X, t)$ is normally written as

$$\partial_t r = -\tfrac{1}{6} \partial_X^3 r + 3r \partial_X r. \tag{8}$$

This takes the form of a Hamiltonian system with Hamiltonian

$$H := \int_{\mathbb{R}} \tfrac{1}{12}(\partial_X r)^2 + \tfrac{1}{2} r^3 \, dX, \qquad J = \partial_X. \tag{9}$$

One easily checks that this is in the form (1), which in this context is

$$\dot{r} = \partial_X \, grad_{\,r} H. \tag{10}$$

The nonlinearity $g(X,r) = \partial_X(3r^2/2)$ is not the only one of interest. In particular the case $\partial_X r^3$ is a Hamiltonian PDE which is also a completely integrable system. Replacing either of the above two equations with a general nonlinear term $g(X,r)$ also results in a Hamiltonian PDE, which is sometimes considered as a model dispersive evolution equation which is not completely integrable.

(iv) The Boussinesq system

Another well known PDE which was originally derived in the study of water waves is the Boussinesq system,

$$\partial_t \begin{pmatrix} p \\ q \end{pmatrix} = \begin{pmatrix} 0 & \partial_X \\ \partial_X & 0 \end{pmatrix} \begin{pmatrix} p + \frac{1}{2}q^2 \\ qq + \partial_X^2 q + pq \end{pmatrix}. \tag{11}$$

This system of equations is a variant of a common one studied by Zakharov [13], and it has been shown to be another example of a completely integrable Hamiltonian PDE in Kaup [10] and Sachs [11]. The Hamiltonian for the system (11) is given by

$$H := \tfrac{1}{2} \int_{\mathbb{R}} p^2 + q^2 - (\partial_X q)^2 + pq^2 \, dX, \tag{12}$$

with a symplectic structure given by the matrix operator

$$J := \begin{pmatrix} 0 & \partial_X \\ \partial_X & 0 \end{pmatrix} \tag{13}$$

which is already in appearance in the above system of equations (11).

We now have a number of examples in hand, many of which stem originally from the study of water waves, that is the fluid dynamical problem of wave propagation in the surface of a body of fluid. A natural question is as to how these systems are related to each other. In particular we note that among these systems the number of dependent variables are different, the number of independent variables is different, and the symplectic structures are also changed from one system to another. In order to address this question, even on the formal level that is given in these lectures, we will undertake a detailed description of the problem of water waves itself from the point of view of the equation as an infinite dimensional Hamiltonian dynamical system.

3 The problem of water waves

The equations of evolution for the free surface of a body of water in the influence of gravity as a restoring force are a classical example of a system of Hamiltonian PDEs for which the structure of the equations as such has led to important developments in fluid dynamics. I will first describe the system of equations in standard Eulerian coordinates, after which the formulation of the problem as a Hamiltonian PDE can be derived. The fluid domain is given by $S(\eta) := \{x \in \mathbb{R}^{d-1}, y \in (-h, \eta(x))\}$, where we are assuming that the free surface is given as the graph of the function η; $\Gamma(\eta) := \{(x,y) : y = \eta(x)\}$. Normally the dimension is taken to be either $d = 2, 3$, although mathematically it makes sense for it to be any integer $d \geq 2$. The force of gravity is take to act vertically, given by $F = -g(0,1)$. One of the unknowns of the problem is the time dependent fluid domain $S(\eta)$ defined in terms of the function $\eta(\cdot, t)$. The other unknowns are the components of the fluid velocity $\mathbf{u}(x, y, t)$ at every point in space and time in the fluid domain.

In $S(\eta)$ the fluid velocity vector field is taken to satisfy the conditions of incompressibility and irrotationality, respectively

$$\nabla \cdot \mathbf{u} = 0, \qquad \nabla \wedge \mathbf{u} = 0.$$

The latter is the condition that the vector field $\mathbf{u}$ is given in the form of a potential flow; $\mathbf{u} = \nabla \varphi$ at each instant of time, while the former states that the potential φ is harmonic in $S(\eta)$;

$$\Delta \varphi = 0.$$

Furthermore, on the solid bottom boundary of $S(\eta)$ the fluid velocity is taken to have no normal component; $N \cdot \mathbf{u} = 0$, hence the potential satisfies Neumann boundary conditions on this component of the domain boundary;

$$N \cdot \nabla \varphi = 0.$$

All of the time dependent and nonlinear content of the problem is thus expressed in the boundary conditions posed on the free surface $\Gamma(\eta)$, namely

$$\partial_t \eta = \partial_y \varphi - \partial_x \eta \cdot \partial_x \varphi \tag{1}$$

$$\partial_t \varphi = -g\eta - \tfrac{1}{2}|\nabla \varphi|^2,$$

known respectively as the kinematic and the Bernoulli conditions. The first boundery condition follows from the fact that a fluid particle which originates on the free surface will remain on the free surface under time evolution, so that a tangent vector T to its trajectory in space–time must always be orthogonal to the space–time normal vector N to the free surface; $N \cdot T = 0$. The Bernoulli condition simply represents an expression of the Euler equations for an inviscid fluid, in integrated form and evaluated on the free surface which itself is a surface of constant pressure.

The energy H of the system of equations for fluid motion with a free surface is straightforward to express, indeed it is the sum of kinetic and potential energy contributions;

$$H = K + P := \int_{\mathbb{R}^{d-1}} \int_h^{\eta(x)} |\mathbf{u}|^2 \, dy \, dx + \int_{\mathbb{R}^{d-1}} \int_h^{\eta(x)} gy \, dy \, dx \qquad (2)$$

$$= \int_{\mathbb{R}^{d-1}} \int_h^{\eta(x)} \tfrac{1}{2} |\nabla \varphi|^2 \, dy \, dx + \int_{\mathbb{R}^{d-1}} \tfrac{g}{2} \eta^2 \, dx - C, \qquad (3)$$

where the constant C is irrelevant to the dynamics and can be neglected. It is useful to rewrite the kinetic energy by integrating by parts.

$$K = \int_{\mathbb{R}^{d-1}} \int_h^{\eta(x)} \tfrac{1}{2} |\nabla \varphi|^2 \, dy \, dx = -\int_{\mathbb{R}^{d-1}} \int_h^{\eta(x)} \tfrac{1}{2} \varphi \Delta \varphi \, dy \, dx$$
$$+ \int_{\mathbb{R}^{d-1}} \tfrac{1}{2} \varphi N \cdot \nabla \varphi \, dS_{\text{bottom}} + \int_{\mathbb{R}^{d-1}} \tfrac{1}{2} \varphi N \cdot \nabla \varphi \, dS_{\text{free surface}}.$$

Because the velocity potential is harmonic and satisfies Neumann bottom boundary conditions, the first two terms of the right hand side vanish. Denoting the boundary values on the free surface $\Gamma(\eta)$ by $\xi(x) = \varphi(x, \eta(x))$, we have then

$$K = \int_{\mathbb{R}^{d-1}} \tfrac{1}{2} \xi N \cdot \nabla \varphi \, dS_{\text{free surface}}.$$

We are taking care to distinguish between φ the potential function itself, and ξ its values on the free surface $\Gamma(\eta)$. The elements of Laplace's equation that remain in this expression are the normal derivative of the potential φ on the free surface. It is useful to describe this quantity in terms of the boundary values $\xi(x)$ and an integral operator on the free surface itself.

Definition 3.1. *(Dirichlet–Neumann operator)* For the fluid domain $S(\eta)$ defined by the function $\eta \in C^1$, give boundary values $\xi(x)$ on the free surface $\Gamma(\eta)$, and consider their harmonic extension $\varphi(x, y)$ to the fluid domain satisfying Neumann bottom boundary conditions. The Dirichlet–Neumann operator is defined by the normal derivative of φ on the free surface, namely

$$G(\eta)\xi(x) = (\partial_y - \partial_x \eta(x) \cdot \partial_x)\varphi(x, \eta(x)) = R(N \cdot \nabla \varphi)(x, \eta(x)), \qquad (4)$$

where $R = \sqrt{1 + |\partial_x \eta|^2}$ is a normalization factor so that $G(\eta)$ is self-adjoint on $L^2(dx)$.

The Hamiltonian (2) can be conveniently written in terms of $G(\eta)$, indeed following [7] we write

$$H = \int_{\mathbb{R}^{d-1}} \tfrac{1}{2} \xi G(\eta) \xi + \tfrac{g}{2} \eta^2 \, dx. \qquad (5)$$

Theorem 3.1. (Zakharov [12]) *There exist canonical variables for the water waves problem* (1), *in which it can be written in the form* (1) *in Darboux coordinates, with Hamiltonian* (5).

Proof. Our derivation of the canonical conjugate variables is based on first principles of mechanics. Given the kinetic energy K and the potential energy P, the Lagrangian for the water waves problem is clearly

$$L = K - P. \tag{6}$$

We should express this in terms of the quantities $(\eta, \dot{\eta})$ (*i.e.* tangent space variables), for which we use the kinematic condition (1),

$$\dot{\eta} = \partial_y \varphi - \partial_x \eta \cdot \partial_x \varphi = G(\eta)\xi.$$

The Lagrangian is thus

$$L(\eta, \dot{\eta}) = \tfrac{1}{2} \int_{\mathbb{R}^{d-1}} \dot{\eta} G \eta) \dot{\eta} - \tfrac{g}{2} \eta^2 \, dx. \tag{7}$$

From this expression the Legendre transform dictates that the canonical conjugate variables are $(\eta, \partial_{\dot{\eta}} L) = (\eta, G^{-1}(\eta)\dot{\eta}) = (\eta, \xi)$. These are precisely the variables presented by Zakharov in [12], in terms of which one may give the water waves Hamiltonian (5). $\square$

Therefore the equations for water waves can be rewritten as a Hamiltonian system in Darboux coordinates;

$$\dot{\eta} = grad_{\xi} H = G(\eta)\xi \tag{8}$$

$$\dot{\xi} = -grad_{\eta} H = -g\eta - grad_{\eta} K.$$

It is interesting to remark that the expressions for K and $grad_{\eta} K$ involve derivatives of the Dirichlet–Neumann operator with respect to perturbations of the domain $S(\eta)$. This idea was already discussed by Hadamard [8, 9] in his Collège de France lectures in 1910 and 1916, in the context of the Green's function for Laplace's equation on a domain in $\mathbb{R}^d$. In these lectures he explicitly mentions the possibility of hydrodynamical applications.

4 The Dirichlet–Neumann operator

Any analysis of the water wave in the above formulation depends upon a detailed knowledge of the Dirichlet–Neumann operator. The fluid domain $S(\eta)$ is given by $\eta(x)$ defining the free surface. Given $\xi(x)$ the boundary values for the velocity potential, then $\varphi(x, y)$ is its harmonic extension to $S(\eta)$ which satisfies the appropriate Neumann bottom boundary conditions. The principal facts about $G(\eta)\xi(x) = \partial_y \varphi(x, \eta(x)) - \partial_x \eta \cdot \partial_x \varphi(x, \eta(x))$ that we use are contained in the lemma.

Proposition 4.1. *Suppose that $\eta \in C^1$. Then $G(\eta)$ satisfies the following properties:*

1. $G(\eta)$ is positive semidefinite.
2. It is self-adjoint (on an appropriatly chosen domain).
3. $G(\eta)$ maps $H^1(\Gamma)$ to $L^2(\Gamma)$ continuously.
4. As an operator $G(\eta) : H^1(\Gamma) \rightarrow L^2(\Gamma)$ it depends analytically upon $\eta \in B_R(0) \subseteq C^1(\Gamma)$, for a nonzero value of R.

The latter item entails questions of the bounded of singular integrals on hypersurfaces, and was proved in the case $d = 2$ by Coifman & Meyer [2], and in the case $d \geq 2$ in [6] using the fundamental results of Christ & Journé [1]. In particular it implies the existence of a convergent Taylor expansion for the operator.

Lemma 4.1. *The Taylor expansion of the Dirichlet–Neumann operator is given by the expression*

$$G(\eta)\xi = \sum_{j \geq 0} G^{(j)}(\eta)\xi \tag{1}$$

where each $G^{(j)}(\eta)$ is homogeneous of degree j in η. Explicitely,

$$G^{(0)}\xi(x) = |D_x|\tanh(h|D_x|)\xi(x) \tag{2}$$

$$G^{(1)}(\eta)\xi(x) = D_x \cdot \eta D_x - G^{(0)}\eta G^{(0)}\xi(x) \tag{3}$$

$$G^{(2)}(\eta)\xi(x) = \tfrac{1}{2}(G^{(0)}\eta^2 D_x^2 + D_x^2 \eta^2 G^{(0)} - 2G^{(0)}\eta G^{(0)}\eta G^{(0)})\xi(x). \tag{4}$$

The terms $G^{(j)}(\eta)$ in the Taylor expansion are polynomial expressions in the quantities D_x and $G^{(0)}$ of order $j+1$, however for $\eta \in C^1$ these terms are nevertheless bounded from $H^1 \rightarrow L^2$. I is because of the form of the operator which is related to a multiple commutator; $[\eta, \ldots j \times \ldots [\eta, D_x^j]] = (-1)^j j!(\partial_x \eta)^j$. With regard to this series for the Dirichlet – Neumann operator, the water waves Hamiltonian itself is analytic on an appropriately chosen subset of, and possesses a Taylor series expansion about the equilibrium solution $(\eta, \xi) = 0$, namely

$$H(\eta, \xi) = \int_{\mathbb{R}^{d-1}} \tfrac{1}{2}\xi G^{(0)}\xi + \tfrac{g}{2}\eta^2 \, dx + \sum_{j \geq 3} \tfrac{1}{2} \int_{\mathbb{R}^{d-1}} \xi G^{(j-2)}(\eta)\xi \, dx$$

$$= \sum_{j \geq 2} H^{(j)}(\eta, \xi), \tag{5}$$

where $H^{(j)}(\eta, \xi)$ is homogeneous of degree j with respect to the variables (η, ξ).

5 Perturbation theory

Suppose that the Hamiltonian function H depends upon an additional parameter ε; $H(v; \varepsilon) = H^{(0)} + \varepsilon H^{(1)} + \ldots \varepsilon^m H^{(m)} + \varepsilon^{m+1} R(v; \varepsilon)$, for $\varepsilon \in \mathscr{E}$ a space of parameters. It is natural to approximate orbits $v(t; \varepsilon)$ by those of the truncated problem

$$\dot{v} = Jgrad_v(H^{(0)} + \varepsilon H^{(1)} + \ldots \varepsilon^m H^{(m)}) \tag{1}$$

$$v(0) = v_0, \qquad v(t) = v(t;\varepsilon,m)$$

The solution $v(t) = v(t;\varepsilon,m)$ clearly depends upon both ε and the degree m of the Taylor series approximation, and there is the natural expectation that, at least for finite time intervals, the solutions $v(t;\varepsilon,m)$ of (1) approximate the solutions of the full problem (1), with a better approximation given with larger m. Indeed, for C^2 Hamiltonians this is the case.

Proposition 5.1. *Suppose that the Hamiltonian $H \in C^{2,m+1}(M \times \mathscr{E})$. Then, at least for bounded time intervals $|t| \leq T_0$, approximate orbits $v(t;\varepsilon,m)$ of (1) are ε^m close to orbits of the full Hamiltonian system (1).*

Our intentions are to discuss Hamiltonian systems in infinite dimensional Hilbert spaces, and in particular Hamiltonian partial differential equations, which we have already pointed out are rarely given by smooth Hamiltonian vector fields. Therefore the above proposition is not applicable. Nonetheless it serves as a basic guiding principle to the problems we are aiming to discuss. It is also true that one can often do better than Proposition 5.1, and in some cases the length of the time interval of validity of this approximation may be longer, or indeed very much longer. However the only improvement on this statement that can be made at this level of generality is that, if the Lyapunov exponents of both section 1 (1) and (1) are bounded, then for any $m' < m$, approximating orbits remain $\varepsilon^{m'}$ close to true orbits for times $|t| \leq T_\varepsilon$, with $T_\varepsilon \sim \log(1/\varepsilon)$.

6 The calculus of transformations

Given a Hamiltonian system

$$\dot{v} = Jgrad_v H \tag{1}$$

posed on a phase space M, we will subject it to transformations of variables of M. Consider two phase spaces M_1 and M_2 with a symplectic form on M_1 given in terms of J_1. Let $H_1 : M_1 \to \mathbb{R}$ be a Hamiltonian. A transformation

$$\tau : M_1 \to M_2, \qquad v \mapsto w = \tau(v) \tag{2}$$

gives rise to a Hamiltonian defined on M_2, namely $H_2(w) = H_2(\tau(v)) = H_1(v)$. The Hamiltonian vector field $X_{H_1} = J_1 grad_v H_1$ is transformed as follows

$$\dot{w} = \partial_v \tau(v)\dot{v} = \partial_v \tau(v)J_1 grad_v H_1(v),$$

while on the other hand

$$grad_v H_1(v) = (\partial_v \tau)^T grad_w H_2(\tau(v)).$$

Equating the expressions, one observes the following:

Proposition 6.1. *The vector field* $X_{H_1} = J_1 \, grad_{\,v} H_1$ *is transformed to*

$$\dot{w} = \partial_v \tau(v) J_1 (\partial_v \tau)^T grad_{\,w} H_2(\tau(v)). \qquad (3)$$

We denote $J = \partial_v \tau(v) J_1 (\partial_v \tau)^T$ which can be used to define a symplectic structure on M_2. When M_2 already has a symplectic structuree represented by J_2, and the transformation $w = \tau(v)$ is such that $J_2 = \partial_v \tau(v) J_1 (\partial_v \tau)^T$, then τ is called *canonical*. In particular when $M_1 = M_2$ and $J_1 = J_2$ is given in Darboux coordinates, these are the usual canonical transformations which play a special rôle in the subject of Hamiltonian mechanics.

Examples of transformations. While the subject of canonical transformations and their generating functions is basic knowledge in finite dimensional Hamiltonian systems, it is less developed in the study of PDE and other infinite dimensional cases. In the following paragraphs we will work through some of the more elementary transformations that occur in Hamiltonian PDE, putting them into context. Furthermore we will make use of particular parameter families of such transformations in order to introduce a small parameter into the Hamiltonian. In this way the principle outlined in Section 5 can be invoked, with the result that we have a natural approximation procedure for solutions through a (albeit formal) series expansion of the Hamiltonian. This procedure and its general context has been worked out in a number of papers that have appeared over the span of several years, by the author along with M. Groves [3], P. Guyenne & H. Kalisch [4] and P. Guyenne, et al. [5].

Initially, the setting is that $M = L^2(\mathbb{R}^{d-1})^2$ will be considered the phase space, with

$$v = \begin{pmatrix} \eta \\ \xi \end{pmatrix} \in M, \qquad \langle v_1 | v_2 \rangle = \int_{\mathbb{R}^{d-1}} \eta_1 \eta_1 + \xi_1 \xi_2 \, dx \qquad (4)$$

$$J = \begin{pmatrix} 0 & I \\ -I & 0 \end{pmatrix} \qquad (5)$$

which is the case of Darboux coordinates.

(i) Amplitude scaling

Consider the elementary transformations $\tau : v \mapsto w$, where

$$w = \begin{pmatrix} \eta' \\ \xi' \end{pmatrix} = \begin{pmatrix} \alpha \eta \\ \beta \xi \end{pmatrix} = \tau(v), \qquad (6)$$

for $\alpha, \beta \in \mathbb{R}^+$. The Jacobian of the transformation τ is given by

$$\partial_v \tau = \begin{pmatrix} \alpha I & 0 \\ 0 & \beta I \end{pmatrix}$$

therefore the symplectic form induced by the transformation is

$$J_1 = \partial_v \tau J \partial_v \tau^T = \alpha\beta J,\tag{7}$$

with the Darboux operator J given in (5). The effects of such transformations are easily restored to the usual Darboux coordinates through a time change $t' = \alpha^{-1}\beta^{-1}$.

The small amplitude regime of the water wave problem is introduced by an amplitude scaling which is a transformation of this form. Namely one sets

$$\begin{pmatrix} \varepsilon^2\eta' \\ \varepsilon\xi' \end{pmatrix} = \begin{pmatrix} \eta \\ \xi \end{pmatrix}, \qquad \varepsilon \ll 1,\tag{8}$$

which is to say that we are seeking solutions for which the amplitude η of a solution is small, and represented in its asymptotic regime by an order one quantity η' times ε^2, and similarly for $\xi = \varepsilon\xi'$. The resulting change of symplectic form is that

$$J_1 = \varepsilon^{-3}J,$$

which is equivalent to a rescaling to a slow time variable. The effect on the water waves Hamiltonian Section 3 (5) and its Taylor expansion Section 4 (5) is that

$$H_1 = \int_{\mathbb{R}^{d-1}} \tfrac{1}{2}\varepsilon^2 \xi' G^{(0)} \xi' + \tfrac{g}{2}\varepsilon^4 \eta'^2 \, dx + \sum_{j\geq 3} \tfrac{1}{2}\int_{\mathbb{R}^{d-1}} \varepsilon^{2+2j}\xi' G^{(j-2)}(\eta')\xi' \, dx.$$

In particular a small parameter has been introduced into the Hamiltonian $H_1 = H_1(\eta',\xi';\varepsilon)$, for which one may consider approximations by its Taylor series. For instance, up to order $\mathcal{O}(\varepsilon^4)$

$$\varepsilon^2 H_1^{(2)} + \varepsilon^4 H_1^{(4)} = \varepsilon^2\left(\int \tfrac{1}{2}\xi' G^{(0)}\xi' \, dx\right) + \varepsilon^4\left(\int \tfrac{g}{2}\eta'^2 + \tfrac{1}{2}\xi' G^{(1)}(\eta')\xi' \, dx\right),$$

where we recall that $G^{(1)}(\eta') = D_x \eta' D_x - G^{(0)}\eta' G^{(0)}$.

(ii) Spatial scaling

The long wave regime of the water waves problem highlights solutions whose typical wavelength is asymptotically long; it is represented through a small parameter introduced into the problem by the spational scaling

$$x \mapsto X := \varepsilon x.\tag{9}$$

The resulting transformation of phase space M is thus

$$\tau : v(x) \mapsto w(X) = v(X/\varepsilon) = \tau(v)(X).\tag{10}$$

The Jacobian of the transformation on a vector field $V(x) \in T_v(M)$ is

$$\partial_v \tau(v) V(X) = \frac{d}{ds}\Big|_{s=0}\Big(v(X/\varepsilon) + sV(X/\varepsilon)\Big) = V(X/\varepsilon).$$

The transpose is slightly less obvious, we compute it using the identity;

$$\langle V_1 | \partial_v \tau V_2 \rangle = \int_{\mathbb{R}^{d-1}} V_1(X) V_2(X/\varepsilon), dX \tag{11}$$

$$= \int_{\mathbb{R}^{d-1}} V_1(\varepsilon x) V_2(x)\, \varepsilon^{d-1} dx = \langle (\partial_v \tau)^T V_1 | V_2 \rangle. \tag{12}$$

Therefore $(\partial_v \tau)^T V(x) = \varepsilon^{d-1} V(\varepsilon x)$, and the induced symplectic form is

$$J_2 = \partial_v \tau J (\partial_v \tau)^T = \varepsilon J, \tag{13}$$

at least if we are working with the Darboux symplectic structure. Thus, modulo a rescaling of time, this recovers the original symplectic form.

It is necessary to study the effect that this transformation has on the Hamiltonian.

Lemma 6.1. *Let* $\tau(v)(X) = v(X/\varepsilon) = w(X)$ *be be the transformation in question, and let* $m(D_x)$ *be a Fourier multiplier oprerator*

$$(m(D_x)v)(x) = \frac{1}{(2\pi)^{d-1}} \int\!\!\int_{\mathbb{R}^{2(d-1)}} e^{ik\cdot(x-x')} m(k) v(x')\, dx' dk. \tag{14}$$

Under τ*, the operator is transformed to*

$$\tau(m(D_x)v)(X) = (m(\varepsilon D_X)\tau(v))(X). \tag{15}$$

Proof. This is the fact that that cotangent variables (x,k) of pseudo-differential operators are transformed symplectically under changes of variables. Indeed one calculates

$$\tau(m(D_x)v)(X) = \frac{1}{(2\pi)^{d-1}} \int\!\!\int_{\mathbb{R}^{2(d-1)}} e^{ik\cdot(X/\varepsilon-x')} m(k) v(x')\, dx' dk$$

$$= \frac{1}{(2\pi)^{d-1}} \int\!\!\int_{\mathbb{R}^{2(d-1)}} e^{ik\cdot(X/\varepsilon-X'/\varepsilon)} m(k) v(X'/\varepsilon)\, \frac{dX' dk}{\varepsilon^{d-1}}$$

$$= \frac{1}{(2\pi)^{d-1}} \int\!\!\int_{\mathbb{R}^{2(d-1)}} e^{iK\cdot(X-X')} m(\varepsilon K) v(X'/\varepsilon)\, dX' dK$$

$$= m(\varepsilon D_X)\tau(v)(X). \qquad \square$$

Considering the water wave Hamiltonian, the Dirichlet–Neumann operator

$$G^{(0)}(D_x) = |D_x| \tanh(h|D_x|)$$

is transformed to

$$G^{(0)}(\varepsilon D_X) = \varepsilon |D_x| \tanh(\varepsilon h |D_x|) \sim \varepsilon^2 h |D_X|^2 - \frac{\varepsilon^4 h^3}{3} |D_X|^4 + \ldots \qquad (16)$$

Using this expression, the Hamiltonian (5) becomes

$$H_2 = \varepsilon^4 \int_{\mathbb{R}^{d-1}} \left(\tfrac{1}{2} \xi (h |D_X|^2 \xi + \tfrac{g}{2} \eta^2) \right) + \frac{\varepsilon^2}{2} \left(\xi(-\frac{h^3}{3} |D_X|^4 \xi) + \xi D_X \cdot \eta D_X \xi \right) \frac{dX}{\varepsilon^{d-1}}$$

$$+ \varepsilon^7 R_2^{(2)}. \qquad (17)$$

(iii) Surface elevation–velocity coordinates

It is often convenient to write the Euler equations in terms of the variables $w = (\eta, u)$, $u = \partial_x \xi$ instead of $v = (\eta, \xi)$. That is, the second variable represents a velocity instead of a potential; in this case it essentially represents the horizontal velocity of the fluid at the free surface $\Gamma(\eta)$. We restrict our discussion of these *surface elevation–velocity* coordinates to the case of two dimensions, for simplicity. That is,

$$w = (\eta, u) = \tau(v) = (\eta, \partial_x \xi). \qquad (18)$$

The Jacobian of the transformation is

$$\partial_v \tau(v) = \begin{pmatrix} I & 0 \\ 0 & \partial_x \end{pmatrix}$$

and the induced symplectic form is represented by the operator

$$J_2 = \begin{pmatrix} 0 & -\partial_x \\ -\partial_x & 0 \end{pmatrix} = \partial_v \tau J (\partial_v \tau)^T, \qquad (19)$$

where J is in Darboux coordinates. One recognizes this as the operator representing the Boussinesq symplectic form (13), up to a trivial change of sign.

Returning to the Hamiltonian (17), and phrasing it in surface elevation–velocity coordinates, we have

$$H_2 = \varepsilon^3 \int_{\mathbb{R}} \left(\tfrac{h}{2} u^2 + \tfrac{g}{2} \eta^2 \right) + \frac{\varepsilon^2}{2} \left(\frac{-h^3}{3} (\partial_X u)^2 + \eta u^2 \right) dX + \mathcal{O}(\varepsilon^7), \qquad (20)$$

while using the operator J_2 of (19) when expressing Hamilton's equations (1). The truncated system (20) up to order $\mathcal{O}(\varepsilon^5)$ is precisely the Boussinesq system (11) (modulo adjusting the value of several constants and the sign change $(p,q)^T = (\eta, -u)^T$).

(iv) Moving reference frame

It is part of the theory of nonlinear waves to work in coordinate systems which move with the characteristic speed of solutions, namely

$$v'(x,t) := v(x - tc, t), \tag{21}$$

for appropriate choices for the velocity c. However this transformation does not at first glance fit into the setting of the transformation theory described above, as the time variable is distinguished, and (21) mixes the rôles of the spatial and temporal variables. We observe however that in the problems under discussion the *momentum*

$$I(\eta, \xi) := \int_{\mathbb{R}} \eta(x) \partial_x \xi(x) \, dx \tag{22}$$

is a conserved quantity, as can be seen from its Poisson bracket with the Hamiltonian

$$\{I, H\} := \langle \operatorname{grad}_v I \,|\, J \operatorname{grad}_v H \rangle = 0. \tag{23}$$

Therefore their respective flows commute; $\varphi_t^H \circ \varphi_s^I(v) = \varphi_s^I \circ \varphi_t^H(v)$. The Hamiltonian flow of the momentum

$$\partial_s \begin{pmatrix} 0 & I \\ -I & 0 \end{pmatrix} \operatorname{grad}_v I = \begin{pmatrix} -\partial_x \eta \\ -\partial_x \xi \end{pmatrix} \tag{24}$$

is simply constant unit speed translation

$$\varphi_s^I(v)(x) = v(x - s).$$

Thus the flow along the diagonal is clearly $\varphi_t^H \circ \varphi_{tc}^I(v) = \varphi_t^{H+cI}$. Therefore the Hamiltonian flow of $H(v) + cI(v)$ is the Hamiltonian flow of $H(v)$ observed in a coordinate frame translating with velocity c.

In the context of the water wave problem the characteristic velocity is $c_0 := \sqrt{gh}$; to study the problem of water waves in our present point of view, we are to look at the flow of the system whose Hamiltonian is $H_2 + \sqrt{gh}I$.

Writing the momentum in surface elevation–velocity coordinates and scaling the coordinates appropriately, we find that

$$I = \varepsilon^3 \int_{\mathbb{R}} u\eta \, dX, \tag{25}$$

and therefore

$$H_2 + \sqrt{gh}I = \varepsilon^3 \int_{\mathbb{R}} \tfrac{1}{2} \left(hu^2 + 2\sqrt{gh}u\eta + g\eta^2 \right) + \tfrac{\varepsilon^2}{2} \left(\tfrac{-h^3}{3}(\partial_X u)^2 + \eta u^2 \right) dX$$

$$= \varepsilon^3 \int_{\mathbb{R}} \tfrac{1}{2} (\sqrt{h}u + \sqrt{g}\eta)^2 + \tfrac{\varepsilon^2}{2} \left(\tfrac{-h^3}{3}(\partial_X u)^2 + \eta u^2 \right) dX. \tag{26}$$

(v) Characteristic coordinates

Focusing on the first term H_2 of the Hamiltonian, it is a common situation to have it in the quadratic form

$$H_2^{(2)} = \tfrac{1}{2} \int_{\mathbb{R}} Au^2 + B\eta^2 \, dX,$$

with $A, B > 0$. Hamilton's equations (1) for $H_2^{(2)}$ alone are the wave equation

$$\partial_t \begin{pmatrix} \eta \\ u \end{pmatrix} = \begin{pmatrix} 0 & -\partial_X \\ -\partial_X & 0 \end{pmatrix} grad_v H_2^{(2)} = \begin{pmatrix} 0 & -A \\ -B & 0 \end{pmatrix} \begin{pmatrix} \partial_X \eta \\ \partial_X u \end{pmatrix}. \tag{27}$$

We seek a transformation of coordinates $(r,s)^T = \tau(\eta,u)^T$ which will accomplish three things.

1. It should diagonalize the symplectic form J_2;

$$J_3 := \partial_v \tau \begin{pmatrix} 0 & -\partial_X \\ -\partial_X & 0 \end{pmatrix} (\partial_v \tau)^T = \begin{pmatrix} \partial_X & 0 \\ 0 & -\partial_X \end{pmatrix}. \tag{28}$$

2. It should transform the Hamiltonian to normal form

$$H_3^{(2)} = \tfrac{1}{2} \int_{\mathbb{R}} \sqrt{AB}(r^2 + s^2) \, dX. \tag{29}$$

3. And it should transform the wave equation (27) to characteristic form

$$\partial_t \begin{pmatrix} r \\ s \end{pmatrix} = \begin{pmatrix} C & 0 \\ 0 & -C \end{pmatrix} \begin{pmatrix} \partial_X r \\ \partial_X s \end{pmatrix}. \tag{30}$$

The transformation $w = \tau(v) = Tv$, where T is the matrix

$$T = \begin{pmatrix} \sqrt[4]{\frac{B}{4A}} & -\sqrt[4]{\frac{A}{4B}} \\ \sqrt[4]{\frac{B}{4A}} & \sqrt[4]{\frac{A}{4B}} \end{pmatrix}$$

accomplishes all three of these goals, with the result that $C = \sqrt{AB}$.

In the case of the water wave Hamiltonian H_3, we have $A = h$ and $B = g$, so that

$$\begin{pmatrix} r \\ s \end{pmatrix} = \begin{pmatrix} \sqrt[4]{\frac{g}{4h}} & -\sqrt[4]{\frac{h}{4g}} \\ \sqrt[4]{\frac{g}{4h}} & \sqrt[4]{\frac{h}{4g}} \end{pmatrix} \begin{pmatrix} \eta \\ u \end{pmatrix}, \tag{31}$$

and in these terms, the relevant Hamiltonian approximation which is to be valid up to $\mathcal{O}(\varepsilon^5)$ is given by

$$H_2 + \sqrt{gh}I = \varepsilon^3 \int_{\mathbb{R}} \sqrt{gh}s^2 \, dX \tag{32}$$

$$+ \varepsilon^5 \int_{\mathbb{R}} -\frac{h^3}{6}\left(\sqrt{\frac{g}{4h}}\right)(\partial_X r - \partial_X s)^2 + \frac{1}{4\sqrt{2}} \sqrt[4]{\frac{g}{h}}(r^3 - r^2 s - rs^2 + s^3) \, dX.$$

Now restrict this Hamiltonian to the hypersurface $M_1 := \{s = 0\} \subseteq M$, denoting it by H_4;

$$H_4 = \varepsilon^5 \int_{\mathbb{R}} -\frac{h^3}{6}\left(\sqrt{\frac{g}{4h}}\right)(\partial_X r)^2 + \frac{1}{4\sqrt{2}}\sqrt[4]{\frac{g}{h}} r^3 \, dX. \tag{33}$$

The subspace M_1 is a *symplectic* subspace of M, possessing the symplectic form $J_4 = \partial_X$, it being the restriction of the symplectic form J_3 of (28). This is unlike the situation in Darboux coordinates, in which M_1 would be a Lagrangian subspace. The equations of motion (1) for r on M_1, or at least in an $o(\varepsilon^2)$ neighborhood of it, are thus

$$\begin{aligned}
\partial_t r &= \partial_X \, grad_{\, r} H_4 \\
&= \varepsilon^2 \partial_X (c_1 \partial_X^2 r + c_2 r^2),
\end{aligned} \tag{34}$$

with $c_1 = \frac{h^3}{3}\sqrt{\frac{g}{4h}}$ and $c_2 = \frac{3}{4\sqrt{2}}\sqrt[4]{\frac{g}{h}}$. This is precisely the KdV equation given in (8), modulo a simple change of time scale $\partial_t = \varepsilon^2 \partial_\tau$ ($\tau = \varepsilon^2 t$), and with a few extra but unimportant constants that could have been normalized in the above calculation with some forethought.

We have seen that a formal calculation, using basic transformations and a small parameter have given us the KdV equation as an approximation of the equations of water waves. It has been a research program to understand the rigorous aspects of this correspondence between solutions of the full Euler's equations and solutions of the KdV or of other model long wave equations. However at this point none of the rigorous results follow along the lines of the above concatenation of transformations. We believe that such an approach would be a rewarding line of work.

References

1. M. Christ & J.-L. Journé. Polynomial growth estimates for multilinear singular integral operators. *Acta Math.* **159** (1987), no. 1–2, 51–80.
2. R. Coifman & Y. Meyer. Nonlinear harmonic analysis and analytic dependence. Pseudodifferential operators and applications (Notre Dame, Ind., 1984), 71–78, Proc. Sympos. Pure Math., 43, Amer. Math. Soc., Providence, RI (1985).
3. W. Craig & M. Groves. Hamiltonian long-wave approximations to the water-wave problem. *Wave Motion* **19** (1994), no. 4, 367–389.
4. W. Craig, P. Guyenne & H. Kalisch. Hamiltonian long-wave expansions for free surfaces and interfaces. *Comm. Pure Appl. Math.* **58** (2005), no. 12, 1587–1641.
5. W. Craig, P. Guyenne, D. P. Nicholls & C. Sulem. Hamiltonian long-wave expansions for water waves over a rough bottom. *Proc. R. Soc. Lond. Ser. A Math. Phys. Eng. Sci.* **461** (2005), no. 2055, 839–873.
6. W. Craig, U. Schanz & C. Sulem. The modulational regime of three-dimensional water waves and the Davey-Stewartson system. *Ann. Inst. H. Poincaré Anal. Non Linéaire* **14** (1997), no. 5, 615–667.
7. W. Craig & C. Sulem. Numerical simulation of gravity waves. *J. Comput. Phys.* **108** (1993), no. 1, 73–83.
8. J. Hadamard. *Leçons sur le calcul des variations*. Hermann, Paris, 1910; paragraph §249.
9. J. Hadamard. Sur les ondes liquides. *Rend. Acad. Lincei* **5** (1916), no. 25, pp. 716–719.

10. D. Kaup. A higher-order water-wave equation and the method for solving it. *Progr. Theoret. Phys.* **54** (1975), no. 2, 396–408.

11. R. Sachs. On the integrable variant of the Boussinesq system: Painlevé property, rational solutions, a related many-body system, and equivalence with the AKNS hierarchy. *Phys. D* **30** (1988), no. 1–2, 1–27.

12. V. E. Zakharov. Stability of periodic waves of finite amplitude on the surface of a deep fluid. *J. Appl. Mech. Tech. Phys.* **9** (1968), pp. 1990–1994.

13. V. E. Zakharov. On stochastization of one-dimensional chains of nonlinear oscillators. *Zh. Eksp. Teor. Fiz.* **65** (1973), pp. 219–225.

Three theorems on perturbed KdV

Sergei B. Kuksin[1]

Abstract This short paper is based on a lecture, given at the NATO Advanced Study Institute on Hamiltonian dynamical systems (Montréal, 2007). Its goal is to discuss three theorems on the long-time behaviour of solutions of a perturbed KdV equation under periodic boundary conditions. These theorems are infinite-dimensional analogies of three classical results on small perturbations of an integrable finite-dimensional system:

- The KAM theorem
- The first-order averaging theory for Hamiltonian perturbations
- The Khasminskii averaging theory for random perturbations

The three theorems raise many new questions, some of which are mentioned below.

We stress that the three theorems are infinite-dimensional *analogies* of some finite-dimensional statements. That is, for nearly integrable nonlinear PDEs (under periodic boundary conditions) we do not know any result which is essentially infinite-dimensional. There are no doubts that such results exist. To find them is a big challenge.

1 KdV equation

Consider the KdV equation under zero-mean value periodic boundary conditions:

$$\dot{u} + u_{xxx} - 6uu_x = 0, \qquad x \in \mathbb{T}^1 = \mathbb{R}/2\pi, \qquad \int u\,dx \equiv 0. \tag{KdV}$$

It can be written in the form

$$\dot{u} = \frac{\partial}{\partial x}\,\frac{\delta}{\delta u(x)}\,H_{KdV}, \qquad H_{KdV} = \int(\frac{1}{2}u_x^2 + u^3)\,dx, \tag{1}$$

[1] Centre de Mathématiques Ecole Polytechnique, 91128 Palaiseau, France
e-mail: s.b.kuksin@math.polytechnique.fr

W. Craig (ed.), Hamiltonian Dynamical Systems and Applications, 85–92.
© 2008 *Springer Science + Business Media B.V.*

and hence (KdV) is a Hamiltonian PDE. Due to Novikov and Lax [8, 11] and McKean-Trubowitz [9] it is integrable.

The KdV equation

$$\dot{u} + u_{xxx} - 6uu_x = 0 \tag{2}$$

may be considered under other boundary conditions; e.g., under the L^2-boundary conditions

$$x \in \mathbb{R}, \qquad u(t,\cdot) \in L^2(\mathbb{R}) \qquad \forall t, \tag{L^2}$$

or under quasi-periodic boundary conditions

$$x \in \mathbb{R}, \qquad u(t,x) \qquad \text{is quasiperiodic in } x. \tag{QP}$$

Equation (2) + (L^2) is integrable. This is a simpler dynamical system than (KdV) due to the phenomenon of radiation. There is a number of results on this equation and its perturbations. Equation (2) + (QP) also is an integrable system, but nothing is known about its perturbations.

1.1 Integrability of (KdV)

Now we discuss what does it exactly mean that the equation (KdV) is integrable. Denote

$$Z = \{u(x) \in L^2 \mid \int u\,dx = 0\}, \qquad \|\cdot\| - \text{the } L^2\text{-norm in } Z.$$

(KdV) defines in the space Z a dynamical system with infinitely-many analytic integrals of motion $I_1, I_2, \ldots$. The functions I_j may be choosen to be non-negative and such that:

- For a vector $I = (I_1, I_2, \ldots)$, where the numbers $I_j \geqslant 0$ decay with j sufficiently fast, denote
$$T_I = \{u \in Z \mid I_j(u) = I_j \ \forall j\}.$$

  Then T_I is an analytic torus in Z and
$$\dim T_I = \sharp\{j \mid I_j > 0\}.$$

  The r.h.s. is called *the number of open gaps*.
- The set $\mathscr{T}^{2n} = \cup\{T_I \mid I_1 \geqslant 0, \ldots, I_n \geqslant 0, 0 = I_{n+1} = \ldots\}$
  is a smooth $2n$-manifold, called "the n-gap manifold". Obviously it is (KdV)-invariant.
- Each torus T_I carries a cyclic coordinate q such that in the (I,q)-variables the equation (KdV) becomes
$$\dot{I} = 0, \quad \dot{q} = W(I). \tag{3}$$

The *frequency map* $W : \mathbb{R}_+^\infty \to \mathbb{R}^\infty$ is analytic.

- (KdV), restricted to any manifold $\mathcal{T}^{2n}$, is Liouville–Arnold integrable.
- The variables (I,q) form action-angles for (KdV) in Z. That is, $\omega_2 = dI \wedge dq$, $H_{KdV} = h_{KdV}(I_1,I_2,\dots)$ and $W(I) = \nabla h_{KdV}(I)$. Here ω_2 is the symplectic form corresponding to the Hamiltonian form (1) of (KdV).

The coordinates (I,q) are singular when some of the $I_j = 0$. This is a serious disadvantage if we try to use them for analytical studies of (KdV).

Below in this paper we discuss the following

PROBLEM: What happens to the tori T_I, the manifolds $\mathcal{T}^{2n}$ and solutions (2) under small perturbations of the equation?

A crucial step in the study of this problem is to introduce in the vicinity of a torus T_I a new coordinate system, free from the disadvantages of the action-angle coordinates (I,q), and such that the Hamiltonian H_{KdV}, written in these variables, is a "nice" function. Such coordinate systems exist and are given by normal forms.

1.2 Normal forms for (KdV)

NF$_1$ (SK, [6]). In the vicinity of any n-gap torus T_I in Z there exist analytic coordinates

$$(\varphi, p, y),\, \varphi \in \mathbb{T}^n,\, p \in \mathbb{R}^n,\, y = (y^+, y^-) = (y_1^+, y_1^-; y_2^+, y_2^-; \dots) \in Y = Z \ominus R^{2n}$$

such that

- The symplectic form is $\omega_2 = dp \wedge d\varphi + dy^+ \wedge dy^-$;
- $\mathcal{T}^{2n} = \{y = 0\}$;
- In these coordinates the equation (KdV) reads as follows;

$$\dot\varphi = \nabla_p H,\quad \dot p = -\nabla_\varphi H,\quad \dot y = J\nabla_y H,$$

$$H = h(p) + \tfrac{1}{2}\langle A(p)y, y\rangle + h_3(\varphi, p, y),$$

where $J(y^+, y^-) = (-y^-, y^+)$, $h_3 = O(\|y\|^3)$, A is an analytic operator-valued function of p and h, h_3 are analytic scalar functions of (φ, p, y).

This normal form exists for "all" other integrable PDE (see [7]). It is sufficient for KAM-purposes.

NF$_2$ ("Birkhoff coordinates", Thomas Kappeler 1991–2001, see in [4]). In Z there exist analytic coordinates $y = (y^+, y^-)$, $y = (y_1^+, y_1^-; y_2^+, y_2^-; \dots)$, such that

- $\omega_2 = dy^+ \wedge dy^-$;
- $I_j = \tfrac{1}{2}((y_j^-)^2 + (y_j^+)^2)$ and $q_j = \mathrm{Arctan}(y_j^-/y_j^+)$ for $j = 1, 2, \dots$;
- The Hamiltonian H_{KdV}, written in the y-variables, is $h_{KdV}(I_1, I_2, \dots)$.

This normal form is more powerful, but less general. It is needed for averaging theorems for perturbed KdV.

2 KAM-theory

Consider a perturbed KdV Hamiltonian $H_\varepsilon(u) = H_{KdV}(u) + \varepsilon \int f(u(x),x)\,dx$. The corresponding Hamiltonian equation is

$$\dot{u} + u_{xxx} - 6uu_x - \varepsilon \tfrac{\partial}{\partial x} f'_u(u(x),x) = 0. \tag{1}$$

Question: In what sense do the tori T_I and solutions (2) persist in (1)?

The answer is known only for finite-gap tori; i.e., when $I = (\hat{I},0,\dots)$, $\hat{I} \in \mathbb{R}^n_+$. We will write $T_I = T_{\hat{I}}$. This is an n-dimensional torus.

Theorem 1 (SK, see [6]). *For most of $\hat{I} \in \mathbb{R}^n_+$ eq. (1) has an analytic invariant n-torus $T_{\hat{I}}^\varepsilon \subset Z$ which is $C\sqrt{\varepsilon}$-close to $T_{\hat{I}}$. It is filled in with time-quasiperiodic solutions*

$$\dot{\varphi} = \hat{W}^\varepsilon(\hat{I}); \quad |\hat{W}^\varepsilon - \hat{W}| \leqslant C\varepsilon, \quad \hat{W} = (W_1,\dots,W_n).$$

The corresponding linearised equations are reducible to constant coefficients and are linearly stable.

Here "for most of $\hat{I} \in \mathbb{R}^n_+$" means the following. Any compact set $K \subset \mathbb{R}^n_+$ contains a Borel subset K_ε such that $\mathrm{meas}(K \setminus K_\varepsilon) \to 0$ when $\varepsilon \to 0$ and for any $\hat{I} \in K_\varepsilon$ the theorem's assertions hold.

"Proof". In the variables (φ, p, y) from the normal form NF_1, the Hamiltonian of the perturbed equation is

$$H_\varepsilon = (h(p) + \tfrac{1}{2}\langle A(p)y,y\rangle + h_3(\varphi,p,y)) + \varepsilon H_{perturb}(\varphi,p,y),$$

and the torus $T_{\hat{I}}$ is formed by the points $\{(\varphi,\hat{I},0),\ \varphi \in \mathbb{T}^n\}$. Scale variables (φ,p,y) near this torus by ε:

$$\varphi = \varphi, \quad p = \hat{I} + \varepsilon^{2/3}p, \quad y = \varepsilon^{1/3}y.$$

In the scaled variables H_ε becomes

$$H_\varepsilon = \mathrm{const} + \omega \cdot p + \tfrac{1}{2}\langle A(\hat{I})y,y\rangle + \varepsilon^{1/3}\tilde{h}(\varphi,p,y;\hat{I},\varepsilon),$$

where $\tilde{h}$ is an analytic function and

$$\omega = \omega(\hat{I}) = \nabla_{\hat{I}} h(\hat{I},0,\dots).$$

When $\hat{I}$ varies in $\mathbb{R}^n_+$, ω varies in the domain $\Omega = \nabla h(\mathbb{R}^n_+)$. So H_ε may be regarded as a function $H_\varepsilon(\varphi,p,y;\omega,\varepsilon)$, depending on the n-dimensional parameter $\omega \in \Omega$.

If $H_\varepsilon(\varphi, p, y; \omega, \varepsilon)$ as a function of ω satisfies suitable nondegeneracy conditions, then the infinite-dimensional KAM-theory for systems with parameters applies to H_ε and implies the assertions. The nondegeneracy condition was checked in [1]. So the theorem holds. $\qquad\square$

For a real proof see [6] and the books [4, 7].

Now some remarks:

◇ Theorem 1 is a general result which also holds for perturbations of other integrable Hamiltonian PDE, see [7].

◇ The time-quasiperiodic solutions of the perturbed (KdV) (1) that are constructed, with $n = 1, 2, \ldots$ correspond to "few" initial data, occupying a "small" subset of Z which cannot be explicitly described.

◇ Still these solutions become dense in the phase-space Z as $\varepsilon \to 0$.[1]

3 Averaging: Hamiltonian perturbations

Consider again the perturbed (KdV) (1). This is a Hamiltonian perturbation of an integrable Hamiltonian system. For perturbations of *finite-dimensional* Hamiltonian systems the actions $I(u(t))$ of any solution u remain almost constant over a time interval of duration $e^{\varepsilon^{-\kappa}}$, $\kappa > 0$, under the Nekhoroshev's steepness condition, see [10]. For the KAM-solutions of equation (1) of Section 2, constructed in Theorem 1, the change of actions is $\leqslant C\varepsilon$ for all t. But these solutions correspond to very special initial data. How do the actions of a general solution of equation (1) of Section 2 behave?

Statement. Let $u_\varepsilon(t)$ be a solutions of (1) such that $u_\varepsilon(0) = u_0 \in \mathscr{T}^{2n}$. Then

$$|I(u_\varepsilon(t)) - I(u_0)| \leqslant \theta_n(\varepsilon) \qquad \forall t \leqslant T_\varepsilon = \varepsilon^{-1}, \qquad \text{(S)}$$

where $\theta_n \to 0$ as $\varepsilon \to 0$.

Theorem 2 (T. Kappeler and SK, see [3]). There exist positive constants $\delta_1, \delta_2, \ldots$ such that (S) holds if $u_0 \in \mathscr{T}^{2n}$ and

$$\|u_0\| \leqslant \delta_n. \qquad \text{(1)}$$

Most likely the result is true without the smallness assumption (1). Much more difficult (and more important) related questions are the following:

• Is the result true with $\theta_n(\varepsilon) = \theta(\varepsilon, \|u_0\|)$? (where $\|u_0\|$ is arbitrary).
• Does (S) hold with $T_\varepsilon = \varepsilon^{-a}$, $a \geqslant 2$?
(We can prove (S) for some $a > 1$, but not for $a = 2$.)

---

[1] This means that for any open set $Q \subset Z$ there exists $\varepsilon_0 > 0$ such that for each $\varepsilon < \varepsilon_0$ the perturbed equation has a time-quasiperiodic solution which passes through Q.

4 Averaging: case of non-Hamiltonian perturbations

4.1 Deterministic perturbations

Consider a non-Hamiltonian perturbation of (KdV). Say, the following equation, often considered in the literature:

$$\dot{u} + u_{xxx} - 6uu_x = \varepsilon u_{xx}, \quad u(0) = u_0$$

(the boundary conditions as in (KdV) are assumed). Write it in the (I,q)-variables:[2]

$$\dot{I} = \varepsilon F(I,q), \qquad \dot{q} = W(I) + \varepsilon G(I,q). \tag{1}$$

Consider the averaged equation for $I(t)$:

$$\dot{I} = \varepsilon \langle F \rangle (I), \quad \langle F \rangle (I) = \int_{T_I} F(I,q) \, dq; \qquad I(0) = I(u_0), \tag{2}$$

where dq is the Haar measure on T_I.

Averaging Principle: If $(I(t), q(t))$ is a solution of (1) and $J(t)$ is a solution of (2), then

$$|I(t) - J(t)| \leqslant \theta(\varepsilon) \quad \forall 0 \leqslant t \leqslant \varepsilon^{-1},$$

where $\theta \to 0$ with ε.

   To prove the Averaging Principle is a big open problem due to the following two difficulties:

 (i) KdV-dynamics on some tori T_I is resonant
(ii) the averaged equation (2) is not well-posed

   To avoid the first difficulty it is natural to introduce randomness.

4.2 Random perturbations

Now consider a randomly perturbed KdV:

$$\dot{u} + u_{xxx} - 6uu_x = \varepsilon u_{xx} + \sqrt{\varepsilon}\,\eta(t,x),$$

$$\eta(t,x) = \frac{\partial}{\partial t} \sum_{j \in \mathbb{Z}_0} b_j \beta_j(t) e_j(x). \tag{3}$$

Here $\mathbb{Z}_0$ is the set of all non-zero integers and

  • All $b_j > 0$ and decay fast when $j \to \infty$

---

[2] Concerning properties of the functions F, G and their smoothness see [5].

- $\{\beta_j(t)\}$ are independent standard Wiener processes
- $\{e_j(x)\}$ is the standard trigonometric basis for the space of periodic functions

The scaling factor $\sqrt{\varepsilon}$ in the r.h.s. is natural since only with this scaling do solutions of equation (3) remain of order one when $t \to \infty$ and $\varepsilon \to 0$.

Using Ito's formula we write the corresponding equation for the vector $I(u(t)) = I^\omega(t)$:

$$\dot{I} = \varepsilon F(I,q) + \sqrt{\varepsilon} \sum_j G_j(I,q) \frac{\partial}{\partial t} \beta_j(t).$$

Average it:

$$\dot{I} = \varepsilon \langle F \rangle(I) + \sqrt{\varepsilon} \sum \langle G_j \rangle(I) \frac{\partial}{\partial t} \beta_j(t).$$

Here $\langle F \rangle$ is the same as in (2) and $\langle G_j \rangle(I)$, $j \in \mathbb{Z}_0$, are the column-vectors, forming the infinite matrix $\langle G \rangle(I)$. The latter is defined as a symmetric square root of the matrix

$$\int_{T_I} G(I,q) G^t(I,q)\,dx,$$

where the matrix $G(I,q)$ is formed by the columns $G_j(I,q)$.

Theorem 3 (SK and A. Piatnitski, see [5]). Let $J(t)$ be a solution of the averaged equation such that $J(0) = I(u_0)$. Then

$$\mathrm{dist}\{\mathscr{D}(I(u(t))), \mathscr{D}(J(t))\} \leqslant \theta(\varepsilon) \quad \forall 0 \leqslant t \leqslant \varepsilon^{-1},$$

where $\theta \to 0$ with ε. Here $\mathscr{D}(I(u(t)))$ is the law of the random variable $I(u(t)) \in \mathbb{R}^\infty_+$, i.e. a Borel measure in $\mathbb{R}^\infty_+$ (this space is given a weighted l_1-distance, see [5]), and *dist* is the Lipschitz-dual distance in the space of Borel measures.

This result is an infinite-dimensional analogy of finite-dimensional averaging theorems due to Khasminskii and Freidlin–Wentzell (see [2]).

Theorem 3 illustrates well.

Principle: Introducing randomness to a nonlinear PDE we simplify the equation.

References

1. A. I. Bobenko and S. B. Kuksin, *Finite-gap periodic solutions of the KdV equation are non-degenerate*, Phys. Lett. A **161** (1991), no. 3, 274–276.
2. M. Freidlin and A. Wentzell, *Random Perturbations of Dynamical Systems*, 2nd ed., Springer, New York, 1998.
3. T. Kappeler and S. Kuksin, *Paper under preparation*.
4. T. Kappeler and J. Pöschel, *KAM & KdV*, Springer, 2003.
5. S. B. Kuksin and A. L. Piatnitski, *Khasminskii-Whitham averaging for randomly perturbed KdV equation*, Preprint (2006), see at `www.ma.hw.ac.uk/~kuksin/rfpdeLim.html`.
6. S. B. Kuksin, *The perturbation theory for the quasiperiodic solutions of infinite-dimensional Hamiltonian systems and its applications to the Korteweg – de Vries equation*, Math. USSR Sbornik **64** (1989), 397–413.

7. ———, *Analysis of Hamiltonian PDEs*, Oxford University Press, Oxford, 2000.
8. P. D. Lax, *Periodic solutons of the KdV equations*, Commun. Pure Appl. Math. **28** (1975), 141–188.
9. H. McKean and E. Trubowitz, *Hill's operator and hyperelliptic function theory in the presence of infinitely many branching points*, Comm. Pure Appl. Math. **29** (1976), 143–226.
10. N. N. Nekhoroshev, *Exponential estimate of the stability of near integrable Hamiltonian systems*, Russ. Math. Surveys **32** (1977), no. 6, 1–65.
11. S. P. Novikov, *A periodic problem for the Korteweg-de-Vries equation, I*, Funct. Anal. Appl. **8** (1974), 236–246.

Groups and topology in the Euler hydrodynamics and KdV

Boris Khesin[1]

Abstract We survey applications of group theory and topology in fluid mechanics and integrable systems. The main reference for most facts in this paper is [1], see also details in [4].

1 Euler equations and geodesics

1.1 The Euler hydrodynamics equation

Consider an incompressible fluid occupying a domain M in $\mathbb{R}^n$. The fluid motion is described by a velocity field $v(t,x)$ and a pressure field $p(t,x)$ which satisfy the classical Euler equation:

$$\partial_t v + (v \cdot \nabla)v = -\nabla p, \tag{1}$$

where $\operatorname{div} v = 0$ and the field v is tangent to the boundary of M. The function p is defined uniquely modulo an additive constant by the condition that v has zero divergence at any moment t.

The flow $(t,x) \mapsto g(t,x)$ describing the motion of fluid particles is defined by its velocity field $v(t,x)$:

$$\partial_t g(t,x) = v(t,g(t,x)), \ g(0,x) = x.$$

The acceleration of particles is given by $\partial_t^2 g(t,x) = (\partial_t v + (v \cdot \nabla)v)(t,g(t,x))$, according to the chain rule, and hence the Euler equation (1) is equivalent to

$$\partial_t^2 g(t,x) = -(\nabla p)(t,g(t,x)).$$

---

[1] Department of Mathematics, University of Toronto, ON M5S 2E4, Canada
e-mail: khesin@math.toronto.edu

W. Craig (ed.), Hamiltonian Dynamical Systems and Applications, 93–102.

The latter form of the Euler equation (for a smooth flow $g(t,x)$) says that the acceleration of the flow is given by a gradient and hence it is L^2-orthogonal to the set of volume-preserving diffeomorphisms, which satisfy the incompressibility condition $\det(\partial_x g(t,x)) = 1$. More precisely, it is L^2-orthogonal to the tangent to this set, the space of divergence-free fields. In other words, the fluid motion $g(t,x)$ is a geodesic line on the set of such diffeomorphisms of the domain M with respect to the induced L^2-metric. Note that this metric is invariant with respect to reparametrizing the fluid particles, i.e. it is *right-invariant* on the set of volume-preserving diffeomorphisms (a reparametrization of the independent variable is the *right* action of a diffeomorphism).

More generally, the Euler equation describes an ideal incompressible fluid filling an arbitrary Riemannian manifold M, see [1, 5]. It defines the geodesic flow on the group of volume-preserving diffeomorphisms of M. It turns out that the group-geodesic point of view, developed in [1] is quite fruitful for topological and qualitative understanding of the fluid motion, as well as for obtaining various quantitative results related to stability and first integrals of the Euler equation.

1.2 Geodesics on Lie groups

In [1] V. Arnold suggested a general framework for the Euler equations on an arbitrary group, which we recall below. In this framework the Euler equation describes a geodesic flow with respect to a suitable one-sided invariant Riemannian metric on the given group.

More precisely, consider a (possibly infinite-dimensional) Lie group G, which can be thought of as the configuration space of some physical system. (Examples from [1]: the group $SO(3)$ for a rigid body and the group $\mathrm{SDiff}(M)$ of volume-preserving diffeomorphisms for an ideal fluid filling a domain M.) The tangent space at the identity of the Lie group G is the corresponding Lie algebra $\mathfrak{g}$. Fix some (positive definite) quadratic form, the energy, on $\mathfrak{g}$. We consider right translations of this quadratic form to the tangent space at any point of the group (the "translational symmetry" of the energy). This way the energy defines a right-invariant Riemannian metric on the group G. The geodesic flow on G with respect to this energy metric represents the extremals of the least action principle, i.e., the actual motions of our physical system. (For a rigid body one has to consider left translations.)

To describe a geodesic on the *Lie group* with an initial velocity $v(0) = \xi$, we transport its velocity vector at any moment t to the identity of the group (by using a right translation). This way we obtain the evolution law for $v(t)$, given by a (nonlinear) dynamical system $dv/dt = F(v)$ on the *Lie algebra* $\mathfrak{g}$ (Fig. 1).

Theorem 1.1 *The system on the Lie algebra* $\mathfrak{g}$*, describing the evolution of the velocity vector along a geodesic in a right-invariant metric on the Lie group* G*, is called the* Euler equation *corresponding to this metric on G.*

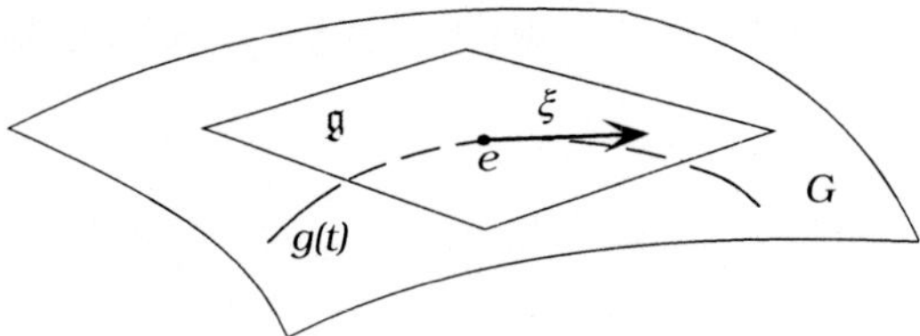

Fig. 1 The vector ξ in the Lie algebra $\mathfrak{g}$ is the velocity at the identity e of a geodesic $g(t)$ on the Lie group G

1.3 Geodesic description for various equations

A similar Arnold-type description via the geodesic flow on a Lie group can be given to a variety of conservative dynamical systems in mathematical physics. Below we list several examples of such systems to emphasize the range of applications of this approach. The choice of a group G (column 1) and an energy metric E (column 2) defines the corresponding Euler equations (column 3).

Group	Metric	Equation
$SO(3)$	$< \omega, A\omega >$	Euler top
$SO(3)\dotplus\mathbb{R}^3$	quadratic forms	Kirchhoff equations for a body in a fluid
$SO(n)$	Manakov's metrics	$n-$dimensional top
$\mathrm{Diff}(S^1)$	L^2	Hopf (or, inviscid Burgers) equation
Virasoro	L^2	KdV equation
Virasoro	H^1	Camassa $-$ Holm equation
Virasoro	$\dot{H}^1$	Hunter $-$ Saxton (or Dym) equation
$\mathrm{SDiff}(M)$	L^2	Euler ideal fluid
$\mathrm{SDiff}(M)\dotplus\mathrm{SVect}(M)$	L^2+L^2	Magnetohydrodynamics
$\mathrm{Maps}(S^1, SO(3))$	H^{-1}	Landau $-$ Lifschits equation

In some cases these systems turn out to be not only Hamiltonian, but also bi-hamiltonian. More detailed descriptions and references can be found in the book [4].

2 Topology of steady flows

2.1 Arnold's classification of steady fluid flows

The stationary Euler equation in the domain M has the form

$$(v \cdot \nabla)v = -\nabla p$$

on a divergence-free vector field v. In 3D this equation can be rewritten as follows:

$$v \times \mathrm{curl}\,v = -\nabla\alpha,$$

i.e. the cross-product of the fields v and $\mathrm{curl}\,v$ is a potential vector field. Here $\alpha = p + |v|^2/2$ is called the *Bernoulli function*. (Another way to express this is to say that the field v commutes with its vorticity $\mathrm{curl}\,v$. The latter commutativity condition is valid in any dimension.)

Theorem 2.1 [2,3] *Let M be a three-dimensional manifold without boundary. Then all non-critical level sets of α are 2-tori. Furthermore, both fields v and $\mathrm{curl}\,v$ are tangent to these levels and define there the $\mathbb{R}^2$-action.*

On a manifold M with boundary, the α-level sets are either 2-tori or annuli. On tori the flow lines are either all closed or all dense, and on annuli all flow lines are closed.

The proof of the theorem is based on the observation that v is always tangent to the level sets of α, i.e. the function α is a first integral of the equation. On non-critical sets one has $\nabla\alpha \neq 0$, which implies that $v \neq 0$. Thus the α-level sets are two-dimensional orientable surfaces which admit a non-vanishing tangent vector field. Thus these surfaces must be tori, since their Euler characteristic is 0. For M with boundary, the α-level sets could intersect the boundary, in which case they are annuli, see Fig. 2.

Remark 2.1. [2,3] *(i)* Analyticity assumptions on M and v imply that there is a finite number of cells between the critical levels of α, which are foliated by tori or annuli.

(ii) The $\mathbb{R}^2$-action on tori is given by two commuting vector fields v and $\mathrm{curl}\,v$. In particular, locally around a non-critical level of α there are coordinates $\{\phi_1, \phi_2, z\}$ such that the α-levels are given by $\{z = const\}$ and

$$v = v_1(z)\partial_{\phi_1} + v_2(z)\partial_{\phi_2},$$

$$\mathrm{curl}\,v = w_1(z)\partial_{\phi_1} + w_2(z)\partial_{\phi_2}.$$

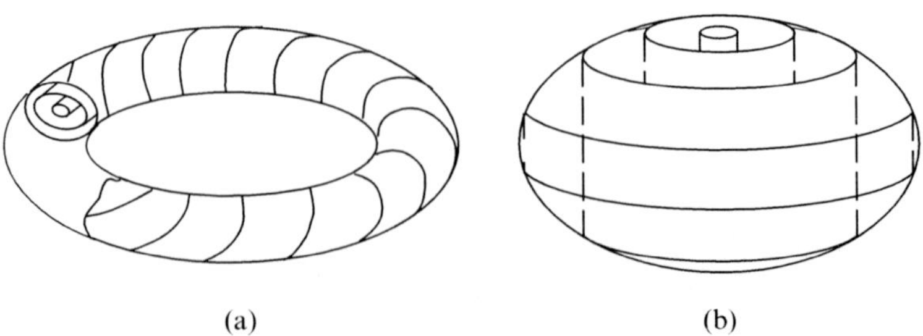

(a)   (b)

Fig. 2 The flow lines of steady flows typically lie on tori or annuli: see the cases of M without boundary (a) and with boundary (b)

This way a steady 3D flow looks like a completely integrable Hamiltonian system with two degrees of freedom.

(iii) It could happen that $\nabla\alpha = 0$ everywhere, i.e. $\alpha = const$. Then

$$v \times \mathrm{curl}\,v = 0,$$

and hence v is collinear with curl$v$ at every point. Such fields are called *force-free*.

If $v \neq 0$ everywhere, we can express curl$v$ as curl$v = \kappa(x)v$ for a smooth function $\kappa(x)$ on M. Then κ is a first integral of our dynamical system given by the field v. Indeed,

$$0 = div(\mathrm{curl}\,v) = div(\kappa v) = \nabla\kappa \cdot v.$$

Again, the vector field v is tangent to the level sets of κ. On these sets there is only the $\mathbb{R}$-action.

(iv) Another interesting case is when $\kappa(x) = const$. Then

$$\mathrm{curl}\,v = \lambda v,$$

i.e. v is an eigen field for the curl operator: $\mathrm{curl}\,\xi = \lambda\xi$. Such fields are called *Beltrami fields* (or flows). One famious example is given by the so called ABC flows on a 3D-torus, which exhibit chaotic behavior and draw special attention in fast dynamo constructions:

$$v_x = A\sin z + C\cos y$$

$$v_y = B\sin x + A\cos z$$

$$v_z = C\sin y + B\cos x$$

(v) There is a two-dimensional version of the above Arnold's theorem. Any area-preserving field in dimension 2 is a Hamiltonian field (with possibly a multivalued Hamiltonian function): $v = \mathrm{sgrad}\,\psi$. This Hamiltonian function ψ is called the *stream function* for the field v. The condition that v is a steady flow, i.e. that it commutes with its vorticity curl$v$, amounts in 2D to the fact that the stream function ψ and its Laplacian $\Delta\psi$ have the same level curves. In other words, locally there is a function $F : \mathbb{R} \to \mathbb{R}$ such that $\Delta\psi = F(\psi)$.

2.2 Variational principles for steady flows

The stationary solutions of the Euler equation come by as extremals from two different variational principles [3,8].

i) The magneto-hydrodynamic ("MHD") variational principle: consider the energy functional

$$E(v) = \int_M |v|^2 d^3x$$

on divergence-free vector fields v on a 3D manifold M. Then extremals of the energy functional among the *fields diffeomorphic* to a given one are singled out by the same condition as the steady Euler flows: such fields must commute with their vorticities. (This problem on conditional extremum corresponds to the restriction of the energy E to the *adjoint orbits* of the diffeomorphism group.)

ii) The ideal hydrodynamic ("IHD") principle: steady fields are extremal fields for the energy functional among the fields with *diffeomorphic vorticities*, i.e. among isovorticed fields. (The latter corresponds to the energy restriction to the *coadjoint orbits* of the same group.) In this sense these principles are dual to each other, but give the same sets of extremal fields.

3 Euler equations and integrable systems

3.1 Hamiltonian reformulation of the Euler equations

The differential-geometric description of the Euler equation as a geodesic flow on a Lie group has a Hamiltonian reformulation. Fix the notation $E(v) = \frac{1}{2}\langle v, Av \rangle$ for the energy quadratic form on $\mathfrak{g}$ which we used to define the Riemannian metric. Identify the Lie algebra and its dual with the help of this quadratic form. This identification $A : \mathfrak{g} \to \mathfrak{g}^*$ (called the *inertia operator*) allows one to rewrite the Euler equation on the dual space $\mathfrak{g}^*$.

It turns out that the Euler equation on $\mathfrak{g}^*$ is Hamiltonian with respect to the natural Lie–Poisson structure on the dual space [1]. Moreover, the corresponding Hamiltonian function is minus the energy quadratic form lifted from the Lie algebra to its dual space by the same identification: $-E(m) = -\frac{1}{2}\langle A^{-1}m, m \rangle$, where $m = Av$. Here we are going to take it as the *definition* of the Euler equation on the dual space $\mathfrak{g}^*$. (The minus is related to the consideration of a right-invariant metric on the group. It changes to plus for left-invariant metrics.)

Definition 3.1 (see, e.g., [4]) The Euler equation on $\mathfrak{g}^*$, corresponding to the right-invariant metric $E(m) = \frac{1}{2}\langle Av, v \rangle$ on the group, is given by the following explicit formula:

$$\frac{dm}{dt} = -\mathrm{ad}^*_{A^{-1}m}m, \tag{1}$$

as an evolution of a point $m \in \mathfrak{g}^*$. Here ad^* is the coadjoint operator, dual to the operator defining the structure of the Lie algebra $\mathfrak{g}$.

Below we explain the meaning of this operator in the case of the Virasoro algebra, "responsible" for several equations of mathematical physics.

3.2 The Virasoro algebra and the KdV equation

Definition 3.2. The Virasoro algebra $vir = \mathrm{Vect}(S^1) \oplus \mathbb{R}$ is the vector space of pairs which consist of a smooth vector field on the circle and a number. This space is equipped with the following commutation operation:

$$\left[\left(f(x)\frac{\partial}{\partial x}, a\right), \left(g(x)\frac{\partial}{\partial x}, b\right)\right] = \left((f'(x)g(x) - f(x)g'(x))\frac{\partial}{\partial x}, \int_{S^1} f'(x)g''(x)\,dx\right),$$

for any two elements $(f(x)\partial/\partial x, a)$ and $(g(x)\partial/\partial x, b)$ in vir.

The bilinear skew-symmetric expression $c(f,g) := \int_{S^1} f'(x)g''(x)dx$ is called the *Gelfand–Fuchs 2-cocycle*.

There exists a Virasoro group, an extension of the group of smooth diffeomorphisms of the circle, whose Lie algebra is the Virasoro algebra vir. Fix the L^2-energy quadratic form in the Virasoro Lie algebra:

$$E\left(f(x)\frac{\partial}{\partial x}, a\right) = \frac{1}{2}\left(\int_{S^1} f^2(x)dx + a^2\right).$$

Applying the construction of Section 1 to the Virasoro group, one can equip this group with a (right-invariant) Riemannian metric and consider the corresponding Euler equation, i.e., the equation of the geodesic flow generated by this metric on the Virasoro group.

Theorem 3.1 [9] *The Euler equation corresponding to the geodesic flow (for the above right-invariant metric) on the Virasoro group is a one-parameter family of the Korteweg–de Vries (KdV) equations:*

$$\partial_t u + u\partial_x u + c\partial_x^3 u = 0;\ \partial_t c = 0$$

on a time-dependent function u on S^1. Here c is a (constant) parameter, the "depth" of the fluid.

Proof. The space vir^* can be identified with the set of pairs

$$\left\{(u(x)(dx)^2, c)\,\big|\,u(x) \text{ is a smooth function on} S^1, c \in \mathbb{R}\right\}.$$

Indeed, it is natural to contract the quadratic differentials $u(x)(dx)^2$ with vector fields on the circle, while the constants are to be paired between themselves:

$$\left\langle\left(v(x)\frac{\partial}{\partial x}, a\right), (u(x)(dx)^2, c)\right\rangle = \int_{S^1} v(x) \cdot u(x)dx + a \cdot c.$$

The coadjoint action of a Lie algebra element $(f\partial/\partial x, a) \in vir$ on an element $(u(x)(dx)^2, c)$ of the dual space vir^* is

$$\mathrm{ad}^*_{(f\partial/\partial x, a)}(u(dx)^2, c) = (2(\partial_x f)u + f\partial_x u + c\partial_x^3 f, 0).$$

It is obtained from the identity

$$\langle[(f\frac{\partial}{\partial x},\boldsymbol{a}),(g\frac{\partial}{\partial x},\boldsymbol{b})],(u(dx)^2,c)\rangle = \langle(g\frac{\partial}{\partial x},\boldsymbol{b}),\mathrm{ad}^*_{(f\frac{\partial}{\partial x},\boldsymbol{a})}(u(dx)^2,c)\rangle,$$

which holds for every pair $(g\frac{\partial}{\partial x},\boldsymbol{b}) \in vir$.

The quadratic energy functional E on the Virasoro algebra vir determines the "tautological" inertia operator $A : vir \rightarrow vir^*$, which sends a pair $(u(x)\partial/\partial x, c) \in vir$ to $(u(x)(dx)^2, c) \in vir^*$.

In particular, it defines the quadratic Hamiltonian on the dual space vir^*,

$$E(u(dx)^2, c) = \frac{1}{2}(\int u^2 dx + c^2)$$

$$= \frac{1}{2}\langle(u\frac{\partial}{\partial x},c),(u(dx)^2,c)\rangle = \frac{1}{2}\langle(u\frac{\partial}{\partial x},c),\boldsymbol{A}(u\frac{\partial}{\partial x},c)\rangle.$$

The corresponding Euler equation for the right-invariant metric defined by E on the group (according to the general formula (1) above) is given by

$$\frac{\partial}{\partial t}(u(dx)^2,c) = -\mathrm{ad}^*_{A^{-1}(u(dx)^2,c)}(u(dx)^2,c).$$

Making use of the explicit formula for the Virasoro coadjoint action ad^* for

$$(f\partial/\partial x, a) = A^{-1}(u(dx)^2, c) = (u\partial/\partial x, c),$$

we obtain the required Euler equation:

$$\partial_t u = -2(\partial_x u)u - u\partial_x u - c\partial_x^3 u = -3u\partial_x u - c\partial_x^3 u, \partial_t c = 0.$$

The coefficient c is preserved in time, and the function u satisfies the KdV equation. QED.

3.3 Equations–relatives and conservation laws

For different metrics on the Virasoro group, other interesting equations can appear from the same scheme. The Euler equation on the Virasoro group with respect to the right-invariant H^1-metric gives the Camassa–Holm equation:

$$\partial_t u - \partial_{xxt} u = -3u\partial_x u + 2(\partial_x u)\partial_{xx} u + u\partial_{xxx} u + c\partial_x^3 u,$$

see [7]. Similarly, the homogeneous $\dot{H}^1$–metric gives the Hunter–Saxton equation (an equation in the Dym hierarchy):

$$\partial_{xxt} u = -2(\partial_x u)\partial_{xx} u - u\partial_{xxx} u,$$

see [6].

Remark 3.1. It turns out that all these three equations (KdV, CH, and HS) are bihamiltonian systems, and hence admit an infinite family of conservation laws. The corresponding Hamiltonian (or Poisson) structures are naturally related to the Virasoro algebra.

For instance, for the KdV equation these conserved quantities can be expressed in the following way. Consider the KdV equation on $(u(x)(dx)^2, c)$ as an evolution of Hill's operator $c\frac{d^2}{dx^2} + u(x)$. The monodromy $M(u)$ of this operator is a 2×2-matrix with the unit determinant. Look at the following function of the monodromy for a family of Hill's operators:

$$h_\lambda(u) := \log(\operatorname{trace} M(u - \lambda^2)),$$

where $M(u - \lambda^2)$ is the monodromy of the Hill operator $\frac{d^2}{dx^2} + u(x) - \lambda^2$.

Now, the expansion of the function h_λ in λ produces the first integrals of the KdV equation:

$$h_\lambda(u) \approx 2\pi\lambda - \sum_{n=1}^{\infty} c_n h_{2n-1}(u)\lambda^{1-2n},$$

where

$$h_1 = \int_{S^1} u(x)\,dx, \quad h_3 = \int_{S^1} u^2(x)\,dx, \quad h_5 = \int_{S^1} \left(u^3(x) - \frac{1}{2}(u_x(x))^2\right)\,dx, \quad \ldots$$

and $c_1 = 1/2, c_n = (2n-3)!!/(2^n n!)$ for $n > 1$. One can recognize here the familiar form of higher KdV integrals. Their appearance in this expansion is due to the fact that the trace of the monodromy $M(u)$ is a Casimir function for the Virasoro algebra, while the coefficients in a Casimir expansion provide a hierarchy of conserved charges for any bihamiltonian systems, see [6] for more details on the KdV and other related equations.

Acknowledgements This paper presents the lecture notes for author's minicourse on Mathematical Fluid Dynamics, given at the Fields Institute in September 2003. I am obliged to Walter Craig for the invitation to give this course, encouragement and careful notes. I am also grateful to the University of Geneva and Anton Alekseev for hospitality during the preparation of this survey. The present work was partially sponsored by an NSERC research grant.

References

1. Arnold, V.I. (1966) Sur la géométrie différentielle des groupes de Lie de dimension infinie et ses applications à l'hydrodynamique des fluides parfaits. *Ann. Inst. Fourier* **16**, 316–361.
2. Arnold, V.I. (1973) The asymptotic Hopf invariant and its applications. *Proc. Summer School in Diff. Equations at Dilizhan*, Erevan (in Russian); English transl.: *Sel. Math. Sov.* **5** (1986), 327–345.
3. Arnold, V.I. (1974) Mathematical methods in classical mechanics. English transl.: *Graduate Texts in Mathematics, v.60. Springer, New York, 1989*, 516 pp.

4. Arnold, V.I. & Khesin, B.A. (1998) Topological methods in hydrodynamics. *Applied Mathematical Sciences, vol. 125, Springer, New York,* pp. xv+374.
5. Ebin, D. & Marsden, J. (1970) Groups of diffeomorphisms and the notion of an incompressible fluid. *Ann. of Math.* **92(2)**, 102–163.
6. Khesin, B. & Misiołek, G. (2003) Euler equations on homogeneous spaces and Virasoro orbits. *Advances in Math.* **176**, 116–144.
7. Misiołek, G. (1998) A shallow water equation as a geodesic flow on the Bott-Virasoro group. *J. Geom. Phys.* **24(3)**, 203–208; Classical solutions of the periodic Camassa-Holm equation. *Geom. Funct. Anal.* **12(5)** (2002), 1080–1104.
8. Moffatt, H.K. & Tsinober, A. (1992) Helicity in laminar and turbulent flow. *Annual Review of Fluid Mechanics* **24**, 281–312.
9. Ovsienko, V.Yu. & Khesin, B.A. (1987) Korteweg-de Vries super-equation as an Euler equation. *Funct. Anal. Appl.* **21(4)**, 329–331.

Infinite dimensional dynamical systems and the Navier–Stokes equation

C. Eugene Wayne[1]

Abstract In this set of lectures I will describe how one can use ideas of dynamical systems theory to give a quite complete picture of the long time asymptotics of solutions of the two-dimensional Navier–Stokes equation. I will discuss the existence and properties of invariant manifolds for dynamical systems defined on Banach spaces and review the theory of Lyapunov functions, again concentrating on the aspects of the theory most relevant to infinite dimensional dynamics. I will then explain how one can apply both of these techniques to the two-dimensional Navier–Stokes equation to prove that any solution with integrable initial vorticity will will be asymptotic to a single, explicitly computable solution known as an Oseen vortex equations.

1 First lecture: infinite dimensional dynamical systems

In this first lecture I recall some common techniques used in finite dimensional dynamical systems and discuss their generalization to the infinite dimensional context needed for applications to partial differential equations. The two main tools we will use in these lectures will by invariant manifolds and Lyapunov functions. We will use the former to analyze the behavior of systems near stationary solutions and the latter to obtain more global information about solutions. Good general references for this material are [11] and [12].

We begin by recalling a very simple situation. Suppose that one has a system of n ordinary differential equations

$$\frac{d\mathbf{x}}{dt} = \tilde{f}(\mathbf{x}) , \quad \mathbf{x} \in \mathbb{R}^n .\tag{1}$$

Suppose further that the origin is a fixed point of this this system of equations. If we want to analyze the behavior of solutions near zero an obvious approach is to linearize the equation, i.e. we write

---

[1] Department of Mathematics and Statistics, Boston University

W. Craig (ed.), Hamiltonian Dynamical Systems and Applications, 103–141.
© 2008 *Springer Science + Business Media B.V.*

$$\tilde{f}(\mathbf{x}) = \tilde{f}(0) + \left(D_0\tilde{f}\right)\mathbf{x} + \mathscr{O}(|\mathbf{x}|^2) \equiv L\mathbf{x} + f(\mathbf{x}) \ . \tag{2}$$

In this last equality we have used the fact that $\tilde{f}(0) = 0$ (since the origin is a fixed point) and defined the $n \times n$ matrix $L = D_0\tilde{f}$, i.e. the Jacobian matrix of $\tilde{f}$ at the fixed point. The function $f(x)$ collects the nonlinear terms in the equation – in particular, $f(\mathbf{x}) = \mathscr{O}(|\mathbf{x}|^2)$ for $\mathbf{x}$ near zero. If $\mathbf{x}$ is very small then the terms $\mathscr{O}(|\mathbf{x}|^2)$ should be much smaller than the linear terms in $\mathbf{x}$ suggesting that a good approximation to the solutions of (1) should be given by

$$\frac{d\mathbf{x}}{dt} = L\mathbf{x} \ . \tag{3}$$

This equation is easily solved – if L has n linearly independent eigenvectors $\{\mathbf{v}_j\}_{j=1}^n$, with eigenvalues $\{\lambda_j\}_{j=1}^n$, then any solution of (3) can be written as

$$\mathbf{x}(t) = c_1 e^{\lambda_1 t}\mathbf{v}_1 + \cdots + c_n e^{\lambda_n t}\mathbf{v}_n \ , \tag{4}$$

for some choice of constants c_j.

Remark 1.1. The constants c_j are determined by the initial conditions. If $\{\mathbf{w}_j\}_{j=1}^n$ are the *adjoint*-eigenvectors of L, normalized so that $\langle \mathbf{w}_j, \mathbf{v}_k \rangle = \delta_{j,k}$ then we have

$$c_j = \langle \mathbf{w}_j, \mathbf{x}(0) \rangle \ .$$

(Here $\langle \cdot, \cdot \rangle$ is the inner product on $\mathbb{R}^n$.) Hence for later use we note that we will want to know not only eigenvectors for the linear part of equations we study but also adjoint eigenvectors.

From (4) we see that we can split $\mathbb{R}^n$ into a direct sum of three subspaces – the stable subspace, $\mathbb{E}^s$, the center subspace $\mathbb{E}^c$ and the unstable subspace, $\mathbb{E}^u$, which are respectively the spectral subspaces associated with the eigenvalues whose real parts have negative, zero, or positive real parts. Note that any solution with initial condition in $\mathbb{E}^s$ approaches the origin as $t \to \infty$ while any solution with initial condition in $\mathbb{E}^u$ approaches the origin as $t \to -\infty$.

An obvious question is to what extent this structure survives when we include the nonlinear terms that were omitted in (3). We certainly don't have explicit solutions like those in (4) any longer but geometrical structures analogous to the stable, center and unstable subspaces do persist, at least in a neighborhood of the fixed point – this is the content of the invariant manifold theorems. We state these informally for the moment, reserving a more formal treatment until we discuss the corresponding results for infinite dimensional systems below. Suppose $\tilde{f} \in C^1(\mathbb{R}^n)$. Then there exists a neighborhood of the origin $B_r \subset \mathbb{R}^n$ and functions h^s defined on B_r such that

$$h^s : B_r \cap \mathbb{E}^s \to \mathbb{E}^c \oplus \mathbb{E}^u$$

The function h^s is C^1, and its graph, known as the local stable manifold $\mathscr{W}_{loc}^s$ is locally invariant (i.e. for any initial condition in $\mathscr{W}_{loc}^s$ the corresponding solution of (1) remains in $\mathscr{W}_{loc}^s$ for as long as it remains in the domain of definition of h^s.).

Furthermore, any solution which remains in $\mathcal{W}^s$ for all $t \geqslant 0$ approaches the origin as $t \to \infty$. In addition the local stable manifold is locally unique – no other manifold in a neighborhood of the origin shares all these properties.

Analogous results hold for the local unstable manifold. However, things are slightly more delicate for the center-manifold. In particular, one no longer has uniqueness. There are simple examples of systems of ordinary differential equations with infinitely many local center manifolds.

One property that makes center manifolds particularly important and interesting is that one can show that there exists a neighborhood of the origin (which we can assume to be B_r, without loss of generality) such that any solution which remains in this neighborhood for all $t \in \mathbb{R}$ must lie in the local center manifold. This implies that any periodic orbits or additional fixed points in a neighborhood of the origin must lie in the center-manifold. If one is looking specifically for periodic orbits, say, this can lead to a very big simplification since it permits one to reduce the search from the original system of n equations to a system whose dimension equals that of the center manifold which is often much less than n. Such a reduction is even more important in the context of partial differential equations where it frequently results in reduction from an infinite dimensional set of equations to one whose dimension is small and finite.

We next turn to a discussion of the appropriate generalization of these invariant manifold theorems to partial differential equations. Suppose that we consider a (system) of partial differential equations

$$\frac{\partial u}{\partial t} = \mathcal{L}u + f(u, \nabla u) \,, \tag{5}$$

where $u = u(x,t) \in \mathbb{R}^n$, $x \in \mathcal{D} \subset \mathbb{R}^d$ and $t \geqslant 0$, $\mathcal{L}$ is a linear, differential operator and f is a nonlinear term depending on u and its (first order) partial derivatives. One could also consider quasilinear partial differential equations but in these lectures we restrict attention to this semilinear case.

Following the intuition gained from the finite dimensional case above we would like to compare solutions of this equation to those of the linear equation

$$\frac{\partial u}{\partial t} = \mathcal{L}u \tag{6}$$

There are many additional difficulties that are encountered in treating this infinite dimensional case in comparison with the finite dimensional case discussed above. Some of these difficulties are only technical and reflect the more complicated analysis necessary in an infinite dimensional setting. However, other problems represent qualitative differences in the behavior of the partial differential equations *vis-a-vis* ordinary differential equations. Among the problems that must be overcome are:

1. The spectrum of $\mathcal{L}$ may no longer consist only of eigenvalues as in (3) but may now contain continuous spectrum.
2. Since the operator $\mathcal{L}$ will in general be unbounded it may not be possible to define solutions for $t < 0$ for general initial conditions – in this case discussing the behavior of solutions as $t \to -\infty$ is clearly problematic!

3. If the continuous spectrum approaches the imaginary axis there may be not clear splitting between the center subspace and the stable and unstable subspaces. This problem, often called the lack of a spectral gap, is particularly common when studying problems defined on unbounded spatial domains.

4. One cannot in general hope that the nonlinearity f in (5) will be C^1 - indeed due to the presence of derivatives of u in the nonlinear term it often even fails to map the Banach space in which solutions lie back into itself. This will be the case, for example in the Navier–Stokes equations which are the subject of the third and fourth lectures in this series.

Many authors have addressed the question of the existence and properties of invariant manifolds for partial differential equations. In contrast to the case of ordinary differential equations where it is more or less clear what the "right" assumptions on the vectorfield are and what the "correct" conclusion ought to be, this is by no means so clear in the case of partial differential equations. In particular, depending on the context one may wish to make either stronger or weaker assumptions about the linear part of the equation (which affect, for instance, the smoothing properties of the semi-group associated with (6), or even whether the linear part defines a semi-group). These assumptions then entail making either different assumptions on the nonlinear term, or changing (typically, weakening) the results one hopes to obtain. For examples of typical results in this context see [1], [13] or [16]. One general principle which emerges from this collection of results is that if (5) and (6) define semi-flows then it is often easier to work with the semi-flow than with the differential equation itself. This is because the semi-flow already incorporates any smoothing properties that the equation may possess. By working with the semi-flow, Chen, Hale and Tan (CHT) [4] have given a very general form of the invariant manifold theorem, applicable to many partial differential equations. It is their result that I will use in subsequent lectures and which I now state.

From now on, we assume that the partial differential equation (5) defines a semi-flow Φ^t on some Banach space X. Then (CHT) make the following assumptions:

(H.1) $\Phi^t(u)$ is continuous for (t,u) in $\mathbb{R}^+ \times X$ and there exist positive constants q and D such that

$$\sup_{0 \leqslant t \leqslant q} \mathrm{Lip}(\Phi^t) = D < \infty$$

where

$$\mathrm{Lip}(\Phi^t) \equiv \sup_{u,v \in X} \frac{\|\Phi^t(u) - \Phi^t(v)\|}{\|u - v\|}.$$

(H.2) For some $\tau \in (0,q]$, one can decompose Φ^τ as

$$\Phi^\tau = S + R$$

where S is a bounded linear operator from X to itself and R is globally Lipshitz.

(H.3) There exist subspaces X_1 and X_2 such that $X = X_1 \oplus X_2$, and continuous projections $P_i : X_i \to X_i$, $i = 1, 2$ which are invariant with respect to S. Also S commutes with P_i. If $S_i = S|_{X_i}$, then S_i has bounded inverse and there exist constants

C_i and α_i such that $\alpha_1 > \alpha_2 \geqslant 0$ and

$$\|S_1^{-k}P_1\| \leqslant C_1\alpha_1^{-k}$$
$$\|S_2^{k}P_2\| \leqslant C_2\alpha^{k}$$

(H.4) The constant C_i and α_i from (H.3) are related to the nonlinearity in such a way that

$$\left(\frac{(\sqrt{C_1}+\sqrt{C_2})^2}{\alpha_1 - \alpha_2}\right)\mathrm{Lip}(R) < 1 \ .$$

Remark 1.2. For later use we fix two additional constants γ_1 and γ_2 with $\alpha_2 < \gamma_2 < \gamma_1 < \alpha_1$ such that

$$\mathrm{Lip}(R)\left(\frac{C_1}{\alpha_1 - \gamma_1} + \frac{C_2}{\gamma_2 - \alpha_2}\right) = 1 \ .$$

Note that by making $\mathrm{Lip}(R)$ sufficiently small we can make γ_1 arbitrarily close to α_1 and γ_2 arbitrarily close to α_2.

Remark 1.3. Before stating the conclusions of the (CHT) theorem we comment briefly on the meaning of these hypotheses. The hypothesis (H.1) implies that (5) defines a well-behaved semi-flow. This hypothesis typically rules out applying these results to elliptic equations, for example. Hypothesis (H.2) is just an assumption that the semi-group splits nicely into its linear and nonlinear parts. Hypothesis (H.3) implies a "spectral gap" for the linear part of the semigroup. The spectrum of S_2 must lie inside a circle of radius α_2 and the spectrum of S_1 must lie outside a circle of radius α_2. Note however, that there is no assumption that S_2 is invertible – we do not assume that the original partial differential equation is solvable "backwards" in time for general initial data. Finally, hypothesis (H.4) requires that the nonlinear term must be small, in the appropriate sense, with respect to the spectral gap.

We now state the theorem of (CHT) which we will use later:

Theorem 1.1. *Suppose that (H.1)-(H.4) hold. Then there exists a globally Lipshitz map $g : X_1 \rightarrow X_2$ such that the graph of g*

$$G = \{u_1 + g(u_1)\,|\,\boldsymbol{u}_1 \in X_1\}$$

satisfies:

(i) (Invariant Manifold) The restriction of Φ^t to G can be extended to a Lipshitz flow on G.

(ii) (Lyapunov exponents) Any negative semi-orbit $\|u(t)\|_{t\leqslant 0} \subset X$ that satisfies

$$\lim_{t\to\infty}\frac{1}{|t|}\log\|u(t)\| < -\frac{1}{\tau}\log\gamma_2 \tag{7}$$

must be contained in G. In particular, if $\gamma_2 < 1$, any fixed point of Φ^t must lie in G.

(iii) (Invariant Foliation) There exists a continuous map $h : X \times X_2 \to X_1$ such that if $v \in G$, then $h(v, P_2 v) = P_1 v$ and the set

$$M_v = \{h(v, w) + w \mid \mathbf{w} \in X_2\}$$

passing through v satisfies $\Phi^t(M_v) \subset M_{\Phi^t(v)}$ and

$$M_v = \left\{ w \in X \mid \limsup_{t \to \infty} \frac{1}{t} \log |\Phi^t(w) - \Phi^t(v)| \leqslant \frac{1}{\tau} \log \gamma_2 \right\} .$$

Remark 1.4. We did not discuss the finite dimensional analogue of point (iii) but roughly speaking the fibers M_v of this foliation contain all points whose asymptotic behavior it the same as that v – i.e. we can characterize the asymptotics of all points (sufficiently close to the fixed point) by those of points on the invariant manifold. Note that the estimate on the rate of convergence of points in the fiber toward the invariant manifold also implies as a corollary that all solutions near the invariant manifold approach (assuming that $\gamma_2 < 1$.)

Remark 1.5. Note that in Hypothesis (H.1) we assume that there is a global bound on the Lipshitz constant of the semi-flow. (Here, I mean global in X, not in time.) This is rarely true in practice but this hypothesis is why the manifold constructed here is not constrained to a neighborhood of the fixed point but rather is defined for all $u_1 \in X_1$. In practice we "cut off" the nonlinear terms in the equation outside a small neighborhood of the fixed point in order to allow this hypothesis to be verified and this will make the applications of this theorem "local" in character.

Remark 1.6. If the term R in the decomposition of the semiflow is nonlinear in the sense that $R(0) = 0$ and $DR(0) = 0$ then the function g whose graph defines the invariant manifold has the same property – namely $g(0) = 0$ and $Dg(0) = 0$.

This invariant manifold theorem will be our main tool to investigate the local behavior of solutions of partial differential equations in the later lectures. However we will also want to consider more global questions. For those, we will make use of *Lyapunov functions*. Here, the transition from the finite dimensional to infinite dimensional setting involves fewer changes than in the case of the invariant manifold theorems so we work directly with the infinite dimensional case without first reviewing the finite dimensional results. The presentation here largely follows that of D. Henry in [11] – see that work, or [12] for more details.

Let Φ^t be a semi-flow on a Banach space X. We want to characterize the long-time behavior of solutions of the differential equation defining Φ^t and with that in mind make the following two definitions:

Definition 1.1. Given $\mathbf{u}_0 \in X$, we define the *forward orbit* of $\mathbf{u}_0$ as:

$$\mathcal{O}^+(\mathbf{u}_0) = \{\Phi^t(\mathbf{u}_0) \mid t \geqslant 0\} .$$

Definition 1.2. The *omega limit set* of a point $\mathbf{u}_0$ is the set of all points which the forward orbit of $\mathbf{u}_0$ approaches arbitrarily closely as t tends to infinity. More precisely,

$$\omega(\mathbf{u}_0) = \{\mathbf{u} \in X \,|\, \text{there exists } \{t_n\} \subset \mathbb{R} \text{ such that } \lim_{n \to \infty} t_n = \infty$$

$$\text{and } \lim_{n \to \infty} \|\Phi^{t_n}(\mathbf{u}_0) - \mathbf{u}\| = 0\} \ .$$

Exercise 1.1. Suppose that $\tilde{\mathbf{u}} \in \mathcal{O}^+(\mathbf{u}_0)$. Show that $\omega(\mathbf{u}_0) = \omega(\tilde{\mathbf{u}})$. Thus we can refer without ambiguity not just to $\omega(\mathbf{u}_0)$ but also $\omega(\mathcal{O}^+(\mathbf{u}_0))$.

Exercise 1.2. Show that if $\mathbf{u}^*$ is an element of the ω-limit set of $\mathbf{u}_0$ and if $\mathcal{O}^+(\mathbf{u}_0) \subset K$, a compact subset of X, then the orbit of $\mathbf{u}^*$ is defined for all $t \in \mathbb{R}$ and the entire orbit of $\mathbf{u}^*$ is contained in $\omega(\mathbf{u}_0)$.

One might worry that the omega-limit set was empty, but this turns out not to be the case, at least not if the forward orbit remains in a compact set:

Proposition 1.1. *If $\mathcal{O}^+(\mathbf{u}_0) \subset K$, a compact subset of X. then $\omega(\mathbf{u}_0)$ is non-empty and invariant* (i.e. *if $\mathbf{u}^* \in \omega(\mathbf{u}_0)$, then $\Phi^t(\mathbf{u}^*) \in \omega(\mathbf{u}_0)$ for all $t \in \mathbb{R}$.*)

The proof of this proposition is not difficult – see [11] for details. The only slightly surprising point is that the omega-limit set is invariant in both forward and backward time, even though we do not know (or expect) that the semi-group itself is defined for $t \leqslant 0$ for general initial conditions.

A key tool for investigating omega-limit sets are Lyapunov functions.

Definition 1.3. If X is a Banach space, a Lyapunov function for the semi-flow Φ^t is a continuous, real-valued function Ψ such that

$$\limsup_{t \to 0^+} \frac{\Psi(\Psi^t(\mathbf{u})) - \Psi(\mathbf{u})}{t} < 0 \text{ for all } \mathbf{u} \in X \ .$$

This means that Ψ is non-increasing along orbits of Φ^t.

Remark 1.7. Note that if the limit in Definition 1.3 exists it is just the derivative of Ψ along the trajectory with initial condition $\mathbf{u}$ so a common way of verifying that a given function is a Lyapunov function is to show that its derivative is non-positive along solutions.

A key tool we will use in Lecture 4 is the LaSalle Invariance Principle:

Proposition 1.2. *Let Ψ be a Lyapunov function for the semi-flow Φ^t. Define* $\mathbb{E} = \{\mathbf{u} \in X \,|\, \frac{d\Psi}{dt} \circ \Phi^t(\mathbf{u})|_{t=0} = 0\}$ *If $\mathcal{O}^+(\mathbf{u}_0)$ is contained in a compact subset of X then $\omega(\mathbf{u}_0) \subset \mathbb{E}$.*

Because of the importance of this result for our applications we sketch its proof:

Proof. By the compactness of the forward orbit and continuity of Ψ we know that there exists some finite M such that

$$V(\Phi^t(\mathbf{u}_0)) \geqslant M$$

for all $t \geqslant 0$. Since Ψ is monotonic along the orbit of $\mathbf{u}_0$ we therefore conclude that

$$\lim_{l\to\infty}\Psi(\Phi^l(\mathbf{u}_0)) = \Psi^\infty$$

for some Ψ^∞. If $\mathbf{w}\in\omega(\mathbf{u}_0)$ the definition of the omega-limit set, plus the continuity of Ψ imply that there exists a sequence of times $\{t_n\}$ approaching infinity such that $\lim_{n\to\infty}\Psi(\Phi^{t_n}(\mathbf{u}_0)) = \Phi(\mathbf{w})$, from which we conclude that $\Phi(\mathbf{w}) = \Psi^\infty$. But then, by the invariance of the omega-limit set we have

$$\Psi(\Phi^l(\mathbf{w})) = \Psi^\infty \text{ for all } t\in\mathbb{R},$$

which implies that $\mathbf{w}\in\mathbb{E}$. $\quad\square$

Example 1.1. We finish this lecture with an example of a somewhat unusual Lyapunov function which will play a role later in this series of talks. Consider the dynamical system defined by the partial differential equation

$$\frac{\partial w}{\partial\tau} = \mathscr{L}w\,,\ w = w(\xi,\tau)\,;\ \xi\in\mathbb{R}^d,\ \tau\geqslant 0 \tag{8}$$
$$w(\xi,0) = w_0(\xi)$$

where $\mathscr{L}w = \Delta_\xi w + \frac{1}{2}\nabla\cdot(\xi w)$. The reason for considering this unusual equation will be explained in Lecture 2 but for the moment assume two facts about the evolution:

1. The solutions of (8) obey the maximum principle. In particular, if $w_0(\xi)\geqslant 0$ then $w(\xi,\tau) > 0$ for all ξ for any $\tau > 0$.
2. If $w(\xi,0)\in L^1(\mathbb{R}^d)$ the ω-limit set of the corresponding trajectory exists.

The reason that equation (8) obeys the maximum principle will be explained in Lecture 2.

We next show that the L^1 norm is a Lyapunov function in this case.

Lemma 1.1. *Let $w_0\in L^1(\mathbb{R}^d)$ and let $w(\xi,\tau)$ be the solution of* (8) *with this initial condition. Then the function*

$$\Phi(w)(\tau) = \int_{\mathbb{R}^d}|w(\xi,\tau)|d\xi$$

is non-increasing along trajectories. More precisely, $\Phi(w)(\tau)\leqslant\Phi(w_0)$ for every $\tau > 0$ and equality holds if and only if w_0 does not change sign.

Proof. Define $w_0^+(\xi) = \max(w_0(\xi),0)$, $w_0^-(\xi) = -\min(w_0(\xi),0)$ Let $w^\pm(\xi,\tau)$ be the solutions of (8) with initial conditions $w_0^\pm$ respectively. Note that from the form of the equation we see immediately that the equation conserves the integral of the solution. Thus

$$\int_{\mathbb{R}^d}w(\xi,\tau)d\xi = \int_{\mathbb{R}^d}w_0(\xi)d\xi$$

and

$$\int_{\mathbb{R}^d}w^\pm(\xi,\tau)d\xi = \int_{\mathbb{R}^d}w_0^\pm(\xi)d\xi\,.$$

Now note that if w_0 does change sign, $w_0^{\pm}$ are both non-trivial. Furthermore they have disjoint support. However, by the maximum principle, $w^{\pm}(\xi,\tau)$ will both be positive for all ξ whenever $\tau > 0$. Thus,

$$\int_{\mathbb{R}^d} |w(\xi,\tau)|\,d\xi = \int_{\mathbb{R}^d} |w^+(\xi,\tau) - w^-(\xi,\tau)|\,d\xi$$

$$< \int_{\mathbb{R}^d} \left(w^+(\xi,\tau) + w^-(\xi,\tau)\right)d\xi = \int_{\mathbb{R}^d} \left(w_0^+(\xi) + w_0^-(\xi)\right)d\xi = \int_{\mathbb{R}^d} |w_0(\xi)|\,d\xi,$$

which shows that Φ decreases along orbits if w_0 changes sign. The fact that Φ is constant when w_0 is everywhere non-negative or non-positive is easier and left as an exercise.

Note that if we combine this Lemma with the LaSalle Invariance Principle we immediately have

Corollary 1.1. *Any point in the ω-limit set of a solution of (8) must be either everywhere positive, everywhere negative, or identically zero.*

This Corollary may not seem very strong at first glance since one might think that all solutions just tend toward zero. However, this can be ruled out by the fact that solutions conserve the integral of the initial condition and conditions on the decay of solutions at infinity – thus, if the integral of the initial data is non-zero, we can conclude that the ω-limit set is either everywhere positive or everywhere negative, a fact which will be important in the last lecture in this series.

2 Second lecture: invariant manifolds for partial differential equations on unbounded domains

In this lecture we examine the application of invariant manifold theorems to some partial differential equations on unbounded spatial domains. For concreteness we focus primarily on the family of semi-linear heat equations:

$$\frac{\partial u}{\partial t} = \Delta u - u|u|^{p-1}\,, \quad p > 1 \tag{1}$$

$$u = u(x,t),\ t \geqslant 0,\ x \in \mathbb{R}^d\,.$$

The long-time behavior of solutions of this equation have been intensively studied and not surprisingly the value of the exponent p in the nonlinear term plays an essential role in this behavior. The dynamical systems approach described below gives a very simple explanation of this p-dependence.

Remark 2.1. There are a host of other applications of invariant manifold theorems to partial differential equations – see the references [1], [13] or [16] for a small sampling. We focus on this particular family of equations both because it will serve as a good "warm up" for treating the Navier–Stokes equations later and also because it illustrates one way of dealing with lack of a spectral gap which often arises in treating problems on unbounded spatial domains.

If one uses the Duhamel formula to convert (1) to an integral equation it is not difficult to show that this equation defines a smooth semigroup, at least for small initial data. However, if one tries to apply the invariant manifold theorem of Chen, Hale and Tan (CHT) described in the previous lecture one immediately runs into the problem that one cannot split the Banach space in the way described in hypotheses (H.3) and (H.4). The reason for this is a lack of a spectral gap and the origin of this problem is seen immediately even for the case of the linear heat equation

$$\frac{\partial u}{\partial t} = \Delta u , \quad u = u(x,t),\ t \geqslant 0,\ x \in \mathbb{R}^d . \tag{2}$$

For this equation we can immediately write down a representation of the semi-group. It is particularly easy to analyze in terms of the Fourier transform

$$\hat{u}(k,t) = \frac{1}{(2\pi)^{d/2}} \int_{\mathbb{R}^d} u(x,t) e^{-ix \cdot k} dx \tag{3}$$

If we are given initial conditions $u(x,0) = u_0 \in L^2(\mathbb{R}^d)$ then the solution of (2) can be written as:

$$\hat{u}(k,t) = e^{-|k|^2 t} \hat{u}_0(k) \tag{4}$$

Since the semigroup in this case is just a multiplication operator we see that its spectrum equals the closed interval $[0,1]$. Since there is no gap in the spectrum there is no way to split the space of initial conditions in the way required by the (CHT) theorem. Thus, there is no easy way to identify subspaces of our Banach space which correspond to solutions with particular decay properties. A way to circumvent this problem emerges if one recalls the form of the fundamental solution of the heat equations:

$$G(x,t) = \frac{1}{(4\pi t)^{d/2}} e^{-|x|^2/(4t)} . \tag{5}$$

Examining this solution we see that x appears in a special way – namely as the combination $x/\sqrt{t}$ and this suggests that it might be more natural to study (2) not in terms of the independent variables (x,t) but rather in terms of the new variable $\xi = x/\sqrt{t}$. With this in mind we introduce new dependent and independent variables through the definition:

$$u(x,t) = \frac{1}{(1+t)^{\alpha/2}} w(\frac{x}{\sqrt{1+t}}, \log(1+t)) \tag{6}$$

$$\xi = \frac{x}{\sqrt{1+t}} , \quad \tau = \log(1+t)$$

Note that in defining the new variables (often called "scaling" or "similarity" variables) we have defined $\xi = x/\sqrt{1+t}$ rather than $x/\sqrt{t}$ simply to avoid the singularity at $t = 0$. This can be thought of as simply changing the origin of the time axis and since our equation is autonomous it has no effect on the problem. Also, the exponent α which occurs in the definition of w will be chosen in a way convenient to each of the problems considered. For the moment, in our discussion of the linear heat equation we will take $\alpha = d$.

If we rewrite (2) in terms of these new variables we find that

$$\frac{\partial w}{\partial \tau} = \mathscr{L}w, \quad w = w(\xi, \tau), \quad \xi \in \mathbb{R}^d \tag{7}$$

$$\mathscr{L}w = \Delta_\xi w + \frac{1}{2}\nabla \cdot (\xi w) \,.$$

At first sight, this may not seem like an improvement as we have traded the heat equation for an apparently more complicated equation. However, as we will see this form of the equation has the advantage that a gap in the spectrum appears which separates the slowly decaying modes from the more rapidly decaying ones and allows us to apply the invariant manifold theorem of the preceeding lecture.

Remark 2.2. Note that (7) is precisely the equation considered in Example 1.1 at the end of the previous lecture. Since this equation is just the heat equation rewritten in new variables it is clear that solutions of this equation will inherit a maximum principle from the maximum principle satisfied by the heat equation.

To see why and how this spectral gap forms, consider the eigenvalue problem for $\mathscr{L}$ – for simplicity, we consider the case of $d = 1$ though the following results are true in any dimension:

$$\mathscr{L}\phi = \lambda \phi \tag{8}$$

If we take the Fourier transform of this equation we find

$$-|k|^2 \hat{\phi}(k) - \frac{1}{2}k\frac{d\hat{\phi}}{dk}(k) = \lambda \phi(k) \tag{9}$$

This first order equation can be solved with the aid of integrating factors and we find that for any λ one has a solution

$$\hat{\phi}^\lambda(k) = A^+|k|^{-2\lambda}e^{-|k|^2}\Theta(k) + A^-|k|^{-2\lambda}e^{-|k|^2}\Theta(-k) \,, \tag{10}$$

where $\Theta(k)$ is the Heaviside function. (Note that the singularity at the origin means we can have different constants A^+ and A^- depending on whether k is positive or negative.) Thus, we have a solution of the eigenvalue equation for any value of λ so one might at first think that the spectrum of $\mathscr{L}$ is the whole complex plane. However, note that if λ is real and positive, $\hat{\phi}^\lambda$ is singular at the origin and thus whether or not $\hat{\phi}^\lambda$ is an eigenfunction depends on what function space we are working on. This observation reminds us that in general the spectrum of an operator depends on its domain of definition and as we will see that is very true of the operator $\mathscr{L}$.

It has long been known that the time decay properties of parabolic equations are often connected with the spatial decay properties of their solutions. With this in mind we define the family of weighted Sobolev spaces:

$$L^2(m) = \{f \in L^2(\mathbb{R}^d) \,|\, \|f\|_m < \infty\} \tag{11}$$

$$\|f\|_m = \left(\int_{\mathbb{R}^d} (1 + |\xi|^2)^m |f(\xi)|^2 d\xi \right)^{1/2} \tag{12}$$

$$H^s(m) = \{\partial^\alpha f \in L^2(m) \,|\, \text{for all } \alpha = (\alpha_1, \ldots, \alpha_d) \text{ with } |\alpha| \leqslant s\} \tag{13}$$

One standard property of these spaces which is very convenient for our subsequent use is that Fourier transformation is an isomorphism from $H^s(m)$ to $H^m(s)$. Thus, if we consider the spectrum of the operator $\mathscr{L}$ on the space $L^2(m)$, the point λ will be in the spectrum if the function $\hat{\phi}^\lambda \in H^m(0)$ – i.e. if the function $\hat{\phi}^\lambda$ is in the "ordinary" Sobolev space H^m. Clearly $\hat{\phi}^\lambda$ is sufficiently smooth and rapidly decaying to be in H^m for any m, provided we stay away from the origin. Thus, ϕ^λ will be in H^m provided it, and all of its derivatives of order m or less are square integrable in some small neighborhood of the origin.

From the form of $\hat{\phi}^\lambda$ we see that the cases with $\lambda = -n/2$ a non-positive half integer are "special". In this case, if we choose A^+ and A^- appropriately we find that $\phi^\lambda(k) = k^n e^{-|k|^2}$ is a solution of the eigenvalue equation. Since this function is entire and rapidly decaying $\hat{\phi}^\lambda$ is any Sobolev space H^m and thus the points $\{-\frac{n}{2} | n = 0, 1, 2 \ldots\}$ are in the spectrum of $\mathscr{L}$ when considered on any of the spaces $L^2(m)$. Furthermore, the corresponding eigenfunctions are given by the inverse Fourier transform of $k^n e^{-|k|^2}$ which implies $\phi^{n/2}(\xi) = C_n \frac{d}{d\xi^n} e^{-\xi^2/4}$. Of particular importance in our subsequent discussions will be the Gaussian eigenfunction of $\lambda = 0$, $\phi^0(\xi) = \frac{1}{\sqrt{4\pi}} e^{-\xi^2/4}$, with the prefactor chosen so that ϕ^0 has integral one.

For other values of λ, the most singular behavior of $\hat{\phi}^\lambda$ and its derivatives will occur for the derivative of highest order and we see that near $k = 0$ one has

$$\frac{d^m \hat{\phi}^\lambda}{dk^m}(k) \sim |k|^{-2\lambda - m} . \tag{14}$$

This expression will be square integrable provided $2(2\mathrm{Re}(\lambda) + m) < 1$, i.e. if

$$\mathrm{Re}(\lambda) < \frac{1}{4} - \frac{m}{2} \tag{15}$$

Thus we have shown

Proposition 2.1. *Fix $m > 1$ and $d = 1$ and let $\mathscr{L}$ be the operator in (7) acting on its maximal domain in $L^2(m)$. Then*

$$\sigma(\mathscr{L}) \supset \left\{ \lambda \in \mathbb{C} \,|\, \mathrm{Re}(\lambda) \leqslant \frac{1}{4} - \frac{m}{2} \right\} \cup \left\{ -\frac{n}{2} \,\Big|\, n = 0, 1, 2, \ldots \right\} .$$

In fact, as mentioned above, this result also holds for dimensions greater than 1. Furthermore, in addition to the eigenvalues computed above one might have additional parts to the spectrum but it turns out that this is all of the spectrum in this case and one can prove that in dimension d one has:

Theorem 2.1. *([7], Theorem A.1) Fix $m > 1$ and let $\mathscr{L}$ be the operator in (7) acting on its maximal domain in $L^2(m)$. Then*

$$\sigma(\mathscr{L}) = \left\{\lambda \in \mathbb{C} \,|\, \mathrm{Re}(\lambda) < \frac{d}{4} - \frac{m}{2}\right\} \cup \left\{-\frac{n}{2} \,|\, n = 0, 1, 2, \dots\right\}.$$

Although the spectral picture above gives valuable intuition about the behavior of the semigroup $e^{\tau\mathscr{L}}$, for later applications we will need more precise estimates on its properties. In particular, recall that the heat equation has strong smoothing properties (i.e. solutions of the heat equation with "rough" initial data are infinitely differentiable for all $t > 0$) and we will need to know to what extent these smoothing properties survive when we introduce scaling variables.

Remark 2.3. Note that it is not automatic that the semigroup $e^{\tau\mathscr{L}}$ will be smoothing. From the spectral picture in Theorem 2.1 we see that the operator $\mathscr{L}$ is not sectorial in any of the $L^2(m)$ spaces. Thus in contrast to the heat equation semigroup, $e^{\tau\mathscr{L}}$ is *not* an analytic semigroup.

In addition to the smoothing properties of the semigroup for our later applications we will need to know what the spectral projection operators onto the various spectral subspaces of $\mathscr{L}$ are. From the discussion in Lecture 1 we expect these to be given by eigenfuctions of the adjoint operator $\mathscr{L}^\dagger$. Formally the adjoint operator has the form

$$\mathscr{L}^\dagger \psi = \Delta_\xi \psi - \frac{1}{2}\xi \cdot \nabla_\xi \psi \tag{16}$$

If we specialize to one dimension again for simplicity the eigenvalue equation for $\mathscr{L}^\dagger$ is

$$\mathscr{L}^\dagger \psi = \psi'' - \frac{1}{2}\xi \psi' = \lambda \psi. \tag{17}$$

This is Hermite's equation and thus, we find that the spectral projections are defined in terms of the Hermite polynomials. If $\alpha = (\alpha_1, \dots, \alpha_d) \in \mathbb{N}^d$ we define

$$H^\alpha(\xi) = \frac{2^{|\alpha|}}{\alpha!} e^{|\xi|^2/4} \partial_\xi^\alpha \left(e^{-|\xi|^2/4}\right) \tag{18}$$

and then the projection P_n onto the eigenspace corresponding to the eigenvalues $\lambda_k = -\frac{k}{2}$, $k = 0, 1, \dots, n$ is defined by

$$(P_n f)(\xi) = \sum_{|\alpha| \leqslant n} \left(\int_{\mathbb{R}^d} H^\alpha(\xi') f(\xi') d\xi'\right)^{1/2} \phi_\alpha(\xi) \tag{19}$$

$$(Q_n f)(\xi) = ((1 - P_n)f)(\xi) \tag{20}$$

We make two remarks about these projection operators that will be useful later.

Remark 2.4. The projection P_0 onto the zero eigenspace is simply

$$(P_0 f)(\xi) = \left(\int_{\mathbb{R}^d} f(z) dz\right) \phi^0(\xi), \tag{21}$$

i.e. the projection of a function f onto the zero eigenspace is just given by the product of the Gaussian, ϕ^0, with the integral of f. In particular, any function of mean zero lies in the complementary subspace.

Remark 2.5. Following up on the preceding remark we see that a function f lies in the range of Q_n if and only if

$$\int_{\mathbb{R}^d} \xi^\alpha f(\xi)d\xi = 0 , \tag{22}$$

for all $\alpha = (\alpha_1,\ldots,\alpha_d) \in \mathbb{N}^d$ with $|\alpha| \leqslant n$.

We now state our main technical estimate on the semigroup $e^{\tau\mathcal{L}}$.

Proposition 2.2. *Fix* $n \in \mathbb{N} \cup \{-1\}$ *and fix* $m > n+1+\frac{d}{2}$ *For all* $\alpha \in \mathbb{N}^d$, *there exists* $C > 0$ *such that*

$$\|\partial^\alpha(e^{\tau\mathcal{L}}Q_n f)\|_m \leqslant \frac{C}{(1-e^{-\tau})^{|\alpha|/2}}e^{-(\frac{n+1}{2})\tau}\|f\|_m , \tag{23}$$

for all $f \in L^2(m)$ *and all* $\tau > 0$.

Proof. For the details of the proof we refer to [7], Appendix A. However, we note that the decay rate is exactly what we expect from the spectral picture in Theorem 2.1. The more delicate smoothing properties (quantified by the estimates of the derivatives of the semigroup) are obtained from the explicit integral representation of the semigroup which we easily obtain by noting that $e^{\tau\mathcal{L}}w_0$ is the solution of (7) with the initial condition w_0 which when combined with (6) gives

$$(e^{\tau\mathcal{L}}w_0)(\xi) = w(\xi,\tau) = e^{\frac{d}{2}\tau}u(\xi e^{\tau/2},e^\tau - 1) \tag{24}$$

and we then use the integral representation of u in terms of w_0 which follows from the fact that u solves the heat equation with initial condition w_0. $\square$

We now consider the implications of this result for the invariant manifold theorem. Recall that the problem with applying the invariant manifold theorem directly to the heat equation was that the semi-group had no spectral gap. If we now consider the semigroup defined by (7) then we see that the modes corresponding to the eigenvalues $\lambda = -\frac{n}{2}$ will decay like $e^{-\frac{n}{2}\tau}$ while modes lying in the half plane of essential spectra will all decay at least with a rate $e^{(\frac{d}{4}-\frac{m}{2})\tau}$ and by choosing m appropriately we can separate the decay rate of these modes from the most slowly decaying ones. In particular, if we choose $m > d/2$ we expect that as t tends toward infinity solutions of (7) will approach a point on the eigenspace corresponding to the eigenvalue zero. Thus, we expect solutions of (7) to behave as

$$w(\xi,t) \sim \frac{C_0}{(4\pi)^{d/2}}e^{-|\xi|^2/4}$$

as t tends toward infinity, which just reflects, in these new variables, the fact that solutions of the heat equation tend toward a Gaussian profile as t tends toward infinity.

(Note that this is consistent with the conclusion of Example 1.1 where we showed that the ω-limit set of non-zero solutions of (7) should be either everywhere positive or everywhere negative.)

We now turn to the nonlinear equation (1). We want to apply the results of (CHT) from the first lecture and to do that we need to study the semi-flow defined by this equation. We begin by rewriting the equation in terms of the scaling variables, (6). It is convenient if the resulting equation is autonomous and in order to insure that this is the case we pick the exponent α in the prefactor of w to be $\alpha = \frac{2}{p-1}$. For what comes later it will be convenient to consider the exponent p itself to be one of the dependent variables with a trivial time evolution. With this choice of exponent, and introducing an equation for p, (1) is transformed into

$$\frac{\partial w}{\partial \tau} = \mathscr{L}w + \left(\frac{1}{p-1} - \frac{d}{2} \right) w - |w|^{p-1}w \,, \tag{25}$$

$$\frac{dp}{d\tau} = 0 \,.$$

Here, $\mathscr{L}$ is exactly the same operator studied in connection with the heat equation and the change in the exponent α simply introduces the additional constant term $\left(\frac{1}{p-1} - \frac{d}{2} \right) w$ which just shifts the entire spectrum of $\mathscr{L}$ by that constant amount. Indeed, for simplicity in what follows we will focus particularly on the behavior of p close to the value $\frac{d+2}{d}$ – i.e. close to the value for which this additional term vanishes. With this in mind, we exchange the variable p for the variable η defined by $p = 1 + \frac{2}{d+2\eta}$ so that $\left(\frac{1}{p-1} - \frac{d}{2} \right) w = \eta w$, and recalling that η (as was p) is considered to be one of the dependent variables this term can be considered a part of the nonlinearity! Thus, after these changes, we finally rewrite (1) in the form

$$\frac{\partial w}{\partial \tau} = \mathscr{L}w + \eta w - |w|^{\frac{2}{d+2\eta}} w \,, \tag{26}$$

$$\frac{d\eta}{d\tau} = 0 \,.$$

Now, to verify the hypotheses of the invariant manifold theorem of (CHT) we study the semiflow defined by this system of equations. The evolution of η is trivial so we focus on the first component of the semiflow which we can write with the aid of Duhamel's formula as

$$\Phi^{\tau}(w_0) = w(t) = e^{\tau \mathscr{L}} w_0 + \int_0^{\tau} e^{(\tau - s)\mathscr{L}} \left(\eta w(s) - |w(s)|^{\frac{2}{d+2\eta}} w(s) \right) ds \,, \tag{27}$$

where we have suppressed the dependence of w on ξ to avoid overburdening the notation.

We now discuss the various hypotheses in the (CHT) theorem. The first is that Φ^{τ} should be globally Lipshitz. This is not true of (27) due to the growth of the nonlinear term when w becomes large. This is a standard problem with the application of invariant manifold theorems even in the context of ordinary differential equations and we handle it here in the same way it is usually handled in that setting, namely by "cutting off" the nonlinear term. Let $\chi(x)$ be a smooth, positive function on $\mathbb{R}$ satisfying

$$\chi(x) = \begin{cases} 1, & |x| < 1 \\ 0, & |x| > 2 . \end{cases} \tag{28}$$

Then define

$$\Phi_r^\tau(w_0) = w(t) = e^{\tau\mathscr{L}} w_0 + \int_0^\tau e^{(\tau-s)\mathscr{L}}$$
$$\times \left(\chi \left(\frac{\|w(x)\|_m}{r} \right) \left(\eta w(s) - |w(s)|^{\frac{2}{d+2\eta}} w(s) \right) \right) ds , \tag{29}$$

With this definition the nonlinear term vanishes if $\|w(s)\|_m$ is larger than $2r$ but $\Phi_r^\tau(w_0)$ is equal to $\Phi^\tau(w_0)$ for all solutions that remain within a ball of radius r in $L^2(m)$.

Remark 2.6. Note that the cutoff function $\chi\left(\frac{\|w(x)\|_m}{r}\right)$ is a smooth function on $L^2(m)$. It is always possible to find such a smooth cutoff function on a Hilbert space, but there are natural Banach spaces on which no such smooth cutoff function exists. This can cause problems for certain applications of invariant manifold theorems in infinite dimensional settings.

It is now a standard exercise to verify that:

(N.1) $\Phi_r^\tau(w_0)$ is well defined for $w_0 \in L^2(m)$.
(N.2) The nonlinear term

$$\mathscr{R}_r^\tau(\eta, w_0) = \int_0^\tau e^{(\tau-s)\mathscr{L}} \left(\chi \left(\frac{\|w(x)\|_m}{r} \right) \left(\eta w(s) - |w(s)|^{\frac{2}{d+2\eta}} w(s) \right) \right) ds$$

is globally Lipshitz with Lipshitz constant bounded by $C_{\mathscr{R}}(\eta + r^{\frac{2}{d+2\eta}})$ for some constant $C_{\mathscr{R}}$. Thus, the Lipshitz constant can be made arbitrarily small for η and r sufficiently small.

These two observations are sufficient to verify hypotheses (H.1) and (H.2) of the of the (CHT) theorem. (We can choose the constants $q = \tau = 1$ and set $\Lambda = e^{\tau\mathscr{L}}$ and $\mathscr{R} = \mathscr{R}_r^1$.

We next verify hypothesis (H.3). Here we must make a choice. Given any $n = 0, 1, 2, \ldots$ we could, by choosing m appropriately set $X_1 = P_n L^2(m)$ and $X_2 = Q_n L^2(m)$. We would then obtain an invariant manifold tangent at the origin to the eigenspace corresponding to the eigenvalues $\{-\frac{k}{2} | k = 0, 1, \ldots, n\}$. The long-time behavior of solutions close to the origin could then be determined up to corrections which go to zero at least as fast as $e^{-\gamma\tau}$ with $\gamma > n/2$ just by studying the asymptotics of solutions of the finite dimensional system of *ordinary* differential equations which results from restricting (26) to this invariant manifold.

For now we focus on the simplest possible case, namely we will assume that $m > \max(1, d/2)$ and take $X_1 = P_0 L^2(m)$. In this case X_1 is the one-dimensional subspace spanned by $\phi^0(\xi) = \frac{1}{(4\pi)^{d/2}} e^{-|\xi|^2/4}$. Next we find

$$\Lambda_1 = P_0 e^{\tau \mathscr{L}} P_0 = \mathbf{1} \, , \tag{30}$$

the identity operator and we can take the constants $C_1 = \alpha_1 = 1$ in hypothesis (H.3). Then

$$\Lambda_2 = Q_0 e^{\tau \mathscr{L}} Q_0 \tag{31}$$

and from Proposition 2.2 we see that (H.3) holds for $\alpha_2 = e^{-(\frac{m}{2} - \frac{d}{2})} < 1$ and for some $C_2 > 0$.

Finally, condition (H.4) is satisfied since by remark (N.2) the Lipshitz constant of $\mathscr{R}$ can be made arbitrarily small for η and r sufficiently small.

Since we are considering η to be one of the dependent variables we should also consider the evolution of η – however, this evolution is trivial and hence we can just apply the (CHT) theorem to $\Phi^{\tau=1}$ for each value of η small, treating η as a parameter.

Applying the (CHT) theorem we conclude

Proposition 2.3. *Fix $m > \max(1, d/2)$. There exists $r_0 > 0$ and $\eta_0 > 1$ such that if $|\eta| < \eta_0$ and $0 < r < r_0$ there exists a globally Lipshitz map $g : P_0 L^2(m) \to Q_0 L^2(m)$ with $g(0) = Dg(0) = 0$ such that the submanifold*

$$W_c = \{\alpha \phi^0 + g(\alpha \phi^0) | \alpha \in \mathbb{R}\}$$

has the following properties:

(i) (Invariance) Φ^τ leaves W_c invariant.
(ii) (Fixed Points) If $\{w(t)\}_{\tau \leqslant 0}$ is a negative semi-orbit with $\|w(\tau)\|_m \leqslant r_0$ for all $\tau \leqslant 0$, then $w(\tau) \in W_c$ for all τ.
(iii) (Attractivity) Fix μ such that $0 < \mu < (\frac{d}{4} - \frac{m}{2})$. There exists C and r_2, positive constants, such that for $\tilde{w}_0 \in L^2(m)$ with $\|\tilde{w}_0\|_m < r_2$, there exists a unique $w_0 \in W_c$ such that

$$\|\Phi^\tau(\tilde{w}_0) - \Phi^\tau(w_0)\|_m \leqslant Ce^{-\mu\tau} \, .$$

Remark 2.7. The "Fixed Points" and "Attractivity" parts of the conclusions of this theorem follow respectively from the "Lyapunov Exponents" and "Invariant Foliation" parts of the (CHT) theorem if we use the fact that in this problem we can choose $\gamma_2 < 1$. In particular, $\mu = -\log \gamma_2$.

Note that since, for $\|w(\tau)\|_m < r$ the semiflow Φ_r^τ coincides with Φ^τ, the semiflow for (26) the rescaled heat equation will also have a *local* invariant manifold which attracts all solutions in some sufficiently small neighborhood of the origin.

We conclude this lecture by considering the implications of this manifold for the long-time behavior of solutions of (26). From the "Attractivity" part of Proposition

2.3, the long-time behavior of small solutions of (26) will (up to higher order corrections) be the same as those of solutions lying on the manifold W_c, and the long-time behavior of solutions lying on this manifold can be determined by solving the single *ordinary* differential equation that results from restricting the original partial differential equation to this manifold. If $w(\xi, \tau)$ lies on W_c we can write

$$w(\xi, \tau) = \alpha(\tau)\phi^0(\xi) + g(\alpha(\tau)\phi^0(\xi)) \, . \tag{32}$$

Inserting this representation of w into (26) gives

$$\dot{\alpha}(\tau)\phi^0(\xi) + \alpha(\tau)Dg(\alpha(\tau)\phi^0(\xi))\phi^0(\xi) = \alpha(\tau)(\mathscr{L}\phi^0)(\xi) - |\alpha(\tau)\phi^0(\xi) \tag{33}$$
$$+ g(\alpha(\tau)\phi^0(\xi))|^{\frac{2}{d+2\eta}} (\alpha(\tau)\phi^0(\xi) + g(\alpha(\tau)\phi^0(\xi)))$$

We now reduce this to an ordinary differential equation for $\alpha(\tau)$ by noting that $\mathscr{L}\phi^0 = 0$ and then applying the projection operator P_0 to both sides of the equation. This yields:

$$\dot{\alpha}(\tau) = \eta\alpha(\tau) - \int_{\mathbb{R}} |\alpha(\tau)\phi^0(\xi) + g(\alpha(\tau)\phi^0(\xi))|^{\frac{2}{d+2\eta}}$$
$$\times (\alpha(\tau)\phi^0(\xi) + g(\alpha(\tau)\phi^0(\xi)))d\xi \tag{34}$$

For the moment the only thing we need to know about the complicated nonlinear term is that since $g(0) = 0$, and $Dg(0) = 0$ (by explicit computation of the equation satisfied by the invariant manifold), for for α and η small it behaves like $C_L|\alpha|^{\frac{2}{d+2\eta}}\alpha$, where $C_L = \int_{\mathbb{R}} (\phi^0(\xi))^p d\xi$.

From this equation is clear that varying η (or equivalently p) leads to a bifurcation at $\eta = 0$. From now on, for simplicity we assume that $d = 1$, though the computations can be carried through in a similar way for higher dimensions. Note that in $d = 1$, $\eta = 0$ corresponds to the exponent $p = 3$. To better understand the bifurcation that results when we vary η we first consider solutions of (34) when $\eta < 0$ which corresponds to $p > 3$. In this case the origin is an attractive fixed point for (34) and for any solutions with $\alpha(0)$ sufficiently small we have

$$\alpha(\tau) \sim C_0 e^{-\eta\tau} \, ,$$

for some $C_0 > 0$. From this we immediately conclude that solutions on the invariant manifold W_c behave for large times like

$$w(\xi, \tau) = C_0 e^{-\eta\tau}\phi^0(\xi) + g(C_0 e^{-\eta\tau}\phi^0(\xi)) \tag{35}$$
$$= C_0 e^{-\eta\tau}\phi^0(\xi) + \mathscr{O}(e^{-2\eta\tau}) \, ,$$

where the last equality reflects the fact that since $g(0) = Dg(0) = 0$, the terms $g(C_0 e^{-\eta\tau}\phi^0(\xi))$ will decay faster than $e^{-\eta\tau}$. Furthermore by the "Attractivity" part of Proposition 2.3 all small solutions will behave like (35) to leading order. Thus we have:

Corollary 2.1. *All sufficiently small solutions of* (26), *behave asymptotically like*

$$w(\xi,\tau) = C_0 e^{-\eta\tau}\phi^0(\xi) + \mathcal{O}(e^{-2\eta\tau}) \, .$$

for some constant C_0.

Note that from this corollary it appears that the decay rate of these solutions depends on p through the exponent η. However, if we revert to our original variables we see that solutions of the original equation (1) behave as

$$u(x,t) = \frac{1}{(1+t)^{\alpha/2}}w(\frac{x}{\sqrt{1+t}},\log(1+t)) = \frac{1}{(1+t)^{\frac{1}{(p-1)}}}w(\frac{x}{\sqrt{1+t}},\log(1+t)) \quad (36)$$

$$= \frac{1}{(1+t)^{\frac{1}{(p-1)}}}\left(C_0(1+t)^{-\eta}\phi^0(\frac{x}{\sqrt{1+t}}) + \mathcal{O}((1+t)^{-2\eta}) \right)$$

$$= \frac{C_0}{\sqrt{1+t}}\phi^0(\frac{x}{\sqrt{1+t}}) + \ldots$$

Note that the leading order behavior here is the same as the leading order asymptotic behavior of solutions of the linear heat equation. Thus, for $p > 3$, all small solutions of (1) behave as if the nonlinear term was absent - such nonlinear terms are often referred to as "irrelevant".

Let's now consider what happens if $p < 3$ (or $\eta > 0$). In this case the origin is *unstable* and the fixed point at the origin undergoes a pitchfork bifurcation and a pair of new fixed points appears at $\pm\alpha^* \approx \pm(\eta/C_L)^{\frac{1}{p-1}}$. These fixed points are *stable* (at least for η sufficiently small) at hence all non-zero solutions in W_c will approach one of them. Define

$$w^*(\xi;p) = \alpha^*\phi^0(\xi) + g(\alpha^*\phi^0(\xi)) \, .$$

Then, small solutions of (1) will behave like

$$w(\xi,\tau) \approx w^*(\xi) \quad (37)$$

for τ large.

Remark 2.8. In fact, there are some solutions which will approach the origin even when $\eta < 0$. Those are the solutions that lie in the stable manifold of the origin. However, these solutions for a manifold of codimension-one and hence "most" solutions will behave as in (37).

If we again revert to the original variables we find

Corollary 2.2. *For $p < 3$, all sufficiently small solutions of* (1) *except for those lying in the codimension one stable manifold of the origin, behave like*

$$u(x,t) = \frac{1}{(1+t)^{\frac{1}{p-1}}} w^*(\frac{x}{\sqrt{1+t}}) + \dots .$$

Thus, we see that for $p < 3$ the situation is quite different from that for $p > 3$ since both the rate of decay of the long-time asymptotics and the functional form of the limiting solution depend on the nonlinear term.

Exercise 2.1. Determine the behavior of the long-time asymptotics of solutions when $p = 3$ – the "critical" value of the nonlinear term.

Remark 2.9. By considering the manifolds tangent to the spectral subspaces corresponding to more than just the zero eigenvalue – say to the eigenvalues $\{0, -\frac{1}{2}\}$ or $\{0, -\frac{1}{2}, -1\}$, etc. one can derive more refined estimates of the long-time behavior of the solutions.

Summing up this lecture, we have found a way, at least in some parabolic partial differential equations, to create a spectral gap which allows us to apply invariant manifold theorems to problems on unbounded spatial domains. These theorems can then give detailed information about the long-time asymptotics of solutions. The drawback is that these results are local in nature – in the present examples they apply only to "small" solutions. As we will see in the fourth lecture in this series that restriction can sometimes be lifted by combining these results with Lyapunov functionals which give more global control over the solutions.

3 Third lecture: an introduction to the Navier–Stokes equations

In this section we will discuss the Navier–Stokes equations which describe the velocity of a viscous, incompressible fluid. The focus of this lecture will be the origin of the equations, their representation in terms of both the velocity and vorticity of the fluid and the existence of solutions in the two-dimensional case. In the final lecture in this series we will look in greater detail at the long-time behavior of solutions of two-dimensional Navier–Stokes equations. A more detailed look at the Navier–Stokes equation, but with a similar point of view can be found in the lecture notes of Gallagher and Gallay [6]. For more discussion of the physical origin of these equations one can consult [5].

The Navier–Stokes equations arise from applying Newton's law to determine the motion of a small "blob" of fluid. Assume that the "blob" is a cube of side length Δx, centered at the point $x \in \mathbb{R}^d$, where for physical relevance we restrict to the cases $d = 2$ or 3. If $\mathbf{u}(x,t)$ is the fluid's velocity measured in the laboratory frame of reference, then Newton's Law implies

$$\frac{d}{dt}(\text{momentum}) = \text{applied forces}.$$

If the density of the fluid is ρ, then the momentum will be $\pi(x,t) = \rho(x,t)\mathbf{u}(x,t)\Delta V$, where ΔV is the volume of the little cube of fluid. To simplify the discussion we will assume that the density ρ is constant and check *a posteriori* that this is consistent with the equations of motion. We'll also ignore the factor of ΔV since it will occur in each term and can be cancelled out.

To compute the change in moment of our fluid blob we need to take account of the fact that the fluid is being advected along by its own velocity. Thus,

$$\frac{d\pi}{dt}(x,t) = \lim_{\Delta t \to 0} \frac{\pi(x + \mathbf{u}(x,t)\Delta t, t + \Delta t) - \pi(x,t)}{\Delta t}$$

$$= \mathbf{u}(x,t)\cdot\nabla\pi(x,t) + \frac{\partial\pi}{\partial t}(x,t).$$

This expression is known as the convective derivative of the momentum. Thus, returning to Newton's law, we have

$$\frac{\partial\pi}{\partial t}(x,t) + \mathbf{u}(x,t)\cdot\nabla\pi(x,t) = \text{applied forces}.$$

What are the forces that act on the fluid element?

- Forces due to pressure: $f_{\text{pressure}} = -\nabla p(x,t)$, where p is the pressure in the fluid.
- External forces: we will ignore these.
- Viscous forces: These involve modeling internal properties of the fluid. We will take a standard model which says $f_{\text{visc}} = \alpha\Delta\mathbf{u}$, for some constant α.

Inserting these forces into Newton's law we arrive at the system of partial differential equations:

$$\frac{\partial\pi}{\partial t}(x,t) + \mathbf{u}(x,t)\cdot\nabla\pi(x,t) = \alpha\Delta\mathbf{u}(x,t) - \nabla p(x,t) \tag{1}$$

Assuming that the density is constant, this is a system of d equations, but it contains $d+1$ unknowns – the d components of the velocity, plus the pressure. We need one further equation linking the pressure and momentum in order to close the system. This remaining equation is derived from the property of conservation of mass. If we look at the equation for the change in the amount of mass in a region V, then we see that by conservation of mass, any change in the mass in the region (given by $\int_V(\partial_t\rho)dV$) must be counterbalanced by a flux of mass through the boundary (given by $-\int_{\partial V}\rho\mathbf{u}\cdot\hat{\mathbf{n}}dS$). Equating these two expressions, applying the divergence theorem and using the fact that the region V was arbitrary leads to the conservation equation

$$\frac{\partial\rho}{\partial t}(x,t) + \nabla\cdot(\mathbf{u}(x,t)\rho(x,t)) = 0$$

If we now impose the incompressibility condition $\nabla \cdot (\mathbf{u}(x,t)\rho(x,t)) = 0$ we see that $\rho(x,t) = \rho(x,0)$. In particular, if the density is initially constant it will remain so for all time and the incompressibility condition simplifies to $\nabla \cdot (\mathbf{u}(x,t)) = 0$. Then we have a system of $d + 1$ nonlinear partial differential equations:

$$\rho \frac{\partial \mathbf{u}}{\partial t}(x,t) + \rho \mathbf{u}(x,t) \cdot \nabla \mathbf{u}(x,t) = \alpha \Delta \mathbf{u}(x,t) - \nabla p(x,t) \tag{2}$$

$$\nabla \cdot \mathbf{u}(x,t) = 0 \ .$$

Remark 3.1. Note that given a solution $\mathbf{u}$ of (2) one can recover the pressure by taking the divergence of the momentum equation and using the incompressibility equation from which one finds:

$$\Delta p = -\rho \nabla \cdot (\mathbf{u} \cdot \nabla \mathbf{u}) \ ,$$

so the pressure is obtained as a solution of Poisson's equation.

Remark 3.2. The coefficients in (2) can be simplified somewhat. Suppose that we introduce some fixed length scale L, velocity scale V and reference density $\overline{\rho}$. If we define new, dimensionless variables via $\tilde{x} = x/L$, $\tilde{\mathbf{u}} = \mathbf{u}/V$, $\tilde{t} = (tV)/L$, and $\tilde{\rho} = \rho/\overline{\rho}$, then a simple exercise shows that in terms of the new variables (2) is replaced by:

$$\frac{\partial \tilde{\mathbf{u}}}{\partial \tilde{t}} + \tilde{\mathbf{u}} \cdot \nabla \tilde{\mathbf{u}} = \tilde{\alpha} \Delta \tilde{\mathbf{u}}(x,t) - \frac{1}{\tilde{\rho}} \nabla \tilde{p}(x,t)$$

where $\tilde{\alpha} = \frac{\alpha}{\rho V^2 L}$, $\tilde{p} = p/(\overline{\rho} L)$, and all derivatives are computed with respect to the new variables. These changes of variables are particularly convenient if we study this equation on the domain $\mathbb{R}^d$ since in this case the rescaling has no effect on the domain and if we choose the length scale $L = \alpha/(\rho V^2)$, all coefficients in the equation become equal to one. From now on we will assume that we have made these changes of variables and drop the tildes to avoid burdening the notation.

Remark 3.3. A related quantity is the dimensionless ratio of the inertial forces to the viscous forces given by

$$\mathrm{Re} = \frac{(\rho V^2/L)}{(\alpha V/L^2)} = \frac{\rho V L}{\alpha}$$

known as the Reynolds number.

The remainder of this lecture will be devoted to studying the initial value problem for (2) – namely given some initial velocity distribution $\mathbf{u}(x,0) = \mathbf{u}_0(x)$, show that the equation (2) has a unique solution and describe the properties of this solution. Proving that (2) has a unique, smooth solution for all initial data is a very famous problem. Basically, two alternatives have developed so far:

- Give up smoothness and uniqueness and simply try to show that there is some (weak) solution to the problem. This approach dates back to the work of Leray.

- Attempt to show that the initial value problem is well posed, at the expense of specializing the problem somehow - perhaps considering "small" initial data, or restricting the domain on which the problem is posed.

I will adopt the second approach in these lectures by focusing on the two-dimensional problem. When studying the two-dimensional Navier–Stokes equation defined in the entire plane it turns out to be simpler to work not directly with the velocity field but rather with the *vorticity* of the fluid. The vorticity is defined by the curl of the velocity field – i.e. $\omega(x,t) = \nabla \times \mathbf{u}(x,t)$ and in general it is a vector field, just like the velocity. However, in two dimensions

$$\omega(x,t) = \nabla \times (u_1(x_1,x_2,0,t), u_2(x_1,x_2,0,t), 0) = (0,0,\omega(x_1,x_2,t))$$

so we see that only one component of the vorticity is non-zero and thus we may treat it as a scalar. If we take the curl of the Navier–Stokes equation we find that (in general dimension d)

$$\frac{\partial \omega}{\partial t} - \omega \cdot \nabla \mathbf{u} + \mathbf{u} \cdot \nabla \omega = \Delta \omega \; . \tag{3}$$

Note that one advantage of the vorticity formulation of the problem is that the pressure term drops out entirely.

Remark 3.4. The term $\omega \cdot \nabla \mathbf{u}$ is known as the "vorticity stretching term". It allows for a certain "self amplification" of the vorticity. Note that in two dimensions this term is zero since $\omega \cdot \nabla \mathbf{u} = \omega \partial_{x_3} \mathbf{u}(x_1,x_2,0) = 0$. The absence of this term is another reason why the two-dimensional Navier–Stokes (or vorticity) equation is easier to treat than the three dimensional one.

From now on we will restrict our attention to the two-dimensional vorticity equation and consider the initial value problem

$$\frac{\partial \omega}{\partial t} + \mathbf{u} \cdot \nabla \omega = \Delta \omega, \; t > 0, x \in \mathbb{R}^2 \tag{4}$$

$$\omega(x,0) = \omega_0(x) \; .$$

The principle difficulty in studying (4) is the presence of the velocity, $\mathbf{u}$ in this equation. We must reconstruct the velocity from the vorticity – however, this leads to a somewhat complicated, nonlocal nonlinearity. Recalling that the vorticity is the curl of the velocity and that the velocity is incompressible, we can reconstruct the velocity using the Biot–Savart law

$$\mathbf{u}(x) = \frac{1}{2\pi} \int_{\mathbb{R}^2} \frac{(x-y)^{\perp}}{|x-y|^2} \omega(y,t) dy \; . \tag{5}$$

Here, for any two-dimensional vector $x = (x_1,x_2)$ we define $x^{\perp} = (-x_2,x_1)$.

Exercise 3.1. Verify that the Biot–Savart law does give an incompressible velocity field whose curl is the vorticity.

In order to control the solutions of (4) (and to verify the hypotheses of the (CHT) theorem) we need estimates which relate the norm of the velocity to the vorticity. A collection of such estimates is derived in [7], Appendix B, but as an example of the sort of estimates one needs we prove:

Lemma 3.1. *Let* **u** *be the velocity field associated to the vorticity* ω *by the Biot–Savart law. Fix* $1 < q < 2$. *Then if*

$$\frac{1}{q} - \frac{1}{p} = \frac{1}{2}$$

there exists $C = C(p,q)$ *such that*

$$\|\mathbf{u}\|_{L^p(\mathbb{R}^2)} \leqslant C\|\omega\|_{L^q(\mathbb{R}^2)}$$

Remark 3.5. Define the L^p norm of a vector valued function as the sum of the L^p norms of the components.

Proof. Recall the Hardy–Little–Sobolev Inequality

$$\int_{\mathbb{R}^d} f(x) \left(\int_{\mathbb{R}^d} \frac{1}{|x-y|^\lambda} g(y)dy \right) dx \leqslant N(p,q,\lambda,d)\|f\|_{L^s(\mathbb{R}^d)}\|g\|_{L^q(\mathbb{R}^d)} \ ,$$

provided $\frac{1}{s} + \frac{1}{q} + \frac{\lambda}{d} = 2$. Note that

$$|u_j(x,t)| \leqslant \frac{1}{2\pi} \int_{\mathbb{R}^2} \frac{1}{|x-y|} |\omega(y,t)|dy \equiv h(x) \ ,$$

so that $\|u_j(x,t)\|_{L^p(\mathbb{R}^2)} \leqslant \|h\|_{L^p(\mathbb{R}^2)}$. Let $f = h^{p-1}$. Then applying the HLS inequality we find

$$\|h\|_{L^p(\mathbb{R}^2)} \leqslant N\|h^{p-1}\|_{L^s(\mathbb{R}^2)}\|\omega\|_{L^q(\mathbb{R}^2)} \ ,$$

Take $s = \frac{p}{p-1}$. Then $\|h^{p-1}\|_{L^s(\mathbb{R}^2)} = \|h\|_{L^p(\mathbb{R}^2)}^{p-1}$ and hence

$$\|h\|_{L^p(\mathbb{R}^2)} \leqslant N\|\omega\|_{L^q(\mathbb{R}^2)} \ ,$$

with $\frac{1}{q} - \frac{1}{p} = \frac{1}{2}$. $\square$

Exercise 3.2. Use the Biot–Savart law to prove that

$$\|\mathbf{u}\|_{L^\infty(\mathbb{R}^2)} \leqslant C(\|\omega\|_{L^1(\mathbb{R}^2)} + \|\omega\|_{L^\infty(\mathbb{R}^2)}) \ .$$

We now have the tools we need to prove the existence and uniqueness of solutions of the two-dimensional vorticity equation. This is a story with a long history but the approach I describe below was first developed by Ben–Artzi, [2]. My presentation of this approach is close to that of [6]. The first question that arises it what space we should work in. Noting that (4) conserves the total vorticity suggests that the space

$L^1(\mathbb{R}^2)$ is appropriate and it turns out that in this space all initial conditions lead to unique global solutions. More precisely one has

Theorem 3.1. *There exists $C > 0$ such that for any $\omega_0 \in L^1(\mathbb{R}^2)$, the initial value problem* (4) *has a unique solution $u \in C(\mathbb{R}^+; L^1(\mathbb{R}^2))$.*

Proof. The proof basically consists of two steps:

1. Show that given $\omega_0 \in L^1(\mathbb{R}^2)$, the initial value problem has a unique solution for some interval of time T_0. Furthermore, for any positive time this solution is in $L^p(\mathbb{R}^2)$ for all $1 \leqslant p \leqslant \infty$.
2. Show that if the initial condition $\omega_0 \in L^1(\mathbb{R}^2) \cap L^\infty(\mathbb{R}^2)$ then one has a unique solution for all time.

Note that these to points taken together suffice to prove the theorem since given an initial condition $\omega_0 \in L^1(\mathbb{R}^2)$ we first solve the initial value problem for some short time. We then take this solution at some positive time t_0 as our new initial condition and the resulting solution exists for all time.

I'll look in detail at the second part of the proof – details of the first part can be found in [2] and [6]. As a first step we rewrite (4) as an integral equation, just as we did with the semi-linear heat equation in Lecture 2.

$$\omega(t) = \Phi^t(\omega_0) = e^{t\Delta}\omega_0 + \int_0^t e^{(t-s)\Delta}\mathbf{u}(s)\cdot\nabla\omega(s)ds, \tag{6}$$

where $e^{t\Delta}$ denotes the semigroup defined by the heat equation. The proof of the theorem now follows by showing that (6) has a fixed point in an appropriate Banach space.

Remark 3.6. Before beginning the fixed point argument, however, we note that *if* (4) has a solution, the solution has the following important property. In two dimensions, since the vorticity is a scalar, it satisfies the maximum principle. As a consequence not only is the $L^1(\mathbb{R}^2)$ norm a non-increasing function of time (remember Example 1.1) but in fact by a similar argument one finds that $\|\omega(t)\|_{L^p(\mathbb{R}^2)} \leqslant \|\omega_0\|_{L^p(\mathbb{R}^2)}$ for all $1 \leqslant p \leqslant \infty$.

Returning to (6) we write this equation as

$$\omega(t) = \mathscr{F}(\omega)(t) = e^{t\Delta}w_0 + \mathscr{N}(w, w)(t) \tag{7}$$

where

$$\mathscr{N}(\tilde{\omega}, \omega)(t) = \int_0^t e^{(t-s)\Delta}\tilde{\mathbf{u}}(s)\cdot\nabla\omega(s)ds, \tag{8}$$

and $\tilde{\mathbf{u}}$ is the velocity field associated to the vorticity $\tilde{\omega}$ by the Biot–Savart law. Note that $\tilde{\mathbf{u}}$ is a linear function of $\tilde{\omega}$ so $\mathscr{N}$ is a bilinear operator. We'll study the fixed point problem for $\mathscr{F}$ on the Banach space

$$X_T^* = \{f \in C([0, T] : L^1(\mathbb{R}^2) \cap L^\infty(\mathbb{R}^2))\}$$

with norm $\|f\|_* = \sup_{0 \leqslant t \leqslant T}(\|f(t)\|_{L^1(\mathbb{R}^2)} + \|f(t)\|_{L^\infty(\mathbb{R}^2)})$.

We first note that if $\omega_0 \in L^1(\mathbb{R}^2) \cap L^\infty(\mathbb{R}^2)$ then the linear term in (7) is an element of X_T^*. This follows immediately from the estimates:

Lemma 3.2. *For any* $\alpha = (\alpha_1, \alpha_2) \in \mathbb{N}^2$ *and* $1 \leqslant p \leqslant q \leqslant \infty$ *there exists* $C = C(p, q, \alpha)$ *such that*

$$\|\partial^\alpha(e^{t\Delta} f)\|_{L^p(\mathbb{R}^2)} \leqslant \frac{C}{t^{\frac{|\alpha|}{2} + \left(\frac{1}{q} - \frac{1}{p}\right)}} \|f\|_{L^q(\mathbb{R}^2)} .$$

Proof. The proof of this lemma follows easily by applying Young's inequality to the explicit integral representation for the heat semigroup. $\square$

The key estimate is the following bound on the nonlinear term:

Lemma 3.3. *There exists* $C > 0$ *such that for any* $\tilde{\omega}$ *and* ω *in* X_T^*,

$$\|\mathcal{N}(\tilde{\omega}, \omega)\|_* \leqslant C\sqrt{T}\|\tilde{\omega}\|_* \|\omega\|_* .$$

Assuming for the moment that the lemma holds we proceed as follows. Given the estimates of the two preceeding lemmas a standard application of the contraction mapping theorem shows that (7) has a unique fixed point in X_T^* provided

$$4C\sqrt{T}\|\omega\|_* \leqslant 1 . \tag{9}$$

However, this estimate is problematic since it involves the fixed point itself and hence makes it difficult to get a good estimate of the time of existence of the solution (which we want ultimately to show is infinity.) We now make use of Remark 3.6. from which we conclude that $\|\omega\|_* \leqslant (\|\omega_0\|_{L^1(\mathbb{R}^2)} + \|\omega_0\|_{L^\infty(\mathbb{R}^2)})$. But if we couple this observation with (9) we see that we obtain a unique solution of (4) for all times $0 \leqslant t \leqslant T$ such that

$$T = \left(\frac{1}{4C(\|\omega_0\|_{L^1(\mathbb{R}^2)} + \|\omega_0\|_{L^\infty(\mathbb{R}^2)})}\right)^2 . \tag{10}$$

In order to show that this solution actually exists for all time we now repeat this procedure, taking as our new initial condition $\tilde{\omega}_0 = \omega(T)$. This new solution (which is the continuation of our original solution) exists for at least a time

$$\tilde{T} = \left(\frac{1}{4C(\|\tilde{\omega}_0\|_{L^1(\mathbb{R}^2)} + \|\tilde{\omega}_0\|_{L^\infty(\mathbb{R}^2)})}\right)^2 . \tag{11}$$

However, since

$$(\|\tilde{\omega}_0\|_{L^1(\mathbb{R}^2)} + \|\tilde{\omega}_0\|_{L^\infty(\mathbb{R}^2)}) = (\|\omega(T)\|_{L^1(\mathbb{R}^2)} + \|\omega(T)\|_{L^\infty(\mathbb{R}^2)})$$
$$\leqslant (\|\omega_0\|_{L^1(\mathbb{R}^2)} + \|\omega_0\|_{L^\infty(\mathbb{R}^2)})$$

we see that $\tilde{T} \geqslant T$ and hence we can repeat this argument indefinitely, extending our solution for arbitrarily long times.

Thus, the only remaining step in the proof that we have unique global solutions for initial conditions in X_T^* is to prove Lemma 3.3. We begin by showing that the $L^1(\mathbb{R}^2)$ norm of $\mathcal{N}$ is uniformly bounded.

$$\|\mathcal{N}(\tilde{\omega}, \omega)(t)\|_{L^1(\mathbb{R}^2)} = \|\int_0^t e^{(t-s)\Delta} \nabla \cdot (\tilde{\mathbf{u}}(s)\omega(s))ds\|_{L^1(\mathbb{R}^2)} \tag{12}$$

$$\leqslant C \int_0^t \frac{1}{\sqrt{t-s}} \|\tilde{\mathbf{u}}(s)\omega(s)\|_{L^1(\mathbb{R}^2)} ds \,,$$

where the last inequality used Lemma 3.2 to bound the linear semigroup. By Hölder's inequality

$$\|(\tilde{\mathbf{u}}(s)\omega(s))\|_{L^1(\mathbb{R}^2)} \leqslant \|\tilde{\mathbf{u}}(s)\|_{L^4(\mathbb{R}^2)} \|\omega(s)\|_{L^{4/3}(\mathbb{R}^2)} \,,$$

while Lemma 3.1 implies that $\|\tilde{\mathbf{u}}(s)\|_{L^4(\mathbb{R}^2)} \leqslant C\|\tilde{\omega}(s))\|_{L^{4/3}(\mathbb{R}^2)}$. Combining these estimates we find

$$\|\mathcal{N}(\tilde{\omega}, \omega)(t)\|_{L^1(\mathbb{R}^2)} \leqslant C \int_0^t \frac{1}{\sqrt{t-s}} \|\tilde{\omega}(s)\|_{L^{4/3}(\mathbb{R}^2)} \|\omega(s)\|_{L^{4/3}(\mathbb{R}^2)} ds$$

$$\leqslant C\sqrt{T} \|\tilde{\omega}\|_* \|\omega\|_* \,. \tag{13}$$

A similar bound on the $L^\infty(\mathbb{R}^2)$ norm of $\mathcal{N}$ completes the proof. We again begin by using the bound in Lemma 3.2:

$$\|\mathcal{N}(\tilde{\omega}, \omega)(t)\|_{L^\infty(\mathbb{R}^2)} = \|\int_0^t e^{(t-s)\Delta} \nabla \cdot (\tilde{\mathbf{u}}(s)\omega(s))ds\|_{L^1(\mathbb{R}^2)} \tag{14}$$

$$\leqslant C \int_0^t \frac{1}{\sqrt{t-s}} \|\tilde{\mathbf{u}}(s)\omega(s)\|_{L^\infty(\mathbb{R}^2)} ds \,,$$

But by Exercise 3.2 we have $\|\tilde{\mathbf{u}}(s)\|_{L^\infty(\mathbb{R}^2)} \leqslant \|\tilde{\omega}\|_*$ and by interpolation $\|\omega(s)\|_{L^2(\mathbb{R}^2)} \leqslant \|\tilde{\omega}\|_*$, hence

$$\|\mathcal{N}(\tilde{\omega}, \omega)(t)\|_{L^\infty(\mathbb{R}^2)} \leqslant C \int_0^t \frac{1}{\sqrt{t-s}} ds \|\tilde{\omega}\|_* \|\omega\|_* \,.$$

which completes the proof of Lemma 3.3 and concludes this section.

4 Fourth lecture: the long-time asymptotics of solutions of the two-dimensional Navier–Stokes equation

In this section we combine the methods developed in the first two lectures to describe the long-time behavior of solutions of the two-dimensional Navier–Stokes equation. We prove that any solution whose initial vorticity distribution is integrable will tend, as time goes to infinity, toward an Oseen vortex, a simple, explicitly computable solution of the Navier–Stokes equations in two-dimensions. We also give a detailed discussion of the long-time behavior of solutions whose total vorticity is small. The material in this lecture is largely joint work of Th. Gallay and myself and for more details the reader can consult the original papers [7] and [8].

Throughout this lecture we will consider the Navier–Stokes equation in the vorticity representation

$$\frac{\partial \omega}{\partial t} = \Delta_x \omega - \mathbf{u} \cdot \cdot \omega, \tag{1}$$

$$\omega = \omega(x,t) \in \mathbb{R}, \quad x \in \mathbb{R}^2, t \geqslant 0.$$

where $\mathbf{u}$ is the velocity field associated with the vorticity ω via the Biot–Savart law. As discussed in the preceding lecture the vorticity formulation is particularly convenient in two-dimensions where the vorticity is a scalar function. Furthermore as in Lecture 2 we will study solutions of (1) in the weighted Hilbert spaces $L^2(m)$ and the vorticity has the advantage that if the initial vorticity distribution lies in one of these spaces the solution of (1) will remain in this space for all time, whereas that is not in general true of the velocity field. (This fact is not immediately apparent but is discussed and proven in [7].)

We begin, as we did in Lecture 2 by considering solutions of (1) in a neighborhood of the origin. Given the similarity between the vorticity equation and (1) we introduce scaling variables as we did in that case, namely we set:

$$\omega(x,t) = \frac{1}{(1+t)} w\left(\frac{x}{\sqrt{1+t}}, \log(1+t)\right) \tag{2}$$

$$\xi = \frac{x}{\sqrt{1+t}}, \quad \tau = \log(1+t)$$

Note that this corresponds to taking the exponent α in (6) equal to $\alpha = d = 2$. We still need to decide how to rescale the velocity field. Since the vorticity is a derivative of the velocity with respect to x, and since each x derivative results in an extra factor of $\frac{1}{\sqrt{1+t}}$, this suggests that the velocity should scale as

$$\mathbf{u}(x,t) = \frac{1}{\sqrt{1+t}} \mathbf{v}\left(\frac{x}{\sqrt{1+t}}, \log(1+t)\right). \tag{3}$$

Strong evidence that this is the "correct" scaling can be seen from the fact that with the rescaled velocity and vorticity fields defined by (2) and (3) $\mathbf{v}$ and w are still related via the Biot–Savart law, namely:

$$\mathbf{v}(\xi,\tau) = \frac{1}{2\pi} \int_{\mathbb{R}^2} \frac{x-y)^\perp}{|x-y|^2} w(\eta,\tau) d\eta, \tag{4}$$

which we leave as an exercise for the reader to check.

Inserting (2) and (3) into (1) we find that

$$\frac{\partial w}{\partial \tau} = \mathcal{L} w - \mathbf{v} \cdot \nabla w \,. \tag{5}$$

Here, $\mathcal{L}$ is the same operator that we studied in Lecture 2 – namely

$$\mathcal{L} w = \Delta_\xi w + \frac{1}{2} \nabla_\xi \cdot (\xi w) \,.$$

Recall that the spectrum of $\mathcal{L}$ when acting on functions in $L^2(m)$ consists of the non-positive half integers, plus a half-plane of spectrum $\{\lambda \in \mathbb{C} | \mathrm{Re}(\lambda) \leqslant \frac{1}{2} - \frac{m}{2}\}$. Thus, for $m > 1$ we expect that there will be a one-dimensional invariant manifold W_c, tangent at the origin to the eigenspace of the (simple) eigenvalue $\lambda = 0$.

Remark 4.1. Verifying the hypotheses (H.1)–(H.4) of the (CHT) invariant manifold theorem requires combining the ideas of Lectures 2 and 3. Since the linear part of (2) is is the same as that of (26) verifying (H.1) and (H.2) is exactly the same as in Lecture 2. Verifying the hypotheses (H.3) and (H.4) on the nonlinearity follows from estimates very similar to those in Lecture 3 where we estimated the semi-group for (1) since the form of the nonlinear terms in (1) are the same as those in (5). In this case one must cut-off the nonlinear term outside a neighborhood of the origin in order to obtain the global estimates required in the (CHT) theorem, but that is again done in a fashion very similar to that in Lecture 2.

Let's next examine the motion on the manifold W_c. As in the case of the nonlinear heat equation in Lecture 2 a point on W_c can be represented as

$$w^c(\xi, \tau) = \alpha(\tau)\phi^0(\xi) + g(\alpha(\tau)\phi^0(\xi)) \tag{6}$$

for some function $g : P_0 L^2(m) \to Q_0 L^2(m)$, where P^0 is the projection onto the eigenspace of $\lambda = 0$ and Q_0 is its complement. If we insert this form into (5) and apply the projection operator P_0 to both sides of the equation we find that

$$\dot{\alpha}(\tau)\phi^0(\xi) = -P^0\left(\mathbf{v}^c(\xi, \tau) \cdot w(\xi, \tau)\right) \,, \tag{7}$$

where $\mathbf{v}^c$ is the velocity field associated to w^c via the Biot–Savart Law. We now note two things:

1. $(P_0 f)(\xi) = (\int_{\mathbb{R}^2} f(\xi)d\xi)\phi^0(\xi)$.
2. The velocity field $\mathbf{v}^c$ is compressible (i.e. $\nabla \cdot \mathbf{v}^c = 0$) and thus we can write $\mathbf{v}^c(\xi, \tau) \cdot w(\xi, \tau) = \nabla \cdot (\mathbf{v}^c(\xi, \tau)w(\xi, \tau))$.

But these two facts imply that

$$P^0\left(\mathbf{v}^c(\xi, \tau) \cdot w(\xi, \tau)\right) = \left(\int_{\mathbb{R}^2} \nabla \cdot (\mathbf{v}^c(\xi, \tau)w(\xi, \tau))d\xi\right) \phi^0(\xi) = 0 \,.$$

and hence that

$$\dot{\alpha}(\tau) = 0 \, .$$

This implies that the center manifold consists entirely of fixed points! In fact, we can identify these fixed points more precisely. If one checks the velocity field corresponding (via the Biot–Savart) law to the vorticity field ϕ^0 one finds that the velocity field is

$$\mathbf{v}^0(x) = \frac{1}{2\pi} \frac{x^\perp}{|x|^2} \left(1 - e^{-|x|^2/4} \right) \, . \tag{8}$$

For the moment, the most important thing to note about this expression is that it is a purely *tangential* velocity field. As a consequence, since the vorticity ϕ^0 depends only on $|x|$, the radial coordinate of x, we see that the nonlinear term in the vorticity equation:

$$\mathbf{v}^0(x) \cdot (\nabla \phi^0)(x) = 0 \, .$$

Thus, since $\mathscr{L}\phi^0 = 0$ we see that the Gaussian vorticity distribution $\alpha\phi^0$ is a stationary solution of the rescaled vorticity equation (5). This family of solutions is known as the family of *Oseen vortices*.

Remark 4.2. Note that in the original, unrescaled variables, the Oseen vortices are not stationary solutions but rather spread and decay in the same way as does the fundamental solution of the heat equation.

Returning now to our discussion of the center manifold we know first of all, from the general theory of invariant manifolds discussed in Lecture 1 that all fixed points near the origin must lie in the center-manifold. Thus, for small α the family of Oseen vortices must be contained in the center-manifold. However, this is a one-dimensional family of solutions and the center-manifold itself is one-dimensional so in fact, the center-manifold in this case consists exactly of the family of Oseen vortices!

Again, appealing to the general theory of invariant manifolds we know that solutions near the origin will be attracted to one of the solutions on the center-manifold. In fact, we can determine which of the Oseen vortices is the limit by noting that the rescaled vorticity equation preserves the total vorticity – i.e. if $w(\xi, \tau)$ is the solution with initial condition $w_0(\xi)$ then

$$\int_{\mathbb{R}^2} w(\xi, \tau)d\xi = \int_{\mathbb{R}^2} w_0(\xi)d\xi \tag{9}$$

for all τ. Thus, as τ goes to infinity, $w(\xi, \tau)$ approaches the vortex $\alpha\phi^0$ whose total vorticity is $\alpha = \int_{\mathbb{R}^2} w_0(\xi)d\xi$. More precisely we find

Proposition 4.1. *Fix* $0 < \mu < \frac{1}{2}$. *There exist positive constants* r_2 *and* C *such that for any initial data with* $\|w_0\|_2 < r_2$ *the solution* $w(\cdot, \tau)$ *with initial conditions* w_0 *satisfies*

$$\|w(\cdot, \tau) - \alpha\phi^0(\cdot)\|_2 \leqslant Ce^{-\mu\tau}$$

where $\alpha = \int_{\mathbb{R}^2} w_0(\xi)d\xi$.

By considering the invariant manifolds corresponding to other of the spectral subspaces, one can make other, more detailed statements about the asymptotics of small solutions. For instance, one thing that had been discovered about solutions of the Navier–Stokes equations was that certain relationships were required to hold between the moments of solutions decaying with particular temporal rates [14]. However, the proofs of these moment conditions provided little insight into the meaning or origin of these relationships. In [7] Gallay and I showed that these moment conditions were the consequence of the requirement that the solution lie on certain invariant manifolds in the phase space and as a consequence were able to give a simple geometrical interpretation of the results on optimal decay rates. Additional uses and consequences of these sorts of invariant manifold theorems are contained in [7].

We turn now from the consideration of small solutions to a study of more general sorts of solutions of the two-dimensional Navier–Stokes equation. The first thing we note is that the Oseen vortices are not limited in size. The family of solutions

$$\mathcal{O}^\alpha(\xi) = \alpha\phi^0(\xi)$$

is an exact, stationary solution of (5) for *all* values of α. Thus, we can extend the local center-manifold to a global manifold in this case. However, we cannot assume that the global center-manifold is locally attractive as is the case for the local center-manifold, so our next task is to analyze the local stability of Oseen vortices of large magnitude.

Begin, by linearizing (5) about the vortex $\mathcal{O}^\alpha$. This leads to the linearized equation

$$\frac{\partial w}{\partial \tau} = \mathcal{L}w - \alpha\Lambda w \tag{10}$$

where the linear operator $\mathcal{L}$ is the one we studied in Lecture 2 and the operator Λ is defined by:

$$\Lambda w = \mathbf{v}^0 \cdot \nabla w + \mathbf{v}^w \cdot \nabla \phi^0 \tag{11}$$

with $\mathbf{v}^0$ the velocity field associated to the vorticity ϕ^0 and $\mathbf{v}^w$ the velocity field associated with the vorticity w.

We now consider the spectrum of the operator $\mathcal{L} - \alpha\Lambda$. The first observation is a bit of basic functional analysis. Note that operator Λ is localized – i.e. the coefficient in each term of Λw decays as $|\xi| \to \infty$. Furthermore it is a first order differential operator while $\mathcal{L}$ is second order. These two facts taken together are sufficient to show that Λ is a relatively compact perturbation of $\mathcal{L}$ and hence the essential spectrum of $\mathcal{L}$ and $\mathcal{L} - \alpha\Lambda$ must coincide. Thus we have

Lemma 4.1. *Fix $m > 1$ and consider the operator $\mathcal{L} - \alpha\Lambda$ acting on its maximal domain in $L^2(m)$. Then*

$$\sigma_{ess}(\mathcal{L}) = \sigma_{ess}(\mathcal{L} - \alpha\Lambda) = \{\lambda \in \mathbb{C} | \mathrm{Re}(\lambda) \leqslant \frac{1-m}{2}\} \, .$$

Remark 4.3. More details on the proof of this lemma and succeeding results in this lecture can be found in [8].

As a consequence of Lemma 4.1 the stability or instability of the Oseen vortices of large norm will be determined by whether or not the isolated eigenvalues of $\mathcal{L} - \alpha\Lambda$ lie in the left or right half plane. One of these eigenvalues can be immediately and explicitly computed and we find:

Lemma 4.2. *The operator $\mathcal{L} - \alpha\Lambda$ has an eigenvalue $\lambda = 0$ with eigenfunction ϕ^0 for all values of α.*

Since the projection of a function f onto this eigenspace is just given by the product of ϕ^0 with the integral of f, the complementary subspace to the zero eigenspace consists of the functions of zero mean. Thus, we can restrict our attention of the space of functions $L_0^2(m) = \{f \mid \int_{\mathbb{R}^2} f(\xi)d\xi = 0\}$. When restricted to this space we have the following result:

Proposition 4.2. *Fix $m > 1$ and $\alpha \in \mathbb{R}$. Then any eigenvalue λ of $\mathcal{L} - \alpha\Lambda$ with eigenfunction in $L_0^2(m)$ satisfies*

$$\mathrm{Re}(\lambda) \leqslant \max\left(-\frac{1}{2}, \frac{1-m}{2}\right).$$

Remark 4.4. Note that this proposition, in combination with the above remark about the zero eigenvalue and the essential spectrum implies that the Oseen vortices are spectrally stable for all values of α. Given this spectral information it follows in a fairly straighforward fashion that the Oseen vortices are locally stable for all values of α – namely given an initial condition of (5) sufficiently close to an Oseen vortex the resulting solution of the vorticity equation will converge to an Oseen vortex as time tends toward infinity.

Remark 4.5. Because we have scaled all other physical parameters to have value one, α can be thought of as the Reynolds number for the problem. Thus, in contrast to many other fluid mechanical situations increasing the Reynolds number in this problem does not lead to instability. In fact, numerical computations [15] indicate that the real parts of most eigenvalues of $\mathcal{L} - \alpha\Lambda$ actually become more negative as α increases so that the increasing Reynolds number actually has a sort of stabilizing effect.

The proof of Proposition 4.2 consists of three steps:

1. By writing out the eigenvalue equation in polar coordinates a straightforward but complicated analysis shows that regardless of the value of α any eigenfunction in $L_0^2(m)$ whose real part is larger than $\frac{1-m}{2}$ must have Gaussian decay as $|\xi| \to \infty$. Thus the eigenfunctions are very strongly localized in space, regardless of the value of α. Given these results we define a new Hilbert space $X = \{w \in L^2(\mathbb{R}^2) \mid w/\sqrt{\phi^0} \in L^2(\mathbb{R}^2)\}$, equipped with the innerproduct

$$(w_1, w_2)_X = \int_{\mathbb{R}^2} \frac{\bar{w}_1(\xi) w_2(\xi)}{\phi^0(\xi)} d\xi \ .$$

We know that the eigenfunctions of $\mathscr{L} - \alpha\Lambda$ lie in X and (since we can continue to ignore the eigenvalue zero) we will study the spectrum on the space $X_0 = X \cap L_0^2(m)$.

2. We next compute the representation of $\mathscr{L}$ in the Hilbert space X which is given by

$$\mathscr{L}^X = (\phi^0)^{(-1/2)} \mathscr{L} (\phi^0)^{(1/2)} = \Delta_\xi - \frac{|\xi|^2}{16} + \frac{1}{2} \ .$$

This operator is the well-known quantum mechanical oscillator and as is well known in quantum mechanics:

(a) $\mathscr{L}^X$ is self-adjoint

(b) The spectrum of $\mathscr{L}^X$ consists only of the eigenvalues $-n/2, n = 0, 1, 2, \ldots$.

The second of these points is not surprising but the fact that $\mathscr{L}$ is self-adjoint in the Hilbert space X will be critical in what follows.

3. The final point is the computation of the representation of Λ in X_0. Writing out the expression for Λ in the X-inner product one finds:

$$(\tilde{w}, \Lambda w)_X = \int_{\mathbb{R}^2} \left(\frac{1}{\phi^0} \tilde{w} \mathbf{v}^0 \cdot \nabla w - \frac{1}{2} \tilde{w} (\mathbf{v} \cdot \xi) \right) d\xi, \tag{12}$$

where we used the fact that $\nabla \phi^0 = -\frac{\xi}{2} \phi^0$. Two easy calculations show that

$$\int_{\mathbb{R}^2} \left(\frac{1}{\phi^0} \tilde{w} \right) \mathbf{v}^0 \cdot \nabla w d\xi = - \int_{\mathbb{R}^2} \frac{1}{\phi^0} w \mathbf{v}^0 \cdot \nabla \tilde{w} d\xi. \tag{13}$$

and

$$\tilde{w}(\mathbf{v} \cdot \xi) + w(\tilde{\mathbf{v}} \cdot \xi) = (\xi_1 \partial_1 - \xi_2 \partial_2)(v_1 \tilde{v}_2 + v_2 \tilde{v}_1) + (\xi_1 \partial_2 + \xi_2 \partial_1)(v_2 \tilde{v}_2 - v_1 \tilde{v}_1). \tag{14}$$

Integrating both sides of the second equation in (13) we see that

$$\int_{\mathbb{R}^2} \tilde{w}(\mathbf{v} \cdot \xi) + w(\tilde{\mathbf{v}} \cdot \xi) d\xi = 0$$

which when combined with (12) and (13) imply that

$$(\tilde{w}, \Lambda w)_X = -(\Lambda \tilde{w}, w)_X \ ,$$

or

Lemma 4.3. *The linear operator Λ is skew-symmetric on X_0.*

Proposition 4.2 now follows from the following property from linear algebra. Namely, suppose that $\mathscr{L}$ is a self-adjoint operator on a Hilbert space X_0 whose spectrum lies in the half line $\lambda \leqslant -\mu < 0$. Then if Λ is skew-adjoint on X_0 any eigenvalue of $\mathscr{L} - \alpha \Lambda$ has real part less than equal or equal to μ. To see why this is so, suppose that

$$(\mathscr{L} - \alpha\Lambda)\phi = \lambda\phi \ .$$

Then

$$\lambda(\phi,\phi)_{X_0} = (\phi,\mathscr{L}\phi)_{X_0} - \alpha(\phi,\Lambda\phi)_{X_0}, \text{while} \tag{15}$$
$$\bar{\lambda}(\phi,\phi)_{X_0} = \overline{(\phi,\mathscr{L}\phi)_{X_0}} - \alpha\overline{(\phi,\Lambda\phi)_{X_0}} = (\mathscr{L}\phi,\phi)_{X_0} - \alpha(\Lambda\phi,\phi)_{X_0}$$
$$= (\phi,\mathscr{L}\phi)_{X_0} + \alpha(\phi,\Lambda\phi)_{X_0}$$

Adding these two expressions together yields

$$\mathrm{Re}(\lambda) = (\phi,\mathscr{L}\phi)_{X_0} \leqslant -\mu \ . \qquad \square$$

Reviewing the picture we have of solutions of the two-dimensional Navier–Stokes equation so far we see that we have a global center manifold, consisting of the family of Oseen vortices which are locally stable for all values of α. The final question that we consider is the behavior of solutions of (5) for arbitrary initial data (i.e. for initial vorticity distributions which are not close to one of the Oseen vortices.)

Given the results of Lecture 3 it is natural to require that the initial vorticity distribution be in $L^1(\mathbb{R}^2)$. We know that the solution with this initial condition exists for all time and thus we can ask what its ω-limit set is. From the first lecture we know that in order to be sure that the ω-limit set exists we need to check whether the trajectory remains in a compact subset of $L^1(\mathbb{R}^2)$. The details needed to establish this fact are presented in [8] but we note two main ideas are that by Rellich's criterion subspaces of $L^1(\mathbb{R}^2)$ that have some smoothness and decay at infinity are compact. In our problem:

- Smoothness comes from the smoothing properties of the semigroup which are preserved by the nonlinearity
- Decay at infinity comes from estimates of the solution of the vorticity equation due to Carlen and Loss [3]

Given that the ω-limit set exists how can we calculate it? We determine the ω-limit set with the aid of two Lypunov functions:

1. The first tells us that the ω-limit set consists of functions that do not change sign – i.e. an element of the ω-limit set of a solution with initial value w_0 is either everywhere non-positive or everywhere non-negative.
2. The second will identify those positive (or negative) functions that can be part of the ω-limit set.

Lyapunov Function No. 1: This Lyapunov function is closely related to Example 1.1 from Lecture 1. Define

$$\Phi(w(\tau)) = \int_{\mathbb{R}^2} |w(\xi,\tau)| d\xi \ .$$

One then has:

Lemma 4.4. *Let $w_0 \in L^1(\mathbb{R}^2)$ and let w be the solution of the rescaled vorticity equation with this initial condition. Then $\Phi(w(\tau)) \leqslant \Phi(w_0)$ for all $\tau \geqslant 0$. Moreover, equality holds if and only if $w_0 \in \Sigma$ where*

$$\Sigma = \left\{ w \in L^1(\mathbb{R}^2) \Big| \int_{\mathbb{R}^2} |w(\xi)| d\xi = \Big| \int_{\mathbb{R}^2} w(\xi) d\xi \Big| \right\} .$$

Proof. This lemma follows from the maximum principle very much along the lines of Example 1.1. Indeed that example established this result for the linear terms in (5). Including the nonlinear terms in the equation causes no essential difficulty and we leave the details of this argument as an exercise for the reader. $\quad\square$

Note that as a corollary of this lemma and the LaSalle Invariance Principle we have

Corollary 4.1. *Let $w_0 \in L^1(\mathbb{R}^2)$ The ω-limit set of the solution with this initial condition must lie in Σ.*

Lyapunov Function No. 2: Since from the preceeding corollary the ω-limit set is contained in set of positive (or negative) functions our second Lyapunov function will be defined only on such functions. This second Lyapunov function is motivated by Lyapunov functions used in kinetic theory where one also wants to prove the convergence of solutions toward Gaussian profiles and is known in that field as the *relative entropy* function. Define $\Sigma_+ = \{w \in \Sigma | w(\xi) \geqslant 0 \text{ almost everywhere}\}$ and define $H : \Sigma_+ \cap L^2(m) \to \mathbb{R}$ by

$$H(w(\tau)) = \int_{\mathbb{R}^2} w(\xi, \tau) \log \left(\frac{w(\xi, \tau)}{\phi^0(\xi)} \right) d\xi .$$

If $m > 3$ the functions w decay fast enough at infinity that one can show:

1. H is defined and continuous on $\Sigma_+ \cap L^2(m)$
2. H is bounded below by $-1/e$

Even more importantly for our purposes, H is decreasing along trajectories and hence a Lyapunov function. Assume for the moment that w is smooth enough that we can differentiate $H(w(\tau))$ by pulling the derivative through the integral sign. (The general case can be handled by approximation by smooth functions.) Then

$$\frac{d}{d\tau} H(w(\tau)) = \int_{\mathbb{R}^2} \left(1 + \log \frac{w}{\phi^0} \right) \partial_\tau w d\xi = \int_{\mathbb{R}^2} \left(1 + \log \frac{w}{\phi^0} \right) (\mathscr{L}w - \mathbf{v} \cdot \nabla w) d\xi .$$

$$(16)$$

We break this last integral into two pieces and consider each piece separately. First note that thanks to the special properties of the Gaussian

$$\mathscr{L}w = \operatorname{div} \left(\phi^0 \nabla \left(\frac{w}{\phi^0} \right) \right)$$

so that

$$\int_{\mathbb{R}^2} \left(1 + \log \frac{w}{\phi^0}\right)(\mathscr{L}w)d\xi = -\int_{\mathbb{R}^2} \phi^0 \left(\nabla(\log \frac{w}{\phi^0})\right) \cdot \nabla(\frac{w}{\phi^0})$$

$$= -\int_{\mathbb{R}^2} w \left|\nabla(\log \frac{w}{\phi^0})\right|^2 d\xi$$

To treat the second term in (16) we first integrate by parts to obtain

$$-\int_{\mathbb{R}^2} \left(1 + \log \frac{w}{\phi^0}\right)(\mathbf{v} \cdot \nabla w) = -\int_{\mathbb{R}^2} \left(1 + \log \frac{w}{\phi^0}\right)(\nabla \cdot (\mathbf{v}w))d\xi \qquad (17)$$

$$= \int_{\mathbb{R}^2} \phi^0 \mathbf{v} \cdot \nabla(\frac{w}{\phi^0})d\xi = \int_{\mathbb{R}^2} \mathbf{v} \cdot \nabla w \, d\xi - \frac{1}{2}\int_{\mathbb{R}^2} (\xi \cdot \mathbf{v})w \, d\xi \ .$$

We claim finally that each of these two last integrals vanish. For the first, this is obvious since $\mathbf{v} \cdot \nabla w = \nabla \cdot (\mathbf{v}w)$. For the second note that $w = \partial_{\xi_1} v_2 - \partial_{\xi_2} v_1$ (where $\mathbf{v} = (v_1, v_2)$) and hence

$$\int_{\mathbb{R}^2} (\xi \cdot \mathbf{v})w \, d\xi = \int_{\mathbb{R}^2} (\xi_1 v_1 + \xi_2 v_2)(\partial_{\xi_1} v_2 - \partial_{\xi_2} v_1)d\xi$$

$$= \int_{\mathbb{R}^2} \xi_1 v_1 \partial_{\xi_1} v_2 d\xi + \int_{\mathbb{R}^2} \xi_2 v_2 \partial_{\xi_1} v_2 d\xi - \int_{\mathbb{R}^2} \xi_1 v_1 \partial_{\xi_2} v_1 d\xi - \int_{\mathbb{R}^2} \xi_2 v_2 \partial_{\xi_2} v_1 d\xi \ .$$

Note that the second and third of these integrals vanish since the second can be rewritten, for example as $\frac{1}{2}\int_{\mathbb{R}^2} \partial_{\xi_1}(\xi_2(v_2)^2)d\xi = 0$ and analogously for the third. In the first and fourth integrals we integrate by parts to obtain

$$-\int_{\mathbb{R}^2} v_1 v_2 d\xi - \int_{\mathbb{R}^2} \xi_1(\partial_{\xi_1} v_1)v_2 d\xi + \int_{\mathbb{R}^2} v_1 v_2 d\xi + \int_{\mathbb{R}^2} \xi_2(\partial_{\xi_2} v_2)v_1 d\xi$$

$$= \int_{\mathbb{R}^2} \xi_1(\partial_{\xi_2} v_2)v_2 d\xi - \int_{\mathbb{R}^2} \xi_2(\partial_{\xi_1} v_1)v_1 d\xi = 0$$

where the next to last equality used the fact that $\mathbf{v}$ is incompressible and the final equality noted that the first integral could be written as $\frac{1}{2}\int_{\mathbb{R}^2} \partial_{\xi_2} \left(\xi_1(v_2)^2\right) d\xi = 0$ and similarly for the second.

Remark 4.6. In fact, one needs to take a little more care with this calculation since for general velocity fields $\mathbf{v}$, integrals like $\int_{\mathbb{R}^2} \xi_2 v_2 \partial_{\xi_1} v_2 d\xi$ may fail to converge. Nonetheless, the entire expression $\int_{\mathbb{R}^2}(\xi \cdot \mathbf{v})w \, d\xi$ is convergent because of cancellations between various terms. The easiest way to take advantage of these cancellations is to rewrite the velocity in terms of the vorticity via the Biot–Savart law and then argue that the integral must vanish by symmetry. (See [8] for details.) However, I think that the present argument with works entirely with the velocity field gives somewhat more intuition into why these terms vanish.

Putting these computations together we see that we have shown:

Lemma 4.5.

$$\frac{d}{d\tau}H(w(\tau)) = -\int_{\mathbb{R}^2} w \left|\nabla(\log\frac{w}{\phi^0})\right|^2 d\xi$$

This lemma implies that the Lyapunov function H is strictly decreasing unless w is a multiple of the Gaussian ϕ^0 and implies, as a immediate corollary:

Corollary 4.2. *Assume that $w_0 \in L^2(m) \cap \Sigma$ with $m > 3$. The $H(w(\tau) \leqslant H(w_0)$ for all points $w(\tau)$ in the forward orbit of w_0 and $H(w(\tau)) = H(w_0)$ for all $\tau \geqslant 0$ if and only if $w_0 = \alpha G$ for some $\alpha \geqslant 0$.*

We can now put together the various pieces of this argument to derive a quite complete picture of the long-time asymptotic behavior of solutions of the two-dimensional Navier–Stokes equations. Suppose we consider any solution of (5) whose initial vorticity $w_0 \in L^2(m)$ with $m > 3$. By Lemma 4.4 we know that any point w^* in the omega limit set of w_0 must lie in the set Σ of functions which do not change sign. Assume, without loss of generality, that $w^*(\xi) > 0$.

From the general theory of Lyapunov functionals we know that the solution of the vorticity equation with initial conditions w^* exists for all time $t \in \mathbb{R}$. Combining this observation with Corollary 4.2 implies that the orbit of w^* consists of the single point $\alpha_0 G$ where $\alpha_0 = \int w_0(\xi)d\xi$ and hence that the omega-limit set of any point $w_0 \in L^2(m)$ with $m > 3$ consists of the Oseen vortex with the same total vorticity.

In fact, using results of Carlen and Loss [3] on the spatial decay rate of solutions of the two-dimensional vorticity equation one can prove that any point in the omega-limit set of a solution whose initial vorticity is in $L^1(\mathbb{R}^2)$ must lie in $L^2(m)$ for all $m > 1$ – in particular it must lie in $L^2(m)$ for some $m > 3$. Then, proceeding as above, we find that the omega-limit set must again consist just of an Oseen vortex. If we now undo the change of variables (2) and (3) we see that solutions $\omega(x,t)$ satisfy:

Theorem 4.1. *If $\omega_0 \in L^1(\mathbb{R}^2)$, the solution $\omega(x,t)$ of (1) satisfies*

$$\lim_{t\to\infty} t^{1-\frac{1}{p}} \left|\omega(\cdot,t) - \frac{\alpha}{t}G(\frac{\cdot}{\sqrt{t}})\right|_p = 0, \text{for} 1 \leqslant p \leqslant \infty, \tag{18}$$

where $\alpha = \int_{\mathbb{R}^2} \omega_0(x)\,dx$. If $\mathbf{u}(x,t)$ is the solution of the two-dimensional Navier–Stokes equation obtained from $\omega(x,t)$ via the Biot–Savart law, then

$$\lim_{t\to\infty} t^{\frac{1}{2}-\frac{1}{q}} \left|\mathbf{u}(\cdot,t) - \frac{\alpha}{\sqrt{t}}\mathbf{v}^G(\frac{\cdot}{\sqrt{t}})\right|_q = 0 \text{, for} 2 < q \leqslant \infty. \tag{19}$$

where $\mathbf{v}^G$ is the velocity field (8) associated to the Oseen vortex.

5 Conclusions

Summing up, we see that the dynamical systems method provides a quite complete view of the long-time asymptotics of general solutions of the two-dimensional Navier–Stokes or vorticity equations.

While one cannot hope to obtain comparably complete information about solutions of the three-dimensional Navier–Stokes equation where even the existence of solutions with general initial data is unproven it turns out that one can use the ideas developed above to understand the existence and stability of some classes of vortex solutions related to the Burgers vortices, an explicit family of solutions of the three-dimensional Navier–Stokes equations believed to be important for turbulent flows [9, 10].

Another interesting an open question is to understand the intermediate time behavior of solutions of the two-dimensional Navier–Stokes equation. While the results proven above imply that eventually one converges to a single vortex solution, numerical simulations imply that the evolution at intermediate time scales is dominated by the interaction and merger of pairs of vortices. A better understanding of this merger process would be very intereresting and also have important applications.

Acknowledgements The work of the author is supported in part by the NSF under grant number DMS-0405724.

References

1. Peter W. Bates and Christopher K. R. T. Jones. Invariant manifolds for semilinear partial differential equations. In *Dynamics reported, Vol. 2, Dynam. Report. Ser. Dynam. Systems Appl.*, pages 1–38. Wiley, Chichester, 1989.
2. Matania Ben-Artzi. Global solutions of two-dimensional Navier-Stokes and Euler equations. *Arch. Rational Mech. Anal.*, 128(4):329–358, 1994.
3. Eric A. Carlen and Michael Loss. Optimal smoothing and decay estimates for viscously damped conservation laws, with applications to the 2-D Navier-Stokes equation. *Duke Math. J.*, 81(1):135–157 (1996), 1995. A celebration of John F. Nash, Jr.
4. Xu-Yan Chen, Jack K. Hale, and Bin Tan. Invariant foliations for C^1 semigroups in Banach spaces. *J. Diff. Equations*, 139(2):283–318, 1997.
5. Charles R. Doering and J. D. Gibbon. *Applied analysis of the Navier-Stokes equations*. Cambridge Texts in Applied Mathematics. Cambridge University Press, Cambridge, 1995.
6. Isabelle Gallagher and Thierry Gallay. Équations de la mechanique des fluides. http://www-fourier.ujf-grenoble.fr/ECOLETE/ecole2005/Note-Gallagher.pdf, 2006.
7. Thierry Gallay and C. Eugene Wayne. Invariant manifolds and the long-time asymptotics of the Navier-Stokes and vorticity equations on $\mathbf{R}^2$. *Arch. Ration. Mech. Anal.*, 163(3):209–258, 2002.
8. Thierry Gallay and C. Eugene Wayne. Global stability of vortex solutions of the two-dimensional Navier-Stokes equation. *Comm. Math. Phys.*, 255(1):97–129, 2005.
9. Thierry Gallay and C. Eugene Wayne. Three-dimensional stability of Burgers vortices: the low Reynolds number case. *Phys. D*, 213(2):164–180, 2006.
10. Thierry Gallay and C. Eugene Wayne. Existence and stability of asymmetric Burgers vortices. *J. Math. Fluid Mech.*, 9(2):243–261, 2007.
11. Daniel Henry. *Geometric theory of semilinear parabolic equations*, volume 840 of *Lecture Notes in Mathematics*. Springer, Berlin, 1981.
12. J. P. LaSalle. *The stability of dynamical systems*. Society for Industrial and Applied Mathematics, Philadelphia, Pa., 1976. With an appendix: "Limiting equations and stability of nonautonomous ordinary differential equations" by Z. Artstein, Regional Conference Series in Applied Mathematics.

13. Alexander Mielke. Locally invariant manifolds for quasilinear parabolic equations. *Rocky Mountain J. Math.*, 21(2):707–714, 1991. Current directions in nonlinear partial differential equations (Provo, UT, 1987).

14. T. Miyakawa and M. Schonbek. On optimal decay rates for weak solutions of the navier-stokes equations in $\mathbb{R}^n$. In *Proceedings of Partial Differential Equations and Applications (Olomouc, 1999)*, vol. 126, pages 443–455. 2001.

15. A. Prochazka and D. I. Pullin. On the two-dimensional stability of the axisymmetric Burgers vortex. *Phys. Fluids*, 7(7):1788–1790, 1995.

16. A. Vanderbauwhede and G. Iooss. Center manifold theory in infinite dimensions. In *Dynamics reported: expositions in dynamical systems*, volume 1 of *Dynam. Report. Expositions Dynam. Systems (N.S.)*, pages 125–163. Springer, Berlin, 1992.

Hamiltonian systems and optimal control

Andrei Agrachev[1]

Abstract Solutions of any optimal control problem are described by trajectories of a Hamiltonian system. The system is intrinsically associated to the problem by a procedure that is a geometric elaboration of the Lagrange multipliers rule. The intimate relation of the optimal control and Hamiltonian dynamics is fruitful for both domains; among other things, it leads to a clarification and a far going generalization of important classical results about Riemannian geodesic flows.

1 Introduction

These are notes of the lectures given in June 25–28, 2007 for the school "Hamiltonian Dynamical Systems and Applications" in Montreal which were written up by Natalia Shcherbakova.

Hamiltonian systems play a very important role in the theory of optimal control since the foundation of the this subject in the middle of the twentieth century. Indeed, the first fundamental result of the theory, the Pontryagin maximum principle, is formulated in Hamiltonian form. As I learned from Ponryagin's collaborators, it was the central role played by the Hamiltonian system that convinced Pontryagin of the importance and universality of his optimality condition; see the pioneering book [5] for the original approach and the books [2, 3, 4] for some of the further developments.

The Pontryagin maximum principle is a natural Hamiltonian form for the first order optimality conditions. In these lectures, we explain the Hamiltonian nature of the second order information on the local structure of the optimal control problem which leads, among other things, to curvature-type invariants of Hamiltonian systems on cotangent bundles. These invariants control Hamiltonian dynamics in a way analogous to the way Riemannian sectional curvature enters into geodesic flows.

---

[1] SISSA, Trieste and MIAN, Moscow
e-mail: agrachev@sissa.it

W. Craig (ed.), Hamiltonian Dynamical Systems and Applications, 143–156.

A more detailed and formal exposition of the constructions and facts presented here can be found in [1].

2 First lecture

Consider two smooth[1] manifolds: M, $\dim M = n$, a *state space* and U, $\dim U = m \leqslant n$, a space of *control parameters*.

A *control system* is a family of ordinary differential equations on M:

$$\dot{q} = f(q,u), \quad q \in M, \ u \in U \ . \tag{1}$$

The vector field f is assumed to be smooth with respect to both variables. For $q \in M$ fixed, the set $f(q,U)$ is called the set of admissible velocities.

Any L_∞-curve $t \mapsto u(t)$, $t \in [0,t_1]$, $u(t) \in U$ is called a control function. Here t_1 is fixed. Substituting the control function $u(t)$ into (1) we get a non-autonomous differential equation

$$\dot{q} = f(q,u(t)).$$

A solution $t \mapsto q(t) \in M$ of this equation is called a trajectory of (1) associated to the control function $u(t)$.

Consider

$$\mathscr{W} = \{(u(\cdot),q(\cdot)) : u(\cdot) \in L_\infty([0,t_1],U), \ \dot{q}(t) = f(q(t),u(t))\},$$

the space of *admissible pairs*.

Remark. $\mathscr{W}$ is a Banach manifold modelled on $\mathbb{R}^n \times L_\infty^m([0,t_1])$. This fact is a direct consequence of the standard theorem on the existence, uniqueness and smooth dependence of the data for solutions to the Cauchy problem of systems of ordinary differential equations.

Define

$$J^{t_1}(u(\cdot),q(\cdot)) = \int_0^{t_1} \varphi(q(t),u(t))\,dt,$$

where φ is a smooth scalar function. J^{t_1} is called the cost functional.

Optimal control problem: Given $q_0,q_1 \in M$ minimize J^{t_1} over admissible pairs $(u(\cdot),q(\cdot))$ such that $q(0) = q_0$, $q(t_1) = q_1$.

This problem generalizes the standard problem of the Calculus of Variations, namely minimize a functional

$$\int_0^{t_1} \varphi(q(t),\dot{q}(t))\,dt \longrightarrow \min$$

---

[1] In these lecture notes smooth objects are C^∞ unless otherwise stated.

over all curves such that $q(0) = q_0$, $q(t_1) = q_1$. This problem can be stated as an optimal control problem by setting $\dot{q} = u$, $u \in T_q M$.

Geometrically speaking, we state the problem as follows; consider a locally trivial bundle $V \to M$ over M whose typical fiber is U: $V = \bigcup_{q \in M} V_q$, $V_q \cong U$. Then $f : V \to TM$ such that $f(V_q) \subset T_q M$. An admissible pair is a curve $t \to v(t)$ in V such that $v(t) \in V_{q(t)}$ and $\dot{q}(t) = f(v(t))$.

Given an admissible pair, we can trivialize V along the trajectory.

In order to describe extremals of an optimal control problem we will use the geometrically elaborated Lagrange multipliers method. This method provides first order optimality conditions. In forthcoming lectures we will also discuss second order conditions and related invariants.

2.1 First order conditions

Optimal control problem is a kind of a constrained optimization problem, where constrains are given by the boundary point conditions. We set $F_t : \mathscr{W} \to M$ such that $F_t(u(\cdot 0, q(\cdot)) = q(t)$. The map F_t has the same smoothness as the field f. Our minimization problem is

$$\min J^{t_1}\Big|_{F_0 = q_0, F_{t_1} = q_1}. \tag{2}$$

We are looking for solutions of the problem among solutions of the equation

$$dJ^{t_1}\Big|_{\ker DF_0 \cap \ker DF_{t_1}} = 0$$

which is equivalent to the equation

$$dJ^{t_1} = \lambda_{t_1} DF_{t_1} - \lambda_0 DF_0, \tag{3}$$

where $\lambda_0 : T_{q_0} M \to \mathbb{R}$ is a linear form on $T_{q_0} M$ and λ_{t_1} is a linear form on $T_{q_1} M$, i.e. $\lambda_0 \in T_{q_0}^* M$, $\lambda_{t_1} \in T_{q_1}^* M$. Covectors λ_0, λ_{t_1} are nothing else but the Lagrange multipliers. The sign "$-$" in front of λ_0 is chosen for convenience at a later step in the development.

We have $DF_t : T_{(u,q)} \mathscr{W} \to T_{q(t)} M$ and

$$\lambda_t DF_t : T_{(u,q)} \mathscr{W} \xrightarrow{DF_t} T_{q(t)} M \xrightarrow{\lambda_t} \mathbb{R}.$$

Proposition 1 *Equation (3) implies that there exists a unique Lipschitz curve $\lambda_t \in T_{q(t)}^* M$, $0 \leqslant t \leqslant t_1$ such that*

$$\lambda_t DF_t - \lambda_0 DF_0 = dJ^t,$$

where $J^t = \int_0^t \varphi(q(\tau), u(\tau)) \, d\tau$.

Proof. The uniqueness follows from the fact that F_t is a submersion. Let us prove the existence. Denote by $(\tilde{u}(\cdot), q(\cdot))$ the reference pair. By (3),

$$dJ^{t_1}_{(\tilde{u},q)} = \lambda_{t_1} DF_{t_1} - \lambda_0 DF_0.$$

Let us fix some $t \in [0, t_1]$. We restrict ourselves to the admissible pairs of the form

$$\mathcal{V}_t = \{(u(\cdot), q(\cdot)) : u(\tau) = \tilde{u}(\tau) \text{ for } \tau \geqslant t\}.$$

Then $F_t(u,q) = \Phi^{t_1}_t \circ F_t$, where $\Phi^{t_1}_t$ is a fixed diffeomorphism. The family $\tau \mapsto \Phi^{\tau}_t$ satisfies $\Phi^{t}_t(q) = q$,

$$\frac{\partial}{\partial \tau} \Phi^{\tau}_t(q) = f\left(\Phi^{\tau}_t(q), \tilde{u}(\tau)\right),$$

i.e. $\Phi^{\tau}_t : q(t) \mapsto q(\tau)$. Restricting to $\mathcal{V}_t$ we obtain at $v \in \mathcal{V}_t$

$$\lambda_{t_1} D_v(\Phi^{t_1}_t \circ F_t) - \lambda_0 D_v F_0 = d_v J^t + d_v(a_t \circ F_t),$$

where $a_t = \int\limits_t^{t_1} \varphi\left(\Phi^s_t(q(t)), \tilde{u}(s)\right) ds$. Hence

$$\lambda_t D_v F_t - \lambda_0 D_v F_0 = d_v J^t,$$

where $\lambda_t = \lambda_{t_1} D_v \Phi^{t_1}_t - d_{q(t)} a_t$.

3 Second lecture

Let us consider the Lipschitz curve $\lambda_t \in T^*_{q(t)} M$ whose existence was proved in Lecture 1. This curve satisfies the following relation;

$$\lambda_t D_{(\tilde{u},q)} F_t - \lambda_0 D_{(\tilde{u},q)} F_0 = d_{(\tilde{u},q)} J^t. \tag{1}$$

Now we are going to derive a differential equation for λ_t. First of all, we can introduce local coordinates in the neighborhood of given $q(t) \in M$; then $\lambda = (p,q)$, $p \in \mathbb{R}^{n*}$, $q \in \mathbb{R}^n$ for any q from the coordinate neighborhood and any λ from the cotangent bundle to this neighborhood. In particular, $\lambda_t = (p_t, q(t))$. Then (1) becomes

$$d_{(\tilde{u},q)}(p_t F_t) - \lambda_0 D_{(\tilde{u},q)} F_0 = d_{(\tilde{u},q)} J^t, \tag{2}$$

where $d_{(\tilde{u},q)}$ is the differential in the Banach space of admissible pairs. After taking derivatives with respect to t we get

$$0 = d_{(\tilde{u},q)}\left(\frac{\partial}{\partial t}(p_t F_t) - \varphi(q(t), \tilde{u}(t))\right)$$
$$= d_{(\tilde{u},q)}\left(\frac{dp}{dt} q(t) + p_t f(q(t), \tilde{u}(t))\right) - \varphi(q(t), \tilde{u}(t)).$$

Let us denote

$$h(p,q,u) = pf(q,u) - \varphi(q,u).$$

The function h is called the Hamiltonian of the optimal control problem. We have

$$d_{(\tilde{u},q)}\left(\dot{p}_t q(t) + h(p_t, q(t), \tilde{u}(t))\right) = 0.$$

Hence we obtain

$$\dot{p}_t + \frac{\partial h}{\partial q} = 0 \quad \frac{\partial h}{\partial u} = 0.$$

Recall now that our original dynamics is given by $\dot{q} = f(q,u) = \frac{\partial h}{\partial p}$. So, we obtain a Hamiltonian system

$$\begin{cases} \dot{p} = -\dfrac{\partial h}{\partial q}(p,q,u) \\[2mm] \dot{q} = \dfrac{\partial h}{\partial p}(p,q,u) \end{cases} \qquad (HS)_u$$

plus one extra equation

$$\frac{\partial h}{\partial u}(p,q,u) = 0.$$

In a regular situation, one can find $u = u(p,q)$ from the last equation, with which we denote;

$$H(p,q) = h(p,q,u(p,q)) \,,$$

and for remaining variables we obtain a standard Hamiltonian system

$$\begin{cases} \dot{p} = -\dfrac{\partial H}{\partial q}(p,q) \\[2mm] \dot{q} = \dfrac{\partial H}{\partial p}(p,q). \end{cases} \qquad (HS)$$

From now on, we will always assume that u can be eliminated from the equation $(HS)_u$.

3.1 Second variation

Let $t \in [0,t_1]$, we are going to study the second derivative of J^t under the constraints

$$\dot{q} = f(q,u), \quad q(0) = q_0, \ q(t) = q_t.$$

The correctness of the Cauchy problem for ordinary differential equations allows us to immediately resolve one of the boundary constraints. From now on, we restrict ourselves to $F_t^{-1}(q_t)$ and slightly change notation. In our new notation,

$$J^t := J^t\big|_{F_t^{-1}(q(t))}, \quad F_0 := F_0\big|_{F_t^{-1}(q_t)}.$$

Since the trajectory $q(\cdot)$ is uniquely defined by the corresponding control $u(\cdot)$, the space of admissible pairs is now reduced to the space of control functions $\{u(\cdot)\}$.

Equation (1) now reads

$$\lambda_0 D_{\tilde{u}} F_0 + d_{\tilde{u}} J^l = 0. \tag{1$'$}$$

The Hessian of $J^l \big|_{F_0^{-1}(q_0)}$ is

$$Hess_{\tilde{u}} \left(J^l \big|_{F_0^{-1}(q_0)} \right) = \left(D_{\tilde{u}}^2 J^l + \lambda_0 D^2 F_0 \right) \big|_{\ker D_{\tilde{u}} F_0}.$$

The Hessian is a quadratic form; we are interested in the Morse index of this form that is the supremum of the dimensions of subspaces where the form is negative definite. In particular, the Morse index equals 0 if and only if the form is nonnegative. We use notation $\mathrm{ind} Q$ for the Morse index of a quadratic form Q; the value of this index is a nonnegative integer or $+\infty$.

Let us express everything in terms of the Lagrange multipliers. After the elimination of one of the boundary constraints, equation (2) reads:

$$\begin{cases} p \dfrac{\partial F_0}{\partial u} + \dfrac{\partial J^l}{\partial u} = 0 \\ q = F_0(u). \end{cases}$$

Linearizing this equation we obtain

$$\xi \frac{\partial F_0}{\partial u} + \left(p \frac{\partial^2 F_0}{\partial u^2} + \frac{\partial^2 J^l}{\partial u^2} \right) v = 0 \qquad \eta = \frac{\partial F_0}{\partial u} v. \tag{LS}$$

Here $(\xi, \eta, v) = (\delta p, \delta q, \delta u)$. Denote

$$\Lambda_t = \{ (\xi, \eta) : \exists v \text{ satisfying } (LS) \} ,$$

the projection of the space of solutions of (LS) into the (p, q)-space.

We made the above computations in coordinates, but the construction of Λ_t (linearization and projection) is intrinsic, and Λ_t is actually a well-defined subspace of $T_{\lambda_0}(T^*M)$. Recall that T^*M is a symplectic manifold endowed with the standard symplectic structure $dp \wedge dq$. Hence $T_{\lambda_0}(T^*M)$ is a symplectic space. It is not hard to check that Λ_t is a Lagrangian subspace of $T_{\lambda_0}(T^*M)$.

The family $t \mapsto \Lambda_t$, $t \in [0, t_1]$ is called the *Jacobi curve*.

We will use Λ_t to calculate the Morse index of the second variation of J^l under constraints. Indeed, system (LS) is of the type

$$\xi A + Bv = 0 , \qquad \eta = Av, \tag{3}$$

where

$$A : V \to \mathbb{R}^n, \quad B = p \frac{\partial^2 F_0}{\partial u^2} + \frac{\partial^2 J^l}{\partial u^2},$$

so that $B : V \to V^*$ and $B^* = B$. Then

$$Hess\left(J^t\big|_{F_0^{-1}(q_0)}\right) : v \mapsto \langle Bv, v \rangle, \quad v \in \ker A. \tag{4}$$

Exercise. Assume that the linear mapping A is surjective and B is non-degenerate. Then the quadratic form (4) is non-degenerate if and only if

$$\Lambda \bigcap \{(\xi, 0) : \xi \in \mathbb{R}^{n*}\} = 0 ,$$

where $\Lambda = \{(\xi, \eta) : \exists v \text{ satisfying}(3)\}$.

Let us follow the evolution of the Morse index of the quadratic form $Hess\left(J^t\big|_{F_0^{-1}(q_0)}\right)$, when t runs from 0 to t_1. The Morse index changes for those t for which Λ_t has a nonzero intersection with $\Lambda_0 = \{(\xi, 0) : \xi \in \mathbb{R}^{n*}\}$. Moreover, the Morse index of $Hess\left(J^t\big|_{F_0^{-1}(q_0)}\right)$ grows monotonically with t simply due to the fact that the past does not depend on the future for our control system.

Definition 3.1. We say that t is conjugate to 0 if $\Lambda_t \cap \Lambda_0 \neq 0$.

Theorem 3.1. *If conjugate times are isolated, then*

$$\mathrm{ind}Hess\left(J^t\big|_{F_0^{-1}(q_0)}\right) - Hess\left(J^{t_0}\big|_{F_0^{-1}(q_0)}\right) = \sum_{t_0 \leqslant t < t_1} \dim(\Lambda_t \cap \Lambda_0)$$

for any $t_0 \in (0, t_1)$.

4 Third lecture

Let us consider the trajectories of the Hamiltonian system corresponding to the Hamiltonian

$$H : T^*M \to \mathbb{R}.$$

We denote by $\mathbf{H}$ the associated Hamiltonian vector field and by $e^{t\mathbf{H}}$ its flow on T^*M. Recall that the trajectories of $\mathbf{H}$ describe the extremals of our optimal control problem together with the associated Lagrange multipliers.

In Lecture 2, we saw that second variation of the cost functional J^t under constraints is related to a family of the Lagrangian subspaces Λ_t, $t \in [0, t_1]$, where Λ_t is the tangent space at λ_0 of the submanifold of T^*M formed by the values at time 0 of the solutions to the Hamiltonian system $\dot{\lambda} = \mathbf{H}(\lambda)$ whose values at time t belong to $T^*_{q(t)}M$.

In other words, we consider the Jacobi curve

$$\Lambda_t = T_{\lambda_0}\left(e^{-t\mathbf{H}}(T^*_{q(t)}M)\right).$$

Our aim is to obtain some information on the conjugate points and on the long-time behavior of the flow without solving differential equations. To do that, we introduce a kind of curvature-type invariants associated to the problem.

4.1 Curves in the Lagrange Grassmannians

Consider a $2n$-dimensional symplectic space Σ. Denote by σ the symplectic form. To any pair of transversal Lagrangian subspaces Λ_0, Δ, we can associate coordinates on Σ such that

$$\Sigma = \Lambda_0 \oplus \Delta = \{(p,q) : p,q \in \mathbb{R}^n\}, \quad \sigma = \sum_{i=1}^{n} dp_i \wedge dq_i$$

and

$$\Lambda_0 = \{(p,0) : p \in \mathbb{R}^n\}, \quad \Delta = \{(0,q) : q \in \mathbb{R}^n\}.$$

Any transversal to Δ n-dimensional subspace Λ has a form

$$\Lambda = \{(p, S_\Lambda p) : p \in \mathbb{R}^n\} ,$$

for some $n \times n$-matrix S_Λ. The subspace Λ is Lagrangian (i. e. $\sigma|_\Lambda = 0$) if and only if $S_\Lambda = S_\Lambda^*$, in other words, if S_Λ belongs to the space $\mathrm{Sym}(n)$ of symmetric $n \times n$-matrices. In particular, it follows that the map $\Lambda \mapsto S_\Lambda$ gives local coordinates on the *Lagrange Grassmannian* $L(\Sigma)$ of all Lagrangian subspaces of Σ.

Now consider a curve $t \mapsto \Lambda_t$ in $L(\Sigma)$ and the corresponding curve $t \mapsto S_t = S_{\Lambda_t}$ in $\mathrm{Sym}(n)$.

Lemma 4.1. *The quadratic form $p \mapsto \langle \dot{S}_t p, p \rangle$ is an intrinsically defined quadratic form on the subspace Λ_t.*

Proof. Pick $\lambda_t \in \Lambda_t$ and insert it in some curve $\tau \mapsto \lambda_\tau \in \Lambda_\tau$. Then $\lambda_\tau = (p_\tau, S_\tau p_\tau)$ and we have:

$$\sigma(\lambda_t, \dot{\lambda}_t) = \sigma\left((p_t, S_t p_t), (\dot{p}_t, S_t \dot{p}_t) + (0, \dot{S}_t p_t)\right) = \langle \dot{S} p_t, p_t \rangle.$$

We see that $\sigma(\lambda_t, \dot{\lambda}_t)$ depends only on λ_t and $\dot{\lambda}_t$ and not on the choice of the curve $\tau \mapsto \lambda_\tau$. Hence $\lambda_t \mapsto \sigma(\lambda_t, \dot{\lambda}_t)$ is a well-defined quadratic form on Λ_t presented by the matrix $\dot{S}_t$. $\square$

Corollary 4.1. *The tangent space $T_\Lambda L(\Sigma)$ to the Lagrange Grassmannian is intrinsically identified with the space of quadratic forms on Λ.*

Definition 4.1. We say that the curve $t \mapsto \Lambda_t$ is monotone if the quadratic forms $p \mapsto \dot{S}_t$ are sign-definite.

4.2 *Curvature-type invariants*

Let us come back to the Jacobi curves in $T_z(T^*M)$, $z \in T^*M$. Recall that $T_z(T^*M) = \Sigma_z$ is a symplectic space, so we are in the situation described above.

We want to consider the curves $t \mapsto \Lambda_z(t)$ such that $\Lambda_z(0) = T_z(T_q^*M)$, where $q = \pi(z)$ and $\pi : T^*M \to M$ is the standard projection. As before, choosing appropriate Darboux coordinates in $T_z(T^*M)$ we have:

$$\Lambda_t = \{(p, S_t p) : p \in \mathbb{R}^n\}, \quad S_0 = 0.$$

By a direct computation one verifies that

$$\dot{S}_0 = -\left.\frac{\partial^2 H}{\partial p^2}\right|_z.$$

From now on, we will deal with Hamiltonians that are convex on fibers, i.e. we will assume that

$$\frac{\partial^2 H(p,q)}{\partial p^2} \geqslant 0, \quad (p,q) \in T^*M.$$

In what follows we identify $\dot{\Lambda} \cong \dot{S}_0$.

Let us show that the convexity assumption on H implies the monotonicity of the Jacobi curves. In other words,

$$\dot{\Lambda}_z(0) \leqslant 0 \,\forall z \quad \Rightarrow \quad \dot{\Lambda}_z(t) \leqslant 0 \,\forall t.$$

Indeed, $\Lambda_z(t) \subset T_z(T^*M)$ and

$$\Lambda_z(t) = e_*^{-t\mathbf{h}}\left(T_{e^{t\mathbf{h}}(z)}(T_{q(t)}^*M)\right);$$

then

$$\frac{d}{dt}\Lambda_z(t) = e_*^{-t\mathbf{H}} \left.\frac{d}{d\varepsilon}\right|_{\varepsilon=0} e_*^{-\varepsilon\mathbf{h}}\left(T_{e^{(t+\varepsilon)\mathbf{h}}(z)}(T_{q(t+\varepsilon)}^*M)\right)$$

$$= e_*^{-t\mathbf{H}} \left.\frac{d}{d\varepsilon}\right|_{\varepsilon=0} \Lambda_{e^{t\mathbf{H}}(z)}(\varepsilon) \leqslant 0.$$

Here we use the fact that the quadratic form $\dot{\Lambda}_t$ is defined intrinsically and its sign does not change under a symplectic transformation of the curve Λ_t.

To any pair of transversal Lagrangian subspaces $\Lambda, \Lambda_0 : \Lambda \cap \Lambda_0 = 0$, $\Lambda, \Lambda_0 \subset \Sigma$, we can associate the projector

$$\pi_{\Lambda\Lambda_0} : \Sigma \to \Lambda_0$$

such that

$$\pi_{\Lambda\Lambda_0}|_\Lambda = 0, \quad \pi_{\Lambda\Lambda_0}|_{\Lambda_0} = Id.$$

Lemma 4.2. *The space $\{\pi_{\Lambda\Lambda_0} : \Lambda \cap \Lambda_0\}$ for a fixed Λ_0 is an affine subspace of the space of linear operators on Σ.*

In local coordinates such that

$$\Lambda_0 = \{(p,0) : p \in \mathbb{R}^n\}, \quad \Lambda = \{(Aq,q) : q \in \mathbb{R}^n\},$$

we have:

$$\pi_{\Lambda\Lambda_0} = \begin{pmatrix} Id & -A \\ 0 & 0 \end{pmatrix}.$$

Assume that the curve $t \mapsto \Lambda_t$ is such that $\Lambda_t \cap \Lambda_\tau = 0$ if $t \neq \tau$ and $|t - \tau|$ is sufficiently small. Pick τ and take coordinates such that the germ of the curve at τ has the form

$$\Lambda_t = \{(p, S_t p) : p \in \mathbb{R}^n\}, \quad S_\tau = 0.$$

Then we obtain:

$$\pi_{\Lambda_t \Lambda_\tau} = \begin{pmatrix} Id & -S_t^{-1} \\ 0 & 0 \end{pmatrix}.$$

If $\det S_t$ has a finite order root at τ, then $\pi_{\Lambda_t \Lambda_\tau}$ admits the Laurent expansion

$$\pi_{\Lambda_t \Lambda_\tau} = \sum_{i=-k}^{m} (t - \tau)^i \pi_\tau^i + o(t - \tau)^{m+1}.$$

The free term of the Laurent expansion π_τ^0 belongs to the described in the Lemma affine space of the projectors (where Λ_0 is substituted by Λ_τ). In other words, $\pi_\tau^0 = \pi_{\Lambda_\tau^\circ \Lambda_\tau}$ for some intrinsically defined Lagrangian subspace Λ_τ°, and $\Lambda_\tau^\circ \cap \Lambda_\tau = 0$ by the construction.

In particular, in the simplest case $k = 1$ the coordinate expression of Λ_τ° has the form:

$$\Lambda_\tau^\circ = \left\{ \left(-\frac{1}{2} \dot{S}_\tau^{-1} \ddot{S}_\tau \dot{S}_\tau^{-1} q, q \right) : q \in \mathbb{R}^n \right\}.$$

Now recall that we identify $\dot{\Lambda}_t$ with a quadratic form on Λ_t, i.e.

$$\dot{\Lambda}_t : \Lambda_t \to \Lambda_t^*,$$

a self-adjoint linear mapping. Moreover, the symplectic form σ gives a non-degenerate pairing of the Lagrangian subspaces Λ_t and Λ_t°, so that $\Lambda_t^* \cong \Lambda_t^\circ$ and $(\Lambda_t^\circ)^* \cong \Lambda_t$. In particular, $\dot{\Lambda}_t$ and $\dot{\Lambda}_t^\circ$ can be treated as the mappings

$$\dot{\Lambda}_t : \Lambda_t \to \Lambda_t^\circ, \quad \dot{\Lambda}_t^\circ : \Lambda_t^\circ \to \Lambda_t.$$

Definition 4.2. The operator $R_\lambda(t) : \Lambda_t \to \Lambda_t$ defined by the formula

$$R_\Lambda(t) = \dot{\Lambda}_t^\circ \circ \dot{\Lambda}_t$$

is called the curvature operator of the curve Λ at point t.

The curvature is a kind of "relative velocity" of the curves Λ_t and $\dot{\Lambda}_t$. In the regular case of a non-degenerate $\dot{\Lambda}_t$, the coordinate form of the curvature is as follows:

$$R_\Lambda(t) = \frac{1}{2}\dot{S}_t^{-1}\,\dddot{S}_t - \frac{3}{4}(\dot{S}_t^{-1}\ddot{S}_t)^2.$$

This is nothing else but the matrix version of the Schwartzian derivative of S_t.

5 Fourth lecture

In this lecture, we consider the curves Λ_t in the Lagrange Grassmannian that is defined for all $t \in \mathbb{R}$.

Theorem 5.1. *If $\dot{\Lambda}(t) \geqslant 0$ (non-negative quadratic form) and $\dot{\Lambda}^\circ(t) \leqslant 0$ (non-positive quadratic form), then no conjugate points occur and*

$$\exists \lim_{t \to \pm\infty} \Lambda(t) = \Lambda(\pm\infty).$$

Remark. The same statement is true also in the case $\dot{\Lambda}(t) \leqslant 0$, $\dot{\Lambda}^\circ(t) \geqslant 0$. What is important is that $\dot{\Lambda}(t)$ and $\dot{\Lambda}^\circ(t)$ have opposite signs.

Let us explain this result in the simplest case $n = 1$. Then Σ is the plane equipped with the area form σ, and the Lagrange Grassmanninan is the oriented real projective line that is actually the oriented circle. A monotone curve $\Lambda(t)$ is simply a monotone curve on the circle. A conjugate point occurs when $\Lambda(t)$ makes a complete revolution. On the other hand, $\Lambda(t)$ never coincides with $\Lambda^\circ(t)$, hence there are no conjugate points and $\exists \lim_{t \to \pm\infty} \Lambda(t)$.

If $n > 1$, then the proof remains essentially the same. Indeed, all pairs of transversal Lagrangian subspaces are equivalent by the action of the symplectic group. We may take coordinates in such a way that

$$\Lambda(t) = \{(p, S_t p) : p \in \mathbb{R}^n\}, \quad \Lambda^\circ(t) = \{(p, S_t^\circ p) : p \in \mathbb{R}^n\}$$

for t close to 0 and $S_0 < S_0^\circ$. The relation $\Lambda(t) \cap \Lambda^\circ(t) = 0$ is equivalent to the inequality $\det(S_t - S_t^\circ) \neq 0$. By the monotonicity assumption,

$$\langle p, \dot{S}_t p \rangle \geqslant 0, \quad \langle p, \dot{S}_t^\circ p \rangle \leqslant 0, \quad \forall p \in \mathbb{R}^n.$$

Hence the inequality

$$\langle p, \dot{S}_t p \rangle < \langle p, \dot{S}_t^\circ p \rangle, \quad \forall p \in \mathbb{R}^n$$

remains true for all positive t and each of two sides of the last inequality monotonically tends to a limit as $t \to +\infty$. The case of $t \to -\infty$ is handled by the same argument, but reversing time.

Let us come back to the Hamiltonian setting. We have $H : T^*M \to \mathbb{R}$, $z \in T^*M$, and $\Lambda_z(t) \subset T_z(T^*M)$ such that

$$\Lambda_z(t) = e_*^{-t\mathbf{H}} \left(T_{e^{t\mathbf{H}}(z)}(T^*_{q(t)}M) \right),$$

where $q(t) = \pi\left(e^{t\mathbf{H}}(z)\right)$. The limits $\Lambda_z(\pm\infty)$ are invariant for the flow $e^{t\mathbf{H}}$. Indeed,

$$e_*^{s\mathbf{H}}\Lambda_z(t) = e_*^{s\mathbf{H}}\Lambda_z(t) \lim_{t\to\pm\infty} e_*^{-t\mathbf{H}} \left(T_{e^{t\mathbf{H}}(z)}(T^*_{q(t)}M) \right)$$

$$= \lim_{t\to\pm\infty} e_*^{(s-t)\mathbf{H}} \left(T_{e^{(t-s)\mathbf{H}}(z)}(T^*_{q(t-s)}M) \right)$$

$$= \Lambda_z(\pm\infty).$$

So we have a pair of invariant Lagrangian distributions $\Lambda_z(\pm\infty)$, $z \in T^*M$. Recall that the curvature $R_\Lambda(t) = \dot{\Lambda}_t^\circ \circ \dot{\Lambda}_t$ is in a certain sense the relative velocity of $\Lambda(t)$ and $\Lambda^\circ(t)$. If we assume strong monotonicity, $\dot{\Lambda}(t) > 0$ or $\dot{\Lambda}(t) < 0$, then $|\dot{\Lambda}(t)|$ defines a Euclidean structure on $\Lambda(t)$ and $R_\Lambda(t)$ is a self-adjoint operator for this Euclidean structure. In particular, the operator $R_\Lambda(t)$ is diagonalizable and all its eigenvalues are real. We say that the curvature is positive or negative e.t.s. if all eigenvalues are like that.

Set $v = \text{sign}(\dot{\Lambda}(t))$. The matrix of the operator $R_\Lambda(t)$ is equal to the matrix of the quadratic form $v\dot{\Lambda}(t)$ in the coordinates where the Euclidean structure $|\dot{\Lambda}(t)|$ is presented by the unit matrix. In particular, assumptions of Theorem 2 are satisfied if the quadratic form $\dot{\Lambda}(t)$ is sign-definite and $R_\Lambda(t) \leqslant 0$.

It is important that the construction of the curvature is intrinsic and thus survives under symplectic transformations. In other words, if A is a linear symplectic transformation of Σ and $A\Lambda : t \mapsto A(\Lambda(t))$, then the operator $R_{A\Lambda}(t)$ is similar to the operator $R_\Lambda(t)$ and has the same eigenvalues.

Definition 5.1. Let $\Lambda_z\Lambda(\cdot)$, $z \in T*M$, be the Jacobi curves of the Hamiltonian field $\mathbf{H}$; then $R_{\lambda_z}(0) \stackrel{def}{=} R_z^H$ is called the curvature operator of $\mathbf{H}$ at z.

Recall that $\dot{\Lambda}_z(0) = -\frac{\partial^2 H}{\partial p^2}(z)$, $z = (p,q)$. If H is strongly convex with respect to p, then $\dot{\Lambda}_z(0) < 0$, $\forall z$. It was proved in Lecture 3 that the inequality $\dot{\Lambda}_z(0) \leqslant 0$, $\forall z$ implies $\dot{\Lambda}_z(t) \leqslant 0$, $\forall t$. We can repeat that proof and see that the result remains valid if one substitute the non-strong inequality by the strong one. In fact, we can see much more if we analyze the proof. Indeed, the germ at t of the curve $\Lambda_z(\cdot)$ is the image of the germ at 0 of the curve $\Lambda_{e^{t\mathbf{H}}(z)}(\cdot)$ under the fixed symplectic transformation $e_*^{-t\mathbf{H}}$. Hence all invariant quantities of these germs are equal. In particular, the operator $R_{\Lambda_z}(t)$ has the same eigenvalues as the curvature operator of $\mathbf{H}$ at the point $e^{t\mathbf{H}}(z)$. This fact is very advantageous because the curvature of $\mathbf{H}$ is just a (rather complicated but quite explicit) differential operator of H, in particular we do not need to solve differential equations in order to compute this curvature.

Combining all together, we obtain

Theorem 5.2. *If the restrictions of H to the fibers T_q^*, $q \in M$, are strongly convex functions and $R_z^H \leqslant 0$, $\forall z \in M$, then the trajectories of the flow $e^{t\mathbf{H}}$ do not have conjugate points and, moreover, this flow possesses two invariant Lagrangian distributions $\Lambda_z(\pm\infty)$, $z \in T^*M$.*

Let us discuss what happens when $R_\Lambda \geqslant 0$. First of all, let us see how changes the curvature if we re-parameterize the curve $\Lambda(t)$.

Remark. Clearly, the presence of conjugate point does not depend on the parameterization.

Let $\varphi : \mathbb{R}^n \to \mathbb{R}^n$ be a change of parameter; we assume that $\dot\varphi(t) \neq 0$. Denote

$$R_\varphi(t) = \frac{1}{2}\dot\varphi^{-1}(t)\,\dddot\varphi(t)(t) - \frac{3}{4}\left(\dot\varphi^{-1}\ddot\varphi(t)\right)^2,$$

the Schwartzian derivative of $\varphi(t)$. Let

$$\Lambda_\varphi : t \to \Lambda(\varphi(t))$$

be the re-parameterized curve.

Proposition 5 $R_{\Lambda_\varphi}(t) = \dot\varphi(t)^2 R_\Lambda(\varphi(t)) + R_\varphi(t)Id.$

This formula (the chain rule) can be checked by direct calculation, which we omit here.

Example. Take $\varphi(t) = \frac{1}{\sqrt{c}}\arctan(\sqrt{c}t)$. Then:

$$R_\varphi(t) = \frac{-c}{(ct^2+1)^2}, \quad \dot\varphi(t) = \frac{1}{ct^2+1},$$

$$R_{\Lambda_\varphi}(t) = \frac{1}{(ct^2+1)^2}\left(R_\Lambda(\varphi(t)) - cId\right).$$

Theorem 5.3 (Comparison theorem). *Let $t \mapsto \Lambda(t)$ be a smooth strongly monotone curve in the Lagrange Grassmannian and c be a nonnegative constant.*

1. If $R_\Lambda(t) \leqslant cId$, $\forall t$, then any pair of conjugate points t_1 and t_2 satisfies the inequality $|t_1 - t_2| \geqslant \frac{\pi}{\sqrt{c}}$.

2. If $\frac{1}{n}\mathrm{trace}R_\Lambda(t) \geqslant c$, then any segment $\left[t, t + \frac{\pi}{\sqrt{c}}\right]$ contains a point conjugate to zero.

Remark. If $c \to 0$ in statement 1, then $|t_1 - t_2| \to \infty$ which correlates with our previous result.

Proof. Statement 1. Re-parameterization and the chain rule reduces everything to the case of non-positive curvature (see above Example).

Statement 2. Assume that $\Lambda(t)$ is transversal to some fixed Lagrangian subspace (for instance, to $\Lambda(0)$) for any $t \in [t_1, t_2]$. Then the segment $\Lambda(t)$, $t \in [t_1, t_2]$ of the curve is contained in the fixed coordinate chart of the Lagrange Grassmannian and can be presented in the matrix form:

$$\Lambda(t) = \{(p, S_t p) : p \in \mathbb{R}^n\}, \quad t \in [t_1, t_2].$$

Hence

$$R_\Lambda(t) = \frac{1}{2}\dot{S}_t^{-1}\,\dddot{S}_t - \frac{3}{4}(\dot{S}_t^{-1}\ddot{S}_t)^2.$$

Now set $W_t = \frac{1}{2}\dot{S}_t^{-1}\ddot{S}_t$; we have, $R_\Lambda(t) = \dot{W}(t) - W(t)^2$. What remain is to find the lower bound for the blow-up time of the solutions to the differential inequality $\mathrm{trace}\dot{W} \geqslant \mathrm{trace}(W^2) + nc$. This is an easy task due to the fact that $(\mathrm{trace}W)^2 \leqslant n\,\mathrm{trace}(W^2)$.

In order to conclude we recall that in the Riemannian case the Hamiltonian flow is the geodesic flow. Actually, the Riemannian structure identifies TM and T^*M. So, in this case $\frac{\partial^2 H}{\partial p^2}$ is the Riemannian metric. The curvature R_z^H, where $z \in T_q^*$, is essentially the sectional curvature at q in the two-dimensional directions which include $z \in T_q^*M \cong T_qM$.

References

1. Agrachev, A.A.: Geometry of optimal control problems and Hamiltonian systems. In: C.I.M.E. Lecture Notes in Mathematics, Springer, Berlin (to appear)
2. Agrachev, A.A., Sachkov, Yu.L.: Control Theory from the Geometric Viewpoint. Springer, Berlin (2004)
3. Gamkrelidze, R.V.: Principles of Optimal Control Theory. Plenum Publishing, New York (1978)
4. Jurdjevic, V.: Geometric Control Theory. Cambridge University Press, Cambridge (1997)
5. Pontryagin, L.S., Boltyanskij, V.G., Gamkrelidze, R.V., Mishchenko, E.F.: The Mathematical Theory of Optimal Processes. Pergamon, Oxfrod (1964)

KAM theory with applications to Hamiltonian partial differential equations

Xiaoping Yuan[1]

Abstract In these notes I present a KAM theorem on the existence of lower dimensional invariant tori for a class of nearly integrable Hamiltonian systems of infinite dimensions, where the second Melnikov's conditions are completely eliminated and the algebraic structure of the normal frequencies is not required. This theorem can be used to construct invariant tori and quasi-periodic solutions for nonlinear wave equations, Schrödinger equations and other equations of any spatial dimensions.

1 Brief history and basic ideas of KAM theory

These lecture notes present a KAM theorem with applications to some nonlinear partial differential equations, such as nonlinear wave equations and Schrödinger equations of higher spatial dimensions. Although it is a powerful tool in dynamical systems, the KAM technique is usually thought to be very complicated, even tedious. I will omit some unimportant details so that the basic idea of the KAM theory can be clearly understood.

Before stating the KAM theorem, let us recall some of the background, taking the nonlinear wave (NLW) equation as an example. One wants to find a periodic solution of NLW equation

$$u_{tt} - u_{xx} + V(x)u + g(x,u) = 0, \tag{1}$$

subject to Dirichlet boundary condition $u(t,0) = u(t,\pi) = 0$, where g is a nonlinear term. In the 1970s using variational methods, Rabinowitz [19] showed that there is a non-constant T-periodic solution $u(t,x) \in L^p(\mathbb{R} \times [0,\pi])$ if $T/\pi \in \mathbb{Q}$, where $p > 2$

[1] School of Mathematical Sciences, Fudan University, Shanghai 200433, China,
Key Laboratory of Mathematics for Nonlinear Science (Fudan University), Ministry of Education
e-mail: xpyuan@fudan.edu.cn; yuanxiaoping@hotmail.com

W. Craig (ed.), Hamiltonian Dynamical Systems and Applications, 157–177.
© 2008 *Springer Science + Business Media B.V.*

depends on g. The condition $T/\pi \in \mathbb{Q}$ guarantees some kind of compactness which is usually required in variational methods. A natural question is

▶ *What happens when $T/\pi \in \mathbb{R} \setminus \mathbb{Q}$?*

From a geometric viewpoint, a periodic solution can be regarded as an invariant torus of 1 dimension, i.e., an invariant closed curve, in some phase space. Thus, another natural question should be

▶ *Are there invariant tori of N-dimension with $N > 1$?*

If there are such tori, and any motion on the tori is quasi-periodic, it follows that there are time quasi-periodic solutions of (1).

The previous questions can not be answered at the present time entirely by variational methods, since the compactness conditions can not be fulfilled. Fortunately, KAM theory can answer these questions. In order to see how the KAM theory adresses them, we write the equation (1) in a discrete form. To this end, we let λ_j^2 and $\phi_j(x)$ be the eigenvalues and eigenfunctions, respectively, of the Sturm–Liouville problem[1]

$$-\frac{d^2 y}{dx^2} + V(x)y = \lambda y, \quad y(0) = y(\pi) \ .$$

Note that $\{\phi_j(x) : j = 1,2,...\}$ is a complete orthogonal system of $L^2([0,\pi])$. For simplicity we assume the nonlinearity $g(x,u) = u^3$ without loss of generality. Since we will search for solutions of small amplitude, we can assume

$$g = \varepsilon u^3$$

where ε is a small parameter. This can be fulfilled by substituting $\sqrt{\varepsilon}u$ for u in (1). Let $v = u_t$. Then (1) reads

$$u_t = v, \quad v_t = u_{tt} = -[-u_{xx} + V(x)u + g(x,u)]$$

Substituting for u and v the expressions

$$u = \sum_{j=1}^{\infty} \frac{q_j}{\sqrt{\lambda_j}} \phi_j(x), \quad v = \sum_{j=1}^{\infty} \sqrt{\lambda_j} p_j \phi_j(x)$$

we get a Hamiltonian system

$$\dot{p}_j = \frac{\partial H}{\partial q_j}, \quad \dot{q}_j = -\frac{\partial H}{\partial q_j}, \quad j = 1,2,...$$

where the Hamiltonian is

$$H = \frac{1}{2} \sum_{j=1}^{\infty} \lambda_j (p_j^2 + q_j^2) + \varepsilon G(q) \qquad (2)$$

---

[1] We assume for simplicity that all eigenvalues are positive.

and the nonlinear term is expressed in terms of the eigenfunction expansion by

$$G(q) = \varepsilon \sum_{i,j,k,l} G_{ijkl} q_i q_j q_k q_l, \quad G_{ijkl} = \frac{1}{\sqrt{\lambda_i \lambda_j \lambda_k \lambda_l}} \int_0^\pi \phi_i \phi_j \phi_k \phi_l \, dx \, .$$

Given a positive integer N and a vector $\xi = (\xi_1, ..., \xi_N) \in \mathbb{R}_+^N$. Let

$$q_j = \sqrt{2(I_j + \xi_j)} \cos \theta_j, \quad p_j = \sqrt{2(I_j + \xi_N)} \sin \theta_j, \quad j = 1, ..., N$$

$$\omega = (\lambda_1, ..., \lambda_N), \hat{q} = (q_1, ..., q_N), \tilde{q} = (q_{N+1}, q_{N+2}, ...).$$

Then the Hamiltonian (2) reads

$$H = (\omega, I) + \frac{1}{2} \sum_{j=N+1}^\infty \lambda_j \left(p_j^2 + q_j^2 \right) + \varepsilon R(I, \theta, \hat{q}) \tag{3}$$

with $R(I, \theta, \hat{q}) = G(\tilde{q}, \hat{q})$. Here R is independent of $\tilde{p} = (p_{N+1}, p_{N+2}, ...)$. More generally, we can assume that R depends on $\tilde{p}$, that is, $R = R(\theta, I, \tilde{q}, \tilde{p})$. Let $z_j = (q_j, p_j)$ and $|z_j|^2 = |q_j|^2 + |p_j|^2$. Then (3) reads

$$H = (\omega, I) + \frac{1}{2} \sum_{j=N+1}^\infty \lambda_j |z_j|^2 + \varepsilon R(I, \theta, z) \tag{4}$$

Write $\Lambda = \mathrm{diag}(\lambda_{j+N} : j = 1, 2, ...)$, $J = \mathrm{diag}(J_j : j = 1, 2, ...)$ with

$$J_j = J = \begin{pmatrix} 0 & 1 \\ -1 & 0 \end{pmatrix} \, .$$

The Hamiltonian vector field is

$$\begin{cases} \dot{\theta} = \frac{\partial H}{\partial I} = \omega + \varepsilon \frac{\partial R}{\partial I} \\[2mm] \dot{I} = -\frac{\partial H}{\partial \theta} = -\varepsilon \frac{\partial R}{\partial \theta} \\[2mm] \dot{z} = J\frac{\partial H}{\partial z} = J\Lambda z + \varepsilon \frac{\partial R}{\partial z} \, . \end{cases} \tag{5}$$

We see that when $\varepsilon = 0$, the manifold

$$\mathcal{T}_0 := \{\theta = \omega t\} \times \{I = 0\} \times \{z = 0\}$$

is an invariant N-torus. KAM theory states that for "most" ω the invariant torus can be preserved if ε is sufficiently small. The basic idea is to seek a symplectic transformation (which is the composition of a series of transformations) with which to kill or to eliminate the perturbation R. However, up to present one has not found a symplectic transformation whish will kill the whole term R. A revised idea is to kill all lower order ($\leqslant 2$) terms (l.o.t.) of R, that is, the linear part of vector field X_R. More precisely, expanding R in a Fourier–Taylor series

$$R = R^\theta(\theta) + R^I(\theta) \cdot I + \langle R^z(\theta), z \rangle + \langle R^{zz}(\theta)z, z \rangle \quad \text{(l.o.t.)}$$

$$+ O(|I|^2 + |I||z| + |z|^3) \quad \text{(h.o.t.)}$$

If we can find a symplectic transform Ψ such that

$$\tilde{H} = H \circ \Psi = (\tilde{\omega}, I) + \frac{1}{2} \sum_j \tilde{\lambda}_j |z_j|^2 + \tilde{R}$$

where

$$\tilde{R} = O(|I|^2 + |I||z| + |z|^3),$$

then (5) reads

$$\begin{cases} \dot{\theta} = \frac{\partial H}{\partial I} = \tilde{\omega} + \varepsilon \frac{\partial R}{\partial I} = \tilde{\omega} + \varepsilon O(|I|) \\ \dot{I} = -\frac{\partial H}{\partial \theta} = -\varepsilon \frac{\partial R}{\partial \theta} = -\varepsilon O(|I|^2 + |I||z| + |z|^3) \\ \dot{z} = J \frac{\partial H}{\partial z} = J\Lambda z + \varepsilon \frac{\partial R}{\partial z} = J\Lambda z + \varepsilon O(|I| + |z|^2). \end{cases} \quad (6)$$

We see that $\mathcal{T}_0$ is still an invariant torus of (6). Thus, $\Psi^{-1}(\mathcal{T}_0)$ is an invariant torus of the original hamiltonian H. In searching for the symplectic transformation Ψ, one will encounter the following small divisors problems:

▲ In order to eliminate the terms $R^\theta(\theta)$ and $(R^I(\theta), I)$, one needs conditions:

$$(k, \omega) \neq 0, \quad \text{for all } 0 \neq k \in \mathbb{Z}^N$$

▲ In order to eliminate the term $(R^z(\theta), z)$, the following Melnikov's first conditions are required:

$$(k, \omega) + \lambda_j \neq 0, \quad \text{for all } k \in \mathbb{Z}^N, j = 1, 2, \ldots$$

▲ In order to eliminate the term $(R^{zz}(\theta)z, z)$, the following Melnikov's second conditions are required:

$$(k, \omega) + \lambda_i \pm \lambda_j \neq 0, \quad \text{for all } k \in \mathbb{Z}^N, j = 1, 2, \ldots,$$

where $k \neq 0$ if $i = j$ and "$\pm$" takes "$-$"

These conditions are usually not fulfilled for all ω. For example, if $\omega \in \mathbb{Q}^N$, then there exists $k \in \mathbb{Z}^N$ such that $(k, \omega) = 0$. The method of addressing this problem is to regard ω as a parameter vector (or, equivalently, assume $\omega = \omega(\xi)$ depends on a parameter vector ξ and $\det(\partial \omega / \partial \xi) \neq 0$). Eliminating those ω which violate the previous conditions, one can prove that the set of remaining parameters has positive measure (in a certain sense). Therefore, one has the following KAM theorem.

Theorem 1 *Assume $\lambda_i \neq \lambda_j$ for $i \neq j$ and R is analytic in some neighborhood of the origin. Then for "most" parameters ω, there exists a symplectic transformation Ψ such that H is changed into $\tilde{H}$, therefore, H possesses an invariant torus $\Psi^{-1}(\mathcal{T}_0)$.*

The finite dimensional version of this theorem is due to Melnikov [13, 14], Eliasson [8] and Pöschel [16]. The infinite dimensional version is due to Kuksin [10, 11], Wayne [20] and Pöschel [17]. Applying Theorem 1 to the nonlinear wave equation (1) we get

Theorem 2 *(Kuksin [11]) Assume the potential $V = V(x,\xi)$ of (1) depends on a parameter vector $\xi \in \mathcal{O} \subset \mathbb{R}^N$ a compact set, with Lebesgue measure 1 such that the Jacobian matrix $\frac{\partial \omega}{\partial \xi} = \frac{\partial(\lambda_1,\ldots,\lambda_N)}{\partial \xi}$ is non-degenerate, then for "most" parameters ξ (i.e., there is a subset $\mathcal{O}_1 \subset \mathcal{O}$ with Measure($\mathcal{O}_1$) tending to zero as $\varepsilon \to 0$ such that for $\xi \in \mathcal{O} \setminus \mathcal{O}_1$) there is an invariant torus for (1). The motion on the torus is quasi-periodic with frequency $\tilde{\omega}$ with $|\omega - \tilde{\omega}| < \varepsilon$.*

Wayne [20] also obtains the existence of the quasi-periodic solutions of (1) when the potential V does not belong to some set of "bad" potentials. In [20], the set of all potentials is given a Gaussian measure and the set of "bad" potentials is proved to be of small measure.

Because parameters are needed in Theorem 1, the potential V is assumed to depend on parameters ξ in [11] or the V itself is regarded as parameters in [20]. An important question is what happens when V does *not* contains any parameters. In this direction, early approaches are due to Bobenko–Kuksin [1] and Pöschel [18]. They assume $V \equiv m$ where the constant $m \neq 0$. In [1], the term $mu + u^3$ is regarded as a perturbation of $\sin u$. Thus, (1) is a perturbation of sine-Gordon equation. The latter are known to be integrable, exhibiting many quasi-periodic solutions. They serves as the starting point of KAM theory in (1). An alternative method is using Birkhoff normal. Observe that for $m > 0$

$$|\lambda_i \pm \lambda_j \pm \lambda_k \pm \lambda_l| \geqslant \frac{cm}{\sqrt{n^2 + m}^3}, \quad n = \min\{|i|,|j|,|k|,|l|\} \tag{7}$$

where c is some absolute constant. This inequality allows Pöschel [18] to extract some parameters from the nonlinear term u^3 through Birkhoff normal form. Once the parameters are obtained, one can apply Theorem 1 to (1).

Theorem 3 *([1, 18]) For $V \equiv m > 0$, (1) possesses many invariant elliptic tori, and thus quasi-periodic solutions.*

According to Remark 7 of [18], when $m \in (0, 1)$ the theorem still holds. In [21] it is shown that (1) possesses many invariant hyperbolic-elliptic tori and quasi-periodic solutions, when $m \in (-\infty, -1) \setminus \mathbb{Z}$. In the case $V \equiv m = 0$, the equation (1) is called *completely resonant* in [18]. In this case, One can see that the inequality (7) is useless. Whether there exists invariant torus is a challenging question, which is proposed or concerned by many authors. See references [7, 12, 15, 18]. Observe that ordinary differential $\ddot{y} + y^3 = 0$ is integrable and all non-zero solutions are periodic and their periods depend on amplitudes or initial values. Those solutions are also the solutions of (1), and they are uniform in the space x. Partial resonances can be overcome if we restrict ourselves to look for invariant tori at the neighborhood of those periodic solutions. Consequently, we have

Theorem 4 *([23]) In the neighborhood of the small solutions of $\ddot{y} + y^3 = 0$, the equation (1) subject to periodic boundary conditions has many invariant tori of any dimension and thus quasi-periodic solutions.*[2]

Another question is what happens when V is given but not constant, such as $V = \sin x, \cos x$. Observe that for a given potential V sufficiently smooth,

$$\lambda_j = j + \frac{[V]}{j} + O(1/j^2)$$

where $[V] = \int_0^\pi V(x)\,dx$. Whereas

$$\lambda_j = \sqrt{j^2 + m} = j + \frac{m}{j} + O(1/j^2)$$

when $V \equiv m \neq 0$. By comparing these two asymptotic formulae and carefully checking the inequality (7), we have

Theorem 5 *([24]) For any given potential V sufficiently smooth and $[V] \neq 0$, the equation (1) subject to Dirichlet boundary conditions has many invariant tori of any dimension and thus quasi-periodic solutions.*

So far, we have a clear comprehension of the invariant tori and quasi-periodic solutions of (1). When Hamiltonian partial differential equations with spatial dimension greater than 1 are considered, a significant new problem arises due to the presence of clusters of normal frequencies of the Hamiltonian systems defined by these PDEs. For example, let us consider the higher dimensional nonlinear wave

$$u_{tt} - \triangle u + mu + g(x, u) = 0, \tag{8}$$

subject to Dirichlet b. c. or periodic b. c., where $\triangle$ is the Laplacian in d-dimensions with $d > 1$. In this case, the eigenvalues λ_j^2 ($j \in \mathbb{Z}^d$) of the eigenvalues of the operator $-\triangle + m$ have formula

$$\lambda_j^2 = |j|^2 + m, \quad j \in \mathbb{Z}^d.$$

It follows that

$$\lim_{n \to +\infty} \{j \in \mathbb{Z}^d : \lambda_j = n^2 + m\}^\sharp = +\infty \tag{9}$$

where $\sharp$ denotes the cardinality of the set. Recall the Melnikov's second conditions

$$(k, \omega) + \lambda_i - \lambda_j \neq 0, \quad i \neq j.$$

By letting $k = 0$, it follows that λ_j is simple, i.e., $\lambda_i \neq \lambda_j$ if $i \neq j$. Therefore, the formula (9) violates seriously Melnikov's second conditions. In 2002, by observing some symmetries in (8), the present author [22] showed that there are many

---

[2] There are many authors who investigate periodic solutions and 2-D quasi-periodic solutions of travelling wave type. These excellent works are less related to KAM theory. I do not present them here, because of limit of space in this talk.

quasi-periodic solutions of traveling wave type for any spatial dimension d. Here the difficulty of small divisors was avoided owing to the symmetries. However, one can not avoid this difficulty in the search for more general solutions. In a series of papers [2] through [6], Bourgain developed another profound approach which was originally proposed by Craig–Wayne in [7], in order to overcome the difficulty that the second Melnikov's conditions can not be imposed. Now this approach is called the C-W-B method. Instead of KAM theory, C-W-B method is based on a generalization of Lyapunov–Schmidt procedure and a technique by Fröhlich and Spencer [9]. The quasi-periodic solutions are constructed directly by Newton iteration. In that direction, one will has to investigate the inverse of a "big" matrix where the small divisors problem arises. The Fröhlich and Spencer technique is used to analyze the inverse. Usually the C-W-B method is very complicated and hard to access. Recently the present author [25] modified the classic KAM technique to avoid the second Melnikov's conditions and succeeded to derive a new KAM theorem which can be applied to many kinds of PDEs including (8). Since the Fröhlich and Spencer technique is not needed there, the new KAM theorem is relatively easy to access. The whole of my lectures are devoted to the following KAM theorem, which appears in [25]:

Theorem 6 *Assume $\lambda_i \geqslant c|i|^{\kappa_1}$ and $\{j : \lambda_j = \lambda_{|j|}\}^\sharp \leqslant \kappa_2$ where c, κ_1, κ_2 are absolute positive constants. Assume R is analytic in some neighborhood of the origin. Then for "most" parameters ω, there exists a symplectic transformation Ψ such that H is changed into $\tilde{H}$, therefore, H possesses an invariant torus $\Psi^{-1}(\mathcal{J}_0)$.*

2 Derivation of the linearized equations

As stated in §1, the key point of KAM theory is to eliminate the (lower order) perturbation by a series of symplectic transformations which are generated by systems of linearized equations. In this section, we will derive the linearized equations. Before doing so, we introduce some notation. Let $\mathscr{H}_p$ be the space of sequences $z = (z_1, z_2) = ((z_{1j}, z_{2j}) \in \mathbb{C}^2 : j \in \mathbb{Z}^d)$ satisfying

$$||z||_p^2 = \sum_{j \in \mathbb{Z}^d} (|z_{1j}|^2 + |z_{2j}|^2)|j|^{2p} < \infty$$

where d is the dimension of the Laplacian and $p > d/2$ is given. It is easy to see that $\mathscr{H}^p$ is a Hilbert space with an inner product corresponding to the norm $||\cdot||_p$. (In fact, $\mathscr{H}^p$ corresponds to the so-called Sobolev space H^p by means of Fourier transform.) Denote by $\mathcal{L}(\mathscr{H}_p, \mathscr{H}_p)$ all bounded linear operators from $\mathscr{H}_p$ to $\mathscr{H}_p$. Introduce the phase space:

$$\mathscr{P} := (\mathbb{C}^n/2\pi\mathbb{Z}^n) \times \mathbb{C}^n \times \mathscr{H}^p,$$

where n is a given positive integer. We endow $\mathscr{P}$ with a symplectic structure

$$d\theta \wedge dy + \sqrt{-1}dz_1 \wedge dz_2 = dx \wedge dy + \sqrt{-1} \sum_{j \in \mathbb{Z}^d} dz_{1j} \wedge dz_{2j}, \quad (\theta, I, z_1, z_2) \in \mathscr{P}.$$

Given $r, s > 0$. Define a domain in $\mathscr{P}$ by

$$D(s,r) = \{(\theta, I, z) \in \mathscr{P} : |\operatorname{Im}\theta| \leqslant s, |I| \leqslant r^2, \|z\|_p \leqslant r\}$$

and

$$D(s) = \{\theta \in \mathbb{C}^n/2\pi\mathbb{Z}^n : |\operatorname{Im}\theta| \leqslant s\}.$$

For $z, \tilde{z} \in \mathscr{H}_p$, define

$$\langle z, \tilde{z}\rangle := \sum_j (z_{1j}\tilde{z}_{1j} + z_{2j}\tilde{z}_{2j}).$$

For a sequence of real numbers $\{\lambda_j : j \in \mathbb{Z}^d\}$, let

$$\Lambda = \operatorname{diag}\left(\lambda_j \begin{pmatrix} 1 & 0 \\ 0 & 1 \end{pmatrix} : j \in \mathbb{Z}^d\right).$$

Note $\langle \cdot, \cdot \rangle$ is not an inner product. Consider a Hamiltonian H defined on $D(s,r)$:

$$H = \underline{N} + R$$

where

$$\underline{N} = (\underline{\omega}, I) + \frac{1}{2}\langle \Lambda z, z\rangle + \frac{1}{2}\langle \underline{B}(\theta)z, z\rangle \tag{1}$$

and $R = R(\theta, I, z) : D(s,r) \to \mathbb{C}$ and $\underline{B} = \underline{B}(\theta) : D(s) \to \mathcal{L}(\mathscr{H}_p, \mathscr{H}_p)$ are analytic. As in §1, write

$$R = R^\theta(\theta) + R^I(\theta) \cdot I + \langle R^z(\theta), z\rangle + \frac{1}{2}\langle R^{zz}(\theta)z, z\rangle \quad \text{(l.o.t.)}$$

$$+ O(|I|^2 + |I|\|z\|_p + \|z\|_p^3) \quad \text{(h.o.t.)} \tag{2}$$

The basic idea is to kill the lower terms (l.o.t.). As stated in §1, in order to eliminate the term $\langle R^{zz}(\theta)z, z\rangle$ we need to assume the second Melnikov conditions, which prevents the KAM theorem from being applied to higher dimensional PDEs. Hence we should modify the basic idea. Following Bourgain [6], we put the term $\langle R^{zz}(\theta)z, z\rangle$ into the "integrable" part $\underline{N}$ rather than to eliminate it. However, doing so will make the problem too complicated. We just want to put a part of $\langle R^{zz}(\theta)z, z\rangle$ into $\underline{N}$. To this end, we introduce a cut-off operator Γ as follows. Given a positive number K large enough. For any an operator or (vector) function f defined on the domain $D(s)$, write

$$f = \sum_{k \in \mathbb{Z}^n} \widehat{f}(k) e^{\sqrt{-1}(k,\theta)}.$$

Define the cut-off operator:

$$(\Gamma f)(\theta) = (\Gamma_K)(\theta) := \sum_{|k| \leqslant K} \widehat{f}(k) e^{\sqrt{-1}(k,\theta)}.$$

The modified KAM procedure consists of the following steps.

Step 1. Averaging and cut-off. Recall (1) and (2).

$$
\begin{aligned}
H = \underline{N} + R \\[2mm]
= (\underline{\omega}, I) + \frac{1}{2}\langle \Lambda z, z\rangle + \langle \underline{B}(\theta)z, z\rangle \\[2mm]
\quad + R^{\theta}(\theta) + R^{I}(\theta) \cdot I + \langle R^{z}(\theta), z\rangle + \frac{1}{2}\langle R^{zz}(\theta)z, z\rangle \ \ (\text{l.o.t.}) \\[2mm]
\quad + O(|I|^2 + |I|\|z\|_p + \|z\|_p^3) \\[2mm]
= (\underline{\omega} + \widehat{R^{I}}(0), I) + \frac{1}{2}\langle \Lambda z, z\rangle + \langle (\underline{B}(\theta) + \Gamma R^{zz}(\theta))z, z\rangle \ \ (:= N) \\[2mm]
\quad + \Gamma R^{\theta} + (\Gamma R^{I} - \widehat{R^{I}}(0)) \cdot I + \langle \Gamma R^{z}, z\rangle \ \ (:= R_1) \\[2mm]
\quad + (1 - \Gamma)R^{\theta} + (1 - \Gamma)R^{I} \cdot I + \langle (1 - \Gamma)R^{z}, z\rangle + \frac{1}{2}\langle (1 - \Gamma)R^{zz}z, z\rangle \ \ (:= R_2) \\[2mm]
\quad + O(|I|^2 + |I|\|z\|_p + \|z\|_p^3) \ \ (:= R_3) \\[2mm]
= N + R_1 + R_2 + R_3
\end{aligned}
\tag{3}
$$

Assume $\Gamma \underline{B}(\theta) = \underline{B}(\theta)$. Let $\omega = \underline{\omega} + \widehat{R^{I}}(0)$ and $B(\theta) = \underline{B}(\theta) + \Gamma R^{zz}(\theta)$. Then

$$
N = (\omega, I) + \frac{1}{2}\langle \Lambda z, z\rangle + \frac{1}{2}\langle B(\theta)z, z\rangle.
$$

Notice that $\Gamma B = B, \Gamma R_1 = R_1$. By the way, we can assume $\widehat{R^{\theta}}(0) = 0$, since any constant added to the Hamiltonian function does not affect the dynamics.

Step 2. Seek a symplectic transformation to eliminate the term R_1. Assume we have a Hamiltonian function of the same form as R_1:

$$
F = F^{\theta}(\theta) + F^{I}(\theta) \cdot I + \langle F^{z}(\theta), z\rangle
$$

where $\widehat{F^{\theta}}(0) = 0$, $\widehat{F^{I}}(0) = 0$ and $\Gamma F = F$. Denote by X_F^t the flow of the vector field X_F corresponding to the Hamiltonian function F. Let $\Psi = X_F^1$, it is a symplectic transformation. Let $(\theta(t), I(t), z(t))$ be a solution of the vector field X_F, that is,

$$
\dot{\theta} = \frac{\partial F}{\partial I}, \quad \dot{I} = -\frac{\partial F}{\partial \theta}, \quad \dot{z} = J\partial_z F.
$$

(Referring to (5).) Then we have

$$
\begin{aligned}
\frac{d}{dt} H \circ X_F^t &= \frac{d}{dt} H(\theta(t), I(t), z(t)) \\[2mm]
&= \frac{\partial H}{\partial \theta}\dot{\theta} + \frac{\partial H}{\partial I}\dot{I} + \frac{\partial H}{\partial z}\dot{z} \\[2mm]
&= \left(\frac{\partial H}{\partial \theta}\frac{\partial F}{\partial I} - \frac{\partial H}{\partial I}\frac{\partial F}{\partial \theta} + \frac{\partial H}{\partial z}J\frac{\partial F}{\partial z} \right)(\theta(t), I(t), z(t)) \\[2mm]
&:= \{H, F\} \circ X_F^t
\end{aligned}
\tag{4}
$$

By abuse of notation, we still denote by (θ,I,z) the new variables $X_F^1(\theta,I,z)$. By Taylor's formula and (4), we get

$$
H \circ \Psi = H \circ X_F^1 = H \circ X_F^0 + \frac{d}{dt} H \circ X_F^t|_{t=0} + \frac{1}{2}\int_0^1 (1-t)\frac{d^2}{dt^2} H \circ X_F^t \, dt
$$

$$
= H + \{H,F\} + \frac{1}{2}\int_0^1 (1-t)\{\{H,F\},F\}\circ X_F^t
$$

$$
= N + R_1 + \{N,F\} + \tilde{R}
$$

$$
= (\omega,I) + \frac{1}{2}\langle \Lambda z,z\rangle + \frac{1}{2}\langle B(\theta)z,z\rangle + \Gamma R^\theta + (\Gamma R^I - \widehat{R^I}(0))\cdot I + \langle \Gamma R^z,z\rangle
$$

$$
+ \{\omega\cdot\partial_\theta F^\theta + (\omega\cdot\partial_\theta F^I)\cdot I + \langle \omega\cdot\partial_\theta F^z + \Lambda J F^z + \Gamma(BJF^z),z\rangle
$$

$$
+ \langle (1-\Gamma)(BJF^z),z\rangle + \langle (\partial_\theta B,F^I)z,z\rangle\} + \tilde{R}
$$

where

$$
\tilde{R} = R_2 + R_3 + \{R_1 + R_2 + R_3,F\} + \frac{1}{2}\int_0^1 (1-t)\{\{H,F\},F\}\circ X_F^t.
$$

If we can find F solving the following linear equations:

$$
\omega\cdot\partial_\theta F^\theta = \Gamma R^\theta, \tag{5}
$$

$$
\omega\cdot\partial_\theta F^I = \Gamma R^I - \widehat{R^I}(0), \tag{6}
$$

$$
\omega\cdot\partial_\theta F^z + \Lambda J F^z + \Gamma(BJF^z) = \Gamma R^z , \tag{7}
$$

and write

$$
B_+ = B + (\partial_\theta B, F^I),
$$

$$
N_+ = (\omega,I) + \frac{1}{2}\langle \Lambda z,z\rangle + \frac{1}{2}\langle B_+(\theta)z,z\rangle,
$$

$$
R_+ = \tilde{R} + \langle (1-\Gamma)(BJF^z),z\rangle ,
$$

then we get

$$
H_+ = H\circ\Psi = N_+ + R_+. \tag{8}
$$

In §3, we will show that if $R = O(\varepsilon)$ in some domain, then $F = O(\varepsilon^{1-})$ in a smaller sub-domain which is the result of excision of some parameters ω of small total measure. Therefore, $R_+ = O(\varepsilon^{4/3})$. Repeating the procedure above m-times, then $R_+ = O(\varepsilon^{(4/3)^m})$. Let $m \to +\infty$. Then we get a Hamiltonian

$$
H_\infty = N_\infty = (\tilde{\omega},I) + \frac{1}{2}\langle (\lambda + B(\theta))z,z\rangle .
$$

This completes the proof of Theorem 6.

3 Solutions of the linearized equations

We are now in position to solve the homological equations (5, 6, 7). Solving (5, 6) is easy. We focus our attention on the solution of (7). Observe that

$$\begin{pmatrix} 0 & 1 \\ -1 & 0 \end{pmatrix} = \sqrt{-1}\, Q_0^* \begin{pmatrix} 1 & 0 \\ 0 & -1 \end{pmatrix} Q_0, \quad Q_0 = \frac{1}{\sqrt{2}} \begin{pmatrix} 1 & \sqrt{-1} \\ 1 & -\sqrt{-1} \end{pmatrix}$$

where $*$ is the conjugate transpose of the matrix. It is easy to see that $Q_0^* = Q_0^{-1}$. Let

$$Q = \mathrm{diag}\left(Q_0 : j \in \mathbb{Z}^d \right), \quad E_\pm = \mathrm{diag}\left(\begin{pmatrix} 1 & 0 \\ 0 & -1 \end{pmatrix} : j \in \mathbb{Z}^d \right).$$

Then $J = \sqrt{-1}\, Q^* E_\pm Q$ and $Q \Lambda Q^* = \Lambda$. Notice that E_0^2 is the identity operator of $\ell^2(\mathbb{Z}^d) \otimes \ell^2(\mathbb{Z}^d)$. Left-multiplying (7) by Q we get

$$\omega E_\pm \partial_\theta (E_\pm Q F^z) + \sqrt{-1}\, Q \Lambda Q^* (E_\pm Q F^z) + \sqrt{-1}\, \Gamma(Q B Q^* (E_\pm Q F^z)) = \Gamma Q R^z.$$

Let $F_z = E_\pm Q F^z, R_z = Q R^z$ and $\mathcal{B} = Q B Q^*$. Note $\Gamma Q = Q \Gamma$. Then

$$\omega E_\pm \partial_\theta F_z + \sqrt{-1}\,\Lambda F_z + \sqrt{-1}\,\Gamma(\mathcal{B} F_z) = \Gamma R_z. \tag{1}$$

Write

$$F_z = \sum_{|k| \leqslant K} \widehat{F}(k) e^{\sqrt{-1}(k,\theta)}, \quad \mathcal{B} = \sum_{|k| \leqslant K} \widehat{\mathcal{B}}(k) e^{\sqrt{-1}(k,\theta)},$$

$$R_z = \sqrt{-1} \sum_{|k| \leqslant K} \widehat{R}(k) e^{\sqrt{-1}(k,\theta)}.$$

Let

$$T = \mathrm{diag}(\pm(k,\omega) + \lambda_j : |k| \leqslant K, j \in \mathbb{Z}^d, k \in \mathbb{Z}^n)$$
$$\widehat{\mathcal{B}} = (\widehat{\mathcal{B}}(k-l) : |k|,|l| \leqslant K, k,l \in \mathbb{Z}^n)$$
$$\widehat{F} = (\widehat{F}(k) : |k|, \leqslant K, k \in \mathbb{Z}^n), \quad \widehat{R} = (\widehat{R}(k) : |k|, \leqslant K, k \in \mathbb{Z}^n).$$

Then (7) can be written as

$$(T + \widehat{\mathcal{B}})\widehat{F} = \widehat{R}. \tag{2}$$

Our goal is to prove that the operator $T + \widehat{\mathcal{B}}$ is invertible and to find its inverse. To this end, we need some assumptions.

A1. *Assume $\omega = \omega(\xi)$, $B = B(\theta,\xi)$ depend smoothly (in the sense of Whitney[3]) on a parameter vector $\xi \in \mathcal{O}$ a compact set with $\mathrm{Meas}(\mathcal{O}) > 0$, and*

$$|\det(\partial \omega/\partial \xi)| \geqslant C > 0, \tag{3}$$

$$\sup_{\theta \in D(s), \xi \in \mathcal{O}} \|B(\theta,\xi)\|_p \ll 1, \quad \sup_{\theta \in D(s), \xi \in \mathcal{O}} \|\partial_\xi B(\theta,\xi)\|_p \ll 1. \tag{4}$$

---

[3] We will not mention this further.

We denote by C a universal positive constant whose value may be different in different places. Let $K^\sharp =$ Cardinality of $\{k \in \mathbb{Z}^n : |k| \leqslant K\}$ and $\underline{\mathscr{H}^p} = \mathscr{H}^p \otimes \mathbb{C}^{K^\sharp}$ with norm

$$||u||_p^2 = \sum_{j=1}^{K^\sharp} ||u_j||_p^2, \ \forall\, u = (u_1,...,u_{K^\sharp}) \in \underline{\mathscr{H}^p} .$$

It follows from (8) that $\widehat{\mathcal{B}}$ is bounded linear operator in $\underline{\mathscr{H}^p}$, and

$$||\widehat{\mathcal{B}}||_p \ll 1, \ ||\partial_\xi \widehat{\mathcal{B}}||_p \ll 1, \ \forall \xi \in \mathcal{O} \tag{5}$$

where $|| \cdot ||_p$ is the operator norm of $\underline{\mathscr{H}^p}$.

A2. *Assume the original Hamiltonian $H = \underline{N} + R$ is real for real arguments.*

It follows that for $(\theta, \xi) \in D(s) \times \mathcal{O}$ with $s = 0$, the operator $B(\theta, \xi)$ is self-adjoint in the space $\ell^2(\mathbb{Z}^d) \times \ell^2(\mathbb{Z}^d)$ (note $\ell^2(\mathbb{Z}^d) \times \ell^2(\mathbb{Z}^d) = \mathscr{H}^p$ with $p = 0$). If we regard B as a matrix of infinite dimension, then $\overline{B(\theta, \xi)}^T = B(\theta, \xi)$ where the bar means conjugate and T means transpose. Recall that $\mathcal{B} = Q B \bar{Q}^T$. We have

$$\overline{\mathcal{B}(\theta, \xi)}^T = \mathcal{B}(\theta, \xi), \ \forall\, (\theta, \xi) \in D(0) \times \mathcal{O}.$$

It follows from this that

$$\overline{\widehat{\mathcal{B}}(k-l)}^T = \widehat{\mathcal{B}}(l-k), \ \forall\, k, l \in \mathbb{Z}^n. \tag{6}$$

Then it follows from (6) that $\widehat{\mathcal{B}}$ is self-adjoint in the space $\underline{\ell^2} := \ell^2(\mathbb{Z}^d) \times \ell^2(\mathbb{Z}^d) \times \mathbb{C}^{K^\sharp}$. Note the matrix $T + \widehat{\mathcal{B}}$ is of infinite dimension. We will reduce the inverse of it to one of a matrix of finite dimension. To this end, we need a third assumption.

A3. *Assume the normal frequencies λ_j's satisfy the following growth conditions:*

$$\lambda_j \geqslant C|j|^\kappa, \ \exists\, \kappa > 0, \ \forall\, j \in \mathbb{Z}^d.$$

Let

$$M = 1 + \left(\frac{\sup_\xi |\omega(\xi)|}{C} K \right)^{1/\kappa} := C K^{1/\kappa} .$$

We see that when $|k| \leqslant K, |j| \geqslant M$,

$$|\pm(k, \omega) + \lambda_j| \geqslant 1 . \tag{7}$$

Write

$$T_1 = \mathrm{diag}(\pm(k, \omega) + \lambda_j : |k| \leqslant K, |j| \leqslant M) ,$$

$$T_2 = \mathrm{diag}(\pm(k, \omega) + \lambda_j : |k| \leqslant K, |j| \geqslant M) .$$

Then $T = T_1 \oplus T_2$. And by (7) we get

$$||T_2^{-1}||_p \leqslant 1.$$

While regarding $\widehat{\mathcal{B}}$ as a matrix of infinite dimension, we denote by $\widehat{\mathcal{B}}_{ij}(k-l)$ the elements of matrix where $i,j \in \mathbb{Z}^d, k,l \in \mathbb{Z}^n, |k|,|l| \leqslant K$. We decompose $\widehat{\mathcal{B}}$ into four blocks as follows:

$$\begin{aligned}
\widehat{\mathcal{B}}_{11} &= (\mathcal{B}_{ij}(k-l) : |k|,|l| \leqslant K, |i|,|j| \leqslant M) \\
\widehat{\mathcal{B}}_{12} &= (\mathcal{B}_{ij}(k-l) : |k|,|l| \leqslant K, |i| \leqslant M, |j| > M) \\
\widehat{\mathcal{B}}_{21} &= (\mathcal{B}_{ij}(k-l) : |k|,|l| \leqslant K, |i| > M, |j| \leqslant M) \\
\widehat{\mathcal{B}}_{22} &= (\mathcal{B}_{ij}(k-l) : |k|,|l| \leqslant K, |i| > M, |j| > M).
\end{aligned}$$

Then

$$\widehat{\mathcal{B}} = \begin{pmatrix} \widehat{\mathcal{B}}_{11} & \widehat{\mathcal{B}}_{12} \\ \widehat{\mathcal{B}}_{21} & \widehat{\mathcal{B}}_{22} \end{pmatrix}.$$

According to (5),

$$||\widehat{\mathcal{B}}_{ij}||_p \ll 1, \ ||\partial_\xi \widehat{\mathcal{B}}_{ij}||_p \ll 1, \ \forall \xi \in \mathcal{O}, \ i,j \in \{1,2\}. \tag{8}$$

By $||T_2^{-1}||_p \leqslant C$ and $||\widehat{\mathcal{B}}_{22}||_p \ll 1$, we get

$$||(T_2 + \widehat{\mathcal{B}}_{22})^{-1}||_p \leqslant || \sum_{j=0}^{\infty} ||(T_2^{-1}\widehat{\mathcal{B}}_{22})^j||_p ||T_2^{-1}|||_p \leqslant C.$$

Set

$$\widehat{\mathcal{B}}_{lef} := \begin{pmatrix} E_1 & 0 \\ -(T_2 + \widehat{B}_{22})^{-1}\widehat{B}_{21} & E_2 \end{pmatrix}$$

$$\mathcal{B}_{rig} := \begin{pmatrix} E_1 & -\widehat{B}_{12}(T_2 + \widehat{B}_{22})^{-1} \\ 0 & E_2 \end{pmatrix},$$

where E_1 (E_2, respectively) is a unit matrix of the same order as that of T_1 (T_2, respectively). From $||(T_2 + \widehat{\mathcal{B}}_{22})^{-1}||_p \leqslant C$, it is easy to verify that

$$||\widehat{\mathcal{B}}_{lef}^{-1}||_p, \ |||\widehat{\mathcal{B}}_{rig}^{-1}||_p \leqslant C$$

Let

$$\mathscr{B}_{11} = \widehat{\mathcal{B}}_{11} - \widehat{\mathcal{B}}_{12}(T_2 + \widehat{\mathcal{B}}_{22})^{-1}\widehat{\mathcal{B}}_{21}$$

Then the inverse of $T + \widehat{\mathcal{B}}$ exists;

$$(T + \widehat{\mathcal{B}})^{-1} = \widehat{\mathcal{B}}_{rig}^{-1} \begin{pmatrix} (T_1 + \mathscr{B}_{11})^{-1} & 0 \\ 0 & (T_2 + \widehat{\mathcal{B}}_{22})^{-1} \end{pmatrix} \widehat{\mathcal{B}}_{lef}^{-1}$$

and

$$||(T + \widehat{\mathcal{B}})^{-1}||_p \leqslant C||(T_1 + \widehat{\mathcal{B}}_{11})^{-1}||_p\ ,$$

provided that the inverse of $T_1 + \widehat{\mathcal{B}}_{11}$ exists.

We are now in position to investigate the inverse of $T_1 + \widehat{\mathcal{B}}_{11}$. First of all, we would like to point out that the matrix $\widehat{\mathcal{B}}_{11}$ is of finite order, and the order is bounded by

$$K^* := K^{2n}M^d \leqslant K^{2n+d\kappa^{-1}}\ .$$

Secondly, it follows from the self-adjointness of $\widehat{\mathcal{B}}$ that the matrix $\widehat{\mathcal{B}}_{11}$ is also self-adjoint. Thirdly, by (8) we have

$$||\widehat{\mathcal{B}}_{11}||_p \ll 1,\ qquad ||\partial_\xi \widehat{\mathcal{B}}_{11}||_p \ll 1\ .$$

Fourthly, since each element of $\widehat{\mathcal{B}}$ is continuously differentiable in $\xi \in \mathcal{O}$, the matrix $\widehat{\mathcal{B}}_{11}$ is also continuously differentiable in $\xi \in \mathcal{O}$. In view of (3), we can regard ω itself as a parameter vector instead of ξ, or we can assume $\omega = \xi$. Without loss of generality, we assume the first entry ω_1 of ω is in the interval $[1,2]$. Then the matrix $T_1 + \widehat{\mathcal{B}}_{11}$ is non-singular if and only if

$$A_1 := \omega_1^{-1}T_1 + \omega_1^{-1}\widehat{\mathcal{B}}_{11}$$

is non-singular, and

$$||(T_1 + \widehat{\mathcal{B}}_{11})^{-1}||_p \leqslant C||A_1^{-1}||_p.$$

Let

$$\Xi\ :\varsigma_1 = 1/\omega_1,\ \varsigma_2 = \omega_2/\omega_1,...,\ \varsigma_n = \omega_n/\omega_1\ .$$

Then it is easy to get

$$\left|\det \frac{\partial(\varsigma_1,...,\varsigma_n)}{\partial(\omega_1,...,\omega_n)}\right| = \omega_1^{-(n+1)} \geqslant C > 0\ .$$

Therefore, we can regard ς as a parameter vector. It is easy to see that

$$\text{Meas } \mathcal{O} \leqslant C\,\text{Meas }\Xi(\mathcal{O}) \leqslant C\,\text{Meas }\mathcal{O}$$

and

$$||\widehat{\mathcal{B}}_{11}(\xi(\varsigma))|| \ll 1\ ,\qquad ||\partial_\varsigma \widehat{\mathcal{B}}_{11}(\xi(\varsigma))|| \ll 1\ . \tag{9}$$

After introducing the parameter ς, we can write

$$A_1 = \text{diag}\left(\pm(k_1 + \sum_{l=2}^{n} k_l\varsigma_l) + \varsigma_1\Omega_j^0 : k = (k_1,...,k_n) \in \mathbb{Z}^n, |k| \leqslant K, |j| \leqslant M\right)$$
$$+ \varsigma_1\widehat{\mathcal{B}}_{11}(\xi(\varsigma)).$$

Since $\widehat{\mathscr{B}}_{11}$ is self-adjoint, so is $A_1 = A_1(\varsigma)$ for any $\varsigma \in \Xi(\mathcal{O})$. Therefore, there are continuously differentiable functions $\mu_1(\varsigma), \cdots, \mu_{K^*}(\varsigma)$ representing the eigenvalues of A_1 for $\varsigma \in \Xi(\mathcal{O})$.

Lemma 2 *There exists a subset $\mathcal{O}_+ \subset \mathcal{O}$ with $Meas(\Xi\mathcal{O}_+) \geqslant (Meas\Xi(\mathcal{O}))(1 - CK^{-1}$ such that for any $\varsigma \in \Xi(\mathcal{O}_+)$,*

$$\mu_j(\varsigma) \geqslant (KK^*)^{-1} > 0.$$

We postpone the proof to the end of this section. Since A_1 is self-adjoint, there exists a matrix-valued function $U(\varsigma)$ of order K^* which depends on ς, such that for every $\varsigma \in \Xi(\mathcal{O}_+)$ the following equalities hold:

$$A_1(\varsigma) = U(\varsigma)\mathrm{diag}(\mu_1(\varsigma), \cdots, \mu_{K^*}(\varsigma))U^*(\varsigma),$$

and

$$U(\varsigma)(U(\varsigma))^* = (U(\varsigma))^*U(\varsigma) = E$$

where E is the unit matrix of order K^* and U^* is the conjugate transpose of U. It follows that for $\varsigma \in \Xi(\mathcal{O}_+)$,

$$||A_1(\varsigma)|| \leqslant \max\{\mu_j : j = 1, ..., K^*\} \leqslant KK^*$$

where $|| \cdot ||$ is the ℓ_2 norm of matrix. Since A_1 is of order K^*,

$$||A_1(\varsigma)||_p \leqslant KK^*(K^*)^p := K^C.$$

Thus,

$$||(T + \widehat{\mathcal{B}})^{-1}||_p \leqslant ||(T_1 + \widehat{\mathscr{B}}_{11})^{-1}||_p \leqslant ||A_1(\varsigma)||_p \leqslant K^C, \ \xi \in \mathcal{O}_+ .$$

Assume that

$$\sup_{D(s) \times \mathcal{O}} ||\partial_\xi^l R^z(\theta, \xi))||_p \leqslant \varepsilon, \ l = 0, 1 .$$

It follows that

$$||\widehat{F}||_p \leqslant K^C||\widehat{R}||_p \leqslant K^C \sup_{D(s) \times \mathcal{O}} ||R^z(\theta, \xi))||_p \leqslant K^C\varepsilon .$$

Note $|||\partial_\xi(T_1 + \widehat{\mathscr{B}}_{11})|||_p \leqslant K$. We have that for any $\xi \in \mathcal{O}_+$,

$$||\partial_\xi(T_1 + \widehat{\mathscr{B}}_{11})^{-1}||_p = ||(T_1 + \widehat{\mathscr{B}}_{11})^{-1}(\partial_\xi(T_1 + \widehat{\mathscr{B}}_{11}))(T_1 + \widehat{\mathscr{B}}_{11})^{-1}||_p \leqslant K^{1+2C} := K^C.$$

It follows that

$$||\partial_\xi(T + \widehat{\mathcal{B}})^{-1}||_p \leqslant K^C, \ \xi \in \mathcal{O}_+.$$

Moreover,

$$||\partial_\xi \widehat{F}||_p \leqslant K^C\varepsilon.$$

Note

$$||\widehat{F}||_p^2 = \sum_{|k| \leqslant K} ||\widehat{F^z}(k)||_p^2 + \sum_{|k| \leqslant K} ||\widehat{F^{\bar{z}}}(k)||_p^2.$$

Then

$$\sup_{D(0) \times \mathcal{O}_+} ||F^u(x,\xi)||_p^2 = \sup_{D(0) \times \mathcal{O}_+} ||\sum_{|k| \leqslant K} \widehat{F^u}(k) e^{\sqrt{-1}(k,x)}||_p^2$$

$$\leqslant \sum_{|k| \leqslant K} ||\widehat{F^u}(k)||_p^2 \leqslant (K^C \varepsilon)^2$$

that is,

$$\sup_{D(0) \times \mathcal{O}_+} ||F^u(x,\xi)||_p \leqslant K^C \varepsilon . \tag{10}$$

Similarly,

$$\sup_{D(0) \times \mathcal{O}_+} ||\partial_\xi F^u(x,\xi)||_p \leqslant K^C \varepsilon .$$

Lemma 3 *We can extend the domain $D(0)$ to $D(s)$ such the above inequalities still hold;*

$$\sup_{D(s) \times \mathcal{O}_+} ||F^u(x,\xi)||_p \leqslant K^C \varepsilon, \qquad \sup_{D(s) \times \mathcal{O}_+} ||\partial_\xi F^u(x,\xi)||_p \leqslant K^C \varepsilon .$$

Proof. Rewrite $F^u = (F^z, F^{\bar{z}}))$ and

$$B = \begin{pmatrix} B^{zz} & B^{z\bar{z}} \\ B^{\bar{z}z} & B^{\bar{z}\bar{z}} \end{pmatrix} .$$

Then the homological equation (7) can be rewritten as

$$-\sqrt{-1}\omega \cdot \partial_\theta F^z + \Omega^0 F^z + \Gamma((\Gamma B^{z\bar{z}})F^z - (\Gamma B^{zz})F^{\bar{z}}) = -\sqrt{-1}\Gamma R^z(x,\xi) ,$$

$$\sqrt{-1}\omega \cdot \partial_\theta F^{\bar{z}} + \Omega^0 F^{\bar{z}} + \Gamma((\Gamma B^{z\bar{z}})F^{\bar{z}} - (\Gamma B^{\bar{z}\bar{z}})F^z) = \sqrt{-1}\Gamma R^{\bar{z}}(x,\xi) .$$

The following equalities can be fulfilled by the assumption that H is real for real argument:

$$\overline{R^z} = R^{\bar{z}}, \ \overline{B^{zz}} = B^{\bar{z}\bar{z}}, \ \overline{B^{z\bar{z}}} = B^{\bar{u}u} = \overline{B^{\bar{z}z}} = B^{u\bar{u}}, \ \theta \in D(0).$$

See [25] for the details. It follows that $\overline{F^z}(\theta) = F^{\bar{z}}(\theta)$ for $\theta \in D(0)$. Note F^z is analytic in $D(s)$. Thus, $\overline{F^z}(\theta) = F^{\bar{z}}(\theta)$ for $\theta \in D(s)$. Let $i^2 = -1$. In the proof of this lemma, we can assume $\omega = (1, ..., 1)$ without loss of generality. An important fact is that λ_j's are positive. When the dimension $n = 1$ of the angle variable θ, the proof shows more clearly our basic idea. Firstly, assume $n = 1$. Let $\theta = it + r$. Arbitrarily fix $r \in \mathbb{R}/2\pi\mathbb{Z}$. Write $F(t) = F^{\bar{z}}(it + r)$. By the second homological equation and using the method of variation of constants,

$$F(t) = F^{\bar{z}}(r) - \int_0^t e^{-\Lambda(t-\tau)}(BF + R^z(it)))d\tau, \ t \in [0,s] \, . \tag{11}$$

Let $||\cdot||_{p,s} = \sup_{D(s)} ||\cdot||_p$ where we write formally

$$BF = \Gamma[\Gamma B^{z\bar{z}}(i\tau))F(i\tau) + (\Gamma B^{z\bar{z}})\overline{F}(it)].$$

Note $||B||_{p,s} \leqslant \delta \ll 1, ||R^z||_{p,s} \leqslant \varepsilon$. And note $||F^z||_p = ||F^{\bar{z}}||_p$. By (10) we have $||F^{\bar{z}}(r)||_p = ||F^z(r)||_p \leqslant K^C \varepsilon$. Therefore, by (11) we get

$$||F(t)||_p \leqslant K^C \varepsilon + \delta ||F^z||_{p,s}, \ t \in [0,s].$$

By the first homological equation, we get

$$||\overline{F(it)}||_p \leqslant K^C \varepsilon + \delta ||F^z||_{p,s}, \ t \in [-s,0].$$

Thus,

$$||F(it)||_p \leqslant K^C \varepsilon + \delta ||F^z||_{p,s}, \ t \in [-s,s].$$

That means

$$||F^{\bar{z}}(it+r)||_p \leqslant K^C \varepsilon + \delta ||F^z||_{p,s}, \ t \in [-s,s], r \in \mathbb{R}/2\pi\mathbb{Z}.$$

This leads to

$$||F^z||_{p,s} \leqslant 2K^C \varepsilon := K^C \varepsilon.$$

That is

$$\sup_{D(s) \times \mathscr{O}_+} ||F^z(x,\xi)||_p \leqslant K^C \varepsilon \, .$$

Now let us consider the dimension $n = 2$. Fix an arbitrary $r \in \mathbb{R}/2\pi\mathbb{Z}$. Let $\theta_1 = it + r$ with $t \in [0,s]$, let $F(t) := F^{\bar{z}}(it+r,\phi)$, and restrict $\phi \in \mathbb{R}$. By the second homological equation, we have

$$F(t) = F^{\bar{z}}(r,\phi) - \int_0^t e^{-\Lambda(t-\tau)} e^{-i(t-\tau)\partial_\phi}(BF + R^z), \ t \in [0,s] \tag{12}$$

For any analytic 2π-periodic function $f : \{x : |\mathrm{Im}\,x| \leqslant s\} \to H^p$, using Cauchy's theorem, we have

$$\sup_{x \in \mathbb{R}/2\pi\mathbb{Z}} ||e^{-i(t-\tau)\partial_x} f(x)||_p \leqslant e^{|t-\tau|/s} ||f||_{p,s}.$$

By (12) we have

$$||F^{\bar{z}}(it+r,\phi)||_p \leqslant ||F^{\bar{z}}(r,\phi)||_p + \int_0^t e^{(t-\tau)/s}[\delta ||F^{\bar{z}}||_{p,s} + ||R^z||_{p,s}]d\tau$$

$$\leqslant K^C \varepsilon + \delta ||F^{\bar{z}}||_{p,s}, \quad t \in [0,s], \phi \in \mathbb{R}.$$

Again by the first homological equation and noting $\overline{F^z} = F^{\bar{z}}$, we have

$$||F^{\bar{z}}(it+r,\phi)||_p \leqslant K^C\varepsilon + \delta||F^{\bar{z}}||_{p,s}, \quad t \in [-s,s], \phi \in \mathbb{R} \tag{13}$$

For any constant c, the line $L = L(c): x - y = c$ is a characteristic line of $\partial_x + \partial_y$. Let

$$F(y) := F(i(y+c)+r_1, iy+r_2), \quad y \in [0,s], r_1, r_2 \text{ fixed.}$$

By the second homological equation, we have

$$F(y) = F^{\bar{z}}(ic+r_1, r_2) - \int_0^y e^{-\Lambda(y-\tau)}(BF + R^z)d\tau, \quad y \in [0,s]. \tag{14}$$

It follows that

$$||F^{\bar{z}}(i(y+c)+r_1, iy+r_2)||_p \leqslant ||F^{\bar{z}}(ic+r_1, r_2)||_p + \delta||F^{\bar{z}}||_{p,s} + \varepsilon, \quad y \in [0,s].$$

Moreover, by the first homological equation, we have

$$||F^{\bar{z}}(i(y+c)+r_1, iy+r_2)||_p \leqslant ||F^{\bar{z}}(ic+r_1, r_2)||_p + \delta||F^{\bar{z}}||_{p,s} + \varepsilon, \quad y+c, y \in [-s,s].$$

By (13),

$$||F^{\bar{z}}(i(y+c)+r_1, iy+r_2)||_p \leqslant K^C\varepsilon + 2\delta||F^{\bar{z}}||_{p,s} + \varepsilon, \quad y+c, y \in [-s,s].$$

Let the line $L(c)$ run over the square $[-s,s]^2$, we have

$$||F^{\bar{z}}||_{p,s} \leqslant K^C\varepsilon + 2\delta||F^{\bar{z}}||_{p,s} + \varepsilon.$$

It follows

$$||F^{\bar{z}}||_{p,s} \leqslant 4K^C\varepsilon := K^C\varepsilon.$$

We will omit the proof of $||\partial_\xi F^{\bar{z}}||_{p,s} \leqslant K^C\varepsilon$. This proof is finished by mathematical induction on n.

Proof of Lemma 2. Let

$$\Xi(\underline{O}_l) = \{\varsigma \in \Xi(O) : |\mu_l| < 1/(KK^*)\}, \quad l = 1,...,K^*.$$

Take an arbitrary $\mu = \mu(\varsigma) \in \{\mu_1(\varsigma),...,\mu_{K^*}(\varsigma)\}$. Let ϕ be the normalized eigenvector corresponding μ. It is easy to prove that $\partial_\varsigma\mu = ((\partial_\varsigma A_1)\phi, \phi)$. By computing $\partial_\varsigma A_1$ and using (9) and Lemma 4, we get

$$\partial_\varsigma\mu = ((\text{diag}(\lambda_j : |k| \leqslant K, |j| \leqslant M))\phi, \phi) + o(1)$$
$$\geqslant \min\{\lambda_j : j \in \mathbb{Z}^d\} + o(1) \geqslant C > 0.$$

It follows that Meas $\Xi(\underline{O}_l) < 1/(KK^*)$. Thus,

$$\text{Meas} \bigcup_{l=1}^{K^*} \Xi(\underline{O}_l) < 1/K.$$

Let

$$\Xi(\underline{0}) = \Xi(0) \setminus \bigcup_{l=1}^{K^*} \Xi(\underline{0}_l).$$

Therefore,

$$\text{Meas } \Xi(\underline{0}) \geqslant \text{Meas } \Xi(0)(1 - O(\frac{1}{K}))$$

and for any $\varsigma \in \Xi(\underline{0})$,

$$|\mu_l(\varsigma)| \geqslant 1/KK^*.$$

This completes the proof of Lemma 2.

4 Applications to partial differential equations in higher dimensions

We will just give the application of Theorem 6 to a nonlinear wave equation in higher dimension. See [25] for other applications such as to the nonlinear Schrödinger equation.

Consider the nonlinear wave equation

$$u_{tt} - \triangle u + M_\sigma u + \varepsilon u^3 = 0, \ \ \theta \in \mathbb{T}^d, \ d \geqslant 1 \tag{1}$$

where $u = u(t, \theta)$ and $\triangle = \sum_{j=1}^d \partial_{\theta_j}^2$ and M_σ is a real Fourier multiplier

$$M_\sigma \cos(j, \theta) = \sigma_j \cos(j, \theta), \ M_\sigma \sin(j, \theta) = \sigma_j \sin(j, \theta), \ \ \sigma_j \in \mathbb{R}, \ j \in \mathbb{Z}^d \ .$$

Pick a set $\underline{\varpi} = \{\varpi_1, ..., \varpi_n\} \subset \mathbb{Z}^d$. Let $\underline{\mathbb{Z}}^d = \mathbb{Z}^d \setminus \{\varpi_1, ..., \varpi_n\}$. Following Bourgain [4], we assume

$$\begin{cases} \sigma_{\varpi_l} = \sigma_l, \ (l = 1, ..., n) \\ \sigma_j = 0, \ j \in \underline{\mathbb{Z}}^d \ . \end{cases}$$

Theorem 7 *([25]) Let* $\omega_l^0 = \sqrt{\lambda_{\varpi_l}} = \sqrt{|\varpi_l|^2 + \sigma_l}, \ \ (l = 1, ..., n)$ *and* $\sigma = (\sigma_1, ..., \sigma_n)$ *and* $\omega^0 = (\omega_1^0, ..., \omega_n^0)$. *Then there is a subset* $\mathcal{O}_0 \subset [1, 2]^d$ *with* $\text{Meas } \mathcal{O}_0 \geqslant (1 - C\varepsilon)$ *such that for any* $\sigma \in \mathcal{O}_0$, *the nonlinear wave equation with small* ε *has a rotational invariant torus of frequency vector* ω *with* $|\omega - \omega^0| = O(\varepsilon)$. *The motion on the torus can be expressed by* $u(t, \theta)$ *which is quasi-periodic (in time) with frequency* ω *and* $u(\cdot, \theta) : \mathbb{R} \to H^p(\mathbb{T}^n)$ *is an analytic map, and thus the solution* $u(t, \theta)$ *is, at least, a sufficiently smooth function of* (t, θ) *if p is taken large enough.*

Acknowledgements The author is very grateful to Prof. W. Craig for his help.

Appendix

Lemma 4 *Assume the matrix $A = A(\varsigma)$ is self-adjoint and smooth in ς. Let $\mu = \mu(\varsigma)$ be any eigenvalue of A and ϕ be the eigenfunction corresponding to μ. Then we have*

$$\partial_\varsigma \mu = ((\partial_\varsigma A)\phi, \phi).$$

Proof. Note ϕ is not necessarily smooth in ς. Consider the difference operator:

$$\triangle f = \triangle_{\varsigma_1 \varsigma_2} f = \frac{f(\varsigma_1) - f(\varsigma_2)}{\varsigma_1 - \varsigma_2}.$$

Apply $\triangle$ to $A\phi = \mu\phi$,

$$(\triangle A)\phi + A(\triangle \phi) = (\triangle \mu)\phi + \mu(\triangle \phi).$$

Taking inner product with ϕ,

$$\langle (\triangle A)\phi, \phi \rangle + \langle A(\triangle \phi), \phi \rangle = \langle (\triangle \mu)\phi, \phi \rangle + \mu \langle \triangle \phi, \phi \rangle.$$

Since A is self-adjoint,

$$\langle A(\triangle \phi), \phi \rangle = \langle (\triangle \phi), A\phi \rangle = \mu \langle \triangle \phi, \phi \rangle.$$

Thus

$$\langle (\triangle A)\phi, \phi \rangle = \langle (\triangle \mu)\phi, \phi \rangle = (\triangle \mu)\langle \phi, \phi \rangle = \triangle \mu.$$

Letting $\varsigma_1 \to \varsigma_1 := \varsigma$, we have $\partial_\varsigma \mu = ((\partial_\varsigma A)\phi, \phi)$.

References

1. Bobenko, A.I. and Kuksin, S.B., The nonlinear Klein-Gordon equation on an interval as a perturbed sine-Gordon equation, *Comment. Math. Helv.* **70** (1995), no. 1, 63–112.
2. Bourgain, J., Construction of quasi-periodic solutions for Hamiltonian perturbations of linear equations and application to nonlinear PDE. *Int. Math. Research Notices* **11** (1994), 475–497.
3. Bourgain, J., Periodic solutions of nonlinear wave equations. Harmonic analysis and partial equations, Chicago Univ. Press (1999), pp. 69–97.
4. Bourgain, J., Quasi-periodic solutions of Hamiltonian perturbations for 2D linear Schrödinger equation. *Ann. Math.* **148** (1998), 363–439.
5. Bourgain, J., Green's function estimates for lattice Schrödinger operators and applications. *Annals of Mathematics Studies* **158** Princeton University Press, Princeton, NJ, (2005).
6. Bourgain, J., On Melnikov's persistence problem, *Math. Res. Lett.* **4** (1997), 445–458.
7. Craig, W. and Wayne, C.E., Newton's method and periodic solutions of nonlinear wave equation, *Commun. Pure. Appl. Math.* **46** (1993), 1409–1501.
8. Eliasson, L.H., Perturbations of stable invariant tori for Hamiltonian systems, *Ann. Scula Norm. Sup. Pisa CL Sci.* **15** (1988), 115–147.
9. Frohlich, J. and Spencer, T., Absence of diffusion in the Anderson tight binding model for large disorder or lower energy. *Commun. Math. Phys.* **88** (1983), 151–184.

10. Kuksin, S.B., Hamiltonian perturbations of infinite-dimensional linear systems with an imaginary spectrum, *Funct. Anal. Appl.*, **21** (1987), 192–205.

11. Kuksin, S.B., Nearly Integrable Infinite-Dimensional Hamiltonian Systems, Lecture Notes in Mathematics, vol. 1556, Springer, Berlin (1993).

12. Kuksin, S.B., Elements of a qualitative theory of Hamiltonian PDEs, Proceedings of ICM 1998, Vol. II, *Doc. Math. J. DMV* (1998), 819–829.

13. Melnikov, V.K., On some cases of conservation of conditionally periodic motions under a small change of the Hamiltonian function, *Dokl. Akad. Nauk SSSR* **165:6** (1965), 1245–1248; *Sov. Math. Dokl.* **6** (1965), 1592–1596.

14. Melnikov, V.K., A family of conditionally periodic solutions of a Hamiltonian system, *Dokl. Akad. Nauk SSSR* **181:3** (1968), 546–549; *Sov. Math. Dokl.* **9** (1968), 882–886.

15. Marmi, S. and Yoccoz, J.-C., Some open problems related to small divisors, (Lecture Notes in Math. 1784). Springer, New York (2002).

16. Pöschel, J., On elliptic lower dimensional tori in hamiltonian systems, *Math. Z.* **202** (1989), 559–608.

17. Pöschel, J., A KAM-Theorem for some nonlinear partial differential equations. *Ann. Sc. Norm. Sup. Pisa* **23** (1996), 119–148.

18. Pöschel, J., Quasi-periodic solution for nonlinear wave equation, *Commun. Math. Helvetici* **71** (1996), 269–296.

19. Rabinowitz, P.H., Free vibrations for a semilinear wave equation. *Commm. Pure Appl. Math.* **31** (1978), 31–68.

20. Wayne, C.E., Periodic and quasi-periodic solutions of nonlinear wave equations via KAM theory, *Comm. Math. Phys*, **127** (1990), no. 3, 479–528.

21. Yuan, X., Invariant manifold of hyperbolic-elliptic type for nonlinear wave equation. *IMMS*, **18** (2003), 1111–1136.

22. Yuan, X., Quasi-periodic solutions of nonlinear Schrödinger equations of higher dimension, *J. Diff. Eq.*, **195** (2003), 230–242.

23. Yuan, X., Quasi-periodic solutions of completely resonant nonlinear wave equations, *J. Differential Equations*, **230** (2006), 213–274.

24. Yuan, X., Invariant tori of nonlinear wave equations with a prescribed potential. *Discrete Continuous Dynamical Sys.*, **16** (2006), no. 3 615–634.

25. Yuan, X., A KAM theorem with applications to partial differential equations of higher dimensions, *Commun. Math. Phys.*, **275** (2007), 97–137.

Four lectures on KAM for the non-linear Schrödinger equation

L.H. Eliasson[1] and S.B. Kuksin[2]

Abstract We discuss the KAM-theory for lower-dimensional tori for the non-linear Schrödinger equation with periodic boundary conditions and a convolution potential in dimension d. Central in this theory is the homological equation and a condition on the small divisors often known as the second Melnikov condition. The difficulties related to this condition are substantial when $d \geqslant 2$.

We discuss this difficulty, and we show that a block decomposition and a Töplitz-Lipschitz-property, present for non-linear Schrödinger equation, permit to overcome this difficuly. A detailed proof is given in [EK06].

1 The non-linear Schrödinger equation

We formulate the equation as an ∞-dimensional Hamiltonian system and as a problem of persistency of lower-dimensional invariant tori.

1.1 The non-linear Schrödinger equation

We consider the Δ-dimensional nonlinear Schrödinger equation

$$-i\dot{u} = -\Delta u + V(x) * u + \varepsilon \frac{\partial F}{\partial \bar{u}}(x, u, \bar{u}) \,,$$

for $u = u(t,x)$ under the periodic boundary condition $x \in \mathbb{T}^d$. The convolution potential $V : \mathbb{T}^d \to \mathbb{C}$ have real Fourier coefficients $\hat{V}(a)$, $a \in \mathbb{Z}^d$, and we shall

---

[1] University of Paris 7, Department of Mathematics, Paris
e-mail: hakane@math.jussieu.se

[2] Heriot-Watt University, Department of Mathematics, Edinburgh
e-mail: Kuksin@math.hw.ac.uk

W. Craig (ed.), Hamiltonian Dynamical Systems and Applications, 179–212.
© 2008 *Springer Science + Business Media B.V.*

suppose it is analytic. (This equation is a popular model for the 'real' NLS equation, where instead of the convolution term $V * u$ we have the potential term Vu.) F is an analytic function in $\mathrm{Re}\, u, \mathrm{Im}\, u$ and x. When $F(x, u, \bar{u}) = (u\bar{u})^2$ this is the cubic Schrödinger equation.

For $\varepsilon = 0$ the equation is linear and has time–quasi-periodic solutions

$$u(t, x) = \sum_{a \in \mathscr{A}} \hat{u}(a) e^{i(|a|^2 + \hat{V}(a))t} e^{i<a,x>},$$

where $\mathscr{A}$ is any finite subset of $\mathbb{Z}^d$ and $|\hat{u}(a)| > 0$. We shall treat $\omega_a = |a|^2 + \hat{V}(a)$, $a \in \mathscr{A}$ as free parameters in some domain $U \subset \mathbb{R}^{\mathscr{A}}$.

For $\varepsilon \neq 0$ we have under general conditions:

If $|\varepsilon|$ is sufficiently small, then there is a large subset U' of U such that for all $\omega \in U'$ the solution u persists as a time–quasi-periodic solution which has all Lyapounov exponents equal to zero and whose linearized equation is reducible to constant coefficients.

In these lectures we shall describe the basic difficulty related to this result – often known as the second Melnikov condition – and the ideas behind its solution. A detailed proof is given in [EK06].

1.2 An ∞-dimensional Hamiltonian system

We write

$$\begin{cases} u(x) = \sum_{a \in \mathbb{Z}^d} u_a e^{i<a,x>} \\ \overline{u(x)} = \sum_{a \in \mathbb{Z}^d} v_a e^{i<-a,x>} \quad (v_a = \bar{u}_a). \end{cases}$$

In the symplectic space $\{(u_a, v_a) : a \in \mathbb{Z}^d\} = \mathbb{C}^{\mathbb{Z}^d} \times \mathbb{C}^{\mathbb{Z}^d}$,

$$i \sum_{a \in \mathbb{Z}^d} du_a \wedge dv_a \,,$$

the equation becomes a Hamiltonian system

$$\begin{cases} \dot{u}_a = i\frac{\partial}{\partial v_a}(h + \varepsilon f) \\ \dot{v}_a = -i\frac{\partial}{\partial u_a}(h + \varepsilon f) \end{cases} \quad a \in \mathbb{Z}^d,$$

with an integrable part

$$h(u, v) = \sum_{a \in \mathbb{Z}^d} (|a|^2 + \hat{V}(a)) u_a v_a$$

plus a perturbation

$$\varepsilon f(u, v) = \varepsilon \frac{1}{(2\pi)^d} \int_{\mathbb{T}^d} F(x, u(x), \overline{u(x)}) dx.$$

The second derivatives of f have a *Töplitz invariance*:

$$\frac{\partial^2 f}{\partial u_{a+c}\partial v_{b+c}} = \frac{\partial^2 f}{\partial u_a \partial v_b}$$

and

$$\frac{\partial^2 f}{\partial u_{a+c}\partial u_{b-c}} = \frac{\partial^2 f}{\partial u_a \partial u_b}$$

(and similar for the second derivatives with respect to v_a, v_b), for any $c \in \mathbb{Z}^d$. This is easy to see for the cubic Schrödinger where

$$f(u,v) = \sum_{a+b-c-d=0} u_a u_b v_c v_d. \tag{1}$$

For example

$$\frac{\partial^2 f}{\partial u_a \partial u_b} = 2 \sum_{c+d=a+b} v_c v_d$$

which clearly have this invariance.

The non-linear Schrödinger is a *real* Hamiltonian system. Indeed if we let

$$\zeta_a = \begin{pmatrix} \xi_a \\ \eta_a \end{pmatrix} = C \begin{pmatrix} u_a \\ v_a \end{pmatrix},$$

with

$$C = \frac{1}{\sqrt{2}} \begin{pmatrix} 1 & 1 \\ -i & i \end{pmatrix}, \tag{2}$$

then, in the symplectic space $\{(\xi_a, \eta_a) =: a \in \mathbb{Z}^d\} = \mathbb{R}^{\mathbb{Z}^d} \times \mathbb{R}^{\mathbb{Z}^d}$,

$$\sum_{a \in \mathbb{Z}^d} d\xi_a \wedge d\eta_a \, ,$$

the equation becomes

$$\begin{cases} \dot{\xi}_a = -\frac{\partial}{\partial \eta_a}(h+\varepsilon f) \\ \dot{\eta}_a = \frac{\partial}{\partial \xi_a}(h+\varepsilon f) \end{cases} \qquad a \in \mathbb{Z}^d,$$

also written $\dot{\zeta}_a = J\frac{\partial}{\partial \zeta_a}(h+\varepsilon f)$, with the integrable part

$$h(\xi,\eta) = \frac{1}{2}\sum_{a \in \mathbb{Z}^d}(|a|^2 + \hat{V}(a))(\xi_a^2 + \eta_a^2)$$

plus the perturbation $\varepsilon f(\xi,\eta)$ which is real, because F is a real function of $\mathrm{Re}\,u$ and $\mathrm{Im}\,u$.

The Töplitz-invariance of the second derivatives can of course be formulated in these coordinates but the description is more complicated (see Sect. 5.2).

1.3 The topology

Let $\mathscr{L}$ be an infinite subset of $\mathbb{Z}^d$. The space

$$l_\gamma^2(\mathscr{L},\mathbb{R}), \quad \gamma \geqslant 0,$$

is the set of sequences of real numbers $\xi = \{\xi_a : a \in \mathscr{L}\}$, such that

$$\|\xi\|_\gamma = \sqrt{\sum_{a\in\mathscr{L}} |\xi_a|^2 \langle a\rangle^{2m_*} e^{2\gamma|a|}} < i \qquad \langle a\rangle = \max(|a|,1).$$

There is a natural identification of $l_\gamma^2(\mathscr{L},\mathbb{R}) \times l_\gamma^2(\mathscr{L},\mathbb{R})$, whose elements are (ξ,η), with $l_\gamma^2(\mathscr{L},\mathbb{R}^2)$, whose elements are $\{\zeta_a = (\xi_a,\eta_a) : a \in \mathscr{L}\}$, and we will not distinguish between them.

We shall assume that $m_* > \frac{d}{2}$. Then, in the phase space $l_0^2(\mathbb{Z}^d,\mathbb{R}^2)$, our Hamiltonian $h + \varepsilon f$ is analytic (in some domain $\mathscr{O}$ in $l_0^2(\mathbb{Z}^d,\mathbb{R})$). To see that f is analytic, consider for example the cubic Schrödinger in the complex variables (1). Using the estimate

$$\sum_a |u_a| \leqslant \sqrt{\sum_a \langle a\rangle^{-2m_*}} \, \|u\|_0,$$

we have

$$|f(u,v)| \leqslant \|u\|_0^2 \|v\|_0^2,$$

and it follows easily that f is analytic.

Let $<,>$ denote the "pairing"

$$<\zeta,\zeta'> = \sum_{a\in\mathbb{Z}^d} \left(\xi_a\xi_a' + \eta_a\eta_a'\right).$$

Since the phase space is a Hilbert space, its first differential

$$l_0^2(\mathscr{L},\mathbb{R}^2) \ni \hat{\zeta} \mapsto <\hat{\zeta},\partial_\zeta f(\zeta)>$$

defines a vector $\partial_\zeta f(\zeta)$, its "*gradient*", (with respect to the pairing), and its second differential

$$l_0^2(\mathscr{L},\mathbb{R}^2) \ni \hat{\zeta} \mapsto \frac{1}{2}<\hat{\zeta},\partial_\zeta^2 f(\zeta)\hat{\zeta}>$$

defines a matrix $\partial_\zeta^2 f(\zeta) : \mathscr{L} \times \mathscr{L} \to gl(2,\mathbb{R})$, its "*Hessian*", (with respect to the pairing), which is symmetric, i.e.

$$^t\!\left(\frac{\partial^2 f}{\partial\zeta_a\partial\zeta_b}(\zeta)\right) = \frac{\partial^2 f}{\partial\zeta_b\partial\zeta_a}(\zeta).$$

For $\zeta \in \mathscr{O} \cap l_\gamma^2(\mathbb{Z}^d,\mathbb{R}^2)$, $\gamma > 0$, the gradient and the Hessian verifies certain properties of *exponential decay*. These properties are most easily seen in the complex variables u,v. For example for the cubic Schrödinger (1) the gradient of f verifies

$$\left|\frac{\partial f}{\partial u_a}\right| \leqslant \text{cte.} \|u\|_\gamma \|v\|_\gamma^2 e^{-\gamma|a|}.$$

(and similar for the derivative with respect to v_a). The Hessian of f verifies

$$\left|\frac{\partial^2 f}{\partial u_a \partial v_c}\right| \leqslant \text{cte.} \|u\|_\gamma \|v\|_\gamma e^{-\gamma|a-c|},$$

and

$$\left|\frac{\partial^2 f}{\partial u_a \partial u_b}\right| \leqslant \text{cte.} \|v\|_\gamma^2 e^{-\gamma|a+b|}$$

(and similar for the second derivative with respect to v_c, v_d).

The exponential decay of the second derivatives can of course be formulated in the real coordinates (ξ, η) but the description is again more complicated (see Sect. 5.2).

1.4 Action-angle variables

Let $\mathscr{A}$ be a finite subset of $\mathbb{Z}^d$ and fix

$$0 < p_a, \quad a \in \mathscr{A}.$$

The $(\#\mathscr{A})$-dimensional torus

$$\tfrac{1}{2}(\xi_a^2 + \eta_a^2) = p_a, \, a \in \mathscr{A}$$
$$\xi_a = \eta_a = 0, \qquad a \in \mathscr{L} = \mathbb{Z}^d \setminus \mathscr{A},$$

is invariant for the Hamiltonian flow when $\varepsilon = 0$. In the symplectic subspace $\mathbb{R}^{\mathscr{A}} \times \mathbb{R}^{\mathscr{A}}$ we introduce, in a neighborhood of this torus, action-angle variables (r_a, φ_a), $a \in \mathscr{A}$,

$$\xi_a = \sqrt{2(p_a + r_a)} \cos(\varphi_a)$$
$$\eta_a = \sqrt{2(p_a + r_a)} \sin(\varphi_a).$$

These coordinates are analytic near $r = 0$ because the p_a's are all positive.

In these coordinates the Hamiltonian equations becomes

$$\begin{cases} \dot\zeta_a = J\frac{\partial}{\partial \zeta_a}(h + \varepsilon f) & a \in \mathscr{L} \\[2mm] \dot r_a = -\frac{\partial}{\partial \varphi_a}(h + \varepsilon f) \\[2mm] \dot\varphi_a = \frac{\partial}{\partial r_a}(h + \varepsilon f) \end{cases} \quad a \in \mathscr{A}$$

with the integrable part

$$h(\xi, \eta, r) = \sum_{a \in \mathscr{A}} \omega_a r_a + \frac{1}{2} \sum_{a \in \mathscr{L}} \Omega_a(\xi_a^2 + \eta_a^2)$$

(modulo a constant), where

$$\omega_a = |a|^2 + \hat{V}(a), \quad a \in \mathscr{A},$$

are the *basic frequencies*, and

$$\Omega_a = |a|^2 + \hat{V}(a), \quad a \in \mathscr{L},$$

are the *normal frequencies* (of the invariant torus). The perturbation $\varepsilon f(\xi, \eta, r, \varphi)$ will be a function of all variables (under the assumption, of course, that the torus lies in the domain of F).

Since $h + \varepsilon f$ is analytic in (some domain in) the phase space $l_0^2(\mathscr{L}, \mathbb{R}^2) \times \mathbb{R}^{\mathscr{A}} \times \mathbb{T}^{\mathscr{A}}$, it extends to a holomorphic function on a complex domain

$$\mathscr{O}^0(\sigma, \mu, \rho) = \begin{cases} \|\zeta\|_0 = \sqrt{\|\xi\|_0^2 + \|\eta\|_0^2} < \sigma \\ |r| < \mu \\ |\operatorname{Im}\varphi| < \rho. \end{cases}$$

1.5 Statement of the result

The Hamiltonian $h + \varepsilon f$ is a standard form for the perturbation theory of lower-dimensional (isotropic) tori with one exception: it is strongly degenerate. We therefore need external parameters to control the basic frequencies and the simplest choice is to let the basic frequencies (i.e. the potential itself) be our free parameters. The parameters will belong to a set

$$U \subset \{\omega \in \mathbb{R}^{\mathscr{A}} : |\omega| \leqslant C_1\}. \tag{3}$$

The potential V will be analytic and

$$|\hat{V}(a)| \leqslant C_2 e^{-C_3|a|}, \ C_3 > 0, \ \forall a \in \mathscr{L}. \tag{4}$$

The normal frequencies will be assumed to verify

$$\begin{cases} |\Omega_a| \geqslant C_4 > 0 \\ |\Omega_a + \Omega_b| \geqslant C_4 \qquad \forall a, b \in \mathscr{L}. \\ |\Omega_a - \Omega_b| \geqslant C_4 \ |a| \neq |b| \end{cases} \tag{5}$$

This is fulfilled, for example, if V is small and $\mathscr{A} \ni 0$, or if V is arbitrary and $\mathscr{A}$ is sufficiently large.

Theorem 1.1. *Under the above assumptions, for ε sufficiently small there exist a subset $U' \subset U$, which is large in the sense that*

$$\operatorname{Leb}(U \setminus U') \leqslant \operatorname{cte}.\varepsilon^{\exp},$$

and for each $\omega \in U'$, a real analytic symplectic diffeomorphism Φ

$$\mathscr{O}^0\left(\frac{\sigma}{2}, \frac{\mu}{2}, \frac{\rho}{2}\right) \to \mathscr{O}^0(\sigma, \mu, \rho)$$

and a vector $\omega' = \omega'(\omega)$ such that $(h_{\omega'} + \varepsilon f) \circ \Phi$ equals (modulo a constant)

$$<\omega, r> + \frac{1}{2} <\zeta, A(\omega)\zeta> + \varepsilon g,$$

where

$$g \in \mathscr{O}(|r|^2, |r|\,\|\zeta\|_0, \|\zeta\|_0^3)$$

and the symmetric matrix $A(\omega)$ has the form

$$\begin{pmatrix} \Omega_1(\omega) & \Omega_2(\omega) \\ {}^t\Omega_2(\omega) & \Omega_1(\omega) \end{pmatrix}$$

with $\Omega_1 + i\Omega_2$ Hermitian and block-diagonal, with finite-dimensional blocks.
Moreover, $\Phi = (\Phi_\zeta, \Phi_r, \Phi_\varphi)$ verifies, for all $(\zeta, \varphi, r) \in \mathscr{O}^0(\frac{\sigma}{2}, \frac{\mu}{2}, \frac{\rho}{2})$,

$$\left\|\Phi_\zeta - \zeta\right\|_0 + \left|\Phi_r - \rho\right| + \left|\Phi_\varphi - \varphi\right| \leqslant \beta\varepsilon,$$

and the mapping $\omega \mapsto \omega'(\omega)$ verifies

$$\left|\omega' - \mathrm{id}\right|_{\mathrm{Lip}(U')} \leqslant \beta\varepsilon.$$

The exponent $\exp$ only depends on the dimensions $d, \#\mathscr{A}, m_*$, the constant $\mathrm{cte.}$ depends on the dimensions and on $C_1, \ldots, C_4$, and the constant β also depends on V and F.

It follows from this theorem that $\Phi(\{0\} \times \{0\} \times \mathbb{T}^{\mathscr{A}})$ is a KAM-torus for the Hamiltonian system of $h + \varepsilon f$, and it implies the result mentioned in Sect. 1.1. We discuss this notion and its consequences in the next section.

Theorem 1.1, as well as a more generalized version, is proven in [EK06].

1.6 KAM-tori

A *KAM-torus* of a Hamiltonian system in $\mathbb{R}^{2\mathscr{L}} \times \mathbb{R}^{\mathscr{A}} \times \mathbb{T}^{\mathscr{A}}$ is a finite-dimensional torus satisfying

(i) *Invariance* – it is invariant under the Hamiltonian flow
(ii) *Linearity* – the flow on the torus is conjugate to a linear flow $\varphi \mapsto \varphi + t\omega$

   A torus with the two properties (i) + (ii) is nothing more and nothing less than a *quasi-periodic solution* when translated into cartesian coordinates. Often, as we shall do in this paper, one also requires

(iii) *Reducibility* – the linearized equations (the "variational equations") on the torus are conjugate to a constant coefficient system of the form

$$\begin{cases} \frac{d\hat{\zeta}}{dt} = JA\hat{\zeta} \\[6pt] \frac{d\hat{r}}{dt} = 0 \\[6pt] \frac{d\hat{\varphi}}{dt} = \beta\hat{r} \end{cases}$$

and JA has a pure point spectrum

If the quasi-periodic solution has property (iii), then questions related to linear stability and Lyapunov exponents "reduce" to a study of a linear system of constant coefficients, which permits (at least for finite-dimensional systems) to answer such questions and to construct higher order normal forms near the torus.

Reducibility is automatic in two cases: if the torus is one-dimensional (and phase-space is finite-dimensional) it is just a periodic solution, and (iii) is a general fact called Floquet theory; if the torus is Lagrangian (i.e. there is not ζ-part), then (iii) follows from (i) + (ii) by a simple integration [dlL01]. In general, however, it is a delicate property which is far from being completely understood.

KAM is a perturbation theory of KAM-tori. Not only is reducibility an important outcome but also an essential ingredient in the proof. It simplifies the iteration since it reduces all approximate linear equations to constant coefficients. But it does not come for free. It requires a lower bound on small divisors of the form

$$(**) \qquad |{<}k,\omega{>} + \Omega_a(\omega) \pm \Omega_b(\omega)|, \quad k \in \mathbb{Z}^{\mathscr{A}}, \ a,b \in \mathscr{L},$$

where $\Omega_a(\omega)$, $a \in \mathscr{L}$ are the imaginary parts of the eigenvalues of $JA(\omega)$ The basic frequencies ω will be fixed during the iteration – that is what parameters are there for – but the normal frequencies will vary. Indeed the $\Omega_a(\omega)$ are perturbations of $|a|^2 + \hat{V}(a)$ which are not known a priori but are determined by the approximation process.[1]

The difficulty associated with the small divisors $(**)$ may be very large. There is a perturbation theory which avoids this difficulty, but to a high cost: the approximate linear equations are no longer of constant coefficients. Moreover it gives persistence of the invariant tori but no reducibility.

1.7 Consequences of Theorem 1

The consequences of the theorem is that $\Phi(\{0,0\} \times \mathbb{T}^{\mathscr{A}})$ is a KAM-torus for $h_{\omega'} + \varepsilon f$. In order to see this it suffices to show that $\{\zeta = r = 0\}$ is a KAM-torus for $k + \varepsilon g$,

---

[1] A lower bound on $(**)$ is strictly speaking not necessary for reducibility. It is necessary, however, in order to have reducibility with a reducing transformation close to the identity.

$$k = <\omega, r> + \frac{1}{2} <\zeta, A(\omega)\zeta> .$$

Since

$$\frac{\partial g}{\partial \zeta} = \frac{\partial g}{\partial \varphi} = \frac{\partial g}{\partial r} = 0$$

for $\zeta = r = 0$, it follows that $\{\zeta = r = 0\}$ is invariant with a flow $\varphi \mapsto \varphi + t\omega$. The linearized equations on this torus become

$$\begin{cases} \dfrac{d\hat{\zeta}}{dt} = JA(\omega)\hat{\zeta} + \varepsilon Ja(\varphi + t\omega, \omega)\hat{r} \\[2mm] \dfrac{d\hat{r}}{dt} = 0 \\[2mm] \dfrac{d\hat{\varphi}}{dt} = \varepsilon <{}^t a(\varphi + t\omega, \omega), \hat{\zeta}> + \varepsilon b(\varphi + t\omega, \omega)\hat{r} \end{cases}$$

where[2] $a(\varphi) = \frac{\partial^2}{\partial r \partial \zeta} g(0,0,\varphi)$ and $b(\varphi) = \frac{\partial^2}{\partial r^2} g(0,0,\varphi)$.

These equations can be conjugated to constant coefficients if the imaginary part of the the eigenvalues of $JA(\omega)$,

$$\pm i\Omega_a(\omega), \quad a \in \mathcal{L},$$

are non-resonant with respect to ω. In order to see this we consider the equations

(i)
$$<\partial_\varphi Z_1(\varphi), \omega> = JAZ_1(\varphi) + \varepsilon Ja(\varphi),$$

which has a unique smooth solution if ω and $\Omega_a(\omega)$, $a \in \mathcal{L}$, verify an appropriate Diophantine condition

(ii)
$$<\partial_\varphi Z_2(\varphi), \omega> = -Z_2(\varphi)JA + \varepsilon {}^t a(\varphi)$$

which has a unique smooth solution under the same condition on ω

(iii)
$$<\partial_\varphi Z_3(\varphi), \omega> = \varepsilon \, {}^t a(\varphi)Z_1(\varphi) + \varepsilon b(\varphi) - \varepsilon\beta$$

which has a smooth solution if ω is Diophantine and if we chose β such that the meanvalue of the right hand side is $= 0$.

If we now take

$$Z(\varphi) = \begin{pmatrix} I & Z_1(\varphi) & 0 \\ 0 & I & 0 \\ Z_2(\varphi) & Z_3(\varphi) & I \end{pmatrix},$$

____

[2] t is used both as the independent time-variable and to denote transposition, without confusion we hope.

then $<\frac{\partial Z}{\partial \varphi}(\varphi), \omega>$

$$= \begin{pmatrix} JA & \varepsilon Ja(\varphi) & 0 \\ 0 & 0 & 0 \\ \varepsilon' a(\varphi) & \varepsilon b(\varphi) & 0 \end{pmatrix} Z(\varphi) - Z(\varphi) \begin{pmatrix} JA & 0 & 0 \\ 0 & 0 & 0 \\ 0 & \varepsilon \beta & 0 \end{pmatrix},$$

so Z conjugates the linearized equations to

$$\begin{cases} \dfrac{d\hat{\zeta}}{dt} = JA(\omega)\hat{\zeta} \\ \dfrac{d\hat{r}}{dt} = 0 \\ \dfrac{d\hat{\varphi}}{dt} = \varepsilon \beta \hat{r} \end{cases}$$

which is constant coefficients.

The conditions on ω will hold if we restrict the set U' arbitrarily little.

If

$$C = \frac{1}{\sqrt{2}} \begin{pmatrix} I & I \\ -iI & iI \end{pmatrix}, \tag{6}$$

then

$$C^{-1} JA(\omega) C = i \begin{pmatrix} {}^t\Omega(\omega) & 0 \\ 0 & -\Omega(\omega) \end{pmatrix},$$

since $\Omega(\omega) = \Omega_1(\omega) + i\Omega_2(\omega)$ is Hermitian. Moreover, there is a unitary matrix $D = D(\omega)$ such that

$$ {}^t\bar{D}\Omega D = \operatorname{diag}(\Omega_a)$$

is a real diagonal matrix, and therefore

$$\begin{pmatrix} D & 0 \\ 0 & \bar{D} \end{pmatrix}^{-1} i \begin{pmatrix} {}^t\Omega & 0 \\ 0 & -\Omega \end{pmatrix} \begin{pmatrix} D & 0 \\ 0 & \bar{D} \end{pmatrix} = i \begin{pmatrix} \operatorname{diag}(\Omega_a) & 0 \\ 0 & -\operatorname{diag}(\Omega_a) \end{pmatrix}$$

So the linearized equations on the torus have only quasi-periodic solutions and, hence, the torus is linearly stable.

1.8 References

For finite dimensional Hamiltonian systems the first proof of persistence of stable (i.e. vanishing of all Lyapunov exponents) lower dimensional invariant tori was obtained in [Eli85, Eli88] and there are now many works on this subjects. There are also many works on reducibility (see for example [Kri99, Eli01]) and the situation in finite dimension is now pretty well understood in the perturbative setting. Not so, however, in infinite dimension.

If $d = 1$ and the space-variable x belongs to a finite segment supplemented by Dirichlet or Neumann boundary conditions, this result was obtained in [Kuk88] (also see [Kuk93, Pös96]). The case of periodic boundary conditions was treated in [Bou96], using another multi-scale scheme, suggested by Fröhlich–Spencer in their work on the Anderson localization [FS83]. This approach, often referred to as the Craig–Wayne scheme, is different from the KAM-scheme described here. It avoids the cumbersome condition $(**)$ but to a high cost: the approximate linear equations are not of constant coefficients. Moreover, it gives persistence of the invariant tori but no reducibility and no information on the linear stability. A KAM-theorem for periodic boundary conditions has recently been proved in [GY05] (with a perturbation F independent of x) and the perturbation theory for quasi-periodic solutions of one-dimensional Hamiltonian PDE is now sufficiently well developed (see for example [Kuk93, Cra00, Kuk00]).

The study of the corresponding problems for $d \geqslant 2$ is at its early stage. Developing further the scheme, suggested by Fröhlich–Spencer, Bourgain proved persistence for the case $d = 2$ [Bou98]. More recently, the new techniques developped by him and collaborators in their work on the linear problem has allowed him to prove persistence in any dimension d [Bou04]. (In this work he also treats the non-linear wave equation.) For another, and simplified, proof of this result see [Yua07].

1.9 Notation

$<,>$ is the standard scalar product in $\mathbb{R}^d$. $\| \; \|$ is an operator-norm or l^2-norm. $|\;|$ will in general denote a supremum norm, with a notable exception: for a lattice vector $a \in \mathbb{Z}^d$ we use $|a|$ for the l^2-norm.

$\mathscr{A}$ is a finite subset of $\mathbb{Z}^d$, and $\mathscr{L}$ is its complement. A matrix on $\mathscr{L}$ is just a mapping $A : \mathscr{L} \times \mathscr{L} \to \mathbb{C}$ or $gl(2,\mathbb{C})$. Its components will be denoted A_a^b. If A_1, A_2, A_3, A_4 are scalar-valued matrices on $\mathscr{L}$, then we identify

$$A = \begin{pmatrix} A_1 & A_2 \\ A_3 & A_4 \end{pmatrix}$$

with a $gl(2,\mathbb{C})$-valued matrix through

$$A_a^b = \begin{pmatrix} (A_1)_a^b & (A_2)_a^b \\ (A_3)_a^b & (A_4)_a^b \end{pmatrix}.$$

The dimension d will be fixed and m_* will be a fixed constant $> \frac{d}{2}$.

$\lesssim$ means $\leqslant$ modulo a multiplicative constant that only, unless otherwise specified, depends on d, m_* and $\#\mathscr{A}$.

The points in the lattice $\mathbb{Z}^d$ will be denoted $a, b, c, \ldots$. Also d will sometimes be used, without confusion we hope.

For two subsets X and Y of a metric space,

$$\text{dist}(X,Y) = \inf_{x \in X, y \in Y} d(x,y).$$

(This is not a metric.) X_ε is the ε-neighborhood of X, i.e.

$$\{y : \text{dist}(y,X) < \varepsilon\}.$$

Let $B_\varepsilon(x)$ be the ball $\{y : d(x,y) < \varepsilon\}$. Then X_ε is the union, over $x \in X$, of all $B_\varepsilon(x)$. If X and Y are subsets of $\mathbb{R}^d$ or $\mathbb{Z}^d$ we let

$$X - Y = \{x - y : x \in X,\ y \in Y\}$$

– not to be confused with the set theoretical difference $X \setminus Y$.

2 The homological equation

Here we shall describe shortly the quadratic iteration and derive the homological equation which is central in KAM.

2.1 Normal form Hamiltonians

This is a real Hamiltonian of the form

$$h = <\omega, r> + \frac{1}{2} <\zeta, A(\omega)\zeta>, \quad \text{(modulo a constant)}$$

where

$$A = \begin{pmatrix} \Omega_1 & \Omega_2 \\ {}^t\Omega_2 & \Omega_1 \end{pmatrix}$$

is block-diagonal matrix with finite-dimensional blocks (we shall say more about these blocks in Sect. 3) and $\Omega(\omega) = \Omega_1(\omega) + i\Omega_2(\omega)$ is Hermitian. Since $\Omega(\omega)$ is Hermitian the eigenvalues of $JA(\omega)$ are

$$\pm i\Omega_a(\omega) \quad a \in \mathscr{L},$$

where the $\Omega_a(\omega)$ are the (necessarly real) eigenvalues of $\Omega(\omega)$. (See the discussion in Sect. 2.2.)

We also suppose $A(\omega)$ to be close to

$$\begin{pmatrix} \text{diag}(|a|^2 + \hat{V}(a) & 0 \\ 0 & \text{diag}(|a|^2 + \hat{V}(a)) \end{pmatrix}$$

and

$$\|\partial_\omega \Omega(\omega)\| \leqslant \frac{1}{4}.$$

This implies that $\Omega_a(\omega)$ is

$$\approx |a|^2 + \hat{V}(a)$$

and $\mathcal{C}^1$ (or Lipschitz) -small in ω.

2.2 The KAM-iteration

Given a normal form Hamiltonian

$$h = <\omega, r> + \frac{1}{2} <\zeta, A(\omega)\zeta>$$

and a perturbation f. Let Tf be the Taylor polynomial

$$f(0,0,\varphi) + <\frac{\partial f}{\partial r}(0,0,\varphi), r> + <\frac{\partial f}{\partial \zeta}(0,0,\varphi), \zeta> + \frac{1}{2} <\zeta, \frac{\partial^2 f}{\partial \zeta^2}(0,0,\varphi)\zeta>$$

of f – it may also depend on ω.

If Tf was $= 0$ then $\{\zeta = r = 0\}$ would be a KAM-torus for $h + f$. But in general we only have

$$Tf \in \mathcal{O}(\varepsilon).$$

Suppose now there exist a Taylor polynomial s of the same form, i.e. $s = Ts$, and a normal form Hamiltonian

$$k = c(\omega) + <\chi(\omega), r> + \frac{1}{2} <\zeta, B(\omega)\zeta>$$

verifying

$$\{h, s\} = -Tf + k, \tag{1}$$

where $\{ \, , \, \}$ is the Poisson bracket associated to the symplectic form $\sum d\xi_a \wedge d\eta_a + \sum dr_a \wedge d\varphi_a$. This equation is known as the *homological equation*.

Let Φ^t be the flow of

$$\begin{cases} \dot{\zeta} = J\frac{\partial s}{\partial \zeta}(\zeta, \varphi, r) \\ \dot{r} = -\frac{\partial s}{\partial \varphi}(\zeta, \varphi, r) \\ \dot{\varphi} = \frac{\partial s}{\partial r}(\zeta, \varphi, r). \end{cases}$$

If $s, k \in \mathcal{O}(\varepsilon)$, then $(\Phi^t - \mathrm{id}) \in \mathcal{O}(\varepsilon)$ and

$$\begin{aligned} (h + f) \circ \Phi^1 &= h + k + \int_0^1 \frac{d}{dt}(h + tf + (1-t)k) \circ \Phi^t dt \\ &= h + k + \int_0^1 (\{h + tf + (1-t)k, s\} + f - k) \circ \Phi^t dt \\ &= h + k + \int_0^1 (\{tf + (1-t)k, s\} + f - Tf) \circ \Phi^t dt \\ &= h + k + [(f - Tf) + f_1]. \end{aligned}$$

So Φ^1 transforms $h+f$ to a new normal form $h = h+k$ plus a new perturbation f'. Since

$$T(f') \in \mathcal{O}(\varepsilon^2),$$

also

$$f' \in \mathcal{O}(\varepsilon^2)$$

when the domain is sufficiently restricted.

If we can solve the homological equation (1), not only for the normal form Hamiltonian h but also for all normal form Hamiltonians h', close to h, then we will be able to make an iteration which will converge to a solution as in Theorem 1.1 if the estimates a good enough. So the basic thing in KAM is to solve and estimate the solution of the homological equation.

It is clear from the discussion above that it is enough to solve a slightly weaker version of the homological equation, namely

$$\{h, s\} = -Tf + k + \mathcal{O}(\varepsilon^2). \tag{2}$$

2.3 *The components of the homological equation*

We write s as

$$S_{01}(\varphi) + \,<S_{02}(\varphi), r> + \,<S_1(\varphi), \zeta> + \frac{1}{2}<\zeta, S_2(\varphi)\zeta>$$

and k as

$$c + <\chi, r> + \frac{1}{2}<\zeta, B\zeta> \,.$$

The homological equation (2) now decomposes into four linear equations. The first two are

$$\begin{cases} <\partial_\varphi S_{01}(\varphi), \omega> = -f(0,0,\varphi) + c + \mathcal{O}(\varepsilon^2) \\ <\partial_\varphi S_{02}(\varphi), \omega> = -\frac{\partial f}{\partial r}(0,0,\varphi) + \chi + \mathcal{O}(\varepsilon^2). \end{cases} \tag{3}$$

In these equations, we are forced to take

$$c = <f(0,0,\cdot)> \quad \text{and} \quad \chi = <\frac{\partial f}{\partial r}(0,0,\cdot)>,$$

where $<g>$ is the mean value

$$\frac{1}{(2\pi)^d} \int_{\mathbb{T}^d} g(\varphi)d\varphi.$$

The other two are

$$<\partial_\varphi S_1(\varphi), \omega> + JAS_1(\varphi) = -\frac{\partial f}{\partial \zeta}(0,0,\varphi) + \mathcal{O}(\varepsilon^2) \tag{4}$$

and

$$<\partial_\varphi S_2(\varphi), \omega> + AJS_2(\varphi) - S_2(\varphi)JA = -\frac{\partial^2 f}{\partial \zeta^2}(0,0,\varphi) + B + \mathcal{O}(\varepsilon^2). \qquad (5)$$

The most delicate of these equations is the last one which is related to reducibility. Let

$$B = \begin{pmatrix} \Omega_1' & \Omega_2' \\ {}^t\Omega_2' & \Omega_1' \end{pmatrix}, \qquad \Omega' = \Omega_1' + i\Omega_2',$$

and

$$F(\varphi) = \frac{\partial^2 f}{\partial \zeta^2}(0,0,\varphi).$$

If we write $\tilde{F}(\varphi) = {}^t C F(\varphi) C$ and $\tilde{S}_2(\varphi) = {}^t C S_2(\varphi) C$, then (5) becomes

$$<\partial_\varphi \tilde{S}_2(\varphi), \omega> - i \begin{pmatrix} 0 & \Omega \\ {}^t\Omega & 0 \end{pmatrix} J\tilde{S}_2(\varphi) + i\tilde{S}_2(\varphi)J \begin{pmatrix} 0 & \Omega \\ {}^t\Omega & 0 \end{pmatrix}$$

$$= -\tilde{F}(\varphi) + \begin{pmatrix} 0 & \Omega' \\ {}^t\Omega' & 0 \end{pmatrix} + \mathcal{O}(\varepsilon^2).$$

This equation decouples into four equations for scalar-valued matrices. These are of the form

$$<\partial_\varphi R(\varphi), \omega> + i(\Omega R(\varphi) + R(\varphi){}^t\Omega) = G(\varphi) + \mathcal{O}(\varepsilon^2), \qquad (6)$$

for the diagonal terms, and of the form

$$<\partial_\varphi R(\varphi), \omega> + i(\Omega R(\varphi) - R(\varphi)\Omega) = G(\varphi) - \Omega' + \mathcal{O}(\varepsilon^2) \qquad (7)$$

for the off-diagonal terms.

The last equation is underdetermined and there are several possible choices of Ω'. One such choice would be $<G>$ which would give an Hermitian matrix, but in general not a block diagonal matrix. So the Hamiltonian $h' = h + k$ would not be on normal form. Instead we shall make a "smaller" choice.

Due to the exponential decay of the second order derivatives of the Hamiltonian (discussed in Sect. 1.3) the matrix G verifies

$$|G(\varphi)_a^b| \lesssim \varepsilon e^{-\gamma|a-b|} \quad a,b \in \mathcal{L},$$

and we can truncate the matrices away from the diagonal at distance

$$\Delta' \approx \log(\frac{1}{\varepsilon}).$$

We then take

$$(\Omega')_a^b = \begin{cases} <G_a^b> & \text{if } |a| = |b|, \ |a-b| \leqslant \Delta' \\ 0 & \text{if not} \end{cases} \qquad (8)$$

Since the left hand side of the equations (3–7) are linear operators with constant coefficients, equations (3–7) + (8) can be solved in Fourier series, and to get a solution we must prove the convergence of these Fourier series and estimate the

solution. This requires good estimates on the small divisors, i.e. the eigenvalues of
the linear operators in the left hand side.

2.4 Small divisors and the second Melnikov condition

Since the equations are to be solved only modulo $\mathcal{O}(\varepsilon^2)$ and since all functions are
analytic in φ, we can truncate all Fourier series at order

$$\Delta' \approx \log\left(\frac{1}{\varepsilon}\right).$$

We want to bound the eigenvalues (in absolute value) in the left hand side from
below by some quantity κ which should be small but much larger than ε, say

$$\kappa = \varepsilon^{\exp}$$

for some small exponent.

For (3), the eigenvalues of the left hand side operator are

$$i <k,\omega> \quad k \in \mathbb{Z}^{\mathscr{A}},\, 0 < |k| \leqslant \Delta'.$$

These are all larger (in absolute value) than κ for $\omega \in U$ except on a small set of
Lebesgue measure

$$\lesssim (\Delta')^{\#\mathscr{A}}\,\kappa.$$

The eigenvalues in (4) are

$$i < k,\omega > +i\Omega_a(\omega) \quad k \in \mathbb{Z}^{\mathscr{A}},\, |k| \leqslant \Delta',\quad a \in \mathscr{L},$$

where the $i\Omega_a(\omega)$:s are the eigenvalues of $A(\omega)$. By the assumption on $A(\omega)$,

$$\Omega_a(\omega) \approx |a|^2 + \hat{V}(a)$$

and is $\mathcal{C}^1$-small in ω. Therefore there are only finitely many eigenvalues which are
not large, and these can be controlled by an appropriate choice of ω.

Equation (6) is treated in the same way.

It is (7) which gives rise to serious problems. If we define Ω' by (8) and take into
account the exponential decay of the matrices, then the eigenvalues of (7) are

$$i(\Omega_a(\omega) - \Omega_b(\omega)) \quad k = 0,\, |a - b| \leqslant \Delta',\, |a| \neq |b|,$$

(which are all $\gtrsim 1$ by assumption (5) of Section 1) and

$$\begin{cases} i <k,\omega> +\Omega_a(\omega) - \Omega_b(\omega)) \\ 0 < |k| \leqslant \Delta', |a - b| \leqslant \Delta'. \end{cases} \tag{9}$$

In one space dimension $d = 1$ we have

$$|\Omega_a(\omega) - \Omega_b(\omega)| \to \infty$$

when $|a| \to \infty$, $|a - b| \leqslant \Delta'$, except for $a = b$. Therefore there are only finitely many eigenvalues which are not large, and these can be controlled by an appropriate choice of ω.

But in dimension $d \geqslant 2$ there are infinitey many eigenvalues which are not large. How to control (9) – known as the second Melnikov condition – is the main difficulty in the proof. But before we turn to this question we shall discuss more closely the normal form.

3 Normal form Hamiltonians

We shall discuss the block-diagonal property and the Töplitz–Lipschitz-property of the normal form Hamiltonians.

3.1 Blocks

In this section $d \geqslant 2$. For a non-negative integer Δ we define an *equivalence relation* on $\mathscr{L}$ generated by the pre-equivalence relation

$$a \sim b \iff \begin{cases} |a|^2 = |b|^2 \\ |a - b| \leqslant \Delta. \end{cases}$$

Let $[a]_\Delta$ denote the equivalence class (*block*) of a, and let $\mathcal{E}_\Delta$ be the set of equivalence classes. It is trivial that each block $[a]$ is finite with cardinality

$$\lesssim |a|^{d-1}$$

that depends on a. But there is also a uniform Δ-dependent bound.

Lemma 1 *Let*

$$d_\Delta = \sup_a (\mathrm{diam}[a]_\Delta).$$

Then

$$d_\Delta \lesssim \Delta^{\frac{(d+1)!}{2}}.$$

Proof. We give the proof in dimension $d = 2$, the general case being treated in Sect. 4 of [EK06].

It suffices to consider the case when there are $a, b, c \in [a]_\Delta$ such that $a - b$ and $a - c$ are linearly independent and

$$|a - b|, |a - c| \leqslant \Delta.$$

(If not, then $[a]_\Delta = \{a,b\}$ and the result is obvious.)

Since $|a|^2 = |b|^2 = |c|^2$ it follows that

$$\begin{cases} <a,a-b> = \tfrac{1}{2}|a-b|^2 \\ <a,a-c> = \tfrac{1}{2}|a-c|^2 \end{cases}$$

Since $a-b$ and $a-c$ are integer-valued independent vectors it follow from this that

$$|a| \lesssim \Delta^3. \qquad \square$$

The blocks $[a]_\Delta$ have a rigid structure when $|a|$ is large. For a vector $c \in \mathbb{Z}^d \setminus 0$ let

$$a_c \in (a + \mathbb{R}c) \cap \mathbb{Z}^d$$

be the lattice point b on the line $a + \mathbb{R}c$ with smallest norm – if there are two such b's we choose the one with $<b,c> \geqslant 0$.

Lemma 2 *Given a and $c \neq 0$ in $\mathbb{Z}^\Delta$. For all t, such that*

$$|a+tc| \geqslant d_\Delta^2(|a_c|+|c|)\,|c|,$$

the set $[a+tc]_\Delta - (a+tc)$ is independent of t and $\perp$ to c.

Proof. It suffices to prove this for $a = a_c$.

Let $b \in [a+tc]_\Delta - (a+tc)$ for some fixed t as in the lemma. This implies that $|b| \leqslant d_\Delta$ and that $|b+a+tc|^2 = |a+tc|^2$. This last equality can be written

$$2t <b,c> +2 <b,a> +|b|^2 = 0.$$

If $<b,c> \neq 0$, then

$$\begin{aligned} |a+tc| &\leqslant |a| + |t <b,c> ||c| \\ &= |a| + |<b,a>||c| + \tfrac{1}{2}|b|^2||c| \\ &\leqslant (1+d_\Delta)|a||c| + \tfrac{1}{2}d_\Delta^2|c|^2), \end{aligned}$$

but this is impossible under the assumption on $a+tc$.

Therefore $<b,c> = 0$ and hence $[a+tc]_\Delta - (a+tc) \perp c$. Moreover $|b+a+sc|^2 = |a+sc|^2$ for all s, so if $|b| \leqslant \Delta$, then

$$[b+a+sc]_\Delta = [a+sc]_\Delta \quad \forall s.$$

To conclude, let $b_0 = a, b_1, \ldots, b_n$ be the elements of $[a]_\Delta$ ordered in such a way that $|b_{j+1} - b_j| \leqslant \Delta$ for all j. Then the preceding argument shows that

$$[b+a+sc]_\Delta = [a+sc]_\Delta \quad \forall s, \forall j. \qquad \square$$

Description of blocks when $d = 2,3$. For $d = 2$, we have outside $\{|a| :\leqslant d_\Delta \approx \Delta^3\}$

$\star$ Rank$[a]_\Delta = 1$ if, and only if, $a \in \tfrac{b}{2} + b^\perp$ for some $0 < |b| \leqslant \Delta$ – then $[a]_\Delta = \{a,a-b\}$

$\star$ $\mathrm{Rank}[a]_\Delta = 0$ otherwise – then $[a]_\Delta = \{a\}$

For $d = 3$, we have outside $\{|a| :\leqslant d_\Delta \approx \Delta^{12}\}$

$\star$ $\mathrm{Rank}[a]_\Delta = 2$ if, and only if, $a \in \frac{b}{2} + b^\perp \cap \frac{c}{2} + c^\perp$ for some $0 < |b|, |c| \leqslant 2\Delta$ linearly independent – then $[a]_\Delta \supset \{a, a-b, a-c\}$

$\star$ $\mathrm{Rank}[a]_\Delta = 1$ if, and only if, $a \in \frac{b}{2} + b^\perp$ for some $0 < |b| \leqslant \Delta$ – then $[a]_\Delta = \{a, a-b\}$

$\star$ $\mathrm{Rank}[a]_\Delta = 0$ otherwise – then $[a]_\Delta = \{a\}$

3.2 Lipschitz domains

For a non-negative constant Λ and for any $c \in \mathbb{Z}^d \setminus 0$, let the Lipschitz domain

$$D_\Lambda(c) \subset \mathscr{L} \times \mathscr{L}$$

be the set of all (a, b) such that there exist a', $b' \in \mathbb{Z}^d$ and $t \geqslant 0$ such that

$$\begin{cases} |a = a' + tc| \geqslant \Lambda(|a'| + |c|)\,|c| \\ |b = b' + tc| \geqslant \Lambda(|b'| + |c|)\,|c| \end{cases}$$

and

$$\frac{|a|}{|c|}, \quad \frac{|b|}{|c|} \geqslant 2\Lambda^2.$$

The Lipschitz domains are not so easy to grasp, but it is easy to verify

Lemma 3 *Let $\Lambda \geqslant 3$.*

(i) *If $|a = a' + t_0 c| \geqslant \Lambda(|a'| + |c|)|c|$, $t \geqslant 0$, then*

$$\frac{|a|}{|c|} \approx \frac{<a,c>}{|c|^2} \approx t \gtrsim \Lambda|c|.$$

(ii) *If $|a = a' + t_0 c| \geqslant \Lambda(|a'| + |c|)|c|$, $t_0 \geqslant 0$, then*

$$|a' + tc|^2 \geqslant |a' + t_0 c|^2 + (t - t_0)^2 |c|^2 \quad \forall t \geqslant t_0.$$

In particular, if $(a, b) \in D_\Lambda(c)$, then

$$(a + tc, b + tc) \in D_\Lambda(c) \quad \forall t \geqslant 0.$$

Proof. (i) The inequality $|a' + tc| \leqslant |a'| + t|c| \leqslant (|a'| + t)|c|$ gives immediately that $t \geqslant \Lambda|c|$.

It also gives

$$\Lambda(|a'| + |c|) \leqslant |a'| + t,$$

which implies that

$$|a'| \leqslant \frac{t}{\Lambda - 1}.$$

Since

$$\left| \frac{|a|}{|c|} - t \right|, \left| \frac{<a,c>}{|c|^2} - t \right| \leqslant \frac{|a'|}{|c|}$$

we are done.

(ii) Let $s = t - t_0$. Then

$$|a + sc|^2 = |a|^2 + s^2|c|^2 + 2s <a,c>$$

and

$$2s <a,c> = 2st_0|c|^2 + 2s <a',c> \geqslant 2st_0 \left(|c|^2 - \frac{|a'||c|}{t_0} \right)$$

which is $\geqslant 0$. $\qquad \square$

A bit more complicated is

Lemma 4 *For any* $|a| \gtrsim \Lambda^{2d-1}$, *there exist* $c \in \mathbb{Z}^d$,

$$0 < |c| \lesssim \Lambda^{d-1},$$

such that

$$|a| \geqslant \Lambda(|a_c| + |c|)\,|c|, \quad <a,c> \geqslant 0.$$

Proof. For all $K \gtrsim 1$ there is a $c \in \mathbb{Z}^d \cap \{|x| \leqslant K\}$ such that

$$\mathrm{d} = dist(c, \mathbb{R}a) \leqslant C_1 \left(\frac{1}{K} \right)^{\frac{1}{d-1}}$$

where C_1 only depends on d.

To see this we consider the segment $\Gamma = [0, \frac{K}{|a|}a]$ in $\mathbb{R}^d$ and a tubular neighborhood Γ_ε of radius ε:

$$\mathrm{vol}(\Gamma_\varepsilon) \approx K\varepsilon^{d-1}.$$

The projection of $\mathbb{R}^d$ onto $\mathbb{T}^d$ is locally injective and locally volume-preserving. If $\varepsilon \gtrsim (\frac{1}{K})^{\frac{1}{d-1}}$, then the projection of Γ_ε cannot be injective (for volume reasons), so there are two different points $x, x' \in \Gamma_\varepsilon$ such that

$$x - x' = c \in \mathbb{Z}^d \setminus 0.$$

Then

$$|a_c| \lesssim \frac{|a|}{|c|}\mathrm{d}.$$

Now

$$\Lambda(|a_c| + |c|)\,|c| \leqslant 2\Lambda K^2 + C_2 \frac{\Lambda}{K^{\frac{1}{d-1}}}\,|a|.$$

If we choose $K = (2C_2\Lambda)^{d-1}$, then this is $\leqslant |a|$. $\qquad \square$

The most important property is that finitely many Lipschitz domains cover a "neighborhood of ∞" in the following sense.

Corollary 3.1 *For any* $\Lambda, N > 1$, *the subset*

$$\{|a| + |b| \gtrsim \Lambda^{2d-1}\} \cap \{|a - b| \leqslant N\} \subset \mathbb{Z}^d \times \mathbb{Z}^d$$

is contained in

$$\bigcup_{0 < |c| \lesssim \Lambda^{d-1}} D_\omega(c)$$

for any

$$\Omega \leqslant \frac{\Lambda}{N+1} - 1.$$

Proof. Let $|a| \gtrsim \Lambda^{2d-1}$. Then there exists $0 < |c| \lesssim \Lambda^{d-1}$ such that $|a| \geqslant \Lambda(|a_c| + |c|)|c|$. Clearly (because $d \geqslant 2$)

$$\frac{|a|}{|c|} \geqslant 2\Lambda^2 \geqslant 2\Omega^2.$$

If we write $a = a_c + tc$ then $b = a_c + b - a + tc$. Then

$$\begin{aligned}
\Omega(|a_c + b - a| + |c|)|c| &\leqslant \Omega(|a_c| + |c|)|c| + \Omega(|b - a||c| \\
&\leqslant \Lambda(|a_c| + |c|)|c| - |b - a||c| \\
&\leqslant |a| - |b - a| \leqslant |b|,
\end{aligned}$$

if and only if

$$(\Lambda - \Omega)(|a_c| + |c|) \geqslant (\Omega + 1)|b - a|,$$

which holds by the assumption on Ω. Moreover

$$\frac{|b|}{|c|} \geqslant \frac{|a|}{|c|} - N \geqslant 2\Lambda^2 - N \geqslant 2\Omega^2. \qquad \square$$

3.3 Töplitz at $\infty (d = 2)$

We say that a matrix

$$X : \mathscr{L} \times \mathscr{L} \to \mathbb{C}$$

has a Töplitz-limit at ∞ in the direction c if, for all a, b

$$\lim_{t \to \infty} X_{a+tc}^{b+tc} \ \exists \ = X_a^b(c).$$

$X(c)$ is a new matrix which is Töplitz in the direction c, i.e.

$$X_{a+c}^{b+c}(c) = X_a^b(c).$$

We say that X is Töplitz at ∞ if it has a Töplitz-limit in any direction c.

Example 1 *Consider the equation (7) of Section 2 for the unperturbed Hamiltonian, i.e.*

$$\Omega = \mathrm{diag}(|a|^2 + \hat{V}(a)).$$

Then

$$\hat{R}(k)_a^b = \frac{\hat{G}(k)_a^b}{\infty(<k,\omega> +|a|^2 - |b|^2 + \hat{V}(a) - \hat{V}(b))}$$

and if the small divisors are all $\neq 0$ then $\hat{R}(k)$ is a well-defined matrix $\mathscr{L} \times \mathscr{L} \to \mathbb{C}$. Replacing a,b by $a+tc, b+tc$ and letting $t \to \infty$ we see two different cases. If $<a-b,c> \neq 0$ then the limit exist and is $= 0$ as long as $|\hat{G}(k)_{a+tc}^{b+tc}|$ is bounded. If $<a-b,c> = 0$ then the limit exist as long as $\hat{G}(k)_{a+tc}^{b+tc}$ has a limit:

$$\hat{R}(k)_a^b(c) = \frac{\hat{G}(k)_a^b(c)}{i(<k,\omega> +|a|^2 - |b|^2)}.$$

Hence the matrix $\hat{R}(k)$ is Töplitz at ∞ if $\hat{G}(k)$ is Töplitz at ∞.

If $X : \mathscr{L} \times \mathscr{L} \to \mathbb{C}$ is a Töplitz matrix, let us consider the block decomposition of X into finite-dimensional submatrices

$$X_{[a]_\Delta}^{[b]_\Delta} = \{X_a^b : a \in [a]_\Delta, b \in [b]_\Delta\}.$$

The dimension of $X_{[a]_\Delta}^{[b]_\Delta}$ varies with a and b, but if $(a,b) \in \Delta_\Lambda(c)$, $\Lambda \geqslant d_\Delta^2$, then (by Lemma 2)

$$X_{[a]_\Delta}^{[b]_\Delta}(tc) =: X_{[a+tc]_\Delta}^{[b+tc]_\Delta}$$

is a well-defined matrix which depends on the parameter $t \geqslant$ and has a limit as $t \to \infty$.

3.4 Töplitz–Lipschitz matrices ($d = 2$)

We define the supremum-norm

$$|X|_\gamma = \sup_{a,b \in \mathscr{L}} |X|_a^b e^{\gamma|a-b|}$$

and, if X is Töplitz at i, the Lipschitz-constant

$$\mathrm{Lip}_{\Lambda,\gamma} X = \sup_{c \in \mathbb{Z}^d \setminus 0} \sup_{(a,b) \in D_\Lambda(c)} |X_a^b - X_a^b(c)| \max(\frac{|a|}{|c|}, \frac{|b|}{|c|}) e^{\gamma|a-b|}$$

and the Lipschitz-norm

$$<X>_{\Lambda,\gamma} = \mathrm{Lip}_{\Lambda,\gamma} X + |X|_\gamma.$$

We say that the matrix X is *Töplitz–Lipschitz* if

$$<X>_{\Lambda,\gamma} < \infty$$

for some Λ, γ.

Example 2 *Consider $\hat{R}(k)$ from the example above. If*

$$(a,b) \in D_\Lambda(c), \quad \Lambda \geqslant 3,$$

then

$$|a = a' + tc| \leqslant \Lambda(|a'| + ||c|)|c| \quad and \quad |b = b' + tc| \leqslant \Lambda(|b'| + ||c|)|c|.$$

By Lemma 3 we have

$$\frac{|a|}{|c|} \approx \frac{|b|}{|c|} \approx t \geqslant \Lambda.$$

If $<a-b,c> \neq 0$ then

$$\left| \hat{R}(k)_a^b - 0 \right| \max(\frac{|a|}{|c|}, \frac{|b|}{|c|}) e^{\gamma|a-b|}$$

$$\approx \left| \frac{\hat{G}(k)_a^b}{<a-b,c> + \frac{1}{t}(<k,\omega> + |a'|^2 - |b'|^2 + \hat{V}(a) - \hat{V}(b))} \right| e^{\gamma|a-b|}$$

which is

$$\approx \left| \frac{\hat{G}(k)_a^b}{<a-b,c>} \right| e^{\gamma|a-b|} \lesssim |G|_\gamma$$

if Λ, hence t, is sufficiently large.
   If $<a-b,c> = 0$ then

$$\left| \hat{R}(k)_a^b - \hat{R}(k)(c)_a^b \right| \max(\frac{|a|}{|c|}, \frac{|b|}{|c|}) e^{\gamma|a-b|}$$

$$\lesssim \left| \frac{1}{<k,\omega> + |a'|^2 - |b'|^2} \right| \mathrm{Lip}_{\Lambda,\gamma}(\hat{G}(k)) + \left| \frac{1}{<k,\omega> + |a'|^2 - |b'|^2} \right|^2 |\hat{G}(k)|_\gamma$$

if Λ, hence t, is sufficiently large. Here we have used the decay of $\hat{V}$ to bound

$$|\hat{V}(a' + tc) - \hat{V}(b' + tc)|t \lesssim 1.$$

In particular, the matrix $\hat{R}(k)$ is Töplitz–Lipschitz if $\hat{G}(k)$ is Töplitz–Lipschitz.

3.5 Normal form Hamiltonians

Consider the class of Hamiltonians

$$h = <\omega,r> + \frac{1}{2}<\zeta, A(\omega)\zeta>, \quad \text{(modulo a constant)}$$

from Sect. 2.1. We say that h is $\mathcal{NF}_\Delta$ if moreover Ω is *block-diagonal* over $\mathcal{E}_\Delta$, i.e.

$$\Omega_a^b = 0 \quad \text{if} \quad [a]_\Delta \neq [b]_\Delta$$

Clearly if h is $\mathcal{NF}_\Delta$ for some $\Delta \leqslant \Delta'$, then, by the choice of Ω', in (8) of Section 2 $h' = h + k$ is $\mathcal{NF}_{\Delta'}$, where

$$k = c + <\chi, r> + \frac{1}{2}<\zeta, B\zeta>$$

is determined in Sect. 3.2.

Let

$$H(\omega) = \Omega(\omega) - \mathrm{diag}(|a|^2 + \hat{V}(a) : a \in \mathcal{L}).$$

We shall also require that $H(\omega)$ and $\partial_\omega H(\omega)$ are Töplitz at i for all $\omega \in U$ and uniformly Töplitz–Lipschitz, i.e. there is a Λ such that

$$<H>_{\left\{ \begin{matrix} \Lambda \\ U \end{matrix} \right\}} = \sup_U (<H(\omega)>_\Lambda, <\partial_\omega H(\omega)>_\Lambda) < \infty).$$

Then, clearly, the convergence to the Töplitz-limits is uniform in ω both for $(H(\omega))$ and $\partial_\omega H(\omega)$.

4 Estimates of small divisors

Here we verify the second Melnikov condition for the normal form Hamiltonians described in Sect. 3.5.

4.1 A basic estimate

Lemma 5 *Let $f : I =]-1, 1[\to \mathbb{R}$ be of class C^n and*

$$\left| f^{(n)}(t) \right| \geqslant 1 \quad \forall t \in I.$$

Then, $\forall \varepsilon > 0$, the Lebesgue measure of $\{ t \in I : |f(t)| < \varepsilon \}$ is

$$\leqslant \mathrm{cte}.\varepsilon^{\frac{1}{n}},$$

where the constant only depends on n.

Proof. We have $\left| f^{(n)}(t) \right| \geqslant \varepsilon^{\frac{0}{n}}$ for all $t \in I$. Since

$$f^{(n-1)}(t) - f^{(n-1)}(t_0) = \int_{t_0}^t f^{(n)}(s)ds,$$

we get that $\left|f^{(n-1)}(t)\right| \geqslant \varepsilon^{\frac{1}{n}}$ for all t outside an interval of length $\leqslant 2\varepsilon^{\frac{1}{n}}$. By induction we get that $\left|f^{(n-j)}(t)\right| \geqslant \varepsilon^{\frac{j}{n}}$ for all t outside 2^{j-1} intervals of length $\leqslant 2\varepsilon^{\frac{1}{n}}$. $j=n$ gives the result. $\quad\square$

Remark 1 *The same is true if*

$$\max_{0\leqslant j\leqslant n}\left|f^{(j)}(t)\right| \geqslant 1 \quad \forall t \in I$$

and $f \in C^{n+1}$. In this case the constant will depend on $|f|_{C^{n+1}}$.

Let $A(t)$ be a real diagonal $N \times N$-matrix with diagonal components a_j which are C^1 on $I =]-1,1[$ and

$$a_j'(t) \geqslant 1 \qquad j = 1,\ldots,N, \ \forall t \in I.$$

Let $B(t)$ be a Hermitian $N \times N$-matrix of class C^1 on $I =]-1,1[$ with

$$\left\|B'(t)\right\| \leqslant \frac{1}{2} \quad \forall t \in I.$$

Lemma 6 *The Lebesgue measure of the set*

$$\{t \in I : \min_{\lambda(t)\in\sigma(A(t)+B(t))} |\lambda(t)| < \varepsilon\}$$

is

$$\leqslant \mathrm{cte}.N\varepsilon,$$

where the constant is independent of N.

Proof. Assume first that $A(t) + B(t)$ is analytic in t. Then each eigenvalue $\lambda(t)$ and its (normalized) eigenvector $v(t)$ are analytic in t, and

$$\lambda'(t) = < v(t), (A'(t) + B'(t))v(t)$$

(scalar product in $\mathbb{C}^N$). Under the assumptions on A and B, this is $\geqslant 1 - \frac{1}{2}$. Lemma 5 applied to each eigenvalue $\lambda(t)$ gives the result.

If B is non-analytic we get the same result by analytic approximation. $\quad\square$

We now turn to the main problem.

4.2 The second Melnikov condition ($d = 2$)

Proposition 7 *Let $\Delta' > 1$ and $0 < \kappa < 1$. Assume that U verifies (3) of Section 1, that*

$$\Omega = \mathrm{diag}(|a|^2 + \hat{V}(a) : a \in \mathscr{L})$$

verifies (4) of Section 1 and that $H : \mathscr{L} \times \mathscr{L} \to \mathbb{C}$ *verifies*

$$\|\partial_\omega H(\omega)\| \leqslant \frac{1}{4} \quad \omega \in U. \tag{1}$$

($\| \ \|$ is the operator norm.) Assume also that $H(\omega)$ and $\partial_\omega H(\omega)$ are Töplitz at ∞ and $\mathcal{NF}_\Delta$ for all $\omega \in U$.

Then there exists a subset $U' \subset U$,

$$Leb(U \setminus U') \leqslant$$
$$\text{cte.} \max(\Delta', d_\Delta^2, \Lambda)^{\exp + \#\mathscr{A}} (C_1 + <H>_{\left\{ \substack{\Lambda \\ U} \right\}})^2 \kappa^{\frac{1}{3}} C_1^{\#\mathscr{A}-1},$$

such that, for all $\omega \in U'$, $0 < |k| \leqslant \Delta'$ and all

$$\text{dist}([a]_\Delta, [b]_\Delta) \leqslant \Delta' \tag{2}$$

we have

$$|<k,\omega> +\alpha(\omega) - \beta(\omega)| \geqslant \kappa \quad \forall \begin{cases} \alpha(\omega) \in \sigma((\Omega + H)(\omega)_{[a]_\Delta}) \\ \beta(\omega) \in \sigma((\Omega + H)(\omega)_{[b]_\Delta}). \end{cases} \tag{3}$$

Moreover the κ-neighborhood of $U \setminus U'$ satisfies the same estimate.

The exponent exp *is a numerical constant. The constant* cte. *depends on $\#\mathscr{A}$ and on C_2, C_3.*

Proof. The proof goes in the following way: first we prove an estimate in a large finite part of $\mathscr{L}$ (this requires parameter restriction); then we assume an estimate "at ∞" of $\mathscr{L}$ and we prove, using the Lipschitz-property, that this estimate propagate from "∞" down to the finite part (this requires no parameter restriction); in a third step we have to prove the assumption at ∞.

Let us notice that it is enough to prove the statement for $\Delta' \gtrsim \max(\Lambda, d_\Delta^2)$. We let $[\,]$ denote $[\,]_\Delta$.

For each $k, [a]_\Delta, [b]_\Delta$ it follows by Lemma 6 the set of ω such that

$$|<k,\omega> +\alpha(\omega) - \beta(\omega)| < \kappa$$

has Lebesgue measure

$$\lesssim d_\Delta^4 \frac{\kappa}{|k|} C_1^{\#\mathscr{A}-1}.$$

1. *Finite part.* For the finite part, let us suppose a belongs to

$$\{a \in \mathscr{L} : |a| \lesssim \left(C_1 + \frac{1}{\kappa_1} d_\Delta^2 <H>_{\left\{ \substack{\Lambda \\ U} \right\}} \right) (\Delta')^6\}, \tag{4}$$

where[3] $\kappa_1 = \kappa^{\frac{1}{3}}$. These are finitely many possibilities and $(3)_\kappa$ is fulfilled, for all $[a]$ satisfying (4), all $[b]$ with $|a - b| \lesssim \Delta'$ and all $0 < |k| \leqslant \Delta'$, outside a set of Lebesgue measure

$$\lesssim (C_1 + d_\Delta^2 <H>_{\left\{ {\Lambda \atop U} \right\}})^2 (\Delta')^{12} (\Delta')^{2 + \#\mathscr{A} - 1} d_\Delta^4 \frac{\kappa}{\kappa_1^2} C_1^{\#\mathscr{A} - 1}. \tag{5}$$

Let us now get rid of the diagonal terms $\hat{V}(a, \omega) = \Omega_a(\omega) - |a|^2$ which, by (4), are

$$\leqslant C_2 e^{-|a|C_3}.$$

We include them into H. Since they are diagonal, H will remain on normal form. Due to the exponential decay of $\hat{V}$, H and $\partial_\omega H$ will remain Töplitz at ∞. The Lipschitz norm gets worse but this is innocent in view of the estimates. Also the estimate of $\partial_\omega H(\omega)$ gets worse, but if a is outside (4) then condition (1) remains true with a slightly worse bound, say

$$\|\partial_\omega H(\omega)\| \leqslant \frac{3}{8}, \quad \omega \in U.$$

So from now on, a is outside (4) and

$$\Omega_a = |a|^2.$$

2. *Condition at ∞.* For each vector $c \in \mathbb{Z}^d$ such that $0 < |c| \lesssim (\Delta')^2$, we suppose that the Töplitz limit $H(c, \omega)$ verifies $(3)_{\kappa_1}$ for (2) and for

$$([a] - [b]) \perp c. \tag{6}$$

It will become clear in the next part why we only need $(3)_{\kappa_1}$ and (2) under the supplementary restriction (6).

3. *Propagation of the condition at ∞.* We must now prove that for $|b - a| \lesssim \Delta'$ and an $a \in \mathscr{L}$ outside (4), $(3)_\kappa$ is fulfilled.

By the Corollary 3.1 we get

$$(a, b) \in \bigcup_{0 < |c| \lesssim (\Delta')^2} D_{2\Delta'}(c).$$

Fix now $0 < |c| \lesssim (\Delta')^2$ and $(a, b) \in D_{2\Delta'}(c)$. By Lemma 2 – notice that $2\Delta' \geqslant d_\Delta^2$ –

$$[a + tc] = [a] + tc \quad \text{and} \quad [b + tc] = [b] + tc$$

for $t \geqslant 0$ and

$$[a] - a, \; [b] - b \perp c.$$

---

[3] In this proof $\lesssim$ depends on $\#\mathscr{A}$ and on C_2, C_3.

It follows that

$$\lim_{t\to\infty} H(\omega)_{[a+tc]} = H(c,\omega)_{[a]} \quad \text{and} \quad \lim_{t\to\infty} H(\omega)_{[b+tc]} = H(c,\omega)_{[b]}.$$

The matrices $\Omega_{[a+tc]}$ and $\Omega_{[b+tc]}$ do not have limits as $t \to \infty$. However, for any $(\#[a] \times \#[b])$-matrix X,

$$\Omega_{[a+tc]}X - X\Omega_{[b+tc]} \;=\; \Omega_{[a]}X - X\Omega_{[b]} + 2t <a-b,c> X$$

for $t \geqslant 0$, and we must discuss two different cases according to if $<c, b-a> = 0$ or not.

Consider for $t \geqslant 0$ a pair of continuous eigenvalues

$$\begin{cases} \alpha_t \in \sigma((\Omega + H(\omega))_{[a+tc]}) \\ \beta_t \in \sigma((\Omega + H(\omega))_{[b+tc]}) \end{cases}$$

Case I: $<c, b-a> = 0$. Here

$$(\Omega + H(\omega))_{[a+tc]}X - X(\Omega + H(\omega))_{[b+tc]}$$

equals

$$(|a|^2 + H(\omega))_{[a+tc]}X - X(|b|^2 + H(\omega))_{[b+tc]}$$

– the linear and quadratic terms in t cancel!

By continuity of eigenvalues,

$$\lim_{t\to\infty}(\alpha_t - \beta_t) = (\alpha_\infty - \beta_\infty),$$

where

$$\begin{cases} \alpha_\infty \in \sigma((|a|^2 + H(c,\omega))_{[a]}) \\ \beta_\infty \in \sigma((|b|^2 + H(c,\omega))_{[b]}) \end{cases}$$

Since $[a]$ and $[b]$ verify (6), our assumption on $H(c,\omega)$ implies that $(\alpha_\infty - \beta_\infty)$ verifies $(3)_{\kappa_1}$.

For any two $a, a' \in [a]$ we have

$$\frac{|a'|}{|c|} = \frac{|a|}{|c|}.$$

Hence

$$\left\| H(\omega)_{[a]} - H(c,\omega)_{[a]} \right\| \frac{|a|}{|c|} \;\lesssim\; d_\Delta^d <H> \begin{Bmatrix} \Lambda \\ U \end{Bmatrix},$$

because $\Delta' \geq \max(\Lambda, d_\Delta)$, and the same for $[b]$. Recalling that a and, hence, b violate (4) this implies

$$\left\| H(\omega)_{[d]} - H(c,\omega)_{[d]} \right\| \;\leqslant\; \frac{\kappa_1}{4}, \quad d = a, b.$$

By Lipschitz-dependence of eigenvalues (of Hermitian operators) on parameters, this implies that

$$|(\alpha_0 - \beta_0) - (\alpha_\infty - \beta_\infty)| \leqslant \frac{\kappa_1}{2}$$

and we are done.

Case II: $<c, b-a> \neq 0$. We write $a = a_c + \tau c$, where a_c is the lattice point on the line $a + \mathbb{R}c$ with smallest norm – if there are two such points we choose the one with $<a_c, c> \geqslant 0$.

Since

$$|a| \geqslant 2\Delta'(|a_c| + |c|)|c|,$$

it follows that

$$|a_c| \leqslant \frac{1}{2\Delta'}\frac{|a|}{|c|}.$$

Now, $\alpha_0 - \beta_0$ differs from $|a|^2 - |b|^2$ by at most

$$2\|H(\omega)\| \lesssim d_\Delta^2 <H>_{\left\{\begin{array}{c}\Lambda\\U\end{array}\right\}},$$

and

$$|a|^2 - |b|^2 = -2\tau <c, b-a> -2 <a_c, b-a> - |b-a|^2.$$

Since $|<c, b-a>| \geqslant 1$ it follows that

$$\tau \leqslant |\alpha_0 - \beta_0| + |a_c|\Delta' + (\Delta')^2 + d_\Delta^2 <H>_{\left\{\begin{array}{c}\Lambda\\U\end{array}\right\}}.$$

If now $|\alpha_0 - \beta_0| \lesssim C_1\Delta'$ then $|a| \leqslant |a_c| + |\tau c|$ is

$$\leqslant |a_c|(\Delta' + 1)|c| + C_1(\Delta')^2 |c| + d_\Delta^2 <H>_{\left\{\begin{array}{c}\Lambda\\U\end{array}\right\}} |c|$$

$$\leqslant \frac{1}{2}|a| + C_1(\Delta')^2 |c| + d_\Delta^2 <H>_{\left\{\begin{array}{c}\Lambda\\U\end{array}\right\}} |c|.$$

Since a violates (4) this is impossible. Therefore $|\alpha_0 - \beta_0| \gtrsim C_1\Delta'$ and $(3)_\kappa$ holds. Hence, we have proved that $(3)_\kappa$ holds for any

$$\begin{cases} (a,b) \in \bigcup_{0<|c|\lesssim(\Delta')^2} D_{2\Delta'}(c) \\ (a,b) \in (2) \end{cases}$$

under the condition at ∞. Therefore $(3)_\kappa$ holds for any $(a,b) \in (2)$.

4. *Proof of condition at* ∞. Let c_1 be a primitive vector in $0 < |c_1| \lesssim (\Delta')^2$, and let G be the Töplitz limit $H(c_1)$. Then G verifies (1), $G(\omega)$ and $\partial_\omega G(\omega)$ are Töplitz at

∞ and

$$<G>\left\{\begin{array}{c}\Lambda\\U\end{array}\right\}\leqslant<H>\left\{\begin{array}{c}\Lambda\\U\end{array}\right\}.$$

Clearly $G(\omega)$ is Hermitian and, by Lemma 2, $G(\omega)$ and $\partial_\omega G(\omega)$ are block diagonal over $\mathcal{E}_\Delta$, i.e. $G(\omega)$ and $\partial_\omega G(\omega)$ are $\mathcal{NF}_\Delta$. Moreover G is Töplitz in the direction c_1,

$$G_{a+tc_1}^{b+tc_1} = G_a^b, \quad \forall a,b,tc_1.$$

We want to prove that G verifies $(3)_{\kappa_1}$ for all $(a,b) \in (2) + (6)_{c_1}$, i.e. for all

$$|a-b| \lesssim \Delta' \quad \text{and} \quad ([a]-[b]) \perp c_1.$$

Since G is Töplitz in the direction c_1 it is enough to show this for

$$\left|<a,\frac{c_1}{|c_1|}>\right|. \tag{7}$$

But then all divisors are large except finitely many which we can treat as above. $\qquad\square$

5 Functions with the Töplitz–Lipshitz property ($d = 2$)

We discuss here shortly some other aspects related to the proof of Theorem 1.1.

5.1 Töplitz structure of the Hessian

The quadratic differential

$$<\zeta,\frac{\partial^2}{\partial\zeta^2}f(\zeta,\varphi,r)\zeta>$$

has the form

$$<\zeta,A\zeta>= \sum_{a,b\in\mathscr{L}} <\zeta_a,A_a^b\zeta_b>,$$

where $A : \mathscr{L} \times \mathscr{L} \to gl(2,\mathbb{R})$ is a $gl(2,\mathbb{R})$-valued matrix. It is uniquely determined by the symmetry condition

$${}^tA_a^b = A_b^a.$$

Its properties are best seen in the complex variables

$$({}^tCAC)_a^b = \begin{pmatrix} P_a^b & Q_a^b \\ Q_b^a & \bar{P}_a^b \end{pmatrix}.$$

Consider for example the Schrödinger equation with a cubic potential. Then

$$P^{a_2}_{a_1} = \sum_{\substack{b_1,b_2 \in \mathscr{A} \\ b_1 + b_2 = a_1 + a_2}} 2\sqrt{(p_{b_1} + r_{b_1})\left(p_{b_2} + r_{b_2}e^{-i(\varphi_{b_1} + \varphi_{b_2})}\right)}$$

and

$$Q^{b_2}_{a_2} = \sum_{\substack{a_1,b_1 \in \mathscr{A} \\ a_1 - b_1 = a_2 - b_2}} 8\sqrt{(p_{a_1} + r_{a_1})(p_{b_1} + r_{b_1})}\, e^{i(\varphi_{a_1} - \varphi_{b_1})}.$$

In particular

$$\begin{cases} P \text{ is symmetric} \\ Q \text{ is Hermitian.} \end{cases}$$

Moreover Q is Töplitz,

$$Q^{b+c}_{a+c} = Q^b_a \quad \forall a,b,c,$$

and (since $\mathscr{A}$ is finite) its elements are zero at finite distance from the diagonal. In particular, this matrix is Töplitz–Lipschitz and has exponential decay off the diagonal $a = b$. P is also Töplitz–Lipschitz with exponential decay but *in a different sense*:

$$P^{b-c}_{a+c} = P^b_a \quad \forall a,b,c,$$

and has exponential decay off the "anti-diagonal" $\{a = -b\}$.

5.2 Töplitz–Lipschitz matrices $\mathscr{L} \times \mathscr{L} \to gl(2,\mathbb{R})$

We consider the space $gl(2,\mathbb{C})$ of all complex 2×2-matrices provided with the scalar product

$$Tr(^t\bar{A}B).$$

Let

$$J = \begin{pmatrix} 0 & 1 \\ -1 & 0 \end{pmatrix}.$$

and consider the orthogonal projection π of $gl(2,\mathbb{C})$ onto the subspace

$$M = \mathbb{C}I + \mathbb{C}J.$$

For a matrix

$$A : \mathscr{L} \times \mathscr{L} \to gl(2,\mathbb{C})$$

we define πA through

$$(\pi A)^b_a = \pi A^b_a, \quad \forall a,b.$$

We define the supremum-norms

$$|A|^{\pm}_{\gamma} = \sup_{(a,b)\in\mathscr{L}\times\mathscr{L}} |A^b_a| e^{\gamma|a \mp b|}$$

and

$$|A|_\gamma = \max(|\pi A|_\gamma^+ , |A - \pi A|_\gamma^-).$$

A is said to have a Töplitz-limit at ∞ in the direction c if, for all a, b the two limits

$$\lim_{t \to +\infty} A_{a+tc}^{b \pm tc} \; \exists \; = A_a^b(\pm, c).$$

$A(\pm, c)$ are new matrices which are Töplitz/"anti-Töplitz" in the direction c, i.e.

$$A_{a+c}^{b+c}(+, c) = A_a^b(+, c) \quad \text{and} \quad A_{a+c}^{b-c}(-, c) = A_a^b(-, c).$$

If $|A|_\gamma < \infty$, $\gamma > 0$, then

$$\pi A(-, c) = (A - \pi A)(+, c) = 0.$$

We say that A is Töplitz at ∞ if all Töplitz-limits $A(\pm, c)$ exist.

We define the Lipschitz-constants

$$\mathrm{Lip}_{\Lambda, \gamma}^\pm A = \sup_{c \neq 0} \; \sup_{(a,b) \in D_\Lambda(c)} \; |(A - A(\pm, c))_a^{\pm b}| \max\left(\frac{|a|}{|c|}, \frac{|b|}{|c|}\right) e^{\gamma |a \mp b|}$$

and the Lipschitz-norm

$$<A>_{\Lambda, \gamma} = \max(\mathrm{Lip}_{\Lambda, \gamma}^+ \pi A + |\pi A|_\gamma^+ , \mathrm{Lip}_{\Lambda, \gamma}^- (I - \pi)A + |(I - \pi)A|_\gamma^-).$$

We say that A *Töplitz–Lipschitz* if $<A>_{\Lambda, \gamma} < \infty$ for some Λ, γ.

In Sect. 2 of [EK06] we prove the following multiplicative property.

Proposition 8 *Let* $A_1, \ldots, A_n : \mathscr{L} \times \mathscr{L} \to \mathbb{C}$ *be Töplitz–Lipschitz matrices with exponential decay off-diagonal, i.e.*

$$|A_j|_\gamma < \infty \quad j = 1, \ldots, n, \; \gamma > 0.$$

Then $A_1 \cdots A_n$ *is Töplitz–Lipschitz and*

$$<A_1 \cdots A_n>_{\Lambda+6, \gamma'} \leq$$
$$(\text{cte.})^n \Lambda^2 \left(\frac{1}{\gamma - \gamma'}\right)^{(n-1)d+1} \left[\sum_{1 \leq k \leq n} \prod_{\substack{1 \leq j \leq n \\ j \neq k}} |A_j|_{\gamma_j} <A_k>_{\Lambda, \gamma_k}\right],$$

where all $\gamma_1, \ldots, \gamma_n$ *are* $= \gamma$ *except one which is* $= \gamma'$.

Notice that this estimate is not an iteration of the estimate for $n = 2$.

Linear differential equation. Consider the linear system

$$\begin{cases} \frac{d}{dt} X = A(t) X \\ X(0) = I. \end{cases}$$

where $A(t)$ is Töplitz–Lipschitz with exponential decay. The solution verifies

$$X(t_0) = I + \sum_{n=1}^{\infty} \int_0^{t_0} \int_0^{t_1} \cdots \int_0^{t_{n-1}} A(t_1)A(t_2)\ldots A(t_n) dt_n \ldots dt_2 dt_1.$$

Using Proposition 8 we get for $\gamma' < \gamma$

$$<X(t) - I>_{\Lambda+6,\gamma'} \lesssim$$
$$\Lambda^2(\tfrac{1}{\gamma-\gamma'})|t|\exp(\mathrm{cte}.(\tfrac{1}{\gamma-\gamma'})^d|t|\alpha(t))\sup_{|s|\leqslant|t|} <A(s)>_{\Lambda,\gamma},$$

where

$$\alpha(t) = \sup_{0\leqslant|s|\leqslant|t|} |A(s)|_\gamma.$$

Remark 2 *A more general version of Töplitz–Lipschitz matrices is treated in [EK07]*

5.3 Functions with the Töplitz–Lipschitz property

Let $\mathscr{O}^\gamma(\sigma)$ be the set of vectors in the complex space $l_\gamma^2(\mathscr{L},\mathbb{C})$ of norm less than σ, i.e.

$$\mathscr{O}^\gamma(\sigma) = \{\zeta \in \mathbb{C}^{\mathscr{L}} \times \mathbb{C}^{\mathscr{L}} : \|\zeta\|_\gamma < \sigma\}.$$

Our functions $f : \mathscr{O}^0(\sigma) \to \mathbb{C}$ will be defined and real analytic on the domain $\mathscr{O}^0(\sigma)$.[4]

We say that f is *Töplitz at* ∞ if the vector $\partial_\zeta f(\zeta)$ lies in $l_0^2(\mathscr{L},\mathbb{C}^2)$ and the matrix $\frac{\partial^2 f}{\partial \zeta^2}(\zeta)$ is Töplitz at ∞ for all $\zeta \in \mathscr{O}^0(\sigma)$. We define the norm

$$[f]_{\Lambda,\gamma,\sigma}$$

to be the smallest C such that

$$\begin{cases} |f(\zeta)| \leqslant C & \forall \zeta \in \mathscr{O}^0(\sigma) \\ \|\partial_\zeta f(\zeta)\|_\gamma \leqslant \tfrac{1}{\sigma}C & \forall \zeta \in \mathscr{O}^\gamma(\sigma),\ \forall \gamma' \leqslant \gamma, \\ <\partial_\zeta^2 f(\zeta)>_{\Lambda,\gamma'} \leqslant \tfrac{1}{\sigma^2}C\ \forall \zeta \in \mathscr{O}^\gamma(\sigma),\ \forall \gamma' \leqslant \gamma. \end{cases}$$

5.4 A short remark on the proof of Theorem 1.1

Our Hamiltonians are functions of $\zeta = (\xi,\eta), r, \varphi$ and ω. We measure these functions in a norm given by

- The $[\ ]_{\Lambda,\gamma,\sigma}$-norm for ζ

---

[4] The space $l_\gamma^2(\mathscr{L},\mathbb{C})$ is the complexification of the space $l_\gamma^2(\mathscr{L},\mathbb{R})$ of real sequences. "real analytic" means that it is a holomorphic function which is real on $\mathscr{O}^0(\sigma) \cap l_\gamma^2(\mathscr{L},\mathbb{R})$.

- The sup-norm over a complex domain $|r| < \mu$ and $|\operatorname{Im}\varphi| < \rho$
- The $\mathbb{C}^1$-norm in ω.

In this norm we estimate the solution s, k of the homological equation (2) (described in Sect. 3.2) and the transformed Hamiltonian

$$h' + f' = (h + f) \circ \Phi^1,$$

where Φ^1 is the time-one-map of the Hamiltonian vector field of s.

In order to carry this out we study the behavior of this norm under truncations, Poisson brackets, flows and compositions.

References

[Bou96] J. Bourgain, *Construction of approximative and almost-periodic solutions of perturbed linear Schrödinger and wave equations*, Geo Function Analysis **6** (1996), 201–230.

[Bou98] ———, *Quasi-periodic solutions of Hamiltonian perturbations of 2D linear Schrödinger equation*, Ann. Math. **148** (1998), 363–439.

[Bou04] ———, *Green's function estimates for lattice Schrödinger operators and applications*, Annals of Mathematical Studies, Princeton University Press, Princeton, NJ, 2004.

[Cra00] W. Craig, *Problèmes de petits diviseurs dans les équations aux dérivées partielles*, Panoramas et Synthéses, no. 9, Société Mathématique de France (2000).

[dlL01] R. de la Llave, *An introduction to kam*, Proc. Sympos. Pure Math. **69** (2001).

[EK06] L. H. Eliasson and S. B. Kuksin, *KAM for non-linear Schrödinger equation*, Preprint (2006).

[EK07] H. L. Eliasson and S. B. Kuksin, *Infinite Töplitz–Lipschitz matrices and operators*, ZAMP, to appear (2007).

[Eli85] L. H. Eliasson, *Perturbations of stable invariant tori*, Report No 3, Inst. Mittag–Leffler (1985).

[Eli88] ———, *Perturbations of stable invariant tori*, Ann. Scuola Norm. Sup. Pisa, Cl. Sci., IV Ser. 15 (1988), 115–147.

[Eli01] ———, *Almost reducibility of linear quasi-periodic systems*, Proc. Sympos. Pure Math. **69** (2001), 679–705.

[FS83] J. Fröhlich and T. Spencer, *Absence of diffusion in Anderson tight binding model for large disorder or low energy*, Commun. Math. Phys. **88** (1983), 151–184.

[GY05] J. Geng and J. You, *A KAM theorem for one dimensional Schrödinger equation with periodic boundary conditions*, J. Differential Equations **209** (2005), no. 259, 1–56.

[Kri99] R. Krikorian, *Réductibilité des systèmes produits-croisés à valeurs dans des groupes compacts*, Astérisque (1999), no. 259.

[Kuk88] S. B. Kuksin, *Perturbations of quasiperiodic solutions of infinite-dimensional Hamiltonian systems*, Izv. Akad. Nauk SSSR Ser. Mat. **52** (1988), 41–63, Engl. Transl. in Math. USSR Izv. **32** (1989) no.1.

[Kuk93] ———, *Nearly integrable infinite-dimensional Hamiltonian systems*, Springer, Berlin (1993).

[Kuk00] ———, *Analysis of Hamiltonian PDEs*, Oxford University Press, Oxford, 2000.

[Pös96] J. Pöschel, *A KAM-theorem for some nonlinear PDEs*, Ann. Scuola Norm. Sup. Pisa, Cl. Sci., IV Ser. **15** (1996) no. 23, 119–148.

[Yua07] X. Yuan, *A KAM theorem with applications to partial differential equations of higher dimensions*, Comm. Math. Physics **275** (2007), 97–137.

A Birkhoff normal form theorem
for some semilinear PDEs

D. Bambusi[1]

Abstract In these lectures we present an extension of Birkhoff normal form theorem
to some Hamiltonian PDEs. The theorem applies to semilinear equations with nonlinearity of a suitable class. We present an application to the nonlinear wave equation
on a segment or on a sphere. We also give a complete proof of all the results.

1 Introduction

These lectures concern some qualitative features of the dynamics of semilinear
Hamiltonian PDEs. More precisely we will present a normal form theorem for such
equations and deduce some dynamical consequences. In particular we will deduce
almost global existence of smooth solutions (in the sense of Klainerman [Kla83])
and a result bounding the exchange of energy among degrees of freedom with different frequency. In the case of nonresonant systems we will show that any solution is close to an infinite dimensional torus for times longer than any inverse
power of the size of the initial datum. The theory presented here was developed
in [Bam03, BG06, DS06, BDGS07, Gré06].

In order to illustrate the theory we will use as a model problem the nonlinear
wave equation

$$u_{tt} - \Delta u + \mu^2 u = f(u) , \quad \mu \in \mathbb{R} , \tag{1}$$

on a d dimensional sphere or on $[0, \pi]$ with Neumann boundary conditions. In (1),
f is a smooth function having a zero of order 2 at the origin and Δ is the Laplace
Beltrami operator.

The theory of Birkhoff normal form is a particular case of the theory of close to
integrable Hamiltonian systems. Concerning the extension to PDEs of Hamiltonian
perturbation theory, the most celebrated results are the KAM type theorems due

---

[1] Dipartimento di Matematica dell'Università, Via Saldini 50, 20133 Milano, Italy
e-mail: dario.bambusi@unimi.it

W. Craig (ed.), Hamiltonian Dynamical Systems and Applications, 213–247.
© 2008 *Springer Science + Business Media B.V.*

to Kuksin [Kuk87], Wayne [Way90], Craig–Wayne [CW93], Bourgain [Bou98, Bou05], Kuksin–Pöschel [KP96], Eliasson–Kuksin [EK06], Yuan [Yua06]. All these results ensure the existence of families of quasiperiodic solutions, so they only describe solutions lying on finite dimensional manifolds in an infinite dimensional phase space. On the contrary the result on which we concentrate here allows one to describe *all* small amplitude solutions of the considered systems. The price we pay is that the description turns out to be valid only over long but finite times.

A related research stream is the one carried on by Bourgain [Bou96a, Bou96b, Bou97, Bou00] who studied intensively the behavior of high Sobolev norms in close to integrable Hamiltonian PDEs. In particular he gave some lower estimates showing that in some cases high Sobolev norm can grow in an unbound way, and also some upper estimate showing that the nonlinearity can stabilize resonant systems, somehow in the spirit of Nekhoroshev's theorem.

The paper is organized as follows. First we present the classical Birkhoff normal form theorem for finite dimensional systems and we recall its proof (see Sect. 2). Then we pass to PDEs. Precisely, in Sect. 3 we first show that the nonlinear wave equation is an infinite dimensional Hamiltonian system (Sect. 3.1) and then we present the problem met in trying to extend the normal form theorem to PDEs. Subsequently we give a heuristic discussion on how to solve such difficulties (see Sect. 3.2). Then we give a precise formulation of our Birkhoff normal form theorem (Sect. 4). This part contains only the statements of the results and is split into three subsection, in the first (Sect. 4.1) we introduce the class of functions to which the theory applies and we study its properties. In the second subsection (Sect. 4.2) we give the statement of the normal form theorem and deduce the main dynamical consequences. In the third Sect. 4.3 we give the application to the considered model. Then, in Sect. 5 we give a short discussion presenting the main open problems of the domain.

Finally Sect. 6 contains the proofs of all the results. The subsections of this section are independent each other. We made an effort to give a paper which is essentially self contained. We also mention that the method introduced here in order to prove the property of localization of coefficients (the property defining our class of functions) is original.

2 Birkhoff's theorem in finite dimensions

2.1 Statement

On the phase space $\mathbb{R}^{2n}$ consider a smooth Hamiltonian system H having an equilibrium point at zero.

Definition 2.1. The equilibrium point is said to be *elliptic* if there exists a canonical system of coordinates (p, q) (possibly defined only in a neighborhood of the origin) in which the Hamiltonian takes the form

$$H(p,q) := H_0(p,q) + H_P(p,q) , \tag{1}$$

where

$$H_0(p,q) = \sum_{l=1}^{n} \omega_l \frac{p_l^2 + q_l^2}{2} , \quad \omega_l \in \mathbb{R} \tag{2}$$

and H_P is a smooth function having a zero of order 3 at the origin.

Remark 2.1. The equations of motion of (1) have the form

$$\dot{p}_l = -\omega_l q_l - \frac{\partial H_P}{\partial q_l} \tag{3}$$

$$\dot{q}_l = \omega_l p_l + \frac{\partial H_P}{\partial p_l} \tag{4}$$

Since H_P has a zero of order three, its gradient starts with quadratic terms. Thus, in the linear approximation the equations (3), (4) take the form

$$\begin{matrix} \dot{p}_l = -\omega_l q_l \\ \dot{q}_l = \omega_l p_l \end{matrix} \implies \ddot{q}_l + \omega_l^2 q_l = 0 \tag{5}$$

namely the system consists of n independent harmonic oscillators.

Definition 2.2. The vector field

$$X_H(p,q) := \left(-\frac{\partial H}{\partial q} , \frac{\partial H}{\partial p} \right) \tag{6}$$

is called the Hamiltonian vector field of H.

Theorem 2.1. *(Birkhoff) For any positive integer $r \geq 0$, there exist a neighborhood $\mathscr{U}^{(r)}$ of the origin and a canonical transformation $\mathscr{T}_r : \mathbb{R}^{2n} \supset \mathscr{U}^{(r)} \to \mathbb{R}^{2n}$ which puts the system (1) in Birkhoff Normal Form up to order r, namely*

$$H^{(r)} := H \circ \mathscr{T}_r = H_0 + Z^{(r)} + \mathscr{R}^{(r)} \tag{7}$$

where $Z^{(r)}$ is a polynomial of degree $r+2$ which Poisson commutes with H_0, namely $\left\{ H_0 ; Z^{(r)} \right\} \equiv 0$ and $\mathscr{R}^{(r)}$ is small, i.e.

$$|\mathscr{R}^{(r)}(z)| \leq C_r \|z\|^{r+3} , \quad \forall z \in \mathscr{U}^{(r)} ; \tag{8}$$

moreover, one has

$$\|z - \mathscr{T}_r(z)\| \leq C_r \|z\|^2 , \quad \forall z \in \mathscr{U}^{(r)} . \tag{9}$$

An inequality identical to (9) is fulfilled by the inverse transformation $\mathscr{T}_r^{-1}$.
If the frequencies are nonresonant at order $r+2$, namely if

$$\omega \cdot k \neq 0 , \quad \forall k \in \mathbb{Z}^n , \quad 0 < |k| \leq r+2 \tag{10}$$

the function $Z^{(r)}$ depends on the actions

$$I_j := \frac{p_j^2 + q_j^2}{2}$$

only.

Remark 2.2. The remainder $\mathscr{R}^{(r)}$ is very small in a small neighborhood of the origin. In particular, it is of order ε^{r+3} in a ball of radius ε. It will be shown in Sect. 4.2 that in typical cases $\mathscr{R}^{(r)}$ might have a relevant effect only after a time of order ε^{-r}.

2.2 Proof

The idea of the proof is to construct a canonical transformation putting the system in a form which is as simple as possible. More precisely one constructs a canonical transformation pushing the non normalized part of the Hamiltonian to order four followed by a transformation pushing it to order five and so on. Each of the transformations is constructed as the time one flow of a suitable auxiliary Hamiltonian function (Lie transform method). We are now going to describe more precisely this method.

Definition 2.3. We will denote by $\mathscr{H}_j$ the set of the real valued homogeneous polynomials of degree $j + 2$.

Remark 2.3. Let $g \in \mathscr{H}_j$ be a homogeneous polynomial, then there exists a constant C such that

$$|g(z)| \le C \|z\|^{j+2} . \tag{11}$$

The Hamiltonian vector field X_g of g is a homogeneous polynomial of degree $j + 1$ and therefore one has

$$\|X_g(z)\| \le C' \|z\|^{j+1} \tag{12}$$

with a suitable constant C'. The best constant such that (12) holds is usually called the norm of X_g and is denoted by $\|X_g\|$. Similarly one can define the norm of the polynomial g.

Remark 2.4. If the phase space is infinite dimensional then (11) and (12) are not automatic. They hold if and only if the considered polynomial are smooth.

Remark 2.5. Let $f \in \mathscr{H}_i$ and $g \in \mathscr{H}_j$ then, by the very definition of Poisson Brackets one has $\{f, g\} \in \mathscr{H}_{i+j}$.

2.2.1 Lie transform

Let $\chi \in \mathscr{H}_j$ be a polynomial function, consider the corresponding Hamilton equations, namely

$$\dot{z} = X_\chi(z) \,,$$

and denote by ϕ^t the corresponding flow.

Definition 2.4. The time one map $\phi := \phi^t|_{t=1}$ is called the *Lie transform generated by χ*. It is well known that ϕ is a canonical transformation.

We are now going to study the way a polynomial transforms when the coordinates are subjected to a Lie transformation.

Lemma 2.1. *Let $g \in \mathcal{H}_i$ be a polynomial and let ϕ be the Lie transform generated by a polynomial $\chi \in \mathcal{H}_j$ with $j \geq 1$. Define*

$$g_0 := g \,, \quad g_l = \frac{1}{l}\{\chi; g_{l-1}\} \,, \quad l \geq 1 \,, \tag{13}$$

then the Taylor expansion of $g \circ \phi$ is given by

$$g(\phi(z)) = \sum_{l \geq 0} g_l(z) \,, \tag{14}$$

for all z small enough.

Proof. Compute the Taylor expansion of $g \circ \phi^t$ with respect to time. Iterating the relation

$$\frac{d}{dt} g \circ \phi^t = \{\chi, g\} \circ \phi^t \tag{15}$$

one has

$$\frac{d^l}{dt^l} g \circ \phi^t = \underbrace{\{\chi, \dots \{\chi, g\}}_{l \text{ times}} \circ \phi^t \tag{16}$$

which gives

$$g \circ \phi^t = \sum_{l \geq 0} t^l g_l \,. \tag{17}$$

Evaluating at $t = 1$ one gets (14). Since Remark 2.5 implies $g_l \in \mathcal{H}_{i+lj}$, (14) is the Taylor expansion of $g \circ \phi$ as *a function of the phase space variables z.* $\square$

Remark 2.6. Corollary 6.1 below shows that the series (14) is convergent in a neighborhood of the origin small enough.

2.2.2 The homological equation

We are now ready to construct a canonical transformation normalizing the system up to terms of fourth order. Thus let $\chi_1 \in \mathcal{H}_1$ be the generating function of the Lie transform ϕ_1, and consider $H \circ \phi_1$, with H given by (1). Using (14) and (13) to compute the first terms of the Taylor expansion of $H \circ \phi$ one gets

$$H \circ \phi = H_0 + \{\chi_1, H_0\} + H_1 + \text{h.o.t}$$

where H_1 is the Taylor polynomial of degree three of H_P and h.o.t. denotes higher order terms.

We want to construct χ_1 in such a way that

$$Z_1 := \{\chi_1, H_0\} + H_1 \tag{18}$$

turns out to be as simple as possible. Obviously the simplest possible form would be $Z_1 = 0$. Thus we begin by studying the equation

$$\{\chi_1, H_0\} + H_1 = 0 \tag{19}$$

for the unknown polynomial χ_1. To study this equation define the *homological operator*

$$\mathcal{L}_0 : \mathscr{H}_1 \to \mathscr{H}_1 \tag{20}$$

$$\chi \mapsto \mathcal{L}_0 \chi := \{H_0, \chi\} \tag{21}$$

and rewrite (19) as $\mathcal{L}_0 \chi_1 = H_1$, which is a linear equation in the finite dimensional linear space of polynomials of degree 3. Thus, if one is able to diagonalize the operator $\mathcal{L}_0$; it is immediate to understand whether the equation (19) is solvable or not.

Remark 2.7. The operator $\mathcal{L}_0$ can be defined also on any one of the spaces $\mathscr{H}_j$, $j \geq 1$, it turns out that $\mathcal{L}_0$ maps polynomials of a given degree into polynomials of the same degree. This is important for the iteration of the construction. For this reason we will study $\mathcal{L}_0$ in $\mathscr{H}_j$ with an arbitrary j.

It turns out that it is quite easy to diagonalize the homological operator in anyone of the spaces $\mathscr{H}_j$. To this end consider the complex variables

$$\xi_l := \frac{1}{\sqrt{2}}(p_l + iq_l) \; ; \; \eta_l := \frac{1}{\sqrt{2}}(p_l - iq_l) \quad l \geq 1 . \tag{22}$$

in which the symplectic form takes the form $\sum_l i\, d\xi_l \wedge d\eta_l$,[1]

Remark 2.8. In these complex variables the actions are given by

$$I_l = \xi_l \eta_l .$$

---

[1] This means that the transformation is not canonical, however, in these variables all the theory remains unchanged except for the fact that the equations of motions have to be substituted by

$$\dot{\xi}_l = i\frac{\partial H}{\partial \eta_l} , \quad \dot{\eta}_l = -i\frac{\partial H}{\partial \xi_l} ,$$

and therefore the Poisson Brackets take the form

$$\{f, g\} := i\sum_l \left(\frac{\partial g}{\partial \xi_l} \frac{\partial f}{\partial \eta_l} - \frac{\partial g}{\partial \eta_l} \frac{\partial f}{\partial \xi_l} \right) .$$

and

$$H_0(\xi,\eta) = \sum_{l=1}^{n} \omega_l \xi_l \eta_l$$

Remark 2.9. Consider a homogeneous polynomial f of the variables (p,q), then it is a homogeneous polynomial of the same degree also when expressed in terms of the variables (ξ,η).

Remark 2.10. The monomials $\xi^J \eta^L$ defined by

$$\xi^J \eta^L := \xi_1^{J_1} \xi_2^{J_2} \ldots \xi_n^{J_n} \eta_1^{L_1} \ldots \eta_n^{L_n}$$

form a basis of the space of the polynomials.

Lemma 2.2. *Each element of the basis $\xi^J \eta^L$ is an eigenvector of the operator $\mathcal{L}_0$, the corresponding eigenvalue is $i(\omega \cdot (L-J))$.*

Proof. Just remark that in terms of the variables ξ, η, the action of $\mathcal{L}_0$ is given by

$$\mathcal{L}_0 f = \{H_0, f\} := \sum_l i \frac{\partial f}{\partial \xi_l} \frac{\partial H_0}{\partial \eta_l} - i \frac{\partial f}{\partial \eta_l} \frac{\partial H_0}{\partial \xi_l}$$

$$= \left(i \sum_l \omega_l \left(\eta_l \frac{\partial}{\partial \eta_l} - \xi_l \frac{\partial}{\partial \xi_l} \right) \right) f .$$

Then

$$\eta_l \frac{\partial}{\partial \eta_l} \xi^J \eta^L = L_l \xi^J \eta^L$$

and thus

$$\mathcal{L}_0 \xi^J \eta^L = i\omega \cdot (L-J) \xi^L \eta^J$$

which is the thesis. □

Thus we have that for each j the space $\mathcal{H}_j$ decomposes into the direct sum of the kernel K of $\mathcal{L}_0$ and its range R. In particular the Kernel is generated by the *resonant monomials*, namely

$$K = \mathrm{Span}(\xi^J \eta^L \in \mathcal{H}_i : (J,L) \in \mathrm{RS}) \tag{23}$$

and

$$\mathrm{RS} := \{(J,L) : \omega \cdot (L-J) = 0\} \tag{24}$$

is the set of the resonant indexes. Obviously the range is generated by the space monomials $\xi^J \eta^L$ with J,L varying in the complement of the resonant set.

Thus it is easy to obtain the following important lemma.

Lemma 2.3. *Let $f \in \mathcal{H}_j$ be a polynomial, write*

$$f(\xi,\eta) = \sum_{J,L} f_{JL} \xi^J \eta^L \tag{25}$$

and define

$$Z(\xi,\eta) := \sum_{(J,L)\in RS} f_{JL}\xi^J\eta^L \, , \quad \chi(\xi,\eta) := \sum_{(J,L)\notin RS} \frac{f_{JL}}{i\omega\cdot(L-J)}\xi^J\eta^L \tag{26}$$

then one has

$$Z = \{\chi,H_0\} + f \, . \tag{27}$$

and

$$\{Z,H_0\} \equiv 0 \, . \tag{28}$$

Motivated by the above lemma we give the following definition.

Definition 2.5. A function Z will be said to be in normal form if, when written in terms of the variables ξ,η, it contains only resonant monomials, i.e. if writing

$$Z(\xi,\eta) := \sum_{(J,L)} Z_{JL}\xi^J\eta^L \, , \tag{29}$$

one has

$$Z_{JL} \neq 0 \Longrightarrow \omega\cdot(L-J) = 0 \, . \tag{30}$$

Remark 2.11. A property which is equivalent to (30) is $\{Z,H_0\} = 0$, which has the advantage of being coordinate independent.

Remark 2.12. If the frequencies are nonresonant, namely if eq. (10) holds, then the set of the indexes (J,L) such that $\omega\cdot(L-J) = 0$ reduces to the set $J = L$. Thus the resonant monomials are only the monomials of the form

$$\xi^J\eta^J = (\xi_1\eta_1)^{J_1}...(\xi_n\eta_n)^{J_n} \equiv I_1^{J_1}...I_n^{J_n} \, . \tag{31}$$

It follows that in the nonresonant case a function Z is in normal form if and only if it is a function of the actions only.

2.2.3 Proof of Birkhoff's theorem

We proceed by induction. The theorem is trivially true for $r = 0$. Supposing it is true for r we prove it for $r+1$. First consider the Taylor polynomial of degree $r+3$ of $\mathscr{R}^{(r)}$ and denote it by $H_{r+1}^{(r)} \in \mathscr{H}_{r+1}$. Let $\chi_{r+1} \in \mathscr{H}_{r+1}$ be the solution of the homological equation

$$\{\chi_{r+1};H_0\} + H_{r+1}^{(r)} = Z_{r+1} \tag{32}$$

with Z_{r+1} in normal form. By Lemma 2.3 such a χ_{r+1} exists. By corollary 6.1 below, χ_{r+1} generates an analytic flow. Use it to generate the Lie transform ϕ_{r+1} and consider $H^{(r+1)} := H^{(r)} \circ \phi_{r+1}$ and write it as follows

$$H^{(r)} \circ \phi_{r+1} = H_0 + Z^{(r)} \tag{33}$$

$$+ Z_{r+1} \tag{34}$$

$$+ (Z^{(r)} \circ \phi_{r+1} - Z^{(r)}) \tag{35}$$

$$+ H_0 \circ \phi_{r+1} - (H_0 + \{\chi_{r+1}; H_0\}) \tag{36}$$

$$+ (\mathscr{R}^{(r)} - H_{r+1}^{(r)}) \circ \phi_{r+1} \tag{37}$$

$$+ H_{r+1}^{(r)} \circ \phi_{r+1} - H_{r+1}^{(r)} \; . \tag{38}$$

define $Z^{(r+1)} := Z^{(r)} + Z_{r+1}$. To prove that the terms (35–38) have a vector field with a zero of order at least $r + 3$ use Lemma 2.1 which ensures that each line is the remainder of a Taylor expansion (in the space variables) truncated at order $r + 3$.

It remains to show that the estimate (9) of the deformation holds. Denote by R_{r+1} a positive number such that $B_{2R_{r+1}} \subset \mathscr{U}_s^{(r)}$, and remark that, by Lemma 6.2, possibly reducing R_{r+1}, one has

$$\phi_{r+1} : B_\rho \to B_{2\rho} \; , \quad \forall \rho \leq R_{r+1}$$

and

$$\sup_{B_\rho} \|z - \phi_{r+1}(z)\| \leq C\rho^{r+2} \; . \tag{39}$$

Define $\mathscr{T}_{r+1} := \mathscr{T}_r \circ \phi_{r+1}$ then one has

$$Id - \mathscr{T}_{r+1} = Id - \mathscr{T}_r \circ \phi_{r+1} = Id - \mathscr{T}_r + \mathscr{T}_r - \mathscr{T}_r \circ \phi_{r+1}$$

and thus, for any $z \in B_\rho$ with ρ small enough, we have

$$\|z - \mathscr{T}_{r+1}(z)\| \leq \|z - \mathscr{T}_r(z)\| + \|\mathscr{T}_r(z) - \mathscr{T}_r(\phi_{r+1}(z))\|$$
$$\leq C_r\rho^2 + \sup_{z \in B_{2\rho}} \|d\mathscr{T}_r(z)\| \sup_{z \in B_\rho} \|z - \phi_{r+1}(z)\|$$
$$\leq C_r\rho^2 + C\rho^{r+2} \leq C_{r+1}\rho^2$$

from which the thesis follows.

3 The case of PDEs

3.1 Hamiltonian formulation of the wave equation

Consider the nonlinear wave equation (1).

It is well known that the energy is a conserved quantity for (1). It is given by

$$H(u,v) := \int_D \left(\frac{v^2}{2} - \frac{u\Delta u}{2} + \frac{\mu^2 u^2}{2} \right) d^d x + \int_D F(u) d^d x \tag{1}$$

where $v := u_t$ and F is such that $-F' = f$, and D is either S^d (d-dimensional sphere) or $[0,\pi]$. The function H is also the Hamiltonian of the system and the corresponding Hamilton equations are given by

$$\dot{u} = \nabla_v H , \quad \dot{v} = -\nabla_u H \tag{2}$$

where $\nabla_u H$ is the L^2 gradient of H with respect to u, defined by

$$\langle \nabla_u H; h \rangle_{L^2} = \mathrm{d}_u H h \quad \forall h \in C^\infty(D) \tag{3}$$

where d_u is the differential with respect to the u variables. ∇_v is defined similarly.

To write (1) in the form (1) we have to introduce the basis of the eigenfunctions of the Laplacian.

In the case of $[0,\pi]$ the eigenfunctions are given by

$$\mathbf{e}_1 := \frac{1}{\sqrt{\pi}} , \quad \mathbf{e}_j := \frac{1}{\sqrt{\pi/2}} \cos((j-1)x) , \quad j \geq 2 \tag{4}$$

and the corresponding eigenvalues of $-\Delta$ are $\lambda_j = (j-1)^2$.

In the case of the d dimensional sphere the eigenvalues λ_j of $-\Delta$ are given by

$$\lambda_j = (j-1)(j+d-2) ; \tag{5}$$

moreover the jth eigenvalue has multiplicity

$$l^*(j) := \binom{j+d-1}{d} .$$

We will denote by $\mathbf{e}_{jl}$ a basis of eigenfunctions of the Laplacian, which is orthonormal in L^2 and such that

$$-\Delta \mathbf{e}_{jl} = \lambda_j \mathbf{e}_{jl} , \quad j \geq 1 , \quad l = 1,...,l^*(j) . \tag{6}$$

For example they can be chosen to be the spherical harmonics.

In both cases define ω_j, p_{jl} and q_{jl} by

$$\omega_j := \sqrt{\lambda_j + \mu^2} \tag{7}$$

$$u = \sum_{jl} \frac{q_{jl}}{\sqrt{\omega_j}} \mathbf{e}_{jl} , \quad v = \sum_{jl} \sqrt{\omega_j} p_{jl} \mathbf{e}_{jl} \tag{8}$$

with the convention that l takes only the value 1 in the case of $[0,\pi]$ (and that, in such a case it will not be written).

Then the Hamiltonian (1) takes the form (1) with

$$H_0 = \sum_j \sum_l \omega_j \frac{p_{jl}^2 + q_{jl}^2}{2} \tag{9}$$

and H_P is given by the second integral in (1) considered as a function of q_{jl}.

3.2 Extension of Birkhoff's theorem to PDEs: heuristic ideas

In this section we will concentrate on the case of the nonlinear wave equation on $[0,\pi]$.

The main difficulty one meets in order to extend the theory of Birkhoff normal form to infinite dimensional systems rests in the denominators one meets in solving the homological equation, namely in the second of equations (26). Indeed, while in the finite dimensional case one has that the set of vectors with integer components having modulus smaller than a given r is finite, this is no longer true in infinite dimensions.

It turns out that typically the denominators in (26) accumulate to zero already at order 4. An example of such a behavior is the following one. Consider $\omega_{j+1} := \sqrt{j^2 + \mu^2}$. For $l \geq 1$ consider the integer vector $K^{(l)}$ whose only components different from zero are given by $K_l = -2$, $K_{l-1} = 1$ $K_{l+1} = 1$; such a vector has modulus 4, and one has

$$
\begin{aligned}
K^{(l)} \cdot \omega &= \omega_{l+1} + \omega_{l-1} - 2\omega_l \\
&= \sqrt{l^2 + \mu^2} + \sqrt{(l-2)^2 + \mu^2} - 2\sqrt{(l-1)^2 + \mu^2} \sim \frac{\mu^2}{l^3} \to 0
\end{aligned}
$$

Thus Birkhoff theorem does not trivially extend to infinite dimensional systems.

However it turns out that in the case of PDEs the nonlinearity has a particular structure. As a consequence it turns out that most of the monomials appearing in the nonlinearity are small and do not need to be eliminated through the normalization procedure. To illustrate this behavior consider the map

$$
H^s([0,\pi]) \ni u \mapsto u^2 \in H^s([0,\pi]) , \tag{10}
$$

which is the first term of the nonlinearity of the nonlinear wave equation (1). The use of Leibniz formula together with interpolation inequality allows one to prove the so called Tame inequality, namely

$$
\left\| u^2 \right\|_s \leq C_s \left\| u \right\|_s \left\| u \right\|_1 . \tag{11}
$$

The key point is that, if u has only high frequency modes then its H^1 norm is much smaller than the H^s norm. Indeed, assume that, for some large M one has

$$
u = \sum_{k \geq M} \hat{u}_k \mathbf{e}_k \tag{12}
$$

then one has

$$
\|u\|_1^2 = \sum_{k \geq M} k^2 |\hat{u}_k|^2 = \sum_{k \geq M} \frac{k^{2s}}{k^{2(s-1)}} |\hat{u}_k|^2 \leq \frac{1}{M^{2(s-1)}} \|u\|_s^2 . \tag{13}
$$

Collecting (13) and (11) one gets

$$\left\| u^2 \right\|_s \leq C_s \frac{1}{M^{s-1}} \|u\|_s^2 \, , \tag{14}$$

which is very small if M and s are large. In order to exploit such a condition one can proceed as follows: given $u \in H^s$ split it into high frequency and low frequency terms, namely write

$$u^S := \sum_{|k|<M} \hat{u}_k \mathbf{e}_k \, , \quad u^L := \sum_{|k| \geq M} \hat{u}_k \mathbf{e}_k \tag{15}$$

then one has

$$u^2 = (u^S)^2 + 2u^S u^L + (u^L)^2 \, , \tag{16}$$

the norms of these terms are bounded respectively by

$$\|u\|_s^2 \, , \quad \|u\|_s^2 \, , \quad \frac{1}{M^{s-1}} \|u\|_s^2$$

from which one sees that the last term can be considered small and is not relevant to the dynamics. Thus one could avoid to eliminate such a term from the nonlinearity. Correspondingly one will not have to consider small denominators involving frequencies with many small indexes.

To be able to exploit the tame property one has to ensure that it persists under the operations involved in the construction of the normal form, namely the computation of Poisson Brackets and the solution of the Homological equation. Now the stability of the tame property under Poisson Brackets is easy to check, while the verification of the stability under solution of the Homological equation is difficult and at present not known. For this reason one has to perform a more careful analysis. It turns out that it is convenient to understand the structure of the coefficients of the nonlinearity that ensure the Tame property, and to show that such a structure is invariant under the construction. The theory we develop is a variant of that developed by Delort and Szeftel in [DS06].

4 A Birkhoff normal form theorem for semilinear PDEs

4.1 Maps with localized coefficients and their properties

Having in mind the case of the nonlinear wave equation in S^d, consider the space ℓ_s^2 of the sequences $q \equiv \{q_{jl}\}_{j \geq 0}^{1 \leq l \leq l^*(j)}$, such that

$$\|q\|_s^2 := \sum_{1 \leq j} |j|^{2s} \sum_{l=1}^{l^*(j)} |q_{jl}|^2 < \infty \, , \tag{1}$$

with a suitable $l^*(j)$.

Then define the projectors Π_j by

$$\Pi_j q := \sum_l q_{jl} \mathbf{e}_{jl} \tag{2}$$

(sum only over l), and the spaces $E_j := \Pi_j \ell_s^2$, which are independent of s.

The spaces $\mathscr{P}_s := \ell_s^2 \times \ell_s^2 \ni (p,q)$ will be used as phase spaces. We will also use the spaces $\mathscr{P}_\infty := \cap_s \mathscr{P}_s$ and $\mathscr{P}_{-\infty} := \cup_{s \in \mathbb{R}} \mathscr{P}_s$. It is useful to treat the p's and the q's exactly on an equal footing so we will use the notation

$$z_{jl} = q_{jl}, \quad z_{-jl} := p_{jl}, \quad j \geq 1,$$

correspondingly we will denote by z the set of all the variables and we will use the projector

$$\Pi_{-j} z := \sum_l p_{jl} \mathbf{e}_{jl}, \quad j \geq 1. \tag{3}$$

Given an element $z \in \mathscr{P}_s$, one can write it as

$$z = \sum_{j \neq 0} \Pi_j z, \tag{4}$$

so that one has

$$\|z\|_s^2 = \sum_{j \neq 0} |j|^{2s} \left\| \Pi_j z \right\|^2 \tag{5}$$

where we defined

$$\left\| \Pi_j z \right\|^2 := \sum_{l=1}^{l^*(j)} z_{jl}^2. \tag{6}$$

Let $f : \mathscr{P}_\infty \to \mathbb{R}$ be a smooth polynomial functions homogeneous of degree r. We can associate to f a symmetric multilinear map $\tilde{f}$, defined by the property

$$f(z) = \tilde{f}(\underbrace{z, ..., z}_{r-\text{times}}) \tag{7}$$

then we can write

$$f(z) = \sum_{j_1,...,j_r} \tilde{f}(\Pi_{j_1} z, ..., \Pi_{j_r} z). \tag{8}$$

We will assume suitable localization properties of the norm of $\tilde{f}(\Pi_{j_1} z, ..., \Pi_{j_r} z)$ as a function of the indexes $j_1, ..., j_r$.

Definition 4.1. Given a multi-index $j \equiv (j_1, ..., j_r)$, let $(j_{i_1}, j_{i_2}, j_{i_3} ..., j_{i_r})$ be a re-ordering of j such that

$$|j_{i_1}| \geq |j_{i_2}| \geq |j_{i_3}| \geq \cdots \geq |j_{i_r}|.$$

We define $\mu(j) := |j_{i_3}|$ and

$$S(j) := \mu(j) + ||j_{i_1}| - |j_{i_2}||. \tag{9}$$

Definition 4.2. Let $f : \mathscr{P}_\infty \to \mathbb{R}$ be a homogeneous polynomial of degree r. Let $\widetilde{f}$ be the associated multilinear form. We will say that f has localized coefficients if there exists $v \in [0, +\infty)$ such that $\forall N \geq 1$ there exists C_N such that $\forall z \in \mathscr{P}_\infty$ and any choice of the indexes $j_1, ..., j_r$ the following inequality holds

$$\left| \widetilde{f}(\Pi_{j_1} z, ..., \Pi_{j_r} z) \right| \leq C_N \frac{\mu(j)^{v+N}}{S(j)^N} \left\| \Pi_{j_1} z \right\| ... \left\| \Pi_{j_r} z \right\| . \tag{10}$$

Definition 4.3. A function $f \in C^\infty(\mathscr{U}, \mathbb{R})$ with $\mathscr{U} \subset \mathscr{P}_\infty$ will be said to have localized coefficients if

 (i) All the terms of its Taylor expansion have localized coefficients.
 (ii) For any s large enough there exists a neighborhood $\mathscr{U}^{(s)}$ of the origin in $\mathscr{P}_s$ such that $X_f \in C^\infty(\mathscr{U}^{(s)}, \mathscr{P}_s)$.

Remark 4.1. In the case of $[0, \pi]$ the property (4.2) turns out to really be a property of the coefficients of the expansion of the nonlinearity on the basis in which the quadratic part is diagonal. To understand this point consider the case of a homogeneous polynomial dependent on q only. Write $q = \sum_j q_j \mathbf{e}_j$, then one has

$$f(q) = \sum_j \widetilde{f}(\mathbf{e}_{j_1}, ..., \mathbf{e}_{j_r}) q_{j_1} ... q_{j_r} =: \sum_j f_{j_1, ..., j_r} q_{j_1} ... q_{j_r} \tag{11}$$

then (10) is equivalent to

$$\left| f_{j_1, ..., j_r} \right| \leq C_N \frac{\mu(j)^{v+N}}{S(j)^N} , \quad \forall N \geq 1 \tag{12}$$

It is useful to extend the definition to polynomial maps taking value in $\mathscr{P}_s$.

Definition 4.4. Let $F : \mathscr{P}_\infty \to \mathscr{P}_{-\infty}$ be a polynomial map of degree r and let $\widetilde{F}$ be the associated multilinear form. We will say that F has localized coefficients if there exists $v \in [0, +\infty)$ such that

$$\left\| \Pi_i \widetilde{F}(\Pi_{j_1} z, ..., \Pi_{j_r} z) \right\| \leq C_N \frac{\mu(i, j)^{v+N}}{S(i, j)^N} \left\| \Pi_{j_1} z \right\| ... \left\| \Pi_{j_r} z \right\| , \tag{13}$$

$$\forall z \in \mathscr{P}_\infty , \quad \forall N \geq 1 \tag{14}$$

Here we denoted by (i, j) the multi-index $(i, j_1, ..., j_r)$.

Remark 4.2. It is easy to see that if a polynomial function has localized coefficients, then its Hamiltonian vector field has localized coefficients.

Remark 4.3. By the very definition of the property of localization of coefficients it is clear that any (finite) linear combination of functions or maps with localized coefficients has localized coefficients.

Remark 4.4. As it will be clear from the theory of Sects. 4.3, and 6.3 it is quite easy to verify the property of localization of the coefficients.

The main properties of polynomials with localized coefficients are their smoothness, their stability under composition, linear combination and solution of the homological equation. In this subsection we will just state the corresponding results that will be proved in the appendix.

First one has that the vector field of a polynomial with localized coefficients has the tame property.

Theorem 4.1. *Let $F : \mathscr{P}_\infty \to \mathscr{P}_{-\infty}$ be a polynomial of degree r with localized coefficients, then there exists s_0 such that for any $s \geq s_0$ it extends to a smooth map from $\mathscr{P}_s$ to itself, moreover the following estimate holds*

$$\|F(z)\|_s \leq C \|z\|_s \|z\|_{s_0}^{r-1} . \tag{15}$$

Corollary 4.1. *Let f be a function with localized coefficients, then the result of theorem 4.1 holds for its vector field.*

The composition of maps with localized coefficients has localized coefficients. Precisely

Theorem 4.2. *Let $f : \mathscr{P}_\infty \to \mathbb{R}$ be a polynomial of degree r_1 with localized coefficients, and let $G : \mathscr{P}_\infty \to \mathscr{P}_{-\infty}$ be a polynomial of degree r_2 with localized coefficients, then the polynomial*

$$\mathrm{d}f(z)G(z) \tag{16}$$

has localized coefficients.

Thus the strategy in order to verify the property of localization of the coefficients is to verify it for a few simple maps and then to use the composition (16) to show that it holds for more general maps. The precise example we have in mind is that where $f(u) = \int u^3$ and $G(u) = u^2$, in which $\mathrm{d}f(u)G(u) = 3 \int u^4$. Hence by iteration one gets that all polynomials in u have localized coefficients if $\int u^3$ has.

Moreover the following corollary holds.

Corollary 4.2. *The Poisson Bracket of two functions with localized coefficients has localized coefficients.*

In order to develop perturbation theory we need a suitable nonresonance condition. This is given by the following definition.

Definition 4.5. Fix a positive integer r. The frequency vector ω is said to fulfill the *property (r–NR)* if there exist $\gamma > 0$, and $\alpha \in \mathbb{R}$ such that for any N large enough one has

$$\left| \sum_{j \geq 1} \omega_j K_j \right| \geq \frac{\gamma}{N^\alpha} , \tag{17}$$

for any $K \in \mathbb{Z}^\infty$, fulfilling $0 \neq |K| := \sum_j |K_j| \leq r+2$, $\sum_{j>N} |K_j| \leq 2$.

It is easy to see that under this condition one can solve the Homological equation. The precise statement is given by the following lemma.

Lemma 4.1. *Let f be a homogeneous polynomial of degree less or equal than r having localized coefficients. Let H_0 be given by (9) and assume that the frequency vector fulfills the condition r-NR. Consider the Homological equation*

$$\{H_0, \chi\} + f = Z. \tag{18}$$

Its solution χ, Z defined by (26) has localized coefficients. In particular χ has localized coefficients.

4.2 Statement of the normal form theorem and its consequences

Using the above results it is very easy to prove a version of the Birkhoff normal form theorem for PDEs.

Definition 4.6. With reference to a system of the form (1) with H_0 given by (9), the quantity

$$J_j := \sum_l \frac{p_{jl}^2 + q_{jl}^2}{2} \tag{19}$$

is called the total action of the modes with frequency ω_j.

Theorem 4.3. *Fix $r \geq 1$, assume that the nonlinearity H_P has localized coefficients and that the frequencies fulfill the nonresonance condition (r-NR), then there exists a finite s_r a neighborhood $\mathcal{U}_{s_r}^{(r)}$ of the origin in $\mathscr{P}_{s_r}$ and a canonical transformation $\mathscr{T} : \mathcal{U}_{s_r}^{(r)} \to \mathscr{P}_{s_r}$ which puts the system in normal form up to order $r+3$, namely*

$$H^{(r)} := H \circ \mathscr{T} = H_0 + Z^{(r)} + \mathscr{R}^{(r)} \tag{20}$$

where $Z^{(r)}$ and $\mathscr{R}^{(r)}$ have localized coefficients and

(i) $Z^{(r)}$ is a polynomial of degree $r+2$ which Poisson commutes with J_j for all j's, namely $\left\{ J_j; Z^{(r)} \right\} \equiv 0$;

(ii) $\mathscr{R}^{(r)}$ has a small vector field, i.e.

$$\left\| X_{\mathscr{R}^{(r)}}(z) \right\|_{s_r} \leq C \|z\|_{s_r}^{r+2}, \quad \forall z \in \mathcal{U}_{s_r}^{(r)}; \tag{21}$$

(iii) One has

$$\| z - \mathscr{T}_r(z) \|_{s_r} \leq C \|z\|_{s_r}^2, \quad \forall z \in \mathcal{U}_{s_r}^{(r)}. \tag{22}$$

An inequality identical to (9) is fulfilled by the inverse transformation $\mathscr{T}_r^{-1}$.

(iv) For any $s \geq s_r$ there exists a subset $\mathcal{U}_s^{(r)} \subset \mathcal{U}_{s_r}^{(r)}$ open in $\mathscr{P}_s$ such that the restriction of the canonical transformation to $\mathcal{U}_s^{(r)}$ is analytic also as a map from $\mathscr{P}_s \to \mathscr{P}_s$ and the inequalities (21) and (22) hold with s in place of s_r.

The proof is deferred to Sect. 6.2.

In order to deduce dynamical consequences we fix the number r of normalization steps; moreover, it is useful to distinguish between the original variables and the variables introduced by the normalizing transformation. So, we will denote by $z = (p,q)$ the original variables and by $z' = (p',q')$ the normalized variables, i.e. $z = \mathcal{T}_r(z')$. More generally we will denote with a prime the quantities expressed in the normalized variables.

Proposition 4.1. *Under the same assumptions of Theorem 4.3, $\forall s \geq s_r$ there exists ε_{*s} such that, if the initial datum fulfills*

$$\varepsilon := \|z_0\|_s < \varepsilon_{*s}$$

then one has

(i)

$$\|z(t)\|_s \leq 4\varepsilon \quad for \quad |t| \leq \frac{1}{\varepsilon^r} \tag{23}$$

(ii)

$$\sum_j j^{2s} \left| J'_j(t) - J'_j(0) \right| \leq C\varepsilon^{M+3} \quad for \quad |t| \leq \frac{1}{\varepsilon^{r-M}} \,, \quad M < r \tag{24}$$

and

$$\sum_j j^{2s} \left| J_j(t) - J_j(0) \right| \leq C\varepsilon^3 \quad for \quad |t| \leq \frac{1}{\varepsilon^r} \,. \tag{25}$$

(iii) *If for each j the space E_j is one dimensional, then there exists a smooth torus $\mathbb{T}_0$ such that, $\forall M \leq r$*

$$d_s(z(t), \mathbb{T}_0) \leq C\varepsilon^{(M+3)/2} \,, \quad for \quad |t| \leq \frac{1}{\varepsilon^{r-M}} \tag{26}$$

where $d_s(.,.)$ is the distance in $\mathscr{P}_s$.

4.3 Application to the nonlinear wave equation

The aim of this section is to verify the assumptions of Theorem 4.3 in the model problems we are considering.

We start by the property of localization of the coefficients. The main step consists in verifying the property for the Hamiltonian function

$$f(u) := \int_D u^3(x)\mathrm{d}x \,; \tag{27}$$

the corresponding multilinear form is given by

$$\tilde{f}(u_1,u_2,u_3) := \int_D u_1(x)u_2(x)u_3(x)\mathrm{d}x \,, \tag{28}$$

so we have to estimate such a quantity when $u_j \in E_{n_j}$, namely the eigenspace of $-\Delta$ corresponding to the eigenvalue λ_{n_j}. We have the following theorem.

Proposition 4.2. *Let E_n be the eigenspace of $-\Delta$ associated to the eigenvalue λ_n, then for any $N \geq 1$ there exists C_N such that one has*

$$\left| \int_D u_{n_1}(x) u_{n_2}(x) u_{n_3}(x) \mathrm{d}x \right| \leq C_N \frac{\mu(n)^{N+\nu}}{S(n)^N} \|u_{n_1}\|_{L^2} \|u_{n_2}\|_{L^2} \|u_{n_3}\|_{L^2} \qquad (29)$$

for all $u_{n_j} \in E_{n_j}$.

A simple strategy to obtain the proof consists in considering the quantity (28) as the matrix elements of index n_1, n_2 (on the basis of the eigenvectors of the Laplacian) of the operator of multiplication by u_{n_3}. The actual proof is deferred to Sect. 6.3.

Corollary 4.3. *The nonlinearity given by the second integral in (1) has localized coefficient.*

Proof. This is a consequence of Proposition 4.2 and of Theorem 4.2. Indeed a term in the Taylor expansion of $\int_D F(u)\mathrm{d}x$ is a multiple of

$$t_k(u) := \int_{S^d} u^k(x)\mathrm{d}x \qquad (30)$$

and one has $t_k(u) = C \mathrm{d}t_2(u) T_{k-1}(u)$, where $T_{k-1}(u) = u^{k-1}$. Then such a quantity has localized coefficients by Theorem 4.2. $\square$

In order to apply Theorem 4.3 to the nonlinear wave equation (1) there remains to verify the nonresonance condition $(r - NR)$. To this end consider the frequencies

$$\omega_{j+1} = \sqrt{j(j+d-1)+\mu^2} \qquad (31)$$

then we have the following

Theorem 4.4. *There exists a zero measure set $S \subset \mathbb{R}$ such that, if $\mu \in \mathbb{R} - S$, then the frequencies (31) fulfill the condition $(r - NR)$ for any r.*

The proof was given in [Bam03] (see also [BG06]), and for the sake of completeness it is repeated in Appendix 8.

Thus the main theorem and its corollaries apply to the nonlinear wave equation both in the case of $[0,\pi]$ and in the case of the d dimensional sphere.

Remark 4.5. A particular consequence of this theory is that it allows one to ensure existence of smooth solutions of the nonlinear wave equation on the sphere for times of order ε^{-r}. It has to be emphasized that when $d > 1$ local existence is ensured only in H^s, with $s > 1$, so that the energy norm is useless in order to deduce estimate of the existence times of solutions. At present the method of Birkhoff normal form is the only one allowing one to improve the times given by the local existence theory.

5 Discussion

First I would like to mention that, as shown in [BG06], Theorem 4.3 is a theorem that allows one to deal with quite general *semilinear* equations in *one space dimension*.

The limitation to semilinear equation is evident in Theorem 4.3. Thus in particular all the equations with nonlinearity involving derivatives are excluded from the present theory. It would be of major interest to have a theory valid also for some quasilinear equations, since most physical models have nonlinearities involving derivatives. Very little is known on quasilinear problems. At present the only known result is that of [DS04] (and a recent extension by Delort), where only one step of normal form was performed for the quasilinear wave equation. It would be very interesting to understand how to iterate the procedure developed in such papers.

The limitation to one-space dimension is more hidden. Actually it is hidden in the nonresonance condition. Indeed its verification is based on the asymptotic behavior of the frequencies: the nonresonance condition is typically satisfied only if the frequencies grow at infinity at least as $\omega_j \sim j$. As it was shown in the example of the nonlinear wave equation on the sphere, the possible multiplicity of the frequencies is not a problem. The theory easily extends to the case where the differences between couples of frequencies accumulate only at a discrete subset of $\mathbb{R}$. The understanding of the structure of the frequencies in higher dimension is surely a key point for the extension of the theory to higher dimensions.

Finally I would like to mention the fact that all known applications of the theory we are considering pertain to equations on compact manifolds, however in principle the theory applies to smooth perturbations of linear system with discrete spectrum. A nice example of such a kind of systems is the Gross Pitaevskii equation. It would be interesting to show that such an equation fulfills the assumption of Theorem 4.3. This could be interesting also in connection with the study of the blow up phenomenon.

6 Proofs

6.1 Proof of the properties of functions with localized coefficients

Lemma 6.1. *Let* $z \in \mathscr{P}_s$ *with* $s > v + 1/2$ *then there exists a constant* C_s *such that*

$$\sum_{j \neq 0} |j|^v \|\Pi_j z\| \leq C_s \|z\|_s \tag{1}$$

Proof. One has

$$\sum_{j \neq 0} |j|^v \|\Pi_j z\| \leq \sum_j |j|^s \frac{\|\Pi_j z\|}{|j|^{s-v}} \leq \sqrt{\sum_j \frac{1}{|j|^{2(s-v)}}} \sqrt{\sum_j |j|^{2s} \|\Pi_j z\|^2}$$

which is the thesis. $\square$

Proof of Theorem 4.1. Write explicitly the norm of $F(z)$. One has

$$\|F(z)\|_s^2 = \sum_l |l|^{2s} \left\| \sum_{j_1,\dots,j_r} \Pi_l \tilde{F}(\Pi_{j_1} z, \dots, \Pi_{j_r} z) \right\|^2 . \tag{2}$$

In what follows, to simplify the notation we will write

$$a_j := \left\| \Pi_j z \right\| .$$

One has

$$\left\| \sum_{j_1,\dots,j_r} |l|^s \Pi_l \tilde{F}(\Pi_{j_1} z, \dots, \Pi_{j_r} z) \right\| \le C \sum_{j_1,\dots,j_r} |l|^s \frac{\mu(j,l)^{\nu+N}}{S(j,l)^N} a_{j_1} \dots a_{j_r} \tag{3}$$

Since this expression is symmetric in $j_1,\dots j_r$ the r.h.s. of (3) is estimated by a constant times the sum restricted to ordered multi-indexes, namely indexes such that $|j_1| \le |j_2| \le \dots \le |j_r|$. Moreover, in order to simplify the notations *we will restrict to the case of positive indexes*. To estimate (3) remark that for ordered multi-indexes one has

$$l\frac{\mu(j,l)}{S(j,l)} \le 2j_r . \tag{4}$$

Indeed, if $l \le 2j_r$ this is obvious ($\mu/S < 1$ by the very definition), while, if $l > 2j_r$ one has $S(j,l) \ge |l - j_r| > l/2$, and therefore

$$l\frac{\mu(j,l)}{S(j,l)} \le \mu(j,l) \le 2j_r .$$

Remark now that, by the definition of S one has

$$S(j,l) \ge \begin{cases} 1 + |j_r - l| & \text{if } l \ge j_{r-1} \\ \mu(j,l) + j_r - j_{r-1} \ge l + j_r - j_{r-1} & \text{if } l < j_{r-1} \end{cases}$$

Thus define $\hat{S}(j,l) := \min\{1 + |j_r - l|, l + j_r - j_{r-1}\}$ and remark that $S(j,l) \ge \hat{S}(j,l)$. Remark also that $\mu(j,l) \le j_{r-1}$. So it follows that (3) is smaller than (a constant times)

$$\sum_{j_1,\dots,j_r} j_r^s \frac{\mu(j,l)^{N'+\nu}}{\hat{S}(j,l)^{N'}} a_{j_1} \dots a_{j_r} \le \sum_{j_1,\dots,j_r} j_r^s \frac{j_{r-1}^{N'+\nu}}{\hat{S}(j,l)^{N'}} a_{j_1} \dots a_{j_r} \tag{5}$$

$$\le \|z\|_{s_1}^{r-2} \sum_{j_{r-1},j_r} j_r^s \frac{j_{r-1}^{N'+\nu}}{\hat{S}(j,l)^{N'}} a_{j_{r-1}} a_{j_r} \tag{6}$$

where we denoted $N' := N - s$ and we used Lemma 6.1; we denoted by s_1 a number such that $s_1 > 1/2$. Inserting in (2) one gets

$$\|F(z)\|_s^2 \le \|z\|_{s_1}^{2(r-2)} \sum_l \left(\sum_{j_{r-1},j_r} j_r^s \frac{j_{r-1}^{N'+v}}{\hat{S}(j,l)^{N'}} a_{j_{r-1}} a_{j_r} \right)^2$$

$$= \|z\|_{s_1}^{2(r-2)} \sum_l \left(\sum_{j_{r-1}} j_{r-1}^{N'+v} a_{j_{r-1}} \sum_{j_r} j_r^s \frac{a_{j_r}}{\hat{S}(j,l)^{N'/2}} \frac{1}{\hat{S}(j,l)^{N'/2}} \right)^2$$

$$\le \|z\|_{s_1}^{2(r-2)} \sum_l \left(\sum_{j_{r-1}} j_{r-1}^{N'+v} a_{j_{r-1}} \sqrt{\sum_{j_r} j_r^{2s} \frac{a_{j_r}^2}{\hat{S}(j,l)^{N'}}} \sqrt{\sum_{j_r} \frac{1}{\hat{S}(j,l)^{N'}}} \right)^2$$

Now the last sum in j_r is finite provided $N' > 1$. Remark now that $\hat{S}(j,l) \ge \check{S}(j_r,l) := \min\{1+|l-j_r|,l\}$ (independent of j_{r-1}), and therefore the above quantity is estimated by a constant times

$$\|z\|_{s_1}^{2(r-2)} \sum_l \left(\sum_{j_{r-1}} j_{r-1}^{N'+v} a_{j_{r-1}} \sqrt{\sum_{j_r} j_r^{2s} \frac{a_{j_r}^2}{\check{S}(j,l)^{N'}}} \right)^2 \tag{7}$$

$$= \|z\|_{s_1}^{2(r-2)} \sum_{j_r} j_r^{2s} a_{j_r}^2 \sum_l \frac{1}{\check{S}(j,l)^{N'}} \left(\sum_{j_{r-1}} j_{r-1}^{N'+v} a_{j_{r-1}} \right)^2 \tag{8}$$

$$\le C \|z\|_s^2 \|z\|_{s_1}^{2(r-2)} \|z\|_{s_0}^2 \tag{9}$$

where s_0 is such that $s_0 > N' + v + 1/2$. Choosing $s_1 \le s_0$ and estimating $\|z\|_{s_1}$ with $\|z\|_{s_0}$ one gets the thesis. $\square$

Proof of Theorem 4.2. First remark that the multilinear form associated to the polynomial $df(z)G(z)$ is given by the symmetrization of

$$r_1 \tilde{f}(z^{(1)}, ..., z^{(r_1-1)}, \tilde{G}(z^{(r_1)}, ..., z^{(r_1+r_2-1)})) . \tag{10}$$

We will estimate the coefficients of this multilinear function. This will give the result. Forgetting the irrelevant constant r_1, the quantity to be estimated is

$$\tilde{f}(\Pi_{j_1} z, ..., \Pi_{j_{r_1-1}} z, \tilde{G}(\Pi_{i_1} z, ..., \Pi_{i_{r_2}} z)) \tag{11}$$

$$= \sum_l \tilde{f}(\Pi_{j_1} z, ..., \Pi_{j_{r_1-1}} z, \Pi_l \tilde{G}(\Pi_{i_1} z, ..., \Pi_{i_{r_2}} z)) \tag{12}$$

$$\le C_{N,N'} \sum_l \frac{\mu^{v_1+N}(j,l)}{S(j,l)^N} \frac{\mu^{v_2+N'}(i,l)}{S(i,l)^{N'}} \|\Pi_{j_1} z\| ... \|\Pi_{i_{r_2}} z\| \tag{13}$$

Thus it is enough to estimate

$$\sum_l \frac{\mu^{v_1+N}(j,l)}{S(j,l)^N} \frac{\mu^{v_2+N'}(i,l)}{S(i,l)^{N'}} \tag{14}$$

This is the heart of the proof.

In order to simplify the notation we will restrict to the case $r_1 - 1 = r_2 = r$. Due to the symmetry of this estimate we will restrict the case of ordered indexes, that can also be assumed to be positive, so that one has $j_r \geq j_{r-1} \geq \ldots \geq j_1$ and similarly for i.

All along this proof we will use the notation

$$\tilde{S}(j) := j_r - j_{r-1} \equiv S(j) - \mu(j)$$

We have to distinguish two cases.

First case $j_r \geq i_r \geq j_{r-1}$.

The proof of this first case is (up to minor changes) equal to that given in [Gré06].

Take $N' = N$, then before estimating (14), we need to estimate the general term of the sum. So we collect a few facts on it.

The main relation we need is

$$\tilde{S}(i,j) \leq \tilde{S}(i,l) + \tilde{S}(j,l) \ . \tag{15}$$

This will be established by writing explicitly all the involved quantities as l varies. So, first remark that $\tilde{S}(i,j) = j_r - i_r$. Then one has

$$\tilde{S}(i,l) = \begin{cases} i_r - i_{r-1} & \text{if } l \leq i_{r-1} \\ |i_r - l| & \text{if } l > i_{r-1} \end{cases} \quad , \quad \tilde{S}(j,l) = \begin{cases} j_r - j_{r-1} & \text{if } l \leq j_{r-1} \\ |j_r - l| & \text{if } l > j_{r-1} \end{cases}$$

which gives

$$\tilde{S}(i,l) + \tilde{S}(j,l) = \begin{cases} i_r - i_{r-1} + j_r - j_{r-1} \geq j_r - j_{r-1} \geq j_r - i_r & \text{if } l \leq i_{r-1} \\ l - i_r + j_r - j_{r-1} \geq j_r - j_{r-1} \geq j_r - i_r \text{if } i_{r-1} < l \leq j_{r-1} \\ |i_r - l| + |j_r - l| \geq |j_k - l| \geq j_k - i_k & \text{if } j_{k-1} < l \leq i_r \\ |i_r - l| + |j_r - l| = j_r - l + l - i_r & \text{if } i_r < l \leq j_r \\ |i_r - l| + |j_r - l| \geq l - i_r \geq j_r - i_r & \text{if } j_r \leq l \end{cases}$$

from this (15) follows.

One also has

$$\mu(j,l) \leq \mu(i,j) \ , \quad \mu(i,l) \leq \mu(i,j) \ . \tag{16}$$

Thus

$$\frac{S(i,j)}{\mu(i,j)} = 1 + \frac{\tilde{S}(i,j)}{\mu(i,j)} \leq 1 + \frac{\tilde{S}(i,l) + \tilde{S}(j,l)}{\mu(i,j)} \leq 1 + \frac{\tilde{S}(i,l)}{\mu(i,l)} + \frac{\tilde{S}(l,j)}{\mu(l,j)} < \frac{S(i,l)}{\mu(i,l)} + \frac{S(l,j)}{\mu(l,j)}$$

From this one has

$$\frac{\mu(i,j)}{S(i,j)} \geq \frac{1}{2} \min\left\{ \frac{\mu(i,l)}{S(i,l)}, \frac{\mu(l,j)}{S(l,j)} \right\} \ . \tag{17}$$

Separate the sum over those l such that $\frac{\mu(i,l)}{S(i,l)} > \frac{\mu(l,j)}{S(l,j)}$ and that over its complement. Let L_1 be the first set. Then one has

$$\sum_{l\in L_1}\frac{\mu^{v_1+N}(j,l)}{S(j,l)^N}\frac{\mu^{v_2+N}(i,l)}{S(i,l)^N}\leq\sum_{l\geq1}2^{N-1-\varepsilon}\mu(j,l)^{v_1}\frac{\mu(i,j)^{N-1-\varepsilon}}{S(i,j)^{N-1-\varepsilon}}\frac{\mu(i,l)^{1+\varepsilon+v_2}}{S(i,l)^{1+\varepsilon}}$$

$$\leq C\frac{\mu(i,j)^{N+v_1+v_2}}{S(i,j)^{N-1-\varepsilon}}\ .$$

Acting in the same way for the case of L_1^c one concludes the proof in the first case.

Second case $j_r\geq j_{r-1}>i_r$. Here it is easy to see that (15) still holds. However, in some cases it happens that the equation

$$\mu(j,l)\leq\mu(i,j) \tag{18}$$

is violated. When (18) holds the proof of the first case extends also to the present case. So let us consider only the case where (18) is violated. We claim that in this case one has

$$\frac{\mu(j,l)}{\tilde{S}(i,l)}\leq 2\mu(i,j)\ . \tag{19}$$

To prove (19) we distinguish two cases

(i) $j_{r-2}\leq i_r\leq j_{r-1}\leq j_r$.
   Then (18) is violated when $i_r<l\leq j_{r-1}$. In this case one has

$$\tilde{S}(i,l)=l-i_r \tag{20}$$

   It follows that

$$\frac{\mu(j,l)}{\tilde{S}(i,l)}\frac{1}{\mu(i,j)}=\frac{l}{l-i_r}\frac{1}{i_r}$$

   which is easily seen to be smaller than 2 (for example write $l=i_r+\delta$, then the relation becomes evident).
(ii) $i_r<j_{r-2}\leq j_{r-1}\leq j_r$. Here (18) is violated when $j_{r-2}<l\leq j_{r-1}$. It is easy to see that also in this case (20) holds. Then

$$\frac{\mu(j,l)}{\tilde{S}(i,l)}\frac{1}{\mu(i,j)}=\frac{l}{l-i_r}\frac{1}{j_r}\leq\frac{l}{l-i_r}\frac{1}{i_r}$$

   from which (19) still follows.

It is now easy to conclude the proof. Take $N'=2N+v_2$, then, using (19) one has

$$\frac{\mu(i,l)^{v_1+2N+v_2}}{S(i,l)^{2N+v_2}}\frac{\mu(j,l)^{N+v_2}}{S(j,l)^N}\leq\frac{\mu(i,l)^{v_1+2N+v_2}}{S(i,l)^N}\left(\frac{\mu(j,l)}{S(i,l)}\right)^{N+v_2}\frac{1}{S(j,l)^N}$$

$$\leq\frac{\mu(i,l)^{v_1+2N+v_2}}{S(i,l)^N}\frac{\mu(i,j)^{N+v_2}}{S(j,l)^N}$$

From this, following the proof given in the first case it is easy to prove that

$$\frac{S(i,j)}{\mu(i,j)} \le \frac{S(i,l)}{\mu(i,l)} + \frac{S(j,l)}{\mu(i,j)}$$

and to conclude the proof in the same way as in the first case. $\square$

Proof of Lemma 4.1. Consider the polynomial f and expand it in Taylor series. Introduce now the complex variables (22). Remark that this is a linear change of variable so it does not change the degree of a polynomial. Remark that the change of variables does not mix the different spaces $E_j \times E_j$. It follows that if a polynomial has localized coefficients in terms of the real variables p, q it has also localized coefficients when written in terms of the complex variables, i.e. it fulfills (10) with z_j which is either ξ_j or η_j. Remark that the converse is also true. Now, Z is the sum of some of the coefficients of f so it is clear that its coefficients are still localized. In order to estimate χ, remark first that, in the particular case where

$$f(z) \equiv \widetilde{f}(\Pi_{j_1}\xi, ..., \Pi_{j_{r_1}}\xi, \Pi_{l_1}\eta, ..., \Pi_{l_{r_2}}\eta)$$

(no summation over j, l) one has

$$\{H_0, f\} = i(\omega_{j_1} + ... + \omega_{j_{r_1}} - \omega_{l_1} - ... - \omega_{l_{r_2}})f \tag{21}$$

It follows that in the case of general f the function χ solving the homological equation can be rewritten as

$$\chi(\xi, \eta) := \sum_{jl} \frac{\widetilde{f}(\Pi_{j_1}\xi, ..., \Pi_{j_{r_1}}\xi, \Pi_{l_1}\eta, ..., \Pi_{l_{r_2}}\eta)}{i(\omega_{j_1} + ... + \omega_{j_{r_1}} - \omega_{l_1} - ... - \omega_{l_{r_2}})} \tag{22}$$

where the sum runs over the indexes such that the denominators do not vanish. Now, it is easy to verify that by condition (r-NR) the denominators are bounded from below by $\gamma/\mu(j,l)^\alpha$. So χ fulfills the estimate (10) with ν substituted by $\nu + \alpha$, if f does with ν. $\square$

6.2 Proof of the Birkhoff normal form Theorem 4.3 and of its dynamical consequences

In this section we will fix s large enough and work in $\mathscr{P}_s$. Here $B_R \subset \mathscr{P}_s$ will denote the open ball of radius R with center at the origin in $\mathscr{P}_s$. Moreover all along this section $\mathscr{H}_j$ will denote the set of homogeneous polynomials of degree $j+2$ **having a Hamiltonian vector field which is smooth as map from $\mathscr{P}_s$ to itself.** Finally, along this section we will omit the index s from the norm, thus we will simply denote $\|.\| := \|.\|_s$.

First we estimate the domain where the Lie transform generated by a polynomial $\chi \in \mathscr{H}_j$, $(j \geq 1)$ is well defined.

Lemma 6.2. *Let* $\chi \in \mathscr{H}_j$, $(j \geq 1)$ *be a polynomial. Denote by* ϕ^t *the flow of the corresponding vector field. Denote also*

$$\bar{t} = \bar{t}(R, \delta) := \inf_{z \in B_R} \sup \left\{ t > 0 : \phi^t(z) \in B_{R+\delta} \text{ and } \phi^{-t}(z) \in B_{R+\delta} \right\}$$

(minimum escape time of $\phi^t(z)$ *from* $B_{R+\delta}$*). Then one has*

$$\bar{t} \geq \frac{\delta}{2\|X_\chi\| R^{j+1}} \tag{23}$$

where $\|X_\chi\|$ *is the norm defined in remark 2.3. Moreover for any* t, *such that* $|t| \leq \bar{t}$ *and any* $z \in B_R$ *one has*

$$\|\phi^t(z) - z\| \leq |t| R^{j+1} \|X_\chi\| \tag{24}$$

Proof. First remark that, by the definition of $\bar{t}$ one has that there exists $\bar{z} \in B_R$ such that $\|\phi^{\pm \bar{t}}(\bar{z})\| = R + \delta$. Assume by contradiction $\bar{t} < \frac{\delta}{2\|X_\chi\| R^{j+1}}$, then, since for any t with $|t| < \bar{t}$ one has $\phi^t(\bar{z}) \in B_{R+\delta}$. It follows that

$$\left\|\phi^{\bar{t}}(\bar{z})\right\| \leq \|\bar{z}\| + \left\|\phi^{\bar{t}}(\bar{z}) - \bar{z}\right\| = \|\bar{z}\| + \left\|\int_0^{\bar{t}} \frac{d}{ds} \phi^s(\bar{z}) ds\right\|$$

$$\leq R + \int_0^{\bar{t}} \left\|X_\chi(\phi^s(\bar{z})) ds\right\| \leq R + |\bar{t}| R^{j+1} \|X_\chi\|,$$

from which $R + \delta \leq R + \delta/2$ which is absurd. $\square$

Since χ is analytic together with its vector field (it is a smooth polynomial), then one has the following corollary.

Corollary 6.1. *Fix arbitrary* R *and* δ, *then the map*

$$\phi : B_\sigma \times B_R \to B_{R+\delta}, \qquad \sigma := \frac{\delta}{2\|X_\chi R^{j+1}\|}$$

$$(t, z) \mapsto \phi^t(z)$$

is analytic. Here, by abuse of notation, we denoted by B_σ *also the ball of radius* σ *contained in* $\mathbb{C}$.

Proof of Theorem 4.3. The proof proceeds as in the finite dimensional case. The only fact that has to be ensured is that at any step the functions involved in the construction have localized coefficients. By Lemma 4.1 the solution χ_{r+1} of the homological equation (32) has localized coefficients. Thus, by Theorem 4.1 its vector field is smooth on a space $\mathscr{P}_{s_{r+1}}$. This determines the index s_{r+1} of the space with minimal smoothness in which the transformation $\mathscr{T}_{r+1}$ is defined. By corollary 6.1

χ_{r+1} generates an analytic flow. As in the finite dimensional case we use it to generate the Lie transform. Then $H^{(r+1)}$ is still given by (35–38). Remark now that given a Hamiltonian function f, the Hamiltonian vector field of $f \circ \phi_{r+1}$ is given by

$$X_{f \circ \phi_{r+1}}(z) = \mathrm{d}\phi_{r+1}^{-1}(\phi_{r+1}(z)) X_f(\phi_{r+1}(z)) \tag{25}$$

so that the Hamiltonian vector fields of the terms (34), (35), (37), (38) are smooth. To ensure the smoothness of the vector field of (36) write

$$\ell(z) := H_0 \circ \phi - H_0 - \{\chi_{r+1}, H_0\}$$

and remark that

$$H_0(\phi_{r+1}(z)) - H_0(z) = \int_0^1 \frac{d}{dt} H_0(\phi_{r+1}^t(z)) dt = \int_0^1 \{\chi_{r+1}, H_0\} (\phi_{r+1}^t(z)) dt$$

$$= \int_0^1 (H_{r+1}^{(r)}(\phi_{r+1}^t(z)) - Z_{r+1}(\phi_{r+1}^t(z)) dt \; ,$$

where we used the homological equation to calculate $\{\chi_{r+1}, H_0\}$. Denote again $G := H_{r+1}^{(r)} - Z_{r+1}$, then one has

$$\ell(z) = \int_0^1 (G(\phi_{r+1}^t(z)) - G(z)) dt \; ,$$

from which the smoothness of the vector field of (36) immediately follows. Since the Taylor expansion of the terms (35–38) can be computed using (13), by corollary 14 one has that all these functions have localized coefficients. Then, as in the finite dimensional case the terms (35–38) have a vector field with a zero of order at least $r+3$ which ensures the estimate of the remainder.

We show now that the normal form $Z^{(r)}$ commutes with all the J_j. To this end remark that, by construction, the normal form contains only resonant monomials, i.e. monomials $\xi^L \eta^J$ with

$$0 = \sum_{jl} \omega_j (J_{jl} - L_{jl}) = \sum_j \omega_j \left(\sum_l (J_{jl} - L_{jl}) \right) \; . \tag{26}$$

Now the nonresonance condition implies

$$\left(\sum_l (J_{jl} - L_{jl}) \right) = 0 \quad \forall j \; .$$

It follows

$$\{J_j, \xi^L \eta^J\} = \mathrm{i} \left[\sum_l (J_{jl} - L_{jl}) \right] \xi^L \eta^J = 0 \tag{27}$$

which is the desired property.

Finally the estimate (22) of the deformation can be obtained exactly as in the finite dimensional case. $\quad\square$

Proof of Proposition 4.1. We start by (i). Assume that ε is so small that $B_{3\varepsilon} \subset \mathscr{U}_s^{(r)}$; perform the normalizing transformation. Remark that, by (22), one has $z_0' \in B_{2\varepsilon} \subset \mathscr{U}_s^{(r)}$. Define $F(z) := \sum_j |j|^{2s} J_j \equiv \|z\|_s^2$, then, as far as $\|z'(t)\|_s \leq 3\varepsilon$ one has

$$
\left| F(z'(t)) - F(z_0') \right| = \left| \int_0^t \left\{ H^{(r)}, F \right\}(z'(s)) ds \right|
$$
$$
\leq \int_0^t \left| \left\{ \mathscr{R}^{(r)}, F \right\}(z'(s)) \right| ds \leq |t| C\varepsilon^{r+3} \leq C\varepsilon^3 \qquad (28)
$$

where the last inequality holds for the times (23). To conclude the proof of (23) it is enough to show that, for the considered times one actually has $z'(t) \in B_{3\varepsilon}$. To this end we follow the scheme of the proof of Lyapunov's theorem: define

$$
\bar{t} := \sup \left\{ t > 0 \ : \ \|z'(t)\|_s < 3\varepsilon \text{ and } \|z'(-t)\|_s < 3\varepsilon \right\}
$$

To fix ideas assume that the equality is realized for $t = \bar{t}$ Assume by contradiction that $\bar{t} < \varepsilon^{-r}$, then one can use (28) which gives

$$
\|z'(\bar{t})\|^2 = 9\varepsilon^2 = F(\bar{t}) \leq F(z_0') + |F(z'(t)) - F(z_0')| \leq 4\varepsilon^2 + C\varepsilon^3 , \qquad (29)
$$

which is impossible for ε small enough.

We come to (ii). First remark that

$$
j_j' = \sum_l \left(-p_{jl}' \frac{\partial \mathscr{R}^{(r)}}{\partial q_{jl}'} + q_{jl}' \frac{\partial \mathscr{R}^{(r)}}{\partial p_{jl}'} \right) ,
$$

so that

$$
\sum_j j^{2s} |J_j'| = \sum_{jl} j^{2s} \left| -p_{jl}' \frac{\partial \mathscr{R}^{(r)}}{\partial q_{jl}'} + q_{jl}' \frac{\partial \mathscr{R}^{(r)}}{\partial p_{jl}'} \right| \qquad (30)
$$
$$
\leq \left(\sum_{jl} j^{2s}(p_{jl}'^2 + q_{jl}'^2) \right)^{1/2} \left(\sum_{jl} j^{2s} \left(\left| \frac{\partial \mathscr{R}^{(r)}}{\partial q_{jl}'} \right|^2 + \left| \frac{\partial \mathscr{R}^{(r)}}{\partial p_{jl}'} \right|^2 \right) \right)^{1/2} \qquad (31)
$$
$$
\leq \|z'\|_s \|X_{\mathscr{R}^{(r)}}(z')\|_s \leq C \|z'\|_s^{r+3} \qquad (32)
$$

which implies (24).

To prove (25) write

$$
|J_j(t) - J_j(0)| \leq |J_j(z(t)) - J_j(z'(t))| + |J'(t) - J'(0)| + |J_j(z_0) - J_j(z_0')| . \qquad (33)
$$

The contribution of the middle term is estimated by (24). To estimate the contribution of the first and the last term write

$$j^{2s}\left|q_{jl}^2 - q_{jl}'^2\right| \le j^{2s}\left(2|q_{jl}'|\,|q_{jl} - q_{jl}'| + |q_{jl} - q_{jl}'|^2\right) \tag{34}$$

adding the corresponding estimate for the p variables and summing over jl one gets the thesis.

We come to (iii). In the considered case J_j reduces to I_j, so the actions are individually conserved. In this proof we omit the index l which would take only the value 1. Denote $\bar{I}_j' := \frac{p_j'^2(0) + q_j'^2(0)}{2}$ and define the torus

$$\mathbb{T}_0' := \left\{z' \in \mathscr{P}_s \ : \ I_j(z') = \bar{I}_j \right\}$$

One has

$$d(z'(t), \mathbb{T}_0') \le \left[\sum_j j^{2s}\left|\sqrt{I_j'(t)} - \sqrt{\bar{I}_j}\right|^2\right]^{1/2} \tag{35}$$

Notice that for $a, b \ge 0$ one has,

$$\left|\sqrt{a} - \sqrt{b}\right| \le \sqrt{|a - b|}\,.$$

Thus, using (32), one has that

$$\left[d(z'(t), \mathbb{T}_0')\right]^2 \le \sum_j j^{2s}|I_j'(t) - \bar{I}_j| \le C\varepsilon^{M+3}$$

Define now $\mathbb{T}_0 := \mathscr{T}_r(\mathbb{T}_0')$ then, since $\mathscr{T}_r$ is Lipschitz one has

$$d(z(t), \mathbb{T}_0) = d(\mathscr{T}_r(z'(t)), \mathscr{T}_r(\mathbb{T}_0')) \le Cd(z'(t), \mathbb{T}_0') \le C\varepsilon^{\frac{M+3}{2}}\,. \qquad \square$$

6.3 Proof of Proposition 4.2 on the verification of the property of localization of coefficients

In this subsection we will prove the property of localization of coefficients for the function $u \mapsto \int u^3$ in the case where the basis used for the definition (10) is the basis of the eigenfunction of general second order elliptic operator. Thus the present theory directly applies also to the case of the equation

$$u_{tt} - u_{xx} + Vu = f(x, u)$$

with Neumann boundary conditions on $[0, \pi]$. The case of Dirichlet boundary conditions can also be covered by a minor variant (indeed in such a case the function $u \mapsto \int u^3$ has to be substituted by the function $u \mapsto \int u^4$.

Thus consider a second order elliptic operator P, which is L^2 self adjoint. This means that we assume that in any coordinate system there exist smooth functions $V_\alpha(x)$, $\alpha \in \mathbb{N}^d$ such that $P = \sum_{|\alpha| \leq 2} V_\alpha \partial^\alpha$, where we used a vector notation for the derivative. Moreover we will assume that

$$\|u\|_{s+2} \leq \|Pu\|_s \ .$$

Then, by L^2 symmetry, one gets

$$\|u\|_s \leq \left\| P^{s/2} u \right\|_0 , \tag{36}$$

(where $P^{s/2}$ is defined spectrally). We will denote by $D(P^k)$ the domain of P^k. Finally, denote by λ_n the sequence of the eigenvalues of P counted without multiplicity (i.e. in such a way that $\lambda_{n+1} > \lambda_n$). We will assume that the eigenvalues of P behave as $\lambda_n \sim n^2$. We will denote by E_n the eigenspace of P relative to λ_n.

Let A be a linear operator which maps $D(P^k)$ into itself for all $k \geq 0$, and define the sequence of operators

$$A_N := [P, A_{N-1}] , \quad A_0 := A . \tag{37}$$

Lemma 6.3. *Let P be as above and let $u_j \in E_{n_j}$. Then, for any $N \geq 0$ one has*

$$|\langle Au_1 ; u_2 \rangle| \leq \frac{1}{|\lambda_{n_1} - \lambda_{n_2}|^N} |\langle A_N u_1 ; u_2 \rangle| \tag{38}$$

Proof. One has

$$\langle A_1 u_1 ; u_2 \rangle = \langle [A, P] u_1 ; u_2 \rangle = \langle APu_1 ; u_2 \rangle - \langle PAu_1 ; u_2 \rangle$$
$$= \lambda_{n_1} \langle Au_1 ; u_2 \rangle - \langle Au_1 ; Pu_2 \rangle = (\lambda_{n_1} - \lambda_{n_2}) \langle Au_1 ; u_2 \rangle$$

Equation (38) follows applying the above equality to the operator $A_N := [P, A_{N-1}]$ and using an induction argument. $\square$

To conclude the proof we have to estimate the matrix elements of A_N, i.e. the r.h.s. of (38). To this end we need a few remarks and lemma.

Remark 6.1. Consider two d-dimensional multi-indexes α and β and define

$$\binom{\alpha}{\beta} := \frac{\alpha!}{\beta!(\alpha - \beta)!} ,$$

with the convention that it is 0 if $\beta_j > \alpha_j$ for some j. One has

$$\partial^\alpha(uv) = \sum_\beta \binom{\alpha}{\beta} \partial^\beta u \partial^{\alpha-\beta} v . \tag{39}$$

Remark 6.2. Let $A := a(x)\partial^\alpha$ and $B := b(x)\partial^\beta$ with a and b smooth functions. Then one has

$$[A,B] = \sum_{\gamma_j \le \alpha_j + \beta_j} \left[a \binom{\beta}{\gamma} \partial^\gamma b - b \binom{\alpha}{\gamma} \partial^\gamma a \right] \partial^{\alpha+\beta-\gamma} . \tag{40}$$

Lemma 6.4. *Choose a coordinate system, let $A = a_0(x)$ be a multiplication operator, then one has*

$$A_N = \sum_{|\alpha| \le N} c_\alpha^{(N)} \partial^\alpha \tag{41}$$

with $c_\alpha^{(N)}$ of the form

$$c_\alpha^{(N)} = \sum_{|\beta| \le 2N - |\alpha|} V_{\alpha\beta}^{(N)}(x) \partial^\beta a_0 \tag{42}$$

and $V_{\alpha\beta}^{(N)}$ which are C^∞ functions depending only on the functions V_α defining the operator P.

Proof. First remark that by (40), the operator A_N is a differential operator of order N. By induction, using (40) one easily sees that the coefficients of such an operator are linear combinations of the derivatives of a_0. To show (42) we proceed by induction. The result is true for $N = 0$. Then use equation (40) to compute

$$\left[V_\alpha \partial^\alpha ; c_\beta^{(N)} \partial^\beta \right] = \sum_{\gamma_j \le \alpha_j + \beta_j} \left[V_\alpha \binom{\beta}{\gamma} \partial^\gamma c_\beta^{(N)} - c_\beta^{(N)} \binom{\alpha}{\gamma} \partial^\gamma V_\alpha \right] \partial^{\alpha+\beta-\gamma} \tag{43}$$

Consider the first term in the square bracket which is the one involving more derivatives of $c_\beta^{(N)}$. Since $c_\beta^{(N)}$ depends on $\partial^\delta a_0$ with $|\delta| \le 2N - |\beta|$, one has that $\partial^\gamma c_\beta^{(N)}$ depends only on the derivatives $\partial^\delta a_0$ with $|\delta| \le 2N - |\beta| + |\gamma|$; in order to conclude the proof we have to show that this is smaller than $2(N+1) - (|\alpha| + |\beta| - |\gamma|)$, a fact which is true since $|\alpha| \le 2$. $\square$

Remark 6.3. Let $u_n \in E_n$ then by (36) one has

$$\|u_n\|_s \le C n^s \|u_n\|_0$$

Remark 6.4. Let $u_n \in E_n$ with $\|u_n\|_0 = 1$, and b_α be a smooth function ($\alpha \in \mathbb{N}^d$), then one has for any $v_0 > d/2$ one has

$$\|b_\alpha \partial^\alpha u_n\|_0 \le C_{v_0} \|b_\alpha\|_{v_0} n^{|\alpha|} \tag{44}$$

Remark 6.5. Let $u_n \in E_n$ with $\|u_n\|_0 = 1$, and let

$$b_\alpha := V_{\alpha\beta}^{(N)}(x) \partial^\beta u_n \tag{45}$$

(with some β) then one has

$$\|b_\alpha\|_{\nu_0} \leq Cn^{\nu_0+|\beta|} \tag{46}$$

with a C that depends on $V^{(N)}_{\alpha\beta}$.

End of the proof of Proposition 4.2. Assume that $n_3 \leq n_2 \leq n_1$ so that $\mu(n) = n_3$ and $S(n) = n_3 + n_1 - n_2$. Write the l.h.s. of (29) as

$$|\langle Au_{n_2}; u_{n_1} \rangle| \tag{47}$$

with A the multiplication operator by u_{n_3}. Using (38) this is smaller than

$$\frac{1}{|n_1^2 - n_2^2|^N} \|A_N u_{n_2}\|_{L^2} \|u_{n_1}\|_{L^2} . \tag{48}$$

To estimate $\|A_N u_{n_2}\|_{L^2}$ we use (41) and estimate each term separately. By Sobolev embedding theorem, one term is estimated by

$$\left\| c_\alpha^{(N)} \partial^\alpha u_{n_2} \right\| \leq C \left\| c_\alpha^{(N)} \right\|_{\nu_0} \|\partial^\alpha u_{n_2}\|$$

$\nu_0 > d/2$. Using (42), (44), (46) one gets

$$\left\| c_\alpha^{(N)} \right\|_{\nu_0} \leq C \|u_{n_3}\|_{2N+\nu_0-|\alpha|} \leq Cn_3^{2N+\nu_0-|\alpha|} \|u_{n_3}\|_{L^2} .$$

where we used the ellipticity of P. We thus get that the l.h.s. of (29) is estimated by

$$C \sum_{|\alpha|\leq N} n_3^{2N+\nu_0-|\alpha|} n_2^{|\alpha|} \frac{1}{|n_1^2 - n_2^2|^N} \|u_{n_1}\|_{L^2} \|u_{n_2}\|_{L^2} \|u_{n_3}\|_{L^2}$$

A part from a constant, the sum of the coefficients in front of the norms is estimated by

$$n_3^{2N+\nu_0} \left(\frac{n_2}{n_3}\right)^N \frac{1}{|n_1^2 - n_2^2|^N} = \left(\frac{n_2}{n_1+n_2}\right)^N \frac{n_3^{\nu_0+N}}{|n_1 - n_2|^N} \leq \frac{n_3^{\nu_0+N}}{|n_1 - n_2|^N} \tag{49}$$

To conclude the proof just remark that $n_3 = \mu$, $S = \mu + (n_1 - n_2)$ and that if $n_3 > n_1 - n_2$ then the inequality (29) is trivially true. On the contrary, if $n_3 \leq (n_1 - n_2)$ the r.h.s. of (49) is smaller than

$$n_3^{\nu_0+N} \frac{2}{(n_3 + |n_1 - n_2|)^N}$$

which concludes the proof. $\quad\square$

6.4 Proof of Theorem 4.4 on the nonresonance condition

The proof follows the proof of Theorem 6.5 of [Bam03] (see also [BG06]). We repeat the main steps for completeness. Fix r once for all and denote by C any constant depending only on r. The value of C can change from line to line. Finally we will denote $m := \mu^2$.

Lemma 6.5. *For any* $K \leq N$, *consider* K *indexes* $j_1 < \ldots < j_K \leq N$; *consider the determinant*

$$
D := \begin{vmatrix}
\omega_{j_1} & \omega_{j_2} & \cdots & \omega_{j_K} \\
\dfrac{d\omega_{j_1}}{dm} & \dfrac{d\omega_{j_2}}{dm} & \cdots & \dfrac{d\omega_{j_K}}{dm} \\
\cdot & \cdot & \cdots & \cdot \\
\cdot & \cdot & \cdots & \cdot \\
\dfrac{d^{K-1}\omega_{j_1}}{dm^{K-1}} & \dfrac{d^{K-1}\omega_{j_2}}{dm^{K-1}} & \cdots & \dfrac{d^{K-1}\omega_{j_K}}{dm^{K-1}}
\end{vmatrix}
\tag{50}
$$

One has

$$
D = C \left(\prod_l \omega_{i_l}^{-2K+1} \right) \left(\prod_{1 \leq l < k \leq K} (\lambda_{j_l} - \lambda_{j_k}) \right) \geq \frac{C}{N^{2K^2}}.
\tag{51}
$$

Proof. One has

$$
\frac{d^j \omega_i}{dm^j} = \frac{(2j-1)!}{2^{j-1}(j-1)!2^j} \frac{(-1)^j}{(\lambda_i + m)^{j-\frac{1}{2}}}.
\tag{52}
$$

Substitute (52) in the r.h.s. of (6.5), factorize from the $l-th$ column the term $(\lambda_{j_l} + m)^{1/2}$, and from the $j-th$ row the term $\frac{(2j-3)!}{2^{j-2}(j-2)!2^j}$. The determinant becomes

$$
C \left[\prod_{l=1}^{K} \omega_{j_l} \right]
\begin{vmatrix}
1 & 1 & 1 & \ldots & 1 \\
x_{j_1} & x_{j_2} & x_{j_3} & \cdots & x_{j_K} \\
x_{j_1}^2 & x_{j_2}^2 & x_{j_3}^2 & \cdots & x_{j_K}^2 \\
\cdot & \cdot & \cdot & \cdots & \cdot \\
\cdot & \cdot & \cdot & \cdots & \cdot \\
x_{j_1}^{K-1} & x_{j_2}^{K-1} & x_{j_3}^{K-1} & \ldots & x_{j_K}^{K-1}
\end{vmatrix}
\tag{53}
$$

where we denoted by $x_j := (\lambda_j + m)^{-1} \equiv \omega_j^{-2}$. The last determinant is a Vandermond determinant given by

$$
\prod_{1 \leq l < k \leq K} (x_{j_l} - x_{j_k}) = \prod_{1 \leq l < k \leq K} \frac{\lambda_{j_k} - \lambda_{j_l}}{\omega_{j_l}^2 \omega_{j_k}^2} = \left(\prod_{1 \leq l < k \leq K} (\lambda_{j_l} - \lambda_{j_k}) \right) \prod_{l=1}^{K} \omega_{j_l}^{-2K}.
\tag{54}
$$

Using the asymptotic of the frequencies one gets also the second of (51). $\square$

Next we need the lemma from appendix B of [BGG85], namely

Lemma 6.6. *Let $u^{(1)},...,u^{(K)}$ be K independent vectors with $\left\|u^{(i)}\right\|_{\ell^1} \leq 1$. Let $w \in \mathbb{R}^K$ be an arbitrary vector, then there exist $i \in [1,...,K]$, such that*

$$|u^{(i)} \cdot w| \geq \frac{\|w\|_{\ell^1} \det(u^{(1)},...,u^{(K)})}{K^{3/2}} \, .$$

Combining Lemmas 6.5 and 6.6 we deduce

Corollary 6.2. *Let $w \in \mathbb{R}^\infty$ be a vector with K components different from zero, namely those with index $i_1,...,i_K$; assume $K \leq N$, and $i_1 < ... < i_K \leq N$. Then, for any $m \in [m_0,\Delta]$ there exists an index $i \in [0,...,K-1]$ such that*

$$\left| w \cdot \frac{d^i \omega}{dm^i}(m) \right| \geq C \frac{\|w\|_{\ell^1}}{N^{2K^2+2}} \tag{55}$$

where ω is the frequency vector.

From [XYQ97] we have.

Lemma 6.7. *(Lemma 2.1 of [XYQ97]) Suppose that $g(\tau)$ is m times differentiable on an interval $J \subset \mathbb{R}$. Let $J_h := \{\tau \in J : |g(\tau)| < h\}$, $h > 0$. If on J, $\left|g^{(m)}(\tau)\right| \geq d > 0$, then $|J_h| \leq Mh^{1/m}$, where*

$$M := 2(2+3+...+m+d^{-1}) \, .$$

For any $k \in \mathbb{Z}^N$ with $|k| \leq r$ and for any $n \in \mathbb{Z}$, define

$$\mathscr{R}_{kn}(\gamma,\alpha) := \left\{ m \in [m_0,\Delta] \; : \; \left| \sum_{j=1}^N k_j \omega_j + n \right| < \frac{\gamma}{N^\alpha} \right\} \tag{56}$$

Applying Lemma 6.7 to the function $\sum_{j=1}^N k_j \omega_j + n$ and using Corollary 6.2 we get as in [Bam99] Lemma 8.4

Corollary 6.3. *Assume $|k| + |n| \neq 0$, then*

$$|\mathscr{R}_{kn}(\gamma,\alpha)| \leq C(\Delta - m_0)\frac{\gamma^{1/r}}{N^\varsigma} \tag{57}$$

with $\varsigma = \frac{\alpha}{r} - 2r^2 - 2$.

Lemma 6.8. *Fix $\alpha > 2r^3 + r^2 + 5r$. For any positive γ small enough there exists a set $\mathscr{I}_\gamma \subset [m_0,\Delta]$ such that $\forall m \in \mathscr{I}_\gamma$ one has that for any $N \geq 1$*

$$\left| \sum_{j=1}^N k_j \omega_j + n \right| \geq \frac{\gamma}{N^\alpha} \tag{58}$$

for all $k \in \mathbb{Z}^N$ with $0 \neq |k| \leq r$ and for all $n \in \mathbb{Z}$. Moreover,

$$\left| [m_0,\Delta] - \mathscr{I}_\gamma \right| \leq C\gamma^{1/r} \, . \tag{59}$$

Proof. Define $\mathscr{I}_\gamma := \bigcup_{nk} \mathscr{R}_{nk}(\gamma, \alpha)$. Remark that, from the asymptotic of the frequencies, the argument of the modulus in (58) can be small only if $|n| \leq CrN$, By (57) one has

$$\left| \bigcup_k \mathscr{R}_{nk}(\gamma, \alpha) \right| \leq \sum_k |\mathscr{R}_k(\gamma, \alpha)| < C \frac{N^r(\Delta - m_0)\gamma^{1/r}}{N^\varsigma} \ ,$$

summing over n one gets an extra factor rN. Provided α is chosen according to the statement, one has that the union over N is also bounded and therefore the thesis holds. $\square$

Lemma 6.9. *For any γ positive and small enough, there exist a set $\mathscr{J}_\gamma$ satisfying, $|[m_0, \Delta] - \mathscr{J}_\gamma| \to 0$ when $\gamma \to 0$, and a real number α' such that for any $m \in \mathscr{J}_\gamma$ one has for $N \geq 1$*

$$\left| \sum_{j=0}^N \omega_j k_j + \varepsilon_1 \omega_j + \varepsilon_2 \omega_l \right| \geq \frac{\gamma}{N^{\alpha'}} \tag{60}$$

for any $k \in \mathbb{Z}^N$, $\varepsilon_i = 0, \pm 1$, $j \geq l > N$, and $|k| + |\varepsilon_1| + |\varepsilon_2| \neq 0$, $|k| \leq r+2$.

Proof. We consider only the case now the case $\varepsilon_1 \varepsilon_2 = -1$ which is the most complicate. One has

$$\omega_j - \omega_l = j - l + a_{jl} \text{ with } |a_{jl}| \leq \frac{C}{l} \tag{61}$$

So the quantity to be estimated reduces to

$$\sum_{j=0}^N \omega_j k_j \pm n \pm a_{jl} \ , \quad n := j - l$$

If $l > 2CN^\alpha/\gamma$ then the a_{jl} term represents an irrelevant correction and therefore the lemma follows from Lemma 6.8. In the case $l \leq 2CN^\alpha/\gamma$ one reapplies the same lemma with $N' := 2CN^\alpha/\gamma$ in place of N and $r' := r+2$ in place of r. $\square$

To obtain theorem 4.4 just define $\mathscr{J} := \bigcap_{r \geq 1} \bigcup_{\gamma > 0} \mathscr{J}_\gamma$ and remark that its complement is the union of a numerable infinity of sets of zero measure.

References

[Bam99] D. Bambusi. On long time stability in Hamiltonian perturbations of non-resonant linear PDEs. *Nonlinearity*, 12:823–850, 1999.

[Bam03] D. Bambusi. Birkhoff normal form for some nonlinear PDEs. *Comm. Math. Physics*, 234:253–283, 2003.

[BDGS07] D. Bambusi, J.M. Delort, B. Grébert, and J. Szeftel. Almost global existence for Hamiltonian semi-linear Klein-Gordon equations with small Cauchy data on Zoll manifolds. *Comm. Pure Appl. Math.*, 2007.

[BG06] D. Bambusi and B. Grébert. Birkhoff normal form for partial differential equations with tame modulus. *Duke Math. J.*, 135(3):507–567, 2006.

[BGG85]   G. Benettin, L. Galgani, and A. Giorgilli. A proof of Nekhoroshev's theorem for the stability times in nearly integrable Hamiltonian systems. *Celestial Mech.*, 37:1–25, 1985.

[Bou96a]   J. Bourgain. Construction of approximative and almost-periodic solutions of perturbed linear Schrödinger and wave equations. *Geometric Functional Anal*, 6:201–230, 1996.

[Bou96b]   Jean Bourgain. On the growth in time of higher Sobolev norms of smooth solutions of Hamiltonian PDE. *Internat. Math. Res. Notices*, (6):277–304, 1996.

[Bou97]   J. Bourgain. On growth in time of Sobolev norms of smooth solutions of nonlinear Schrödinger equations in $\mathbf{R}^D$. *J. Anal. Math.*, 72:299–310, 1997.

[Bou98]   J. Bourgain. Quasi-periodic solutions of Hamiltonian perturbations of 2D linear Schrödinger equation. *Ann. Math.*, 148:363–439, 1998.

[Bou00]   J. Bourgain. On diffusion in high-dimensional Hamiltonian systems and PDE. *J. Anal. Math.*, 80:1–35, 2000.

[Bou05]   J. Bourgain. *Green's function estimates for lattice Schrödinger operators and applications*, volume 158 of *Annals of Mathematics Studies*. Princeton University Press, Princeton, NJ, 2005.

[CW93]   W. Craig and C. E. Wayne. Newton's method and periodic solutions of nonlinear wave equations. *Comm. Pure Appl. Math.*, 46:1409–1498, 1993.

[DS04]   J. M. Delort and J. Szeftel. Long–time existence for small data nonlinear Klein–Gordon equations on tori and spheres. *Internat. Math. Res. Notices*, 37:1897–1966, 2004.

[DS06]   J.-M. Delort and J. Szeftel. Long-time existence for semi-linear Klein-Gordon equations with small Cauchy data on Zoll manifolds. *Amer. J. Math.*, 128(5):1187–1218, 2006.

[EK06]   H. L. Eliasson and S. B. Kuksin. KAM for non-linear Schrödinger equation. *Preprint*, 2006.

[Gré06]   B. Grébert. Birkhoff normal form and Hamiltonian PDES. *Preprint.*, 2006.

[Kla83]   S. Klainerman. On "almost global" solutions to quasilinear wave equations in three space dimensions. *Comm. Pure Appl. Math.*, 36:325–344, 1983.

[KP96]   S. B. Kuksin and J. Pöschel. Invariant Cantor manifolds of quasi-periodic oscillations for a nonlinear Schrödinger equation. *Ann. Math.*, 143:149–179, 1996.

[Kuk87]   S. B. Kuksin. Hamiltonian perturbations of infinite-dimensional linear systems with an imaginary spectrum. *Funct. Anal. Appl.*, 21:192–205, 1987.

[Way90]   C. E. Wayne. Periodic and quasi-periodic solutions of nonlinear wave equations via KAM theory. *Comm. Math. Physics*, 127:479–528, 1990.

[XYQ97]   J. Xu, J. You, and Q. Qiu. Invariant tori for nearly integrable Hamiltonian systems with degeneracy. *Math. Z.*, 226:375–387, 1997.

[Yua06]   X. Yuan. A KAM theorem with applications to partial differential equations of higher dimensions. *Commun. Math. Phys* to appear (Preprint 2006).

Normal form of holomorphic dynamical systems

Laurent Stolovitch[1]

Abstract This article represents the expanded notes of my lectures at the ASI "Hamiltonian Dynamical Systems and applications". We shall present various recent results about normal forms of germs of holomorphic vector fields at a fixed point in $\mathbb{C}^n$. We shall explain how relevant it is for geometric as well as for dynamical purpose. We shall first give some examples and counter-examples about holomorphic conjugacy. Then, we shall state and prove a main result concerning the holomorphic conjugacy of a commutative family of germs of holomorphic vector fields. For this, we shall explain the role of diophantine condition and the notion of singular complete integrability.

1 Definitions and examples

Let us consider the pendulum with normalized constants :

$$\ddot{\theta} + \sin\theta = 0 \tag{1}$$

We would like to understand the behavior of the motion for the small oscillations of the pendulum, that is to say when θ is small. We are tempted to say that $\sin\theta$ is well approximated by θ and then we would like to consider the much simpler equation $(*)\ \ddot{\theta} + \theta = 0$ instead of (1). If we set $\theta_1 = \theta$ and $\theta_2 = \dot{\theta}$, equation $(*)$ can be written as

$$\begin{cases} \dot{\theta}_1 = \theta_2 \\ \dot{\theta}_2 = -\theta_1 \end{cases}.$$

[1]CNRS UMR 5580, Laboratoire Emile Picard, Université Paul Sabatier, 118 route de Narbonne, 31062 Toulouse cedex 9, France
e-mail: stolo@picard.ups-tlse.fr

W. Craig (ed.), Hamiltonian Dynamical Systems and Applications, 249–284.

The dynamic is completely understood. Its trajectories are circles $\theta_1^2 + \theta_2^2 = constant$. Are these information relevant for the understanding of the dynamic of the original problem (1)? Does the closeness of equation $(*)$ to equation (1) imply that they have the same dynamical properties?

In general, both answer are 'No!'. In these lectures, we shall explain these phenomena and how to define a reasonable simplified problem to study : a normal form.

Let us start with a very elementary example of a similar problem. In order to study the iterates of a square complex matrix A of $\mathbb{C}^n$, that is the orbits $\{A^k x\}_{k \in \mathbb{N}}$ for $x \in \mathbb{C}^n$ near the "fixed point" 0, it is very convenient to transform, with the help of a linear change of coordinates P, the matrix A into a Jordan matrix $J = S + N$, with S a diagonal matrix, N an upper triangular nilpotent matrix commuting with $S : PAP^{-1} = S + N$. Using the (block diagonal) structure of $S + N$, it is easy to study its iterates. Since $A^k = P^{-1} J^k P$, we have $A^n x = P^{-1}(J^n y)$ where $x = P^{-1} y$. We thus obtain all informations needed for the study of the iterates of A.

One of the great ideas of Poincaré was to try to proceed in the same way for vector fields. Is it possible to transform a given vector field X, vanishing at the origin of $\mathbb{R}^n$ (resp. $\mathbb{C}^n$), into a "simpler" one with the help of a local diffeomorphism Φ near the origin and which maps the origin to itself? The group of germs of C^k (resp. holomorphic, formal) diffeomorphisms at $0 \in \mathbb{C}^n$ and tangent to $Id_{\mathbb{C}^n}$ at the origin, acts on the space of germs of holomorphic (resp. formal) vector fields at $0 \in \mathbb{C}^n$ by conjugacy : if X is any representative of a germ of vector field X, and ϕ is any representative of a germ of diffeomorphism Φ, then $\Phi_* X$ is the germ of vector field defined by

$$\phi_* X(\phi(x)) := D\phi(x) X(x)$$

where $D\phi(x)$ denotes the derivative of ϕ at the point x. One may first attempt to linearize formally X, that is to find a formal change of coordinates $\hat{\Phi}$, such that $\hat{\Phi}_* X(y) = DX(0)y$. Assume it is so then, one could expect to understand all about the dynamics of X since the flow of the linear vector field $DX(0)y$ is easy to study. Nevertheless, this cannot be the case unless we are able to pull-back these informations by $\hat{\Phi}$, and this requires some "regularity" conditions on $\hat{\Phi}$. Is there a C^k (resp. smooth) linearizing diffeomorphism? When we are working in the analytic category, this regularity condition should be that $\hat{\Phi}$ is holomorphic in a neighborhood of the origin. What happens in this situation?

These ideas have been widely developed by V.I. Arnol'd and his school. Our main reference for this topic is the great book by V.I. Arnol'd [Arn88a]. We refer also to [AA88] which contain a lot of references on this topic. Singularities of mappings are also studied in the same spirit [AGZV85, AGZV88].

1.1 Vector fields and differential equations

Let us consider a germ of vector field X at a point p : in a coordinate chart at p, it can be written $X(z) = \sum_{i=1}^{n} X_i(z) \frac{\partial}{\partial z_i}$. It is equivalent to consider the system of autonomous differential equations :

$$\begin{cases} \dot{z}_1 = X_1(z) \\ \vdots \\ \dot{z}_n = X_n(z) \end{cases}.$$

The **Lie derivative** of a germ of function f along the vector field X is the germ of function

$$\mathcal{L}_X f(z) := \sum_{i=1}^n X_i(z) \frac{\partial f}{\partial z_i}(z).$$

It will also be denoted by $X(f)$.

We will denote by $[X,Y]$ the **Lie bracket** of the vector fields $X = \sum_{i=1}^n X_i \frac{\partial}{\partial z_i}$ and $Y = \sum_{i=1}^n Y_i \frac{\partial}{\partial z_i}$. It is defined to be

$$[X,Y] = \sum_{i=1}^n \left(\sum_{j=1}^n X_j \frac{\partial Y_i}{\partial z_j} - Y_j \frac{\partial X_i}{\partial z_j} \right) \frac{\partial}{\partial z_i}.$$

It is skew-symmetric and satisfies the Jacobi identity :

$$[X,[Y,Z]] + [Y,[Z,X]] + [Z,[X,Y]] = 0.$$

Moreover, if X,Y are vector fields and f a function

$$[X, fY] = f[X,Y] + \mathcal{L}_X(f)Y. \tag{2}$$

Two vector fields X,Y are said to be **commuting pairwise** whenever $[X,Y] \equiv 0$. From the dynamical point of view, let us start at a point p, then let us follow the flow of X during a time t then follow the flow of Y during a time s. Let q be this end point. Let us start at p again but now follow the flow of Y during a time s first then follow the flow of X during a time t. Let q' be this end point. The fact that X and Y commute pairwise means that $q = q'$.

1.1.1 Notations

Let us set some notations which will be used all along this article : let $k \geq 1$ be an integer,

- $\mathscr{P}_n^k$ denotes the $\mathbb{C}$-space of homogeneous polynomial vector fields on $\mathbb{C}^n$ and of degree k
- $\mathscr{P}_n^{m,k}$ denotes the $\mathbb{C}$-space of polynomial vector fields on $\mathbb{C}^n$, of order $\geq m$ and of degree $\leq k$ ($m \leq k$)
- $\widehat{\mathscr{X}}_n^k$ denotes the $\mathbb{C}$-space of formal vector fields on $\mathbb{C}^n$ and of order $\geq k$ at 0
- $\mathscr{X}_n^k$ denotes the $\mathbb{C}$-space of germs of holomorphic vector fields on $(\mathbb{C}^n, 0)$ and of order $\geq k$ at 0
- p_n^k denotes the $\mathbb{C}$-space of homogeneous polynomial on $\mathbb{C}^n$ and of degree k

- $\widehat{\mathcal{M}}_n^k$ denotes the $\mathbb{C}$-space of formal power series on $\mathbb{C}^n$ and of order $\geq k$ at 0
- $\mathcal{M}_n^k$ denotes the $\mathbb{C}$-space of germs of holomorphic functions on $(\mathbb{C}^n, 0)$ and of order $\geq k$ at 0
- $\widehat{\mathcal{O}}_n$ denotes the ring of formal power series in $\mathbb{C}^n$
- $\mathcal{O}_n$ denotes the ring of germs at 0 of holomorphic functions in $\mathbb{C}^n$

1.1.2 Norms

Let $f \in \mathbb{C}[[x_1, \ldots, x_n]]$ be a formal power series : $f = \sum_{Q \in \mathbb{N}^n} f_Q x^Q$. We define $\bar{f}$ as the formal power series $\bar{f} = \sum_{Q \in \mathbb{N}^n} |f_Q| x^Q$. We will say that a formal power series g dominates a formal power series f, if $\forall Q \in \mathbb{N}^n$, $|f_Q| \leq |g_Q|$. In that case, we will write $f \prec g$. More generally, let $q \geq 1$ be an integer and let $F = (f_1, \ldots, f_q)$ and $G = (g_1, \ldots, g_q)$ be elements of $(\mathbb{C}[[x_1, \ldots, x_n]])^q$; we shall say that G dominates F, and we shall write $F \prec G$, if $f_i \prec g_i$ for all $1 \leq i \leq q$. We shall write $\bar{F} = (\bar{f}_1, \ldots, \bar{f}_q)$. We shall say that F is of order $\geq m$ (resp. polynomial of degree $\leq m$), if each of his components is of order $\geq m$ (resp. polynomial of degree $\leq m$).

Let r be an positive number and $(f, F, G) \in \widehat{\mathcal{O}}_n \times \widehat{\mathcal{O}}_n^q \times \widehat{\mathcal{O}}_n^q$, we define

$$|f|_r := \sum_{Q \in \mathbb{N}^n} |f_Q| r^{|Q|} = \bar{f}(r, \ldots, r)$$

and $|G|_r = \max_i |g_i|_r$; these may not be finite. We have the following properties

$$\overline{fG} \prec \bar{f}\bar{G}$$
$$\text{if } F \prec G \text{ then } |F|_r \leq |G|_r$$
$$\overline{\frac{\partial F}{\partial x_k}} = \frac{\partial \bar{F}}{\partial x_k}$$

Let us define $\mathcal{H}_n^q(r) = \{F \in \widehat{\mathcal{O}}_n^q \mid |F|_r < +\infty\}$; $|.|_r$ is norm on this space. Together with the norm $|.|_r$, this space is a Banach space (see [GR71]).

Lemma 1.1.1 *Let $F = \sum_{Q \in \mathbb{N}^n} F_Q x^Q$ an element of $\mathcal{H}_n^q(r)$, then we have the following inequalities :*

$$\|F\|_r \leq |F|_r \tag{3}$$

$$|F|_R \leq \left(\frac{R}{r}\right)^m |F|_r \quad \text{if} \quad ord(F) \geq m, R \leq r \tag{4}$$

$$|DF|_r \leq \frac{d}{r}|F|_r \quad \text{if } F \text{ is a polynomial of degree} \leq d \tag{5}$$

Here $\|F\|_r$ denotes the supremum of $|F(z)|$ on the polydisc $|z_i| < r$, $1 \leq i \leq l$.

Proof. The first inequality comes from the fact that for all x in the polydisc of radius r, we have

$$\left| \sum_{Q \in \mathbb{N}^n} F_Q x^Q \right| \leq \sum_{Q \in \mathbb{N}^n} |F_Q| |x^Q| \leq |F|_r.$$

For the second, we have

$$\sum_{Q \in \mathbb{N}^n,\, |Q| \geq m} |F_Q| R^{|Q|} \leq \sum_{Q \in \mathbb{N}^n,\, |Q| \geq m} \frac{R^{|Q|}}{r^{|Q|}} |F_Q| r^{|Q|} \leq \frac{R^m}{r^m} \sum_{Q \in \mathbb{N}^n,\, |Q| \geq m} |F_Q| r^{|Q|}.$$

For the last one, we have $F = \sum_{Q \in \mathbb{N}^n,\, |Q| \leq d} F_Q x^Q$. Hence, we have

$$\left| \frac{\partial F}{\partial x_j} \right|_r = \left| \sum_{Q \in \mathbb{N}^n,\, |Q| \leq d} F_Q q_j x^{Q - E_j} \right|_r$$

$$= \sum_{Q \in \mathbb{N}^n,\, |Q| \leq d} |F_Q| q_j r^{|Q| - 1}$$

$$\leq \frac{d}{r} |F|_r.$$

We shall often use the estimate $|(DG).F|_r \leq n |DG|_r |F|_r$ whenever $(F, G) \in \mathcal{H}_n^n(r)$.

Lemma 1.1.2 *[Sto00][Prop. 3.1.1] Let $r > 0$, $a \in \mathbb{C}^*$ and $g \in \mathcal{H}_n^1(r)$. We assume that $|g|_r < |a|$. Then*

$$\left| \frac{1}{a + g} \right|_r \leq \frac{1}{|a| - |g|_r}$$

1.2 Normal forms of vector fields

In the sequel, we will assume that the linear part of X at the origin is not nilpotent (see [CS86] normal form with nilpotent linear part) and for the sake of simplicity we even assume that is it semi-simple :

$$S := DX(0)x = \sum_{i=1}^{n} \lambda_i x_i \frac{\partial}{\partial x_i}$$

is a nonzero diagonal vector field. If $Q = (q_1, \ldots, q_n) \in \mathbb{N}^n$, we will write $(Q, \lambda) := \sum_{i=1}^{n} q_i \lambda_i$, $|Q| := q_1 + \cdots + q_n$ and $x^Q := x_1^{q_1} \cdots x_n^{q_n}$.

Proposition 1.2.1 (Poincaré–Dulac normal form) *Let $X = S + R_2$ be a nonlinear perturbation of the linear vector field S. Then there exits a formal change of coordinates $\hat{\Phi}$ tangent to the identity such that*

$$\hat{\Phi}_* X = S + \hat{N},$$

where the nonlinear formal vector field $\hat{N}$ commutes with S : $[S, \hat{N}] = 0$.

By formal change of coordinates $\hat{\Phi}$ tangent to the identity, we mean that there exists formal power series $\hat{\phi}_i(x) = \sum_{Q\in\mathbb{N}^n, |Q|\geq 2} \phi_{i,Q} x^Q \in \mathbb{C}[[x_1,\ldots x_n]]$ of order ≥ 2, such that $\hat{\Phi}_i(x) = x_i + \hat{\phi}_i(x)$, the ith-component of $\hat{\Phi}$.

Let us describe a normal form in local coordinates. First of all, we notice that

$$\left[S, x^Q \frac{\partial}{\partial x_i} \right] = ((Q,\lambda) - \lambda_i) x^Q \frac{\partial}{\partial x_i}.$$

Therefore, such an elementary vector field commute with S if and only if

$$(Q,\lambda) = \lambda_i.$$

This is called a **resonance relation** and $x^Q \frac{\partial}{\partial x_i}$ the associated **resonant vector field**.

Therefore, the formal normal form proposition can be rephrased as : there exists a formal diffeomorphism $\hat{\Phi}$ (which is not unique in general) such that

$$\hat{\Phi}_* X = \sum_{i=1}^{n} \lambda_i x_i \frac{\partial}{\partial x_i} + \sum_{i=1}^{n} \left(\sum_{(Q,\lambda)=\lambda_i} a_{i,Q} x^Q \right) \frac{\partial}{\partial x_i}$$

where the sum is over the multiintegers $Q \in \mathbb{N}^n$, $|Q| \geq 2$ and the index i which satisfy to $(Q,\lambda) = \lambda_i$ and where the $a_{i,Q}$'s are complex numbers.

Example 1.2.2 *Let ζ be a positive irrational number. Let us consider the vector field X*

$$\begin{cases} \dot{x} = x + f(x,y) \\ \dot{y} = -\zeta y + g(x,y) \end{cases}$$

where f,g are smooth functions vanishing at the origin as well as their first derivatives. It is formally linearizable since the only integer solution (q_1, q_2) of $q_1 - \zeta q_2 = 0$ is $(0,0)$. Hence, there are no resonance relation satisfied.

Example 1.2.3 *Let us consider the vector field X*

$$\begin{cases} \dot{x} = 2x + y^2 + f(x,y) \\ \dot{y} = y + g(x,y) \end{cases} \tag{6}$$

where the smooth functions f,g vanish at order 3 at the origin. There is one and only one resonance relation satisfied : $0\lambda_1 + 2\lambda_2 = \lambda_1$. Therefore, X is formally conjugate to the normal form

$$\begin{cases} \dot{x} = 2x + y^2 \\ \dot{y} = y \end{cases}. \tag{7}$$

Example 1.2.4 *Let us consider the analytic vector field X*

$$\begin{cases} \dot{x} = x + f(x,y) \\ \dot{y} = -y + g(x,y) \end{cases} \tag{8}$$

for some holomorphic functions f, g vanishing at first order at the origin. It is clear that the only solutions of the resonance relation $q_1\lambda_1 + q_2\lambda_2 = \lambda_1$ (resp. $q_1\lambda_1 + q_2\lambda_2 = \lambda_2$) are of the form $q_1 = q_2 + 1$ (resp. $q_2 = q_1 + 1$). Thus, the resonant vector fields are generated by $(xy)^l x\frac{\partial}{\partial x}$ and $(xy)^l y\frac{\partial}{\partial y}$ where l is a positive integer. Applying Poincaré-Dulac theorem to equation (8) leads to a formal normal form

$$\begin{cases} \dot{x} = x\hat{F}(xy) \\ \dot{y} = -y\hat{G}(xy) \end{cases} \tag{9}$$

where $\hat{F}, \hat{G}$ are formal power series which values at 0 is 1.

Example 1.2.5 *Let us extend example 1.2.4 by Example 1.2.3 in a four-dimensional system :*

$$\begin{aligned} \dot{w} &= w + e(w,x,y,z) \\ \dot{x} &= -x + f(w,x,y,z) \\ \dot{y} &= 2iy + g(w,x,y,z) \\ \dot{z} &= iz + h(w,x,y,z) \end{aligned}$$

Its formal normal form is of the form

$$\begin{aligned} \dot{w} &= w\hat{F}(wx) \\ \dot{x} &= -x\hat{G}(wx) \\ \dot{y} &= 2i\hat{H}_1(wz)y + \hat{H}_2(wz)z^2 \\ \dot{z} &= i\hat{H}_3(wz)z \end{aligned}$$

Example 1.2.6 *Let us consider the five-dimensional system*

$$\begin{cases} \dot{x}_1 = x_1 + f_1(x) \\ \dot{x}_2 = -x_2 + f_2(x) \\ \dot{x}_3 = -\zeta x_3 + f_3(x) \\ \dot{x}_4 = ix_4 + f_4(x) \\ \dot{x}_5 = ix_5 + f_5(x) \end{cases}$$

where ζ is a positive irrational number. Its normal form is of the form

$$\begin{cases} \dot{x}_1 = x_1\hat{f}_1(x_1x_2) \\ \dot{x}_2 = -x_2\hat{f}_2(x_1x_2) \\ \dot{x}_3 = -\zeta\hat{f}_3(x_1x_2)x_3 \\ \dot{x}_4 = ix_4 + x_4\hat{g}_{1,1}(x_1x_2) + x_5\hat{g}_{1,2}(x_1x_2) \\ \dot{x}_5 = ix_5 + x_4\hat{g}_{2,1}(x_1x_2) + x_5\hat{g}_{2,2}(x_1x_2) \end{cases} \tag{10}$$

where the $\hat{f}_i$'s (resp. $\hat{g}_{i,j}$'s) are formal power series of one variable (resp. vanishing at the origin).

1.2.1 Hamiltonian vector fields

We refer the Arnold book [Arn97] for this section.

To a germ of function $H : (\mathbb{R}^{2n}, 0) \to (\mathbb{R}, 0)$ vanishing at first order at 0, we can associate a germ of vector field X_H of $(\mathbb{R}^{2n}, 0)$ vanishing at the origin. If (x, y) are local coordinates, it is defined to be

$$\dot{x}_j = \frac{\partial H}{\partial y_j}, \quad j = 1, \ldots, n$$

$$\dot{y}_j = -\frac{\partial H}{\partial x_j}, \quad j = 1, \ldots, n.$$

It is called the **Hamiltonian vector field** associated to H. The function H is called the Hamiltonian of X_H.

Definition 1.2.7 *A change of coordinate* $X_j = \phi_j(x, y)$, $Y_j = \psi_j(x, y)$ *is called* **canonical** *if it preserve the symplectic form* $\omega = \sum_{j=1}^{n} dx_j \wedge dy_j$. *In other words,*

$$\sum_{j=1}^{n} dx_j \wedge dy_j = \sum_{j=1}^{n} dX_j \wedge dY_j.$$

If we conjugate an Hamiltonian vector field X_H by a canonical diffeomorphism Φ, we obtain again an Hamiltonian vector field, namely $X_{H \circ \Phi}$. We shall say that the Hamiltonian H is a **Birkhoff normal form** whenever its associated Hamiltonian vector field X_H is a normal form.

Definition 1.2.8 *In symplectic coordinates* (x, y), *we define the* **Poisson bracket** *of the germ of functions to be*

$$\{f, g\} := \sum_{j=1}^{n} \left(\frac{\partial f}{\partial x_j} \frac{\partial g}{\partial y_j} - \frac{\partial g}{\partial x_j} \frac{\partial f}{\partial y_j} \right).$$

It satisfies the following properties :

- $\{.,.\}$ is bilinear and skew-symmetric
- $\{f, \{g, h\}\} + \{g, \{h, f\}\} + \{h, \{f, g\}\} = 0$ (Jacobi identity)
- $\{f, gh\} = \{f, g\}h + \{f, h\}g$ (Leibniz identity)

It is easy to show that
$$[X_H, X_G] = X_{\{H, G\}}.$$

1.3 Examples about linearization

Example 1.3.1 *The normal form* (7) *is topologically conjugate to the linear part*

$$\begin{cases} \dot{x} = 2x \\ \dot{y} = y \end{cases}.$$

By this, we mean there exists an homeomorphism H fixing the origin which maps the trajectories of the normal form to the trajectories of its linearized : $H(\phi_t(x)) = \phi_t^{linearized}(H(x))$ where ϕ_t denotes the flow at time t starting at x. This is a consequence of Hartman–Grobmann theorem. Nevertheless, it can be shown that the normal form is not C^2-conjugate to its linearized at the origin.

Theorem 1.3.2 *[Ste58, Bru95] Assume the linear part S is non-resonant, i.e. there is no resonance relation satisfied. Then any smooth nonlinear perturbation $X = S + R$ of S is smoothly conjugate to its linear part S.*

What happens in the analytic context?

Example 1.3.3 *We borrow this example to J.-P. Françoise [Fra95]. Let us consider a special case of Example 1.2.2. Let us assume that the irrational number ζ is* **Liouvillian**. *By this, we mean that there exists two sequences of positive integers $(p_n), (q_n)$ both tending to infinity with n such that*

$$\left| \zeta - \frac{p_n}{q_n} \right| < \frac{1}{q_n(q_n!)}.$$

The number ζ is too well approximated by rational numbers. Given such a pair of sequence, let us consider the function

$$f(x,y) = \frac{1}{1 - \sum x^{p_n} y^{q_n}}.$$

It is holomorphic in a neighborhood of the origin and $f(0) = 1$. Let us set $S := x\frac{\partial}{\partial x} - \zeta y\frac{\partial}{\partial y}$ and let us consider the germ of holomorphic vector field defined to be $X = f(x,y)S$. Its linear part at the origin is S. Let us find the formal change of coordinate that linearizes it (in this case, it's unique) : $\tilde{x} = x\exp(-V(x,y))$, $\tilde{y} = y\exp(-W(x,y))$. Then,

$$x\exp(-V(x,y)) = \tilde{x} = \mathscr{L}_X(\tilde{x}) = x\mathscr{L}_X(-V(x,y))\exp(-V(x,y))$$
$$+ \exp(-V(x,y))\mathscr{L}_X(x).$$

Here, the first equality comes from the definition, the second comes from that fact that X is linearized in the new coordinates. Therefore, we have that $\mathscr{L}_X(V) = f - 1$ which is equivalent to

$$\mathscr{L}_S(V) = \frac{f - 1}{f} = \sum x^{p_n} y^{q_n}.$$

This equation has the unique solution

$$V = \sum \frac{1}{p_n - \zeta q_n} x^{p_n} y^{q_n}$$

which is divergent at the origin since $\frac{1}{p_n - \zeta q_n} \geq q_n!$.

This example shows that one need an "arithmetical" condition on the **small divisors** $(Q, \lambda) - \lambda_i \neq 0$. The major step in the understanding of the phenomenon is due to C.L Siegel.

Definition 1.3.4 *We shall say that* $\lambda = (\lambda_1, \ldots, \lambda_n)$ *is* **diophantine** *of type* $\nu \geq 0$ *if there exists* $C > 0$ *such that, for all multiindexes* $Q \in \mathbb{N}^n$, $|Q| \geq 2$,

$$|(Q, \lambda) - \lambda_i| > \frac{C}{|Q|^\mu}.$$

We shall say that there **no small divisor** *if there exists a constant* $c >$ *such that*

$$|(Q, \lambda) - \lambda_i| > c.$$

Theorem 1.3.5 *[Sie42] If the linear vector field* $S = \sum_{i=1}^n \lambda_i x_i \frac{\partial}{\partial x_i}$ *is diophantine, then any holomorphic non-linear perturbation of* S *is holomorphically linearizable.*

This arithmetical condition has been weakened by A.D. Brjuno as we shall see below.

1.4 Examples about nonlinearizable vector fields

Let's go back to Example 1.2.3 where we saw that any holomorphic perturbation of order ≥ 3 of the normal form is formally conjugate to it. What about the holomorphy of such a conjugacy?

Theorem 1.4.1 (Poincaré–Picard) *If the linear part* S *has non-polynomial first integral but the constants and if there are no small divisors then any nonlinear perturbation* $X = S + R$ *is holomorphically conjugate to a polynomial normal form in a neighborhood of the origin.*

Remark 1.4.2 *Usually in the literature, the previous theorem is applied for linear part which spectrum is said to lie in the "Poincaré domain". By this, we mean that there exists a line* (D) *in the complex plane which separate the eigenvalues of* S *from the origin (i.e. the eigenvalues are on one and the same side of the line while 0 is in the other side). Thus, if* S *belongs to the Poincaré domain, then it has only constant polynomial first integral. In fact, if* $\mathscr{L}_X(x^Q) = 0$ *then* $(Q, \lambda) = 0$. *This means that the origin is a linear combination of the* λ_i's *with non negative coefficients. Since the spectrum lies on the same and opposite side from the origin of a line, then* $Q = 0$. *Furthermore, there are no small divisors since the projection of the eigenvalues onto the orthogonal line to* (D) *passing through the origin is bounded from below. So do the small divisors.*

Example 1.4.3 *Let us show that Example 1.2.3 falls into the application scope of the theorem. In fact, a monomial (p,q are non-negative)* $x^p y^q$ *is a first integral of* S

if and only if $S(x^p y^q) = (2p+q)x^p y^q = 0$. This implies that $p = q = 0$. Thus, polynomial first integrals of S are just constants. Moreover, there are no small divisors. In fact, both $|2p+q-2|$ and $|2p+q-1|$ are integers so they don't accumulate the origin. Therefore, any holomorphic system (6) is holomorphically conjugate to its normal form (7).

2 Holomorphic normalization

The main progress are due to Brjuno who gave sufficient conditions that ensure that there is a convergent normalizing transformation to a normal form. These conditions are of two different type. The first one is a condition about the rate of accumulation to zero of the small divisors of the linear part. It is weaker that Siegel condition and is called condition (ω). The second one is linked to the nonlinearity of the perturbation we are considering. It is a condition about a formal normal form of the perturbation.

2.1 Theorem of A.D. Brjuno

Let $X = S + R$ be an holomorphic vector field in a neighborhood of its singular point $0 \in \mathbb{C}^n$ with $S = \sum_{i=1}^{n} \lambda_i x_i \frac{\partial}{\partial x_i}$ and R a nonlinear vector field. We assume that the following diophantine condition like is satisfied:

$$(\omega) \quad -\sum_{k \geq 0} \frac{\ln \omega_k}{2^k} < +\infty$$

where $\omega_k = \inf\{|(Q, \lambda) - \lambda_i| \neq 0,\ 1 \leq i \leq n,\ Q \in \mathbb{N}^n, 2 \leq |Q| \leq 2^k\}$.

Theorem 2.1.1 *[Bru72] Let $X = S + R$ be an holomorphic vector field as above. We assume that S satisfies the Bruno condition (ω). If X has formal normal form of the type $\hat{a}.S$ for some formal power series $\hat{a}$ (with $\hat{a}(0) = 1$), then X is holomorphically normalizable.*

In the case of Hamiltonian vector field and under Siegel diophantine condition, this result is due to H. Rüssmann :

Theorem 2.1.2 *[Rüs67] Let $H = \sum_{i=1}^{n} \lambda_i x_i y_i + \cdots$ be an analytic third order perturbation of the quadratic hamiltonian $h = \sum_{i=1}^{n} \lambda_i x_i y_i$. Assume that h satisfies the Siegel condition:*

$$\left| \sum_{j=1}^{n} q_j \lambda_j \right| > \frac{c}{\left(\sum_{j=1}^{n} |q_j| \right)^\mu}$$

for integer vectors $(q_1, \ldots, q_n) \in \mathbb{Z}^n$ *such that* $\sum_{j=1}^n |q_j| > 0$. *Assume that H has a formal Birkhoff normal form of the form* $\hat{F}(h) = \hat{F}(\sum_{i=1}^n \lambda_i x_i y_i)$ *then H is analytically conjugate to a Birkhoff normal form $F(h)$ for some analytic function F.*

We refer to J. Martinet's Bourbaki seminar for a survey on this topic [Mar80].

Example 2.1.3 *Let us apply the previous result to example* (8). *If it has a formal normal form* (9) *with* $\hat{G} = \hat{F}$ *then it is holomorphically normalizable.*

Example 2.1.4 *Let us consider the two-dimensional system*

$$\begin{cases} \dot{x} = x^2 \\ \dot{y} = x + y \end{cases} \tag{1}$$

There is a unique formal diffeomorphism $x = X$, $y = Y + \hat{\psi}(X)$ that transforms the previous systems into its normal form

$$\begin{cases} \dot{x} = x^2 \\ \dot{y} = y \end{cases} \tag{2}$$

In fact, the conjugacy equation leads to

$$\begin{aligned} \dot{y} = x + y &= X + Y + \hat{\psi}(X) \\ &= \dot{Y} + \hat{\psi}'(X)\dot{X} = Y + \hat{\psi}'(X)X^2. \end{aligned}$$

So $\hat{\psi}$ has to solve the **Euler equation**

$$X^2 \hat{\psi}'(X) - \hat{\psi}(X) = X$$

which formal solution is

$$\hat{\psi}(X) = -\sum_{k \geq 1} (k-1)! X^k.$$

This does not converge in a neighborhood of the origin ! The normal form (2) *does not satisfies Brjuno condition: it is not proportional to the linear part $\dot{x} = 0, \dot{y} = y$. Nevertheless, we can show that there exists sectorial normalizations. This means that there exists germs of holomorphic diffeomorphisms defined only in the product of sector with an edge at the origin (in the x plane) and a disc around 0 (in y) which conjugate equation* (1) *into its normal form. This the starting point of a long story that have been developed by J. Martinet and J.-P. Ramis [MR82, MR83] for two-dimensional vector fields and by J. Ecalle, S. Voronin and B. Malgrange for germs of local diffeomorphisms near a fixed point in the complex plane [Eca, Vor81, Mal82, Il'93]. In higher dimension, the theory has been developed by J. Ecalle and L. Stolovitch [Eca92, Sto96]. Recently, the interplay beetwen these "Stokes phenomena" and small divisors phenomena have been investigated by B. Braaksma and L. Stolovitch [BS07]. We refer to [Bal00, Ram93, RS93] for summability theory and Stokes phenomenon.*

2.2 Theorems of J. Vey

On the other hand, Vey proved two theorems about the normalization of family of commuting vector fields satisfying some geometric properties.

Theorem 2.2.1 *[Vey79] Let $X_1,\ldots,X_{n-1}$ be $n-1$ holomorphic vector fields in a neighborhood of $0 \in \mathbb{C}^n$, vanishing at this point. We assume that :*

- *Each X_i is a volume preserving vector field ($\mathscr{L}_{X_i}\omega = 0$ with ω an holomorphic n-differential form)*
- *The 1-jet $J^1(X_1),\ldots,J^1(X_{n-1})$ are diagonal and independent over $\mathbb{C}$ (this means that if there are complex constants c_i such that $\sum_{i=1}^{n-1} c_i J^1(X_i) = 0$, then $c_i = 0$ for all i.)*
- *$[X_i,X_j] = 0$ for all indices i,j*

Then, $X_1,\ldots,X_{n-1}$ are holomorphically and simultaneously normalizable.

Theorem 2.2.2 *[Vey78] Let $X_1,\ldots,X_n$ be n holomorphic vector fields in a neighborhood of $0 \in \mathbb{C}^{2n}$, vanishing at this point. We assume that :*

- *Each X_i is an Hamiltonian vector field*
- *The 1-jet $J^1(X_1),\ldots,J^1(X_n)$ are diagonal and independent*
- *$[X_i,X_j] = 0$ for all indices i,j*

Then, $X_1,\ldots,X_n$ are holomorphically and simultaneously normalizable.

2.3 Singular complete integrability–Main result

We shall present a general result about normalization of commutative family of holomorphic vector fields vanishing at the same point that unifies both Vey's and Brjuno's theorems. At first glance, such unification could seem a little bit weird. In fact, in Vey's theorems, there no assumption about small divisors while in Brjuno's theorem there is one. In Vey's theorem, vector fields satisfy a geometric assumptions (volume preserving or symplectic) whereas in Brjuno's theorem there is an assumption about the formal normal form.

Let us consider the family $S = \{S_1,\ldots,S_l\}$, $l \leq n$, of linearly independent linear diagonal vector fields

$$S_i = \sum_{j=1}^{n} \lambda_{i,j} x_j \frac{\partial}{\partial x_j}.$$

This means that if $\sum_{i=1}^{l} c_i S_i = 0$ for some complex numbers c_i, then all the c_i's are zero. Let us define the sequence of positive numbers

$$\omega_k(S) = \inf\left\{ \max_{1 \leq i \leq l} |(Q,\lambda^i) - \lambda_{i,j}| \neq 0,\ 1 \leq j \leq n,\ Q \in \mathbb{N}^n, 2 \leq |Q| \leq 2^k, \right\},$$

where $\lambda^i = (\lambda_{i,1},\ldots,\lambda_{i,n})$.

Definition 2.3.1 *We shall say that S is diophantine if*

$$(\omega(S)) \qquad -\sum_{k\geq 0} \frac{\ln \omega_k(S)}{2^k} < +\infty$$

Remark 2.3.2 *The family S can be diophantine while none of the S_i's satisfies Brjuno condition (ω). For instance, consider in $(\mathbb{C}^3,0)$ with complex coordinates (x,y,z) the vector fields $S_1 = E_1 - \zeta E_2$ and $S_2 = -\zeta E_1 + E_2$ where ζ is positive irrational number, $E_1 = x\frac{\partial}{\partial x} - y\frac{\partial}{\partial y}$ and $E_2 = y\frac{\partial}{\partial y} - z\frac{\partial}{\partial z}$. Since $1 - \zeta^2 \neq 0$, S_1 and S_2 are linearly independent. The small divisors relative to S_1 or S_2 look like $q_1 - \zeta q_2$ for some relative integers q_1, q_2. Thus, if ζ is a Liouvillian number then neither S_1 nor S_2 will satisfy Brjuno condition. On the other hand, let λ_i (resp. μ_i) be the vector of eigenvalues of S_i (resp. E_i). We have*

$$A := \begin{pmatrix} (Q,\lambda_1) - \lambda_{1,j} \\ (Q,\lambda_2) - \lambda_{2,j} \end{pmatrix} = \begin{pmatrix} 1 & -\zeta \\ -\zeta & 1 \end{pmatrix} \begin{pmatrix} (Q,\mu_1) - \mu_{1,j} \\ (Q,\mu_2) - \mu_{2,j} \end{pmatrix} =: B.$$

Hence, if we denote the matrix by C, we have then $\|B\| \leq \|C^{-1}\|\|A\|$. Therefore, the sequence of the $\|A\|$'s when Q and j vary do not accumulate the origin since the sequence of the $\|B\|$'s does not. So, the family S is diophantine.

Let $\left(\widehat{\mathscr{X}^1_n}\right)^S$ $\left(resp. \left(\widehat{\mathscr{O}_n}\right)^S\right)$ be the formal centralizer of S (resp. the ring of formal first integrals), that is the set of formal vector fields X (resp. formal power series f) such that $[S_i, X] = 0$ (resp. $\mathscr{L}_{S_i}(f) = 0$) for all $1 \leq i \leq l$.

Let $X = \{X_1, \ldots, X_l\}$ be a family of germs of commuting vector fields at the origin such that the linear part of X_i is S_i; that is $[X_i, X_j] = 0$ for all i, j. We shall call X a nonlinear deformation of S.

Definition 2.3.3 *We shall say that a nonlinear deformation X of S is a **normal form** (with respect to S) if*

$$[S_i, X_j] = 0, \quad 1 \leq i, j \leq l.$$

Definition 2.3.4 *We shall say that X, a nonlinear deformation of S, is **formally completely integrable** if there exists a formal diffeomorphism $\hat{\Phi}$ fixing the origin and tangent to the identity at that point which conjugate the family X to normal form of the type*

$$\hat{\Phi}_* X_i = \sum_{j=1}^l \hat{a}_{i,j} S_j, \quad i = 1, \ldots, l \tag{3}$$

where the $\hat{a}_{i,j}$'s belongs to $\widehat{\mathscr{O}}_n^S$.

Proposition 2.3.5 *If X has a formally completely integrable normal form then all its normal form are also formally completely integrable.*

Theorem 2.3.6 *Under the assumptions above, if S is diophantine, then any formally completely integrable nonlinear deformation $X = S + \varepsilon$ of S is holomorphically normalizable.*

This means that there exists a genuine germ of biholomorphism $\Phi : (\mathbb{C}^n, 0) \rightarrow (\mathbb{C}^n, 0)$ tangent to the identity at 0 which conjugate the family X to normal form of the type

$$\Phi_* X_i = \sum_{j=1}^{l} a_{i,j} S_j, \quad i = 1, \ldots, l \tag{4}$$

where the $a_{i,j}$'s are germ of holomorphic invariant functions, i.e. they belong to $\mathcal{O}_n^S$.

Remark 2.3.7 *The theorem doesn't says that neither $\hat{\Phi}$ nor the $\hat{a}_{i,j}$ converge but rather that there is another normalizing diffeomorphism that converges.*

Remark 2.3.8 *One way to use this theorem is to have* **"a magic word in hand"** *(like Hamiltonian, volume preserving, reversible ....) that will implies that the formal normal form is of the good type. This comes from the data of the problem that one wants to solve.*

Corollary 2.3.9 *If S is diophantine and if the holomorphic nonlinear deformation X is formally linearizable then it is holomorphically linearizable, i.e. there exists a holomorphic change of coordinates in which all the X_i's are linear.*

Of course, if one of the S_i's satisfies Brjuno condition (ω) and if the family X is formally linearizable, then it is also holomorphically linearizable. The point of the previous corollary is that none of the S_i's is required to satisfies (ω) in order that S to be diophantine. A result similar to our corollary was obtained by T. Gramchev and M. Yoshino for germs of commuting diffeomorphisms near a fixed point [GY99] under a slightly coarser diophantine condition. The article of J. Moser [Mos90] was the starting point since he was dealing with germ of one-dimensional diffeomorphisms.

2.3.1 Fundamental structures

Proposition 2.3.10 *[Sto00][prop. 5.3.2] With the notation above, $\widehat{\mathcal{O}}_n^S$ is a formal $\mathbb{C}$-algebra of finite type; $\widehat{\mathcal{X}}_n^S$ is a $\widehat{\mathcal{O}}_n^S$-module of finite type.*

This means the following : if the ring of invariants is nor reduced to the constants, then there exists a finite number of monomials $x^{R_1}, \ldots, x^{R_p}$ such that $\widehat{\mathcal{O}}_n^S = \mathbb{C}[[x^{R_1}, \ldots, x^{R_p}]]$. Moreover, there exists a finite number of polynomial vector fields $Y_1, \ldots, Y_m$ such that if X belongs to $\widehat{\mathcal{X}}_n^S$ (i.e. $[S_i, X] = 0$, for all i) then there exists $\hat{a}_1, \ldots \hat{a}_m \in \widehat{\mathcal{O}}_n^S$ such that $X = \hat{a}_1 Y_1 + \cdots + \hat{a}_m Y_m$. The proof is based on Hilbert theorem : in a Noetherian ring, ideals are generated by a finite number of elements.

Let $2 \leq k$ be an integer and let $\mathscr{P}_n^k$ be the space of homogeneous vector fields of $\mathbb{C}^n$ of degree k. Let us consider the map $\rho : \mathbb{C}^l \rightarrow \mathrm{Hom}_{\mathbb{C}}(\mathscr{P}_n^k, \mathscr{P}_n^k)$ defined by

$$\rho(g)(X) = \left[\sum_{i=1}^{l} g_i S_i, X \right]$$

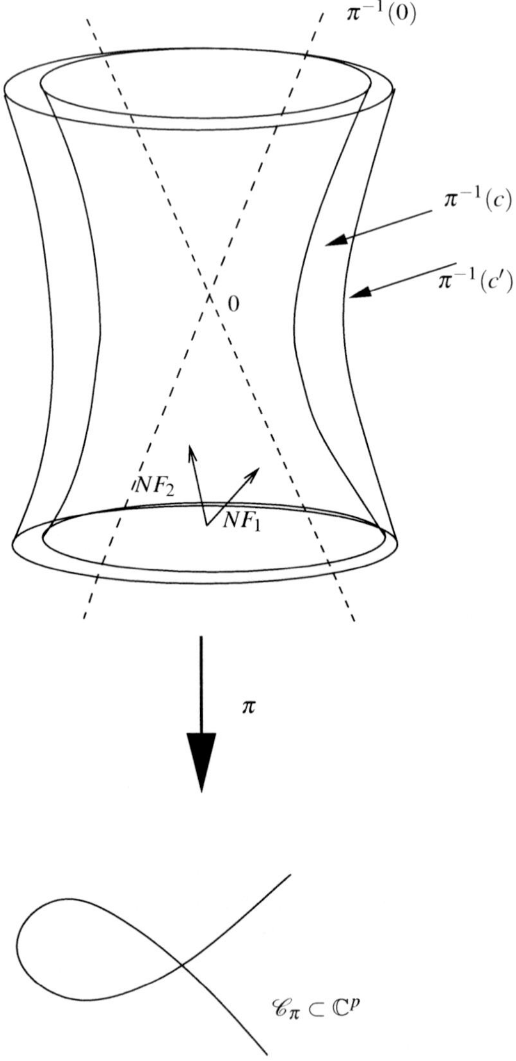

Fig. 1 Singular complete integrability: in the new holomorphic coordinate system, all the fibers (intersected with a fixed polydisc) are left invariant by the vector fields and their motion on it is a linear one

where $g = (g_1, \ldots, g_l)$ and $X \in \mathscr{P}_n^k$ ([.,.] denotes the Lie bracket of vector fields of $\mathbb{C}^n$). It is a **representation** of the commutative Lie algebra $\mathbb{C}^l$ in $\mathscr{P}_n^k$. To such a representation ρ of the abelian Lie algebra $\mathbb{C}^l$ into a finite dimensional vector space M, one can associate the **Chevalley–Koszul complex**

$$0 \to M \xrightarrow{d_0} \mathrm{Hom}_{\mathbb{C}}\left(\mathbb{C}^l, M\right) \xrightarrow{d_1} \mathrm{Hom}_{\mathbb{C}}\left(\wedge^2\mathbb{C}^l, M\right) \xrightarrow{d_2} \cdots \xrightarrow{d_{l-1}} \mathrm{Hom}_{\mathbb{C}}\left(\wedge^l\mathbb{C}^l, M\right) \to 0,$$
(5)

where the differentials d_i are defined in the following way : if $\omega \in \mathrm{Hom}_{\mathbb{C}}\left(\wedge^p\mathbb{C}^l, M\right)$ and $(g_1,\ldots,g_{p+1}) \in (\mathbb{C}^l)^{p+1}$, then

$$d_p(\omega)(g_1,\ldots,g_{p+1}) = \sum_{i=1}^{p+1} (-1)^{i+1}\rho(g_i)\left(\omega(g_1,\ldots,\widehat{g_i},\ldots,g_{p+1})\right)$$
(6)

Here $(g_1,\ldots,\widehat{g_i},\ldots,g_{p+1}) \in (\mathbb{C}^l)^p$ stands for $(g_1,\ldots,g_{i-1},g_{i+1},\ldots,g_{p+1})$. The differentials d_0 and d_1 will be particuliary useful:

$$d_0 U(g) = \rho(g)U, \quad d_1 F(g_1,g_2) = \rho(g_1)F(g_2) - \rho(g_2)F(g_1).$$

The **cohomology spaces** $H^i(\mathbb{C}^l, M)$ are defined to be

$$H^i(\mathbb{C}^l, M) = \mathrm{Ker}\, d_i / \mathrm{Im}\, d_{i-1}, \quad i = 0,\ldots l-1.$$

Let $\alpha = (\alpha_1,\ldots,\alpha_l) \in \mathbb{C}^l$. It defines the complex linear form on $\mathbb{C}^l$, $\alpha(z) = \sum_{i=1}^l \alpha_i z_i$. To such a linear form, we associate the "generalized eigenspace"

$$\mathscr{P}_{n,\alpha}^k = \left\{ X \in \mathscr{P}_n^k \,|\, \forall g \in \mathbb{C}^l,\, [S(g),X] = \alpha(g)X \right\}.$$

In other words, $X \in \mathscr{P}_{n,\alpha}^k$ if and only if $[S_i,X] = \alpha_i X$ for all $1 \le i \le l$. If $\mathscr{P}_{n,\alpha}^k \ne 0$ then α is called a **weight** of S and $\mathscr{P}_{n,\alpha}^k$ is called the associated **weightspace**. There is a decomposition of the space into "generalized eigenspaces", namely the **Fitting decomposition**:

$$\mathscr{P}_n^k = \mathscr{P}_{n,*}^k \oplus \mathscr{P}_{n,0}^k$$

where $\mathscr{P}_{n,*}^k$ is the (finite) direct sum of the weightspaces associated to nonzero weights of S.

2.3.2 Geometric interpretation

In order to illustrate our result, let us first recall the Liouville theorem [Arn97]. Let $H_1,\ldots,H_n$ be smooth functions on a smooth symplectic manifold M^{2n}; let $\pi : M^{2n} \to \mathbb{R}^n$ denotes the map $\pi(x) = (H_1(x),\ldots,H_n(x))$. We assume that, for all $1 \le i, j \le n$, the Poisson bracket $\{H_i,H_j\} = 0$ vanishes. Let $c \in \mathbb{R}^n$ be a regular value of π; we assume that $\pi^{-1}(c)$ is compact and connected. Then there exists a neighborhood U of $\pi^{-1}(c)$ and a symplectomorphism Φ from U to $\pi(U) \times \mathbb{T}^n$ such that, in this new coordinate system, each symplectic vector field X_{H_i} associated to H_i is tangent to the fiber $\{d\} \times \mathbb{T}^n$. It is constant on it and the constant depends only on the fiber.

Let us turn back to our problem and let S be a diophantine family of linearly independent diagonal vector fields of $\mathbb{C}^n$. Let $\widehat{\mathscr{O}}_n^S$ be its ring of formal first integrals. It is a $\mathbb{C}$-algebra of finite type and there are homogeneous polynomials $u_1,\ldots,u_p$ such that $\widehat{\mathcal{O}}_n^S = \mathbb{C}[[u_1,\ldots,u_p]]$. Let $\pi : \mathbb{C}^n \to \mathbb{C}^p$ defined by $\pi(x) = (u_1(x),\ldots,u_p(x))$. Let s be the degree of transcendence of the field of fractions of $\mathbb{C}[u_1,\ldots,u_p]$; it is the maximal number of algebraically independent polynomials among $u_1,\ldots,u_p$. The algebraic relations among $u_1,\ldots,u_p$ define an s-dimensional algebraic variety $\mathcal{C}_S$ in $\mathbb{C}^p$. Hence, π defines a **singular** fibration over $\mathcal{C}_S$. The linear vector fields $S_1,\ldots,S_l$ are tangent and independent on each fiber $\pi^{-1}(b)$ of π; the latter are called **toric variety** because they admit an action of the algebraic torus $\mathbb{C}^*$. Note that we must have $l \leq n - s$. Now, we come to the nonlinear deformation. Let $X = S + \varepsilon$ be a nonlinear deformation of S. Let us assume that it is formally completely integrable. Then, according to our result, there exists a neighborhood U of 0 in $\mathbb{C}^n$ and an holomorphic diffeomorphism Φ on U such that, in the new coordinate system, the vector fields $\Phi_* X_1,\ldots,\Phi_* X_l$ are commuting linear diagonal vector fields on each fiber restricted to U and their eigenvalues depend only on the fiber. Indeed, in this new coordinates, we have $\Phi_* X_i = \sum_{j=1}^l a_{i,j} S_j$ where $a_{i,j} \in \mathcal{O}_n^S$. By definition, these vector fields are all tangent to the fibers of π (therefore, we must have $l \leq n - s$). As consequence $\Phi_* X_i$'s are all tangent to the fibers of π. On each fiber, the functions $a_{i,j}$ are constant so that each $\Phi_* X_i$ reads as a linear diagonal vector field, that is a **linear motion of a toric variety**.

2.3.3 Proper Poincaré extension

The next question that can be asked is the following : under what assumptions can a formally completely integrable nonlinear deformation $X = S + \varepsilon$ of S be extended in an higher dimensional space into another formally completely integrable nonlinear deformation $\hat{S} + \hat{\varepsilon}$ of $\hat{S}$, with the same number of commuting vector fields?

First of all, we shall define a good extension of S in $\mathbb{C}^{n+m}$ to be $\hat{S}_i := S_i \oplus S_i''$, $i = 1,\ldots,l$, where S_i'' is a diagonal linear vector field of $\mathscr{P}_m^1$. Of course, we want the properties of $\hat{S}$ to be derived from those of S; that is, we want $\hat{S}$ to be diophantine as soon as S is and we want that $\mathcal{O}_{n+m}^{\hat{S}} = \mathcal{O}_n^S$. One way to achieve this is to assume that S'' is **Poincaré family relatively to S** : we require that the weights of S all belong to a real linear hyperplane of $\mathbb{R}^{2l}$ whereas the weights of S'' all, but a finite number of them, belong to one and the same side of the hyperplane. Such an extension will be called **proper** if the only weight of S'' which belong to the hyperplane is the zero weight. If $(x_1,\ldots x_{n+m})$ denotes coordinates of $\mathbb{C}^{n+m}$ and if X is a vector field of $(\mathbb{C}^{n+m},0)$, then X'' denotes its projection onto $\frac{\partial}{\partial x_{n+1}},\ldots,\frac{\partial}{\partial x_{n+m}}$.

Definition 2.3.11 *We shall say that a proper Poincaré extension of S in $\mathbb{C}^{n+m}$ is completely integrable if there exists a formal diffeomorphism $\hat{\Phi}$ fixing the origin and tangent to the identity at that point which conjugate the family X to normal form of the type*

$$NF_i := \hat{\Phi}_* X_i = \sum_{j=1}^l \hat{a}_{i,j} S_j + \sum_{j=1}^l \hat{a}_{i,j} S_j'' + D_i'' + Nil_i'' + Res_i'', \quad i = 1,\ldots,l \quad (7)$$

where the $\hat{a}_{i,j} \in \widehat{\mathcal{O}}_n^S$. Here, D_i'' (resp. Nil_i'', Res_i'') denotes a linear diagonal (resp. nilpotent, nonlinear) vector field of $\mathbb{C}^m$ with coefficient in $\widehat{\mathcal{O}}_n^S$ such that the family D'' has the same centralizer as S'' (resp. commuting with the S_i'''s).

In other words, the projection NF'' of the normal form is a polynomial Poincaré normal form of $\mathbb{C}^m$ with coefficients in $\mathcal{O}_n^S$.

Then we have the

Theorem 2.3.12 *[Sto00] Let S be a diophantine family of diagonal linear vector field of $\mathbb{C}^n$. We assume that $\hat{S} = S \oplus S''$ is a proper Poincaré extension of S in $\mathbb{C}^{n+m}$ by S''. Then, any nonlinear deformation of $\hat{S}$ which is formally completely integrable is holomorphically normalizable.*

For one vector field, theses results are due to Brjuno. Let us illustrate this result on Example 1.2.6. Let us define $S = x_1 \frac{\partial}{\partial x_1} - x_2 \frac{\partial}{\partial x_2} - \zeta x_3 \frac{\partial}{\partial x_3}$. Assume that S satisfies Brjuno condition (ω). Let us define $S'' = i x_4 \frac{\partial}{\partial x_4} + i x_5 \frac{\partial}{\partial x_5}$. It is proper Poincaré vector field with respect to S. In fact, all the weights of S are real while those of S'' are purely imaginary. Then nonlinear centralizer of S'' is reduced to zero. First of all assume that in the normal form (10), we have $\hat{f}_1 = \hat{f}_2 = \hat{f}_3$. So that the projection on $\frac{\partial}{\partial x_1}, \ldots, \frac{\partial}{\partial x_n}$ is a formally completely integrable system which nonlinearities are parametrized by $\mathbb{C}^m$. Assume that the formal power series $\hat{g}_{1,1}$ and $\hat{g}_{2,2}$ can be decomposed as $\hat{g}_{i,i} = \hat{f}_{i,i} + \hat{h}_{i,i}$ such that

1.

$$q_1(i + \hat{f}_{1,1}(x_1 x_2)) + q_2(i + \hat{f}_{2,2}(x_1 x_2)) \not\equiv i + \hat{f}_{1,1}(x_1 x_2)$$
$$q_1(i + \hat{f}_{1,1}(x_1 x_2)) + q_2(i + \hat{f}_{2,2}(x_1 x_2)) \not\equiv i + \hat{f}_{2,2}(x_1 x_2)$$

for all $(q_1, q_2) \in \mathbb{N}^2$ such that $q_1 + q_2 \geq 2$. This precisely means that the formal vector field $(i + \hat{f}_{1,1}(x_1 x_2)) x_4 \frac{\partial}{\partial x_4} + (i + \hat{f}_{2,2}(x_1 x_2)) x_5 \frac{\partial}{\partial x_5}$, thought as a vector field of $\mathbb{C}^2$, has the same nonlinear centralizer as S'', that is 0.

2. The vector field

$$\left(\hat{h}_{1,1}(x_1 x_2) x_4 + \hat{g}_{1,2}(x_1 x_2) x_5 \right) \frac{\partial}{\partial x_4} + \left(\hat{g}_{2,1}(x_1 x_2) x_4 + \hat{h}_{2,2}(x_1 x_2) x_5 \right) \frac{\partial}{\partial x_5}$$

is nilpotent and commutes with S''.

Let us a give a geometric interpretation of this last result. Let us consider again the map $\tilde{\pi} : \mathbb{C}^{n+m} \to \mathbb{C}^p$ with $\pi(x) = (x^{R_1}, \ldots, x^{R_p})$. Since, the invariants of $\hat{S}$ are the same as those of S, we have $\tilde{\pi}^{-1}(b) = \pi^{-1}(b) \times \mathbb{C}^m$. Let us apply our result. In a new holomorphic coordinate system at the origin, the projection X_i' on $\mathbb{C}^n$ of the vector field X_i is a completely integrable in the previous sense : it is tangent to any the toric variety $\pi^{-1}(b)$ and its restriction to it is a linear diagonal motion. On the other hand, the projection X_i'' on $\mathbb{C}^m$ is a polynomial normal form (of $\mathbb{C}^m$) which coefficients depend only holomorphically on b.

2.4 How to recover Brjuno's and Vey's theorems from Theorem 2.3.6

Brjuno's theorem correspond precisely to our result for $l = 1$.

Let us prove the volume preserving case of Vey's theorem. Let E be the family of the $n - 1$ linear semi-simple vector fields of $\mathbb{C}^n$ defined to be $E_i = x_i \frac{\partial}{\partial x_i} - x_{i+1} \frac{\partial}{\partial x_{i+1}}$, $1 \le i \le n - 1$. The weights associated to $Q = (q_1, \ldots, q_n) \in \mathbb{N}^n$, $|Q| \ge 2$, $1 \le j \le n$ are $\alpha_{i,Q,j} = q_i - q_{i+1} + \delta_{i,j}\delta_{i+1,j}(-1)^{\delta_{i,j}}$ (the last expression in the sum is 0 if $j \ne i, i+1$, 1 if $j = i + 1$ and -1 if $j = i$). First of all, the values of the nonzero weights of E are integers; thus, they cannot accumulate the origin, so that E is diophantine. Moreover, if we set $u = x_1 \cdots x_n$, then $\widehat{\mathcal{O}}_n^E = \mathbb{C}[[u]]$ and $\left(\widehat{\mathscr{X}}_n^1\right)^E$ is the $\mathbb{C}[[u]]$-module generated by $x_i \frac{\partial}{\partial x_i}$, $1 \le i \le n$. An easy computation shows that $X \in \left(\widehat{\mathscr{X}}_n^1\right)^E$ satisfies to $\mathscr{L}_X(u) = 0$ if and only if X belongs to the $\mathbb{C}[[u]]$-module generated by the E_i's.

Let us write $J^1(X_i) = \sum_{j=1}^n \mu_{i,j} x_i \frac{\partial}{\partial x_j}$. Let us set $\mu^i := (\mu_{i,1}, \ldots, \mu_{i,n})$. Since X_i is volume preserving then, $\mu_{i,1} + \cdots + \mu_{i,n} = 0$; it follows that $J^1(X_i) = \sum_{j=1}^{n-1} a_{i,j} E_j$. By the independence of the 1-jets, the $(n-1) \times (n-1)$ matrix $A_0 = (a_{i,j})$ is invertible. Let us compute the weights of the family of the $J^1(X_i)$'s with respect of those of E. We have

$$
\begin{pmatrix} (Q, \mu^1) - \mu_{1,j} \\ \vdots \\ (Q, \mu^l) - \mu_{l,j} \end{pmatrix} = A_0 \begin{pmatrix} \alpha_{1,Q,j} \\ \vdots \\ \alpha_{l,Q,j} \end{pmatrix}.
$$

Therefore, we have

$$
\|A_0^{-1}\| \begin{vmatrix} (Q, \mu^1) - \mu_{1,j} \\ \vdots \\ (Q, \mu^l) - \mu_{l,j} \end{vmatrix} \ge \begin{vmatrix} \alpha_{1,Q,j} \\ \vdots \\ \alpha_{l,Q,j} \end{vmatrix}.
$$

This means that the family of the $J^1(X_i)$'s is also diophantine.

Since the family can be transformed into a normal form, there exists a formal diffeomorphism $\hat{\Phi}$ such that $\hat{\Phi}^* X_i = \sum_{j=1}^n \hat{F}_{i,j}(u) x_i \frac{\partial}{\partial x_i}$ for some $\hat{F}_{i,j} \in \mathbb{C}[[u]]$. We can assume that $\hat{\Phi}$ is volume preserving; thus the normal forms are also volume preserving. Hence $\mathrm{div}\,(\hat{\Phi}^* X_i) = 0$, that is

$$
\sum_{j=1}^n \frac{\partial x_j \hat{F}_{i,j}(u)}{\partial x_j} = 0 = \left(\sum_{j=1}^n \hat{F}_{i,j}(u)\right) + u \frac{d\left(\sum_{j=1}^n \hat{F}_{i,j}\right)}{du}(u).
$$

An easy computation shows that $\sum_{j=1}^n \hat{F}_{i,j} = 0$. Thus,

$$
\hat{\Phi}^* X_i = \sum_{j=1}^{n-1} \hat{f}_{i,j}(u) E_j = \sum_{j=1}^{n-1} \hat{g}_{i,j}(u) J^1(X_j),
$$

that is X is formally completely integrable. According to our main result, there is an holomorphic diffeomorphism Ψ normalizing X in a neighborhood of the origin. By a classical argument of Vey [Vey79], we can modify holomorphically Ψ so that it becomes volume preserving and still normalizing X.

2.5 Sketch of the proof

Let us give a sketch of the proof of our results. In order to normalize the nonlinear deformation $X = S + \varepsilon$ of S, we shall proceed through a classical Newton method, that is a Nash–Moser induction type.

Let us assume that the nonlinear deformation $X = S + \varepsilon$ is normalized up to order m; we will build a diffeomorphism Φ_m which normalize the deformation up to order $2m$; it is tangent to Id up to order k. Let us show how this works. First of all, we can write the deformation $X_i = NF_i^m + B_i + R_i$, $1 \leq i \leq l$ where NF_i^m is a normal form of degree m, B_i is polynomial of degree $\leq 2m$ and of order $\geq m + 1$ and R_i is of order $\geq 2m + 1$. Let us denote by $B_{i,*}$ (resp. $B_{i,0}$) the projection of B_i onto the sum of the weightspaces associated to a nonzero weight (resp. zero weight) of S in $\mathscr{P}_n^{m+1,2m}$. The compatibility condition (i.e $[X_i, X_j] = 0$ for all $1 \leq i, j \leq l$) shows that, for all $1 \leq i, j \leq l$

$$J^{2m}\left([NF_i^m, B_{j,*}]\right) - [NF_j^m, B_{i,*}]\right) = 0. \tag{8}$$

On the other hand, if we conjugate X_i by a diffeomorphism of the form $\exp(U)$ for some polynomial vector field $U \in \mathscr{P}_n^{m+1,2m}$ and writing $\exp(U)_* X_i = NF_i^m + B_i' + R_i'$ as above, we find out that

$$J^{2m}\left(B_i' - B_i + [NF_i^m, U]\right) = 0$$

The algebraic properties of the weightspaces of S show that, in fact, we have

$$J^{2m}\left(B_{i,*}' - B_{i,*} + [NF_i^m, U_*]\right) = 0.$$

If we assume that the diffeomorphism $\exp(U)$ normalizes simultaneously the X_i's up to order $2m$ then we must have $B_{i,*}' = 0$ for all i. Hence, we have

$$J^{2m}\left(-B_{i,*} + [NF_i^m, U_*]\right) = 0 \quad i = 1, \ldots, l. \tag{9}$$

Let us denote by $\mathscr{P}_{n,*}^{m+1,2m}$ the direct sum of weightspaces associated to a nonzero weight of ρ in $\mathscr{P}_n^{m+1,2m}$. Let us define the linear map

$$\rho_m : \mathbb{C}^l \to \mathrm{Hom}_{\mathbb{C}}\left(\mathscr{P}_{n,*}^{m+1,2m}, \mathscr{P}_{n,*}^{m+1,2m}\right)$$

by $\rho_m(g)(X) = J^{2m}\left(\left[\sum_{i=1}^{l} g_j NF_j^m, X\right]\right)$ if $g = (g_1, \ldots g_l)$. It is well defined and it is a representation of the abelian Lie algebra $\mathbb{C}^l$ into $\mathscr{P}_{n,*}^{m+1,2m}$. To this representation is associated a complex of finite dimensional complex vector spaces; it is the Chevalley-Kozsul complex of this representation. Let us write d_m^i its ith-differential. Therefore, equation (8) reads $d_m^1(B^*) = 0$, that is B_* is a 1-cocycle for this complex. Moreover, equation (9) reads $d_m^0(U) = B_*$, that is B_* is the 0-coboundary of U: it is a **cohomological equation**.

Hence, the Chevalley–Koszul complex of the representation ρ_m plays an important role in our problem. We shall call it the **Newton complex** of order m. According to the discussion above, the first important problem to study is its cohomology. We can show that its 0th-cohomology as well as the 1st-cohomology spaces are zero:

Proposition 2.5.1 *[Sto00][Prop. 7.1.1] We have*

$$H_m^i\left(\mathbb{C}^l, \mathscr{P}_{n,*}^{m+1,2m}\right) = 0, \quad i = 0,1$$

where H_m^i denotes the ith-cohomology space of the Chevalley–Koszul complex associated to ρ_m.

It is not very difficult but rather technical. It leads to the important consequence that, B_* being given as above, there exists a unique $U \in \mathscr{P}_{n,*}^{m+1,2m}$ such that, for all $1 \le i \le l$, $J^{2m}([NF_i^m, U]) = B_{i,*}$; hence, conjugating X_i by $\exp(U)$ normalizes X_i up to order $2m$.

We find out that the formal diffeomorphism defined by $\hat{\Phi} := \lim_{k \to +\infty} \Phi_{2^k} \circ \cdots \circ \Phi_2$ normalizes simultaneously the X_i's where the Φ_{2^k}'s are built as above. In order to prove that $\hat{\Phi}$ is holomorphic in a neighborhood of $0 \in \mathbb{C}^n$, one has to estimate Φ_{2^k}. Here comes the analysis and the major difficulty. To get an estimate of $\Phi_m = \exp(U)$ with $m = 2^k$, we have to estimate U. Hence, we are led naturally to give bounds for the cohomology of the Newton complex : Let $r > 1/2$, the spaces of the Newton complex are provided with norms (depending on a real positive number r) which turn it into a topological complex of vector spaces. By the above algebraic properties, the 0-differential, d_m^0, has a right inverse s on the space of 1-cocycle : if Z is a 1-cocycle of the Newton complex, then $s(Z)$ is the unique element of $\mathscr{P}_{n,*}^{m+1,2m}$ such that $d_m^0(s(Z)) = Z$. Here comes the main assumptions : if the family X is **completely integrable** then there exists constants $d, \eta_1, c(\eta_1)$, such that if $m = 2^k$ and if the r-norms of $NF^m - S$ and $D(NF^m - S)$ are sufficiently small, say $< \eta_1$ (for some $1/2 \le r < 1$) then

$$|s(Z)|_r \le \frac{c(\eta_1)}{\omega_{k+1}^d(S)} |Z|_r; \tag{10}$$

the constant d doesn't depend on η_1 (we recall that $\omega_k(S)$ is the smallest norms of the nonzero weights of S in $\mathscr{P}_n^{2,2^k}$).

Let us describe the way we obtain this estimate. In order to solve the cohomological equation associated the 1-cocycle Z, it is necessary and sufficient to solve the system of l equations $J^{2m}([NF_i^m, U]) = Z_i$, $i = 1, \ldots, l$. We can decompose this equation along the weightspaces of S. In fact, let α be a weight of S and let V belongs to the associated weightspace. Then, by Jacobi identity, we have

$$[S_j, [NF_i^m, V]] = [-NF_i^m, [V, S_j]] - [V, [S_j, NF_i^m]].$$

By assumptions, $NF_i^m = \sum_{j=1}^l a_{i,j}^{m-1} S_j$ where the $a_{i,j}^{m-1}$'s are polynomials invariants for S of degree $\leq m - 1$. Therefore, according to formula (2), $[S_j, NF_i^m] = 0$. Hence, we have $[S_j, [NF_i^m, V]] = [-NF_i^m, [V, S_j]] = \alpha_j [NF_i^m, V]$. It is sufficient to consider for any nonzero weight α of S, the equation with both Z_i's and U in the associated weightspace.

This set of equations can be written in the following matrix form

$$A(x) \begin{pmatrix} \alpha_1 U \\ \vdots \\ \alpha_l U \end{pmatrix} + \begin{pmatrix} D_1(U) \\ \vdots \\ D_l(U) \end{pmatrix} = \begin{pmatrix} Z_1 + \mathfrak{Z}_1 \\ \vdots \\ Z_l + \mathfrak{Z}_l \end{pmatrix}$$

where $A = (a_{i,j}^{m-1})$ is a square $l \times l$ matrix with coefficients in the $\mathbb{C}$-algebra $\mathcal{O}_n^S$ of holomorphic first integrals of the linear part S; $A(0) = Id$; the operators $D_1, \ldots, D_l$ are $\mathcal{O}_n^S$-linear; $\mathfrak{Z}_1, \ldots, \mathfrak{Z}_l$ have order $\geq 2m + 1$. After inverting the matrix A, we obtain l equations $(\alpha_i Id + \tilde{D}_i)(U) = \tilde{Z}_i + \tilde{\mathfrak{Z}}_i$, $i = 1, \ldots, l$. The $\tilde{D}_i$'s (resp. $\tilde{Z}_i, \tilde{\mathfrak{Z}}_i$) are still $\mathcal{O}_n^S$-linear operators and they are linear combination of the D_i's (resp. Z_i, $\mathfrak{Z}_i$) with coefficients in $\mathcal{O}_n^S$. Let us set $\|\alpha\| = \max_{1 \leq j \leq l} |\alpha_j|$ and let i be such that $|\alpha_i| = \|\alpha\| \neq 0$; it is the **"worst small divisor"** of the family.

Let us look through the ith equation; we find out that, at least formally, its solution U is given by

$$U = \frac{1}{\alpha_i} \sum_{k \geq 0} \left(\frac{-1}{\alpha_i} \right)^k \tilde{D}_i^k (\tilde{Z}_i + \tilde{\mathfrak{Z}}_i).$$

This expression does not fancy us since it involves *a priori* infinitely large powers of α_i which can be very small. Thus, instead of using this expression, we shall split the ith equation in an appropriate way. First of all, we shall split the linear diagonal family S in two parts S' and S'' corresponding to the splitting of $\mathbb{C}^n$ as $\mathbb{C}^{n'} \times \mathbb{C}^{n-n'}$; that is, for all $1 \leq i \leq l$,

$$S_i = \underbrace{\sum_{k=1}^{n'} \lambda_{i,k} x_k \frac{\partial}{\partial x_k}}_{S_i'} + \underbrace{\sum_{k=n'+1}^{n} \lambda_{i,k} x_k \frac{\partial}{\partial x_k}}_{S_i''}.$$

The integer n' is such that the linear forms $\{\sum_{i=1}^l \lambda_{i,k} z_i\}_{1 \leq k \leq n'}$ all belong to a real hyperplane H of $\mathrm{Hom}_{\mathbb{C}}(\mathbb{C}^l, \mathbb{C})$ whereas all the linear forms $\{\sum_{i=1}^l \lambda_{i,k} z_i\}_{n'+1 \leq k \leq n}$ all belong (strictly) to one and the same side of H. The integer n' is taken to be the lowest as possible; it may be equal to 0 as well as equal to n. We shall call this splitting, the **analytic splitting** of S. It has been chosen in such a way that the

small divisors as well as the **first integrals** are only due to S'. We show that there is a **separating constant** $Sep(S) > 0$ such that if α is a weight of S which norm is $< Sep(S)$ then it must belong to H (if $n' = n$ we shall set $Sep(S) = +\infty$ in order to have one proof for the theorems). Let X be a vector field of $\mathbb{C}^n$, we shall denote by X' (resp. X'') its projection onto $\frac{\partial}{\partial x_1}, \ldots, \frac{\partial}{\partial x_{n'}}$ (resp. $\frac{\partial}{\partial x_{n'+1}}, \ldots, \frac{\partial}{\partial x_n}$). This being said, let us go back to the study of our equation $(\alpha_i Id + \tilde{D}_i)(U) = \tilde{Z}_i + \tilde{\mathfrak{Z}}_i$. Using the analytic splitting of S as well as the structure of the operator $\tilde{D}_i$, we show that this equation can be written under the following form :

$$U' - \frac{1}{\alpha_i}(P_i(U'))' = \frac{1}{\alpha_i}(\tilde{Z}'_i + \tilde{\mathfrak{Z}}'_i + (Q_i(U'))') \tag{11}$$

$$U'' - \frac{1}{\alpha_i}(Q_i(U''))'' = \frac{1}{\alpha_i}(\tilde{Z}''_i + \tilde{\mathfrak{Z}}''_i + (P_i(U'))'' + (Q_i(U'))''); \tag{12}$$

both P_i and Q_i are $\mathcal{O}_n^S$-linear operators. Let us assume that the weight α is of small norm, that is $< Sep(S)$. Then, we show that $(Q_i(U'))' = 0$ and that, according to the complete integrability assumption, $P'_i \circ P'_i = 0$. Therefore the solution of (11) is given by

$$U' = \left(Id + \frac{P'_i}{\alpha_i}\right)\left(\frac{\tilde{Z}'_i + \tilde{\mathfrak{Z}}'_i}{\alpha_i}\right).$$

Since U' is a polynomial of order $\leq 2m$, then in fact, we have

$$|U'|_r \leq \left|\left(Id + \frac{P'_i}{\alpha_i}\right)\left(\frac{\tilde{Z}'_i}{\alpha_i}\right)\right|_r.$$

An estimate of the operator P'_i will provide the desired estimate of U'. Now, let us study equation (12). Let us denote by $\frac{1}{\alpha_i}\mathfrak{w}_i$ the left handside of this equation. Then, at least formally, we have

$$U'' = \sum_{k \geq 0}\left(\frac{1}{\alpha_i}\right)^k Q_i^k\left(\frac{\mathfrak{w}_i}{\alpha_i}\right).$$

By assumption, NF^m is the m-jet of completely integrable normal form. Therefore, its projection $(NF^m)''$ is the m-jet of a **good deformation** of S''. The point is that **there exists an integer k_0 which do not depend on m and such that** $J^{2m}(Q_i^k(\frac{\mathfrak{w}_i}{\alpha_i})) = 0$ **for all** $k \geq k_0$. The important consequence for the estimates is that the sum above which give U'' is **finite**. Using the estimate of U' which were found above, we can give estimate for $\mathfrak{w}_i$; then using estimate of Q_i, we conclude with estimate of U''. The last case deals with weight α such that $\|\alpha\| \geq Sep(S)$; it it the easiest case.

Now let us give an idea of the induction argument. Let $1/2 \leq r < 1$ and let assume the family of the $X_i = NF_i^m + R_{i,m+1}$'s is normalized up to order $m = 2^k$. Let us assume that the norms $|NF^m - S|_r$ and $|D(NF^m - S)|_r$ are small enough, say $< \eta_1$, and that $|R_i|_r < 1$. The solution of the cohomological equation allows us to normalize the family up to order $2m$: $(\Phi_m)_* X_i = NF_i^{2m} + R_{i,2m+1}$. Using the estimate of this solution, we show that $|NF^{2m} - S|_R$ and $|D(NF^{2m} - S)|_R$ are **still less** than η_1 where $R = \left(\frac{c(\eta_1)}{\omega_{k+1}^d}\right)^{-1/m} m^{-2/m} r < r$ and that $|R_{i,2m+1}|_R < 1$. After a preliminary renormalization, we show that, at each stage, our new objects still satisfy the required assumptions in order to have again the estimate for the solution of the new cohomological equation. Thus, we may proceed again ... Now, because of the diophantine condition, these R are bounded from below by some positive constant Rad. Therefore, at the limit, we have found an holomorphic diffeomorphism in the polydisc of radius Rad centered at $0 \in \mathbb{C}^n$ which normalizes our nonlinear deformation X.

3 Proof of main Theorem 2.3.6

3.1 Bounds for the cohomological equations

Let α be a nonzero weight of S in $\mathscr{P}_n^{m+1,2m}$ and let $\mathscr{P}_{n,\alpha}^{m+1,2m}$ be the associated weight space. As we have seen in proposition 2.5.1, for all $Z \in Z_{N,m}^1(\mathbb{C}^l, \mathscr{P}_{n,\alpha}^{m+1,2m})$, there exists a unique $U \in \mathscr{P}_{n,\alpha}^{m+1,2m}$ such that, for all integer $1 \leq i \leq l$, $J^{2m}([NF_i^m, U]) = Z_i$.

The remaining of this subsection is devoted to the determination of a bound of the norm of this solution under some assumptions. Moreover, we assume that NF^m **is the m-jet of the normal form of a completely integrable deformation of S.** More precisely, we shall prove the

Theorem 3.1.1 *Under the assumptions above, there exists constants $\eta_1 > 0$ and $c_1(\eta_1) > 0$ such that, if $1/2 < r \leq 1$, $m = 2^k$ and $\max(|NF^m - S|_r, |D(NF^m - S)|_r) < \eta_1$, then for any nonzero weight α of S in $\mathscr{P}_n^{m+1,2m}$, for any $Z \in Z_m^1(\mathbb{C}^l, \mathscr{P}_{n,\alpha}^{m+1,2m})$, the unique $U \in \mathscr{P}_{n,\alpha}^{m+1,2m}$ such that $d_{N,m}^0 U = Z$ satisfies the following inequality:*

$$|U|_r \leq \frac{c_1(\eta_1)}{\omega_{k+1}(S)^2} |Z|_r; \tag{1}$$

and d depends only on S.

Proof. The cohomological equation can be written:

$$[NF_i^m, U] = Z_i + \mathfrak{Z}_i, \quad i = 1, \ldots, l.$$

where we have set, for all integers $1 \leq i \leq l$, $\mathfrak{Z}_i := [NF_i^m, U] - J^{2m}([NF_i^m, U])$. By assumptions, we have for all $1 \leq i \leq l$, $NF_i^m = \sum_{j=1}^l a_{i,j}^{m-1} S_j$ where $a_{i,j} \in \mathcal{O}_n^S$ are polynomials of degree $\leq m - 1$ and $a_{i,j}^{m-1}(0) = \delta_{i,j}$. Therefore, we have

$$[NF_i^m, U] = \sum_{j=1}^{l} \left(a_{i,j}^{m-1}[S_j, U] - U(a_{i,j}^{m-1})S_j \right) = Z_i + \mathfrak{Z}_i, \quad i = 1,\dots l \qquad (2)$$

where $U(a_{i,j}^{m-1})$ denotes the Lie derivative of $a_{i,j}^{m-1}$ along U. Let us choose an index $1 \le i \le l$ such that $|\alpha(g_i)| = \|\alpha\| \neq 0$. We recall that U belongs to the α-weightspace of S in $\mathscr{P}_n^{m+1,2m}$; that is, for all $1 \le i \le l$, $[S_i, U] = \alpha_i U$. Therefore, equations (2) can be written into the following matricial form

$$A(x) \begin{pmatrix} [S_1, U] \\ \vdots \\ [S_l, U] \end{pmatrix} + \begin{pmatrix} D_1(U) \\ \vdots \\ D_l(U) \end{pmatrix} = \begin{pmatrix} Z_1 + \mathfrak{Z}_1 \\ \vdots \\ Z_l + \mathfrak{Z}_l \end{pmatrix}$$

where $A = (a_{p,q}^{m-1})_{1 \le p,q \le l}$ and D_i is the $\widehat{\mathscr{O}}_n$-linear map defined by $D_i : U \in \widehat{\mathscr{X}}_n^2 \mapsto -\sum_{j=1}^{l} U(a_{i,j}^{m-1})S_j \in \widehat{\mathscr{X}}_n^2$. Since $A(0) = \mathrm{Id}$, $A(x)$ is formally invertible : if $\tilde{A}^t := (c_{i,j})_{1 \le i,j \le l}$ denotes the transpose of the cofactors matrix of A, then $\hat{A}^{-1} := \frac{1}{\det A}\tilde{A}^t := (b_{i,j})_{1 \le i,j \le l}$ is a matrix which coefficient belong to $\widehat{\mathscr{O}}_n^S$ and satisfy to $\hat{A}^{-1}A = A\hat{A}^{-1} = Id$. It follows that

$$\begin{pmatrix} \alpha_1 U \\ \vdots \\ \alpha_l, U \end{pmatrix} + \begin{pmatrix} \tilde{D}_1(U) \\ \vdots \\ \tilde{D}_l(U) \end{pmatrix} = \hat{A}^{-1} \begin{pmatrix} Z_1 + \mathfrak{Z}_1 \\ \vdots \\ Z_l + \mathfrak{Z}_l \end{pmatrix} \quad \text{where } \tilde{D}_j(U) = \sum_{k=1}^{l} b_{j,k}D_k(U).$$

Here is a key point : equation (2) is overdetermined. To estimate its solution, we select the equation that give the smallest norm a priori. It is the one that correspond to the "biggest" small divisor among the family, that is α_i.

Thus, the ith equation of the cohomological equation can be written

$$U - P_i(U) = \tilde{Z}_i + \tilde{\mathfrak{Z}}_i \qquad (3)$$

Here, we have written

$$\alpha_i \tilde{Z}_i = \sum_{k=1}^{l} b_{i,k} Z_k, \quad \alpha_i \tilde{\mathfrak{Z}}_i = \sum_{k=1}^{l} b_{i,k} \mathfrak{Z}_k,$$

$$P_i(U) = \frac{1}{\alpha_i} \sum_{k=1}^{l} b_{i,k} \sum_{j=1}^{l} U(a_{k,j}^{m-1})S_j.$$

We claim that the operator P_i satisfies to $P_i \circ P_i = 0$. This is due to the fact that $S_q(a_{k,p}^{m-1}) = 0$ since the $a_{k,p}^{m-1}$'s are invariants of S. Hence, we have $(Id - P_i) \circ (Id + P_i) = Id$.

As a consequence, we have

$$U = (Id + P_i)(\tilde{Z}_i + \tilde{\mathfrak{Z}}_i). \qquad (4)$$

Let us give bounds for the operators P_i. To do so, we shall write $A(x) = Id + R(x)$ where $R(0) = 0$; we shall write $R(x) = (r_{i,j}(x))_{1 \leq i,j \leq l}$. Recalling the expression of the determinant, we have $\det(A(x)) = 1 + P(R(x))$ where $P(Z) \in \mathbb{C}[Z_1, \ldots, Z_{l^2}]$ is a polynomial functions of l^2 variables without constant term and of degree l. Since, it vanishes at the origin, there exists $\eta > 0$ such that $|P(Z)|_\eta < 1/2$. It follows that, if $|R(x)|_r = |A(x) - A(0)|_r < \eta$, then $|P(R(x))|_r < 1/2$. By Lemma 1.1.2, if $|A(x) - A(0)|_r < \eta$, then we have

$$\left| \frac{1}{\det A(x)} \right|_r \leq \frac{1}{1 - |P(r_{i,j}(x))|_r} \leq 2$$

$$|\det A(x)|_r \leq 1 + |P(R(x))|_r \leq \frac{3}{2}.$$

We recall that $(c_{i,j})_{1 \leq i,j \leq l} = \tilde{A}^t$ is the transpose of the cofactors matrix of A. Thus, there are universal polynomials of degree $\leq l - 1$, $Q_{i,j}(Z) = \sum_{\substack{S \in \mathbb{N}^{l^2} \\ 1 \leq |S| \leq l-1}} q_{i,j,S} Z^S \in$ $\mathbb{C}[Z_1, \ldots, Z_{l^2}]$ such that $c_{i,j}(x) = Q_{i,j}(A(x))$. It follows that, for all $1 \leq i, j \leq l$, $|c_{i,j}(x)|_r \leq Q(|A|_r)$ where Q is the universal polynomial of one variable defined by

$$Q(t) = \sum_{\substack{S \in \mathbb{N}^{l^2} \\ 1 \leq |S| \leq l-1}} \max_{i,j} |q_{i,j,S}| t^{|S|}.$$

As a consequence, if $|A(x) - A(0)|_r < \eta$, we have

$$|b_{i,j}|_r = \left| \frac{c_{i,j}}{\det A} \right|_r \leq \left| \frac{1}{\det A} \right|_r |c_{i,j}|_r \leq 2Q(|A|_r). \tag{5}$$

In our result, the assumption are about $|NF_i^m - S_i|_r$ and $|D(NF_i^m - S_i)|_r$ and not on the matrix A. So we have to link both estimates. By definition, we have, for all integers $1 \leq i, j \leq l$

$$S_j = \sum_{k=1}^n \lambda_{j,k} x_k \frac{\partial}{\partial x_k}$$

$$NF_i^m = \sum_{j=1}^l a_{i,j}^{k-1} S_j := \sum_{k=1}^n x_k g_{i,k} \frac{\partial}{\partial x_k} \quad \text{with} \quad g_{i,k} = \left(\sum_{j=1}^l \lambda_{j,k} a_{i,j}^{k-1} \right),$$

that is

$$\begin{pmatrix} g_{i,1} \\ \vdots \\ \vdots \\ g_{i,n} \end{pmatrix} = \begin{pmatrix} \lambda_{1,1} & \cdots & \lambda_{l,1} \\ \vdots & & \vdots \\ \vdots & & \vdots \\ \lambda_{1,n} & \cdots & \lambda_{l,n} \end{pmatrix} \begin{pmatrix} a_{i,1}^{k-1} \\ \vdots \\ a_{i,l}^{k-1} \end{pmatrix}.$$

By assumptions, the S_i's ($1 \le i \le l$) are linearly independents over $\mathbb{C}$, the matrix $(\lambda_{j,i})_{\substack{1 \le i \le n \\ 1 \le j \le l}}$ has rank l. Without lost of generality, we may assume that the matrix $L := (\lambda_{j,i})_{1 \le i,j \le l}$ is invertible with inverse $L^{-1} := (\tilde{\lambda}_{i,j})_{1 \le i,j \le r}$.

Let us set $A = (a_{i,j}^{m-1})_{1 \le i,j \le l}$ and $|A|_r = \max_{i,j} |a_{i,j}^{m-1}|_r$, $|D(A)|_r = \max_{i,j,k} \left| \frac{\partial a_{i,j}^{m-1}}{\partial x_k} \right|_r$. Let We have the following

Lemma 3.1.2

$$|A - A(0)|_r \le 2l|L^{-1}||NF^m - S|_r \tag{6}$$

$$|D(A)|_r \le 2l|L^{-1}||D(NF^m - S)|_r \tag{7}$$

We refer to [Sto00][p. 185–186] for a proof.

Let us set $\eta_1 = \eta/(2l|L^{-1}|)$. If $|NF^m - S|_r < \eta_1$, then by (6), we have $|A(x) - A(0)|_r < \eta$ so that $|b_{i,j}|_r \le 2Q(|A|_r)$ by (5). Moreover, we have $|A|_r \le |A(0)| + |A - A(0)|_r \le 1 + \eta$. It follows that $Q(|A|_r) < Q(1 + 2l|L^{-1}|\eta_1)$.

On the other hand, if $|D(NF^m - S)|_r < \eta_1$ then $|U(a_{k,j}^{k-1})|_r \le n|U|_r|D(A)|_r \le n\eta|U|_r$.

We recall (see the section of notations) that, given two elements $Y = (Y_1, \ldots, Y_q)$ and $W = (W_1, \ldots, W_q)$ of $\widehat{\mathcal{O}}_n^q$, we say that Y is dominated by W, and we write $Y \prec W$, if $Y_i \prec W_i$ for all $1 \le i \le q$. Moreover, we shall write $\bar{Y} := (\bar{Y}_1, \ldots, \bar{Y}_q)$. Now, we are able to give estimates for P_i. Since we have

$$P_i(U) = \frac{1}{\alpha_i} \sum_{k=1}^{l} b_{i,k} \sum_{j=1}^{l} U(a_{k,j}^{k-1}) S_j,$$

we obtain

$$P_i(U) \prec \frac{1}{\|\alpha\|} \sum_{k=1}^{l} \bar{b}_{i,k} \sum_{j=1}^{l} \bar{U}(\overline{\bar{a}_{k,j}^{k-1}}) \bar{S}_j.$$

Here, $\bar{S}_j$ stands for $\sum_{k=1}^{n} |\lambda_{j,k}| x_k \frac{\partial}{\partial x_k}$. It follows that if $1/2 < r \le 1$, $\max(|NF^m - S|_r, |D(NF^m - S)|_r) < \eta_1$, then

$$|P_i(U)|_r \le \frac{c_2(\eta_1)}{\|\alpha\|} |U|_r \tag{8}$$

with

$$c_2(\eta_1) = 4l^3 Q(|A(0)| + 2l|L^{-1}|\eta_1) n\lambda |L^{-1}|\eta_1.$$

Here we have set $\lambda = \max_{1 \le i \le l, 1 \le j \le n} |\lambda_{i,j}|$, so that, since $r \le 1$, $|S_j|_r \le \lambda$ for all $1 \le j \le l$.

Let us give the estimate of the solution of the cohomological equation (4). Since U is a polynomial vector field of degree $\le 2m$ then $U \prec (Id + \bar{P}_i)(\bar{Z}_i)$. Hence,

$$|U|_r \le \left| (Id + \bar{P}_i)(\bar{Z}_i) \right|_r$$

$$\leq \left(1 + \frac{c_2(\eta_1)}{\|\alpha\|}\right)|\check{Z}_i|_r$$

$$\leq \left(1 + \frac{c_2(\eta_1)}{\|\alpha\|}\right)\frac{2lQ(1 + 2l|L^{-1}|\eta_1)}{\|\alpha\|}|Z|_r.$$

we are done. $\square$

3.2 Iteration scheme

Let $1/2 < r \leq 1$ be a real number and $\eta_1 > 0$ be the positive number defined in Theorem 3.1.1. For any integer $m \geq [8n/\eta_1] + 1$, let us set

$$\mathscr{N}\mathscr{F}_m(r) = \left\{X \in (\mathscr{P}_n^{1,m})^l \mid \max(|X_i - S_i|_r, |D(X_i - S_i)|_r) < \eta_1 - \frac{8n}{m}\right\}$$

$$\mathscr{B}_{m+1}(r) = \left\{X \in (\mathscr{X}_n^{m+1})^l \mid |X_i|_r < 1\right\}.$$

If $m = 2^k$, for some integer $k \geq 1$, let us define

$$\rho = m^{-1/m}r \quad \text{and} \quad R = \gamma_k m^{-2/m}r \quad \text{where } \gamma_k = \left(\frac{c_1}{\omega_{k+1}^2(S)}\right)^{-1/m}.$$

The core of this section is the following proposition :

Proposition 3.2.1 *With the notations above, let us assume that*
$(NF^m, R_{m+1}) \in \mathscr{N}\mathscr{F}_m(r) \times \mathscr{B}_{m+1}(r)$. *If m is sufficiently large (say $m > m_0$ indepen-dent of r), then there exists a unique $U \in \mathscr{P}_{n,*}^{m+1,2m}$, the sum of nonzero weightspaces of S, such that*

1. $\Phi := (Id + U)^{-1} \in Diff_1(\mathbb{C}^n, 0)$ *is a diffeomorphism such that $D_R \subset \Phi(D_\rho)$*
2. $\Phi^*(\phi + \varepsilon) = NF^{2m} + R_{2m+1}$ *is normalized up to order $2m$*
3. $(NF^{2m}, R_{2m+1}) \in \mathscr{N}\mathscr{F}_{2m}(R) \times \mathscr{B}_{2m+1}(R)$

We have seen that Φ normalizes simultaneously each $X_i = NF_i^m(y) + B_i^{m+1,2m} + C_i(y)$ up to order $2m$. Let us write $\Phi_*X_i(y) = NF_i^{2m}(y) + C_i'(y)$ where both C_i and C_i' is of order $\geq 2m + 1$ whereas $B_i^{m+1,2m}$ is of degree $\leq 2m$ and of order $\geq m + 1$. Using the conjugacy equation and the fact that $NF^m(\Phi^{-1}(y)) - NF^m(y) = \int_0^1 D(NF^m)(y + tU(y))U(y)dt$, we have

$$C_i'(y) = \int_0^1 D(NF_i^m)(y + tU(y))U(y)dt + (B_i^{m+1,2m} + C_i)(\Phi^{-1}(y)) \qquad (9)$$

$$- (NF_i^{2m}(y) - NF_i^m(y)) - D(U)(y)(NF_i^{2m} + C_i')(y).$$

This is expression that we will use in order to prove the third point of the proposition.

3.2.1 Estimate for the diffeomorphism

By assumptions, $NF^m \in \mathcal{N}\mathcal{F}_m(r)$; thus, we can apply Proposition 3.1.1 so that

$$|U|_r \leq \frac{c_1}{\omega_{k+1}^2(S)} |B_*^{m+1,2m}|_r.$$

Since $B_{i,*}^{m+1,2m} \prec \bar{B}_{i,*}^{m+1,2m} + \bar{B}_{i,0}^{m+1,2m} \prec \bar{R}_{i,m+1}$, we have $|B_*^{m+1,2m}|_r < 1$. It follows that $|U|_r \leq \gamma_k^{-m}$.

Lemma 3.2.2 *Under the above hypothesis and if m is large enough (say $m > m_0$), then for all $0 < \theta \leq 1$ and all integer $1 \leq i \leq n$, we have $|y_i + \theta U_i(y)|_R < \rho$. As a consequence, $\Phi(D_\rho) \supset D_R$.*

Proof. We borrow the proof of Bruno [Bru72][p. 203]. It is sufficient to show that $R + |U|_R < \rho$. Since U is of order at least $m+1$ then, by (4) and the inequality above

$$|U|_R \leq \left(\frac{R}{r}\right)^{m+1} |U|_r \leq \left(\gamma_k m^{-2/m}\right)^{m+1} \gamma_k^{-m}$$
$$\leq \gamma_k m^{-2-2/m} \leq m^{-2-2/m}. \tag{10}$$

Since $R = \gamma_k m^{-2/m} r \leq m^{-2/m} r$, it is sufficient to show that $m^{-2/m}(r + m^{-2}) < \rho = m^{-1/m} r$, that is $\frac{m^{-2}}{m^{1/m}-1} < r$. But,

$$\frac{m^{-2}}{m^{1/m}-1} = \frac{m^{-2}}{\exp^{1/m\ln m}-1} \leq \frac{m^{-2}}{1/m\ln m} \leq \frac{1}{m\ln m}$$

since $1 + x \leq \exp x$ for all $x \in \mathbb{R}^+$. But, for $0 < x$ sufficiently large, we have $2 < x\ln x$. Thus, since $1/2 < r$, we obtain the result : $\frac{m^{-2}}{m^{1/m}-1} < \frac{1}{2} < r$.

3.3 Proof of the theorem

In this section, we shall prove our main result. Let $1/2 < r \leq 1$ be a positive number and let us consider the sequence $\{R_k\}_{k \geq 0}$ of positive numbers defined by induction as follow:

$$R_0 = r$$
$$R_{k+1} = \gamma_k m^{-2/m} R_k \quad \text{where} \quad m = 2^k$$

Lemma 3.3.1 *The sequence $\{R_k\}_{k \geq 0}$ converges and there exists an integer m_1 such that for all integer $k > m_1$, $R_k > \frac{R_{m_1}}{2}$.*

Proof. We recall that $\gamma_k = \left(\dfrac{c_1}{\omega_{k+1}^2(S)} \right)^{-1/2^k}$. We have $R_{k+1} = r \prod_{i=1}^k \gamma_i(2^i)^{-2^{1-i}}$, thus

$$\ln R_{k+1} = -2 \sum_{i=1}^k \frac{\ln \omega_{i+1}(S)}{2^i} - \ln c_1 \sum_{i=1}^k \frac{1}{2^i} - 2\ln 2 \sum_{i=1}^k \frac{i}{2^i}$$

the last two sums are convergent and the first is also convergent by assumption. It follows that there exists an integer m_1, such that

$$\prod_{i=m_1+1}^{+\infty} \gamma_i(2^i)^{-2^{1-i}} > 1/2.$$

Thus, if $k > m_1$ then $R_k = R_{m_1} \prod_{i=m_1+1}^k \gamma_i(2^i)^{-2^{1-i}} > \frac{R_{m_1}}{2}$.

Let $X = \{X_i = S_i + \cdots \}_{i=1,\ldots,l}$ be a family of commuting holomorphic vector fields in a neighborhood of the origin in $\mathbb{C}^n$. We may assume that it is holomorphic in a neighborhood of the closed polydisc D_1. Let $m_2 = 2^{k_0} \geq \max(m_0, 2^{m_1})$ where m_0 is the integer defined in Proposition 3.2.1. By a polynomial change of coordinates, we can normalize X up to order m_2 : in these coordinates, X_i can be written $NF_i^{m_2} + R_{i,m_2+1}$. If necessary, we may apply a diffeomorphism $a\mathrm{Id}$ with $a \in \mathbb{C}^*$ sufficiently small so that $(NF^{m_2}, R_{m_2+1}) \in \mathscr{N}\mathscr{F}_{m_2}(1) \times \mathscr{B}_{m_2+1}(1)$. We may define as above the sequence $\{R_k\}_{k \geq k_0}$, with $R_{k_0} = 1$. Thus, for all integer $k > k_0$, we have $1/2 < R_k \leq 1$.

Let us prove by induction on $k \geq k_0$, that there exists a diffeomorphism Ψ_k of $(\mathbb{C}^n, 0)$ such that $\Psi_k^*(NF_i^{m_2} + R_{i,m_2+1}) := NF_i^{2^{k+1}} + R_{i,2^{k+1}+1}$ is normalized up to order 2^{k+1}, $(NF^{2^{k+1}}, R_{2^{k+1}+1}) \in \mathscr{N}\mathscr{F}_{2^{k+1}}(R_{k+1}) \times \mathscr{B}_{2^{k+1}+1}(R_{k+1})$ and

$$|\mathrm{Id} - \Psi_k^{-1}|_{R_{k+1}} \leq \sum_{p=k_0}^k \frac{1}{2^{2p}}.$$

- For $k = k_0$: According to Proposition 3.2.1, there exists a diffeomorphism Φ_{k_0} such that $\Phi_{k_0}^*(NF^{2^{k_0}} + R_{2^{k_0}+1}) = NF^{2^{k_0+1}} + R_{2^{k_0+1}+1}$ is normalized up to order 2^{k_0+1}, $(NF^{2^{k_0+1}}, R_{2^{k_0+1}+1}) \in \mathscr{N}\mathscr{F}_{2^{k_0+1}}(R_{k_0+1}) \times \mathscr{B}_{2^{k_0+1}+1}(R_{k_0+1})$. The estimate $|\mathrm{Id} - \Phi_{k_0}^{-1}|_{R_{k_0+1}} = |U|_{R_{k_0+1}} < 1/2^{2k_0}$ comes from estimate (10)
- Let us assume that the result holds for all integers $i \leq k-1$: by assumptions, $\Psi_{k-1}^*(NF^{m_2} + R_{m_2+1}) = NF^{2^k} + R_{2^k+1}$ is normalized up to order 2^k and $(NF^{2^k}, R_{2^k+1}) \in \mathscr{N}\mathscr{F}_{2^k}(R_k) \times \mathscr{B}_{2^k+1}(R_k)$. Since $1/2 < R_k \leq 1$, we may apply proposition 3.2.1 : there exists a diffeomorphism Φ_k such that $(\Phi_k \circ \Psi_{k-1})^*(NF^{2^{k_0}} + R_{2^{k_0}+1}) = NF^{2^{k+1}} + R_{2^{k+1}+1}$ is normalized up to order 2^{k+1} and $(NF^{2^{k+1}}, R_{2^{k+1}+1}) \in \mathscr{N}\mathscr{F}_{2^{k+1}}(R_{k+1}) \times \mathscr{B}_{2^{k+1}+1}(R_{k+1})$. Let us set $\Psi_k = \Phi_k \circ \Psi_{k-1}$. According to estimate (10), we have $|\mathrm{Id} - \Phi_k^{-1}|_{R_{k+1}} < 1/2^{2k}$. It follows that

$$\left| \mathrm{Id} - \Psi_k^{-1} \right|_{R_{k+1}} \leq \left| (\mathrm{Id} - \Psi_{k-1}^{-1}) \circ \Phi_k^{-1} + (\mathrm{Id} - \Phi_k^{-1}) \right|_{R_{k+1}},$$

$$\leq \left| (\mathrm{Id} - \Psi_{k-1}^{-1}) \circ \Phi_k^{-1} \right|_{R_{k+1}} + \left| (\mathrm{Id} - \Phi_k^{-1}) \right|_{R_{k+1}}.$$

According to proposition $(3.2.1)$, we have $\Phi_k^{-1}(D_{R_{k+1}}) \subset D_{R_k}$. It follows that

$$\left| \mathrm{Id} - \Psi_k^{-1} \right|_{R_{k+1}} \leq \left| (\mathrm{Id} - \Psi_{k-1}^{-1}) \right|_{R_k} + \left| (\mathrm{Id} - \Phi_k^{-1}) \right|_{R_{k+1}}$$

$$\leq \sum_{p=k_0}^{k-1} \frac{1}{2^{2p}} + \frac{1}{2^{2k}};$$

this ends the proof of the induction.

Since $D(1/2) \subset D_{R_k}$ for all integers $k \geq k_0$, then the sequence $\{|\Psi_k^{-1}|_{1/2}\}_{k \geq k_0}$ is uniformally bounded. Moreover, the sequence $\{\Psi_k^{-1}\}_{k \geq k_0}$ converges coefficientwise to a formal diffeomorphism $\hat{\Psi}^{-1}$ (the inverse of the formal normalizing diffeomorphism). Therefore, this sequence converges in $\mathcal{H}_n^n(r)$ (for all $r < 1/2$) to $\hat{\Psi}^{-1}$ (see [GR71]). This means that the normalizing transformation is holomorphic in a neighborhood of $0 \in \mathbb{C}^n$.

4 Miscellaneous results

4.1 Normal forms again

The following theorem is due to Nguyen Tien Zung. It describes a situation of "fully" completely integrable systems : there is "no room" left for small divisors and there are the maximum number of holomorphic first integrals.

Theorem 4.1.1 *[Zun02] Let $X_1 = S + R$ be an holomorphic nonlinear perturbation of a semi-simple linear part S in a neighborhood of the origin in $\mathbb{C}^n$. Assume it belongs to a commutative family a holomorphic vector fields $\{X_1, \dots, X_m\}$ which are assumed to be linearly independent almost everywhere : $X_1 \wedge \cdots \wedge X_l \neq 0$ almost everywhere. Assume that this family has $n - m$ holomorphic common first integrals $\{f_1, \dots, f_{n-m}\}$ which are almost everywhere independent : $df_1 \wedge \cdots \wedge df_{n-m} \neq 0$ almost everywhere. Then, there exists an holomorphic change of coordinates which normalize X_1.*

For the Hamiltonian version of this result see [Zun05]. This result is very related to the following one. As above, let us consider the family S of linear vector fields $S_i = \sum_{j=1}^n \lambda_{i,j} x_j \frac{\partial}{\partial x_j}$, $1 \leq i \leq l$. Let us set $\lambda_i := (\lambda_{i,1}, \dots, \lambda_{i,n})$. Let $r > 0$. We shall say that S satisfies condition (A_r) if $\inf\{|(Q, \lambda^i)| \neq 0, 1 \leq i \leq l\} \geq r^{-|Q|}$.

Proposition 4.1.2 *[Sto97] Let $X_i = \sum_{j=1}^l a_{i,j}(x)S_j$, $1 \leq i \leq l$ be germs of holomorphic vector fields in a neighbourhood of $0 \in \mathbb{C}^n$, commuting pairwise. We assume* *that the matrix $A = (a_{ij})$ of elements $a_{i,j} \in \mathcal{O}_n$ is invertible in a neighbourhood of $0 \in \mathbb{C}^n$. If condition (A_r) is satisfied for some $r > 0$, then the vector fields X_i's are simultaneously and holomorphically normalizable.*

It is very likely that Zung's theorem (at least in some cases) could be obtained in the following way: First of all, apply Artin's theorem in order to transform, by an holomorphic change of coordinates, the first integrals into different monomials generating the first integrals of the S_i's. Then, we are in situation of applying the proposition unless the matrix A is not invertible. In this last case, technics of next result could be applied.

Definition 4.1.3 *We shall say that $s \in span(S_1,\dots,S_l)$ is regular relatively to S whenever*

$$\left(\widehat{\mathscr{X}^1_n} \right)^S = \left(\widehat{\mathscr{X}^1_n} \right)^s.$$

Theorem 4.1.4 *[Sto05b] Let S be a diophantine family of linearly independent linear diagonal vector fields. Let X_1 be a germ of holomorphic vector field with a regular linear part s at the origin. Let $X_2,\dots X_l$ be germs of holomorphic vector fields in a neighborhood of 0 and commuting with X_1. Assume that the family $\{X_1,\dots,X_l\}$ has a normal form (relatively to s) of the type*

$$\hat{\Phi}_* X_i = \sum_{i=1}^{l} \hat{a}_{i,j} S_j$$

where the $\hat{a}_{i,j}$'s belong to $\widehat{\mathcal{O}_n^S}$. Moreover, we assume that the family of the parts of lowest degree of this normal form is free over $\widehat{\mathcal{O}_n^S}$. Then, the family is holomorphically normalizable, i.e. there exists an holomorphic diffeomorphism in a neighborhood of $0 \in \mathbb{C}^n$ transforming the family into a normal form.

One of H. Ito results about normal forms of family of n Hamiltonian vector fields [Ito89] is a particular case the previous results.

Remark 4.1.5 *Contrary to Theorem 2.3.6,* **only X_l is required to have a nonvanishing linear part** *s at the origin. The linear part of the X_i's, $i \geq 2$, could be zero. Moreover, s is* **not supposed to satisfy any diophantine condition**.

4.2 KAM theory

Let us go back to Theorem 2.3.6 and Figure 1 which refer to completely integrable systems. A natural question one may ask is the following: starting from a holomorphic singular completely integrable system in a neighborhood of the origin of $\mathbb{C}^n$ (a common fixed point), we consider a holomorphic perturbation (in some sense) of one its vector fields. Does this perturbation still have invariant varieties in some neighborhood of the origin? Are these varieties biholomorphic to resonant (toric) varieties? To which vector field on a resonant variety does the biholomorphism conjugate the restriction of the perturbation to an invariant variety? Is there a "big set" of surviving invariant varieties?

The aim of article [Sto05a] is to answer these questions which are classical in the Hamiltonian nonsingular setting [Arn63, Arn88b, Bos86, Kol57, Ste69].

4.3 Poisson structures

A **Poisson structure** is defined by a bracket $\{.,.\}$ which is applied to a couple of germs of functions to a germ of function and which satisfies the following properties :

- $\{.,.\}$ est bilinear and skew-symmetric
- $\{f,\{g,h\}\} + \{g,\{h,f\}\} + \{h,\{f,g\}\} = 0$ (Jacobi identity)
- $\{f,gh\} = \{f,g\}h + \{f,h\}g$ (Leibniz identity)

It is equivalent to define a germ of two-vectors field which can be written, in a local chart,

$$P = \frac{1}{2} \sum_{1 \leq i,j \leq N} P_{i,j}(x) \frac{\partial}{\partial x_i} \wedge \frac{\partial}{\partial x_j} = \sum_{1 \leq i < j \leq N} P_{i,j}(x) \frac{\partial}{\partial x_i} \wedge \frac{\partial}{\partial x_j}$$

with $P_{i,j} = -P_{j,i}$ and which satisfies Jacobi's identity

$$\sum_{1 \leq l \leq N} \left(P_{i,l} \frac{\partial P_{j,k}}{\partial x_l} + P_{j,l} \frac{\partial P_{k,i}}{\partial x_l} + P_{k,l} \frac{\partial P_{i,j}}{\partial x_l} \right) = 0$$

for $1 \leq i,j,k \leq N$. We want to study the singularities of analytic Poisson structures, that is points where all the $P_{i,j}$'s vanish. First of all, we are interested in singular point where the linear part of the Poisson structure is not reduced to zero. This linear part defines on the cotangent bundle to M at 0 a structure of Lie algebra. When it is semi-simple, then J. Conn has proved that the Poisson structure is analytically conjugate to its linear part in a neighborhood of the origin [Con84]. Most of the work concerns the linearization problem (and mostly in the smooth case). We refer the book [DZ05] for these aspects and [Arn97][Appendix 14] for dimension 2.

In article [Sto04] we study some analytic Poisson structures which are not even formally linearizable. Our main result is about the holomorphic normalization of some Poisson structures which associated Lie algebra is a skew-product $\mathbb{C}^p \ltimes \mathbb{C}^n$. The proof uses Theorem 2.3.6 in an essential and nontrivial manner.

Recently, P. Lohrmann studied analytic Poisson structures with a vanishing linear part but a nonvanishing quadratic part. J.-P. Dufour and A. Wade have defined a formal normal form for such an object [DW98]. P. Lohrmann proved the **counterpart of Brjuno Theorem 2.1.1 for these Poisson structure:** if the quadratic part satisfies to some diophantine condition and if the formal normal form is "completely integrable" then the Poisson structure is analytically conjugate to a normal form [Loh06, Loh05].

References

[AA88]   D. V. Anosov and V. I. Arnol'd, editors. *Dynamical systems. I*, volume 1 of *Encyclopaedia of Mathematical Sciences*. Springer-Verlag, 1988. Ordinary differential equations and smooth dynamical systems, Translated from the Russian.

[AGZV85]   V. I. Arnol'd, S. M. Gusein-Zade, and A. N. Varchenko. *Singularities of differentiable maps. Vol. I*, volume 82 of *Monographs in Mathematics*. Birkhäuser Boston Inc., 1985.

[AGZV88]   V. I. Arnol'd, S. M. Gusein-Zade, and A. N. Varchenko. *Singularities of differentiable maps. Vol. II*, volume 83 of *Monographs in Mathematics*. Birkhäuser Boston Inc., 1988.

[Arn63]   V.I. Arnold. Proof of a theorem by A.N. Kolmogorov on the persistence of quasiperiodic motions under small perturbations of the hamiltonian. *Russian Math. Surveys*, 18(5):9–36, 1963.

[Arn88a]   V. I. Arnol'd. *Geometrical methods in the theory of ordinary differential equations*, volume 250 of *Grundlehren der Mathematischen Wissenschaften*. Springer-Verlag, second edition, 1988.

[Arn88b]   V.I. Arnold, editor. *Dynamical systems III*, volume 28 of *Encyclopaedia of Mathematical Sciences*. Springer Verlag, 1988.

[Arn97]   V. I. Arnol'd. *Mathematical methods of classical mechanics*, volume 60 of *Graduate Texts in Mathematics*. Springer-Verlag, 1997. Corrected reprint of the second (1989) edition.

[Bal00]   Werner Balser. *Formal power series and linear systems of meromorphic ordinary differential equations*. Universitext. Springer-Verlag, New York, 2000.

[Bos86]   J.-B. Bost. Tores invariants des systèmes dynamiques hamiltoniens (d'après Kolomogorov, Arnol'd, Moser, Rüssmann, Zehnder, Herman, Pöschel,...). In *Séminaire Bourbaki*, volume 133-134 of *Astérisque*, pages 113–157. Société Mathématiques de France, 1986. exposé 639.

[Bru72]   A.D. Bruno. Analytical form of differential equations. *Trans. Mosc. Math. Soc*, 25,131-288(1971); 26,199-239(1972), 1971-1972.

[Bru95]   F. Bruhat. Travaux de Sternberg. In *Séminaire Bourbaki, Vol. 6*, pages Exp. No. 217, 179–196. Soc. Math. France, Paris, 1995.

[BS07]   Boele Braaksma and Laurent Stolovitch. Small divisors and large multipliers. *Ann. Inst. Fourier (Grenoble)*, 57(2):603–628, 2007.

[Con84]   J. Conn. Normal forms for analytic Poisson structures. *Ann. Math.*, 119:577–601, 1984.

[CS86]   R. Cushman and J. A. Sanders. Nilpotent normal forms and representation theory of $\mathrm{sl}(2,\mathbf{R})$. In *Multiparameter bifurcation theory (Arcata, Calif., 1985)*, volume 56 of *Contemp. Math.*, pages 31–51. Amer. Math. Soc., 1986.

[DW98]   J.-P. Dufour and A. Wade. Formes normales de structures de Poisson ayant un 1-jet nul en un point. *J. Geom. Phys.*, 26(1-2):79–96, 1998.

[DZ05]   J.-P. Dufour and Nguyen Tien Zung. *Poisson structures and their normal forms*, volume 242 of *Progress in Mathematics*. Birkhäuser Verlag, Basel, 2005.

[Eca]   J. Ecalle. Sur les fonctions résurgentes. I,II,III Publ. Math. d'Orsay.

[Eca92]   J. Ecalle. Singularités non abordables par la géométrie. *Ann. Inst. Fourier, Grenoble*, 42,1-2(1992),73-164, 1992.

[Fra95]   J.-P. Françoise. *Géométrie analytique et systèmes dynamiques*. Presses Universitaires de France, Paris, 1995.

[GR71]   H. Grauert and R. Remmert. *Analytische Stellenalgebren*. Springer-Verlag, 1971.

[GY99]   T. Gramchev and M. Yoshino. Rapidly convergent iteration method for simultaneous normal forms fo commuting maps. *Math. Z.*, 231:p. 745–770, 1999.

[Il'93]   Yu. S. Il'yashenko, editor. *Nonlinear Stokes phenomena*, volume 14 of *Advances in Soviet Mathematics*. American Mathematical Society, 1993.

[Ito89]   H. Ito. Convergence of birkhoff normal forms for integrable systems. *Comment. Math. Helv.*, 64:412–461, 1989.

[Kol57]   A.N. Kolmogorov. The general theory of dynamical systems and classical mechanics. In *Proceedings of International Congress of Mathematicians (Amsterdam, 1954)*, volume 1, pages 315–333. North-Holland, 1957. English translation in 'Collected Works", Kluwer.

[Loh05]   P. Lohrmann. Sur la normalisation holomorphe de structures de Poisson à 1-jet nul. *C. R. Math. Acad. Sci. Paris*, 340(11):823–826, 2005.

[Loh06]   P. Lohrmann. *Normalisation holomorphe et sectorielle de structures de Poisson.* PhD thesis, Université Paul Sabatier, Toulouse, France, 2006.

[Mal82]   B. Malgrange. Travaux d'Ecalle et de Martinet-Ramis sur les systèmes dynamiques. *Séminaire Bourbaki 1981-1982*, exp. 582, 1982.

[Mar80]   J. Martinet. Normalisation des champs de vecteurs holomorphes. *Séminaire Bourbaki*, 33(1980-1981),564, 1980.

[Mos90]   J. Moser. On commuting circle mappings and simultaneous diophantine approximations. *Math. Z.*, (205):p. 105–121, 1990.

[MR82]   J. Martinet and J.-P. Ramis. Problèmes de modules pour des équations différentielles non linéaires du premier ordre. *Publ. Math. I.H.E.S*, (55):63-164, 1982.

[MR83]   J. Martinet and J.-P. Ramis. Classification analytique des équations différentielles non linéaires résonantes du premier ordre. *Ann. Sci. E.N.S*, 4ème série,16,571-621, 1983.

[Ram93]   Jean-Pierre Ramis. Séries divergentes et théories asymptotiques. *Bull. Soc. Math. France*, 121(Panoramas et Syntheses, suppl.), 1993.

[RS93]   J.-P. Ramis and L. Stolovitch. Divergent series and holomorphic dynamical systems, 1993. Unpublished lecture notes from J.-P. Ramis lecture at the SMS *Bifurcations et orbites périodiques des champs de vecteurs*, Montréal 1992. 57p.

[Rüs67]   H. Rüssmann. Über die Normalform analytischer Hamiltonscher Differentialgleichungen in der Nähe einer Gleichgewichtslösung. *Math. Ann.*, (169):55–72, 1967.

[Sie42]   C.L. Siegel. Iterations of analytic functions. *Ann. Math.*, (43): 807-812, 1942.

[Ste58]   S. Sternberg. On the structure of local homeomorphisms of euclidean n-space. II. *Amer. J. Math.*, (80): 623–631, 1958.

[Ste69]   S. Sternberg. *Celestial Mechanics, Part II.* W.A Benjamin, 1969.

[Sto96]   L. Stolovitch. Classification analytique de champs de vecteurs 1-résonnants de $(\mathscr{B}C^n, 0)$. *Asymptotic Analysis*, (12): 91–143, 1996.

[Sto97]   L. Stolovitch. Forme normale de champs de vecteurs commutants. *C. R. Acad. Sci. Paris Sér. I Math.* (324): 665–668, 1997.

[Sto00]   L. Stolovitch. Singular complete integrabilty. *Publ. Math. I.H.E.S.*, (91): 133–210, 2000.

[Sto04]   L. Stolovitch. Sur les structures de poisson singulières. *Ergod. Th. & Synam. Sys., special volume in the memory of Michel Herman*, (24):1833–1863, 2004.

[Sto05a]   L. Stolovitch. A KAM phenomenon for singular holomorphic vector fields. *Publ. Math. Inst. Hautes Études Sci.*, (102):99–165, 2005.

[Sto05b]   L. Stolovitch. Normalisation holomorphe d'algèbres de type Cartan de champs de vecteurs holomorphes singuliers. *Ann. of Math.*, (161):589–612, 2005.

[Vey78]   J. Vey. Sur certains systèmes dynamiques séparables. *Am. Journal of Math. 100*, pages 591–614, 1978.

[Vey79]   J. Vey. Algèbres commutatives de champs de vecteurs isochores. *Bull. Soc. Math. France,107*, pages 423–432, 1979.

[Vor81]   S.M. Voronin. Analytic classification of germs of conformal mappings $(\mathscr{B}C, 0) \rightarrow (\mathscr{B}C, 0)$ with identity linear part. *Funct. An. and its Appl.*, 15, 1981.

[Zun02]   Nguyen Tien Zung. Convergence versus integrability in Poincaré-Dulac normal form. *Math. Res. Lett.*, 9(2-3):217–228, 2002.

[Zun05]   Nguyen Tien Zung. Convergence versus integrability in Birkhoff normal form. *Ann. of Math. (2)*, 161(1):141–156, 2005.

Geometric approaches to the problem of instability in Hamiltonian systems. An informal presentation

Amadeu Delshams[1], Marian Gidea[2], Rafael de la Llave[3], and Tere M. Seara[4]

Abstract We present (informally) some geometric structures that imply instability in Hamiltonian systems. We also present some finite calculations which can establish the presence of these structures in a given near integrable systems or in systems for which good numerical information is available. We also discuss some quantitative features of the diffusion mechanisms such as time of diffusion, Hausdorff dimension of diffusing orbits, etc.

1 Introduction

The goal of these lectures is to present an overview of some geometric programs to understand instability in Hamiltonian dynamical systems.

Roughly speaking, the problem of instability is to decide whether the effect of small time-dependent perturbations accumulates over time. Relatedly, to show that many orbits of a time-independent Hamiltonian system explore a large fraction of the energy surface.

Instability is a real problem arising in applications. For example, designers of accelerators or plasma confinement devices want to invent devices which are as devoid of instability as possible. Designers of space missions want to find orbits

[1] Departament de Matemàtica Aplicada I, Universitat Politècnica de Catalunya, Diagonal 647, 08028 Barcelona, Spain
e-mail: Amadeu.Delshams@upc.edu

[2] Department of Mathematics, Northeastern Illinois University, Chicago, IL 60625
e-mail: mgidea@neiu.edu

[3] Department of Mathematics, University of Texas, Austin, TX 78712
e-mail: llave@math.utexas.edu

[4] Departament de Matemàtica Aplicada I, Universitat Politècnica de Catalunya, Diagonal 647, 08028 Barcelona, Spain
e-mail: Tere.M-Seara@upc.edu

W. Craig (ed.), Hamiltonian Dynamical Systems and Applications, 285–336.

which can move freely over a wide region of space, but of course, they can only use the intertwining gravitational fields of the nearby celestial bodies. Chemists want to understand how reactions and reconfigurations of molecules take place. As is common with real problems, there are many mathematical formulations depending on the precise mathematical meaning attached to the vague words of the previous paragraph[1] and many techniques which have come to bear on these formulations. For example, the lectures of Professors Cheng, Neishtadt, and Treschev in this volume present other points of view about the problem and will even present different treatments of the same mathematical model.

These lectures can only aim to present informally the ideas behind some of the methods that have been proposed. We do not aim to present all the hypothesis of the results and much less complete proofs. Even when we restrict to geometric methods, we cannot aim to present a complete survey. The subject is progressing very fast. We only hope that these lectures can present an entry point to a portion the literature and indicate what to look for while reading some papers. We just want to present several milestones of the programs and to give some indication of the arguments.

There are two basic steps in all the results presented here. In a first step, we will present several geometric facts that imply that there are orbits that move appreciable lengths. In a second step, we will present some finite calculations which can verify the existence of these objects in quasi-integrable systems or in systems of a special form. Hence, for some systems, deciding that instability happens can be established with a finite computation. This will have the conclusion that some types of diffusion or instability are generic in some sense in some class of systems.

Remark 1.1. It should be emphasized that there are different geometric mechanisms that lead to instability. These mechanisms involve different geometric objects, have different hypothesis and lead to orbits with different characteristics. Several different mechanisms can coexist in the same model. The existence of several mechanisms was documented in some of the heuristic literature. An early paper, which is still worth reading is [LT83].

Remark 1.2. Given the practical importance of the problem of instability, there is a very large numerical and heuristic literature. Even if not easy to read, this literature contains considerable insights and can suggest several theorems. As representative papers of the numerical literature – which we cannot discuss in more detail – we mention [Chi79, Ten82, LT83, ZZN$^+$89, Zas02, GLF05, FGL05, FLG06].

Perhaps the main insight from the numerical literature, is that resonances organize the diffusion (the so called *Arnold web*). This, indeed was one important

---

[1] The previous paragraph contains several imprecise words such as *accumulate, many, explore, large,* etc. There are several rigorous formulations of these ideas. Some of the authors of this paper remember a round table in [Sim99] which included Profs. Arnol'd, Gallavotti, Galgani, Herman, Moser, Simó, Sinai. The panel was asked the question to give a canonical definition of *diffusion* that was preferable to the other definitions then in use. The conclusion was that it was better that each paper contains a precise definition.

The reader is encouraged to compare the precise definitions of diffusion or Arnol'd diffusion used in each paper. See Remark 1.3.

motivation for several of the investigations reported here. On the other hand, we will discuss some mechanisms which do not quite fit in this paradigm. See Sections 2, 7.

Remark 1.3. There are many precise mathematical formulations of what is meant by *diffusion* or *Arnol'd diffusion.* For some authors, the fact that there are whiskered tori as discussed in Section 2 is the key feature. We however take the presence large effects as the key feature. Many papers, for example [HM82] (which we will discuss more fully in Section 4) consider perturbations of size ε of an integrable system and establish existence of whiskered tori with heteroclinic intersections. These chains of whiskered tori, however are rather short and lead only to changes in the action variable of order $\varepsilon^{1/2}$. We, on the other hand, prefer to emphasize the existence of changes of order 1 in the actions, even if they are not accomplished through whiskered tori. A careful discussion of these issues appears in [Moe96].

1.1 Two types of geometric programs

With some simplification, there are two types of geometric programs that we will discuss.

Programs based on invariant objects and their relations

1. Find invariant objects (whiskered tori, normally hyperbolic invariant manifolds, periodic orbits, horseshoes, normally hyperbolic laminations, etc. as well as their stable and unstable manifolds).
2. Prove that if these objects satisfy some appropriate relations (e.g. there is a sequence of whiskered tori such that the unstable manifold of one torus intersects transversally the stable manifold of the next torus) then, there are orbits which move along the chain of invariant objects.

Incipient versions of programs of this type were already present in [Poi99]. The paper which has been more influential in the mathematical literature is [Arn64, Arn63]. The main invariant objects considered in [Arn64] were whiskered tori and their invariant manifolds. We will discuss this paper in some more detail in Section 2. Other early examples of instability were [Sit53, Ale68a, Ale68b, Ale69, Ale81], which were mainly based on hyperbolic and topological properties. The study of instability properties of oscillators was pioneered in [Lit66a, Lit66b, Lev92]. Other papers establishing instability in oscillators are [AO98, Ort97, LY97, Ort04] The papers [Pus77a, Pus77b, Dou89, KPT95, Pus95] study instability in systems with collisions. The papers [Dou88, DLC83] construct examples of instability near elliptic points. The paper [CG98] revived geometric approaches and contained many useful techniques.

To study invariant objects, typically, one finds some representation of them as a function. The condition of invariance is then a functional equation, which is often studied by methods of functional analysis or just numerically or by asymptotic methods. Two very basic methods to study invariance equations are normal hyperbolicity or KAM theory. One often has to supplement them with some preliminary calculations based on averaging or on perturbative calculations.

Programs based on finite orbits "with hooks"

1. Find finite segments of pseudo-orbits such that one segment ends close to the beginning of the following segment.
2. Verify some extra properties of each of the segments.
3. Use these properties to show that there is an orbit that remains close to the whole segment of orbits.

We picturesquely describe the above situation as saying that the segments of orbits have *hooks* so that they can be chained together. The fact that one needs some extra properties of the segments is made clear by the existence of examples – e.g. rigid rotations of the torus – where the conclusions are false.

There are quite a number of mathematical results of this kind. The best known results of this type are, perhaps, the *shadowing* theorems for hyperbolic systems [Shu78]. The hook in this case, is hyperbolicity. For many applications, hyperbolicity is a hard hypothesis to verify – it is often even false! – so that there are many variants. See, for example [Pal00, Pil99] and references there.

For us, the method which so far has proved to be more useful is the method based on *correctly aligned windows*. The basic idea is to use some kind of topological index of the segments of orbits so that one can show that there is an orbit in a neighborhood of the whole chain. One early example, is [Con68, Con78, Eas89]. We will discuss it in Section 5.

One should also mention the variational program started in the 1930s using *broken geodesics* [Mor24, Hed32, Ban88]. The idea was that, if the segments are minimizers of a good variational principle, then, indeed, there are orbits that follow them.[2] Some early implementations of these ideas to the problem of diffusion appear in [Bes96]. More recent applications appear in [BBB03, BCV01]. These methods also have the advantage that they apply to PDE's [RS02, Ang87]. Very deep variational methods that also involve global considerations appeared in [Mat93, Mañ97, CI99].

Remark 1.4. Of course, there are relations between the methods. Even in [Arn64], the invariant objects were used to produce segments of orbits as well as some *obstruction* property which shows that there are orbits that follow the segment. In our

---

[2] The heuristic idea is that, in the space of segments, each of our minimizers is in the center of a ball whose boundary has more action. If we take the whole orbit, the phase space is the product of the phase space of the segments so that the approximate orbit is contained in a ball so that the boundary of a ball has more action.

discussion of applications of the method of correctly aligned windows, we will consider orbits suggested by the invariant objects.[3] Even the more global variational methods of [CY04b, CY04a] start by reducing the problem using the presence of a normally hyperbolic invariant manifold.

One can hope that in the near future there will be even more relations. In particular, the more local variational theories (broken geodesic methods) seem rather close to the geometric methods. One can find relations between variational methods and the windows method is [Moe05].

In these lectures, we will try to present different mechanisms as well as the verification of their presence in some quasi-integrable systems. For the geometric mechanisms we will present in these lectures, the verification of their presence in concrete systems, will involve a rather standard toolkit (averaging theory, the theory of normally hyperbolic manifolds – perturbation theory, λ-lemma –, KAM theory) and some less standard tools such as the scattering map (Section 3.2) and the correctly aligned windows (Section 5). We will omit most of the details, but refer to the literature. The only goal of these lectures is to present a road map to the programs and to indicate the significant mileposts to be reached. Some similar expositions are [DDLLS00, DLS03, dlL06]. The present one incorporates some progress since the previous exposition were written. Fortunately, the new developments have lead to more streamlined proofs.

2 Exposition of the Arnold example

This very explicit example was constructed in [Arn64]. It is, possibly, the best known example in the mathematical literature. Some more detailed expositions of several of the aspects appear in [AA67]. A very complete explanation of the model in [Arn64] and generalizations can be found in [FM01].

In the following paragraphs, we will present the result emphasizing some of the geometric aspects that will play a role in the following. We refer [FM01] for the technical details of many of the proofs. We will emphasize several geometric properties that will play in the future.

Theorem 2.1. *Consider a time-dependent system defined in the action-angle variables* $(I, \Phi) \in \mathbb{R}^2 \times \mathbb{T}^2$ *by:*

$$H(I, \Phi, t) = \frac{1}{2}(I_1^2 + I_2^2) + \varepsilon(\cos \Phi_1 - 1)$$
$$+ \mu\varepsilon(\cos \Phi_1 - 1)(\sin \Phi_2 + \cos t) \,, \tag{1}$$

If $0 < \mu \ll \varepsilon \ll 1$.

Then, there exist orbits of the Hamilton's equation corresponding to (1) *with*

$$|I(T) - I(0)| > 1 \,.$$

---

[3] Strictly speaking, the windowing method only needs that they are approximately invariant.

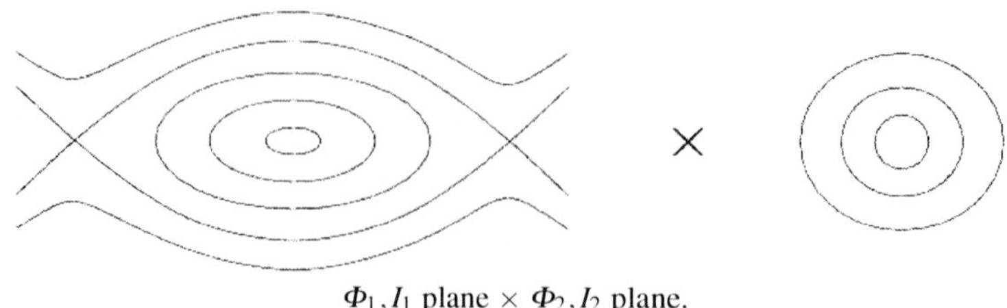

Φ_1, I_1 plane $\times$ Φ_2, I_2 plane.

Fig. 1 Illustration of the dynamics of the time one map of the dynamics of (1) for $\varepsilon > 0, \mu = 0$.

We point out that the Hamiltonian (1) satisfies the conditions of KAM and Nekhoroshev theorems (in spite of being partially degenerate) [Lla01, Nie07] so that the for ε, μ small, the orbits that satisfy the conclusion occupy a small measure (these orbits cannot be in KAM tori) and T has to be very large (by Nekhoroshev's theorem). This gives an idea of the subtleness of the phenomenon.

The system (1) can be easily understood for $\varepsilon > 0, \mu = 0$ since it is a product of two simple systems (a pendulum and a rotator), see Figure 1. We note, in particular that the manifold Λ, obtained by fixing the pendulum variables to the hyperbolic fixed point, (i.e. $(I_1, \Phi_1) = (0,0)$) and letting the (I_2, Φ_2) vary is a normally hyperbolic manifold. Clearly, Λ is topologically an annulus $\mathbb{R} \times \mathbb{T}^1$.

It will be important (for other mechanisms) to remark that the manifold Λ is normally hyperbolic.

The main remark in [Arn64] is that the manifold Λ is foliated by invariant tori (corresponding to fixing I_2). These tori are not normally hyperbolic (perturbations along the I_2 direction do not grow exponentially), but they are *whiskered tori* . That is, tori, whose normal directions contain stable directions (i.e. directions which contract exponentially fast in the future) and unstable directions (i.e. directions that contract exponentially fast in the past). The rates of contraction in the future and in the past are the contracting and expanding eigenvalues of the fixed point of the pendulum. It is easy to see that they are equal to $\lambda = \mp\varepsilon^{1/2}$.

It is shown, in general, that to whiskered tori, one can attach invariant stable (resp. unstable) manifolds consisting of the orbits which converge exponentially fast – with a rate similar to the rate of convergence of the linearization – in the future (resp. in the past). In the uncoupled case that we are considering now, the stable and unstable manifolds can be computed explicitly. The (un)stable manifolds are just the product of the tori and the separatrix of the pendulum. In particular, the stable and unstable manifolds of a torus agree.

Now, we consider $0 < \mu \ll \varepsilon$ and we will treat the term containing μ as a perturbation. In such a case, we can use the general theory of whiskered tori and their manifolds. The application of the general theory to (1) is rather simple because the example has been chosen carefully so that the perturbation and its gradient vanish on Λ. Hence, the family of tori, remains the same. It is part of the general theory that the tori keep being whiskered under the new dynamics and that they have (un)stable manifolds. Furthermore, the manifolds depend smoothly on μ. The first order in the

μ expansion can be computed easily by matching powers in formal expansions[4] and it is not difficult to show that the manifolds of nearby tori intersect transversally. In some ways the result is to be expected since the μ term, even if leaving Λ invariant, is significant in the region occupied by the whiskers. It would be very easy to make perturbations with compact support intersecting the separatrices and which move them.

The construction so far, for any $\delta > 0$ allows to construct a δ pseudo-orbit that moves I_2 by 1. If we start in a torus τ with an irrational rotation, we wait for the appropriate moment, then, jump in its unstable manifold, in such a way that the orbit is also in the stable manifold of another torus τ'. Once we are close enough to τ', we jump into a torus with an irrational rotation – such tori are dense. Then, we can restart again.

Unfortunately, this step does not allow to take the limit $\delta \to 0$ since the orbits change widely. If we make δ smaller, the orbits we constructed have to give more turns till the irrational rotation sets the phase exactly right for the jump.

2.1 The obstruction property

The program of [Arn64] contains an extra step, the *obstruction property* – that constructs a true orbit shadowing some of the pseudo-orbits. Figure 2 depicts schematically some of the pseudoorbits made by joining heteroclinic connections and the orbits shadowing them.

There is a substantial literature on the obstruction argument. We just call attention to the reader that part of the literature includes – sometimes without making it explicit – the assumption that one of the terms in the normal form of the torus vanishes. Some papers rely on normal forms to high order – hence only apply com-

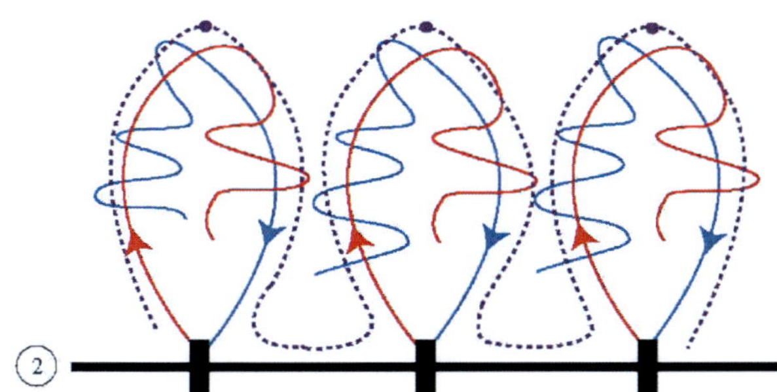

Fig. 2 Illustration of some orbits in the dynamics of (1) for $0 < \mu \ll \varepsilon$. The 2 refers to the fact that Λ is two-dimensional.

---

[4] Of course, matching powers in formal expansions does not justify that the expansions exist. In this case, using the general theory of whiskered tori, we know that these expansions exist. Historically, power matching in cases similar to this one was routinely used many years before it was justified.

fortably to C^∞ or C^ω systems. Others assume that all the tori can be fit in a common system of coordinates. In some papers, the construction depends on the number of tori that the orbit has to explore. Therefore, increasing the number of tori changes substantially the orbit (the time the orbit has to spend in the neighborhood of each tori increases with the total number of tori to be visited). These constructions do not allow to pass to the limit and construct orbits which visit infinitely many tori. Of course, the diversity of arguments is just a reflection of the fact that there are many types of diffusing orbits each with different quantitative and qualitative properties. We cannot survey the rather extensive literature but just call attention on some points to watch for. We certainly hope somebody will write such a survey.

We also note that the obstruction argument is not the only way of constructing orbits which shadow the pseudo-orbits. In this lecture we will discuss the method of correctly aligned windows in other context, which is a topological method – applications to the shadowing of whiskered tori happen in [Rob02, GR04, CG03]. There are also variational methods [Bes96, BBB03, BCV01] for this step.

In practice, the step of constructing the shadowing orbits is what controls the time T in the statement of the result. Many of the methods above lead to different estimates for T and presumably to different orbits. This again reinforces the belief that diffusion is really a superposition of several mechanisms. Here, we will just present some simple argument – we follow closely [DLS00] – which makes more precise some of the ideas in the original papers [AA67]. See also [FM01, FM03, FM00, Cre97]. The main ingredient is a somewhat sharp version of the λ-lemma – for example that in [FM00] – and a point set topology argument. Since no normal forms to higher order are used the method has only modest differentiability requirements. It can also accommodate infinitely long chains. A more elaborate argument along similar lines, but also giving more control on the orbits appears in [DLS06c].

If $\{\mathcal{T}_i\}_{i=1}^\infty$ is a sequence of whiskered tori with irrational rotations and $\{\varepsilon_i\}_{i=1}^\infty$ a sequence of strictly positive numbers, we can find a point P and an increasing sequence of numbers T_i such that

$$\Psi_{T_i}(P) \in N_{\varepsilon_i}(\mathcal{T}_i)$$

where $N_{\varepsilon_i}(\mathcal{T}_i)$ is a neighborhood of size ε_i of the torus $\mathcal{T}_i$. Here Ψ_t represents the flow associated to the system.

To establish this result, note that if $x \in W^s_{\mathcal{T}_1}$, we can find a closed ball B_1, centered at x, and such that

$$\Psi_{T_1}(B_1) \subset N_{\varepsilon_1}(\mathcal{T}_1). \tag{2}$$

By the λ-lemma,

$$W^s_{\mathcal{T}_2} \cap B_1 \neq \emptyset.$$

Hence, there is a closed ball $B_2 \subset B_1$, centered at a point in $W^s_{\mathcal{T}_2}$ such that, besides satisfying (2):

$$\Psi_{T_2}(B_2) \subset N_{\varepsilon_2}(\mathcal{T}_2).$$

Proceeding by induction, we can find a sequence of closed balls

$$B_i \subset B_{i-1} \subset \cdots \subset B_1$$
$$\Psi_{T_j}(B_i) \subset N_{\varepsilon_j}(\mathscr{T}_j), \quad i \leqslant j.$$

Since the closed balls are compact, they have non-empty intersection and any point in the intersection satisfies the desired property.

This argument as presented above does not give estimates on the time needed to transfer. On the other hand, it gives several other information on the orbits. For example, the orbits never leave an ε neighborhood of the segments of $W^{s,u}_{\mathscr{T}_i}$ so that we can be sure that the energy, or the actions, are described, up to errors of size ε by the values along the sequences visited. For future purposes, it is important to point out that the argument only uses that the tori are whiskered and it does not use at all the way that the tori fit together. Later, in Section 4, we will apply this argument to sequences of tori which are not homotopic and that, therefore, cannot be fit in common system of coordinates.

2.2 Some final remarks on the example in [Arn64]

The example (1) is remarkable for many reasons. Here, we just note that the diffusion happens in places where there are no resonances. Indeed, detecting the diffusion numerically in (1) is much harder than in other examples. It is somewhat ironic that much mathematical effort was spent proving instability in models for which the result is indeed very weak.

One feature of the example (1) which is important for the construction is that the second perturbation vanishes identically on a manifold. This is very non-generic and, indeed, it does not happen in many models of interest.[5]

We have done the first order expansion in μ, assuming $\varepsilon > 0$ and fixed. The dependence on ε of this theory is rather complicated. The first order term in the expansion in μ of the angle between the stable and the unstable manifoldso of a torus is of the order $\exp(-A\varepsilon^{-1/2})\mu\varepsilon$. The remainder, on the other hand, is not easy to bound better than $C\mu^2\varepsilon^2$. This is, of course, perfectly fine if $\mu \ll \exp(-A/2\varepsilon^{-1/2})$, but if $\mu = \varepsilon^p$, then, it could happen that the leading order of the perturbation in μ does not give the whole story.

As a consequence, the treatment above – based on just first order perturbation theory in μ can not establish the existence of instability in a whole ball in ε, μ or for $\mu = \varepsilon^p$.

---

[5] One should remark, however, that it does happen in some models of interest. For example [dlLRR07] shows that perturbations which vanish on manifolds, happen naturally in some systems of physical interest such as billiards with moving boundaries and in oscillators provided that they have some symmetries and that an analysis very similar to that of [Arn64] leads to the existence of orbits of unbounded energy in these systems

3 Return to a normally hyperbolic manifold. The two dynamics approach

In the exposition of [Arn64] in the previous section, we have emphasized the normally hyperbolic manifold Λ – which only appeared implicitly in [Arn64].

The reason is that the persistence of normally hyperbolic manifolds holds rather generally as was recognized in the 1960s [Sac65, Fen72, Fen74, HP70, HPS77, Pes04]. Of course, for examples other than the carefully chosen (1), one does not expect that the dynamics in the invariant manifold remains integrable. Indeed, as it is well known (we will present some ideas in Section 4.3) the resonant tori break up under perturbation so that the foliation by invariant tori gets interrupted.

The general theory of normally hyperbolic invariant manifolds establishes not only the persistence of the normally hyperbolic invariant manifolds but also the existence of stable and unstable manifolds and the regularity of the dependence on parameters of these objects. A short summary of the theory of normally hyperbolic invariant manifolds can be found in Appendix A. Of course, this is no substitute for the references above.

The theory of dependence with respect to parameters of normally hyperbolic invariant manifolds, justifies the perturbation theory.

3.1 The basics of the mechanism of return to a normally hyperbolic invariant manifold

The basic assumption is that the stable and unstable manifolds of Λ intersect transversally. This means that there are orbits that leave the manifold but come back. We will refer to these orbits as homoclinic excursions. Note that a simple dimension counting – justified by the regularity given by the theory of normally hyperbolic invariant manifolds – shows that the set of homoclinic excursions is, locally, a manifold of the same dimension as Λ. Hence, we expect that there is an open set $H_- \subset \Lambda$ such that all the points in H_- can make an arbitrarily small jump and, go into the unstable manifold of Λ, perform an homoclinic excursion and come back to Λ. Since this homoclinic excursion moves the orbit far away from Λ it is quite possible that it can be really affected by the perturbation and the action variables can change. In Section 3.2, we will describe some concrete descriptions of these sets.

When the system is conservative, one expects that some of the homoclinic excursions are *favorable* – e.g. the excursion gains energy or action – and others are *unfavorable* – the excursions looses energy or actions. Since there are rather explicit formulas – which we will explain in Sections 3.2 and 4.1, one expects that the points in H_- which lead to favorable or unfavorable excursions are open sets separated by a codimension 1 manifold, which can be calculated as the zero set of a function (in the models discussed in Section 4 perturbative formulas for this function are rather standard).

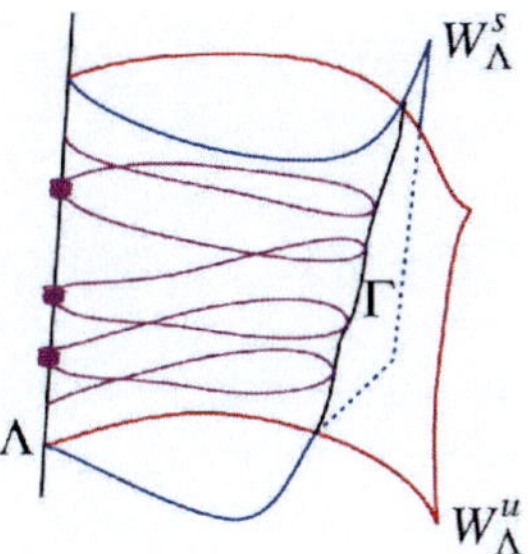

Fig. 3 Illustration of orbits that gain energy by intertwining homoclinic excursions with staying around an invariant manifold.

Note that H_- and the separation between the favorable and unfavorable regions depend very strongly on the perturbation far away from Λ. Hence, we can expect that the dynamics on Λ – which is unaffected by the perturbations away from the manifold – is completely unrelated to the separation between favorable and unfavorable excursions. Hence, unless this separation is invariant under the dynamics in Λ, one can stay around Λ for a carefully chosen time and move into the favorable region. We emphasize that, explicit perturbative computations can give approximations to the manifold separating the favorable from the unfavorable excursions, so that a finite computation can establish that there are orbits in Λ that move into the favorable region.

In this way, for many systems, one can construct pseudo orbits by interleaving orbits that follow a homoclinic excursion and orbits that remain in Λ so that we go from the end of a homoclinic excursion to another favorable excursion. This can be compared to primitive sailing: When the wind is favorable, the boat moves. When the wind turns bad, it moves to the coast and anchors waiting till the wind becomes favorable again.

Of course, if one is interested in true orbits rather than on δ pseudo-orbits with δ arbitrarily small, one still needs an extra step – shadowing or obstruction. Some versions of these arguments are discussed in Sections 2.1 and 5.

To make the above heuristic ideas rigorous, one uses: (a) a tool to describe the homoclinic excursions, which allows explicit computations (b) some explicit description of the dynamics on Λ, (c) some tools to pass from the pseudo-orbits to the orbits.

Of course, the analysis of the dynamics restricted to Λ is just the general problem of dynamical systems. The description of homoclinic intersections will be undertaken in Section 3.2.

We note that the scattering map is not the only way to discuss homoclinic excursions. The paper [Tre02a, Tre02b] introduce the *separatrix map*. We also call attention to [BK05].

3.2 *The scattering map*

The scattering map is a particularly convenient way of describing the behavior on a homoclinic excursion. It was introduced explicitly in [DLS00] as a geometrically natural alternative to Melnikov theory so that issues of domain and monodromy could be discussed in detail. Related ideas for center manifolds were introduced in [Gar00]. A much more systematic theory of the scattering map was developed in [DLS06a].

An orbit is homoclinic if the future and the past converge exponentially fast to Λ. We adopt the same notation as in Appendix A.

We recall that the stable and unstable manifolds can be decomposed into stable manifolds of single points, namely: $W_\Lambda^s = \bigcup_{x \in \Lambda} W_x^s$, $W_\Lambda^u = \bigcup_{x \in \Lambda} W_x^u$. The above decompositions are are foliations because if $x, y \in \Lambda$, $x \neq y$, then $W_x^s \cap W_y^s = \emptyset$, $W_x^u \cap W_y^u = \emptyset$. We will refer to these foliations as $\mathscr{F}_{s,u}$ respectively.

Using the foliations $\mathscr{F}_{s,u}$ we can define the *"wave operators"* Ω_+, Ω_-

$$\Omega_\pm : W_\Lambda^{s,u} \longrightarrow \Lambda \tag{1}$$

defined by

$$x \in W_{\Omega_+(x)}^s \qquad x \in W_{\Omega_-(x)}^u \tag{2}$$

If there is a manifold $\Gamma \subset W_\Lambda^s \cap W_\Lambda^u$ such that Ω_- is a diffeomorphism from Γ to its range $\Omega_-(\Gamma) \equiv H_-^\Gamma$, then we can define $(\Omega_-^\Gamma)^{-1} : H_-^\Gamma \to \Gamma$ and relatedly,

$$s^\Gamma = \Omega_+ \circ (\Omega_-^\Gamma)^{-1} \tag{3}$$

See Figure 4.

This set $H_-\Gamma$ is the set of initial points of trajectories having the property that an small push can make them go through Γ. This is a more precise version of the set H_- wich we discussed in Section 3.1. The set H_-^Γ specifies that the connections go through Γ.

The map $s^\Gamma : H_- \to H_+$, gives an encoding of the homoclinic excursions that pass through Γ. If we consider one such excursion, the orbit is asymptotically close to one orbit in Λ in the past and to another orbit in Λ in the future. The map s^Γ gives the future orbit as a function of the asymptotic orbit in the past.[6] Of course, the scattering map depends very strongly on the manifold Γ we have chosen. Escaping from Λ along different routes will, clearly, have very different effects and the scattering map will be very different. Some examples of celestial mechanics with explicit computations appear in [CDMR06].

Now, we discuss some natural hypothesis that imply that Ω_-^Γ is invertible from its range to Γ and that this is maintained under perturbations and that there is good dependence with respect to parameters. Basically, we will reduce the definitions to

---

[6] This is remarkably similar to the definition of the scattering matrix in the time-dependent scattering theory in quantum mechanics. Indeed, there are many more analogies and we have chosen the notation to reflect them.

Other classical analogues of quantum scattering theory, somewhat different from those considered here, were considered in [Hun68, BT79, Thi83] and in a more general context in [Nel69].

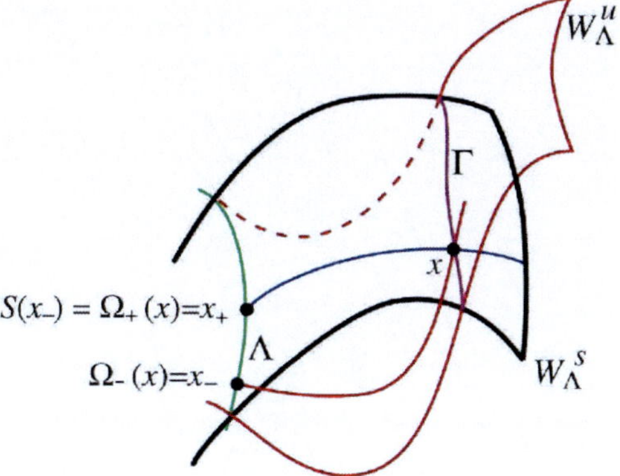

Fig. 4 Illustration of the definition of the scattering map.

transversality conditions so that the implicit function theorem gives the persistence and smooth dependence on parameters.

A natural set of conditions to define scattering map is that for all $x \in \Gamma$,

$$T_x W^s_\Lambda + T_x W^u_\Lambda = T_x M$$
$$T_x W^s_\Lambda \cap T_x W^u_\Lambda = T_x \Gamma \tag{4}$$

$$T_x W^s_{\Omega_+ x} \oplus T_x \Gamma = T_x W^s_\Lambda$$
$$T_x W^u_{\Omega_- x} \oplus T_x \Gamma = T_x W^u_\Lambda \tag{5}$$

The conditions in (4) mean that W^s_Λ, W^u_Λ *"intersect transversally"* along Γ. The first condition in (5) means that Γ is *"transversal to the foliation"* $\{W^s_x\}_{x \in \Lambda}$ inside W^s_Λ. The second equation in (5) means that Γ satisfies an analogous property relative to the unstable foliations. See Figure 5.

If we have (4) for just one x_0, the implicit function theorem tells us that we can find a smooth manifold Γ containing x_0 such that (4) holds for all $x \in \Gamma$. Since the manifold Γ is obtained applying the implicit function theorem, if both W^s_Λ, W^u_Λ, are C^k manifolds in a neighborhood of x, then Γ will also be a C^k manifold.

Similarly, applying the the implicit function theorem, the regularity theory for the manifolds and their smooth dependence on parameters, discussed in Appendix A, we conclude that if f_ε is a C^1 family and f_0 has a Λ_0, Γ_0 satisfying the normal hyperbolicity and transversality conditions, that there is a C^1 family of manifolds Λ_ε which are normally hyperbolic and another family of manifolds Γ_ε satisfying the properties. In the case that we can guarantee that $W^{s,u}_{\Lambda_\varepsilon}$ are $C^{\ell-1}$ families, we obtain that Γ_ε is a $C^{\ell-1}$ family and we can also obtain smooth dependence on parameters for the $\Omega^{\Gamma_\varepsilon}_\pm$ and for the scattering map.[7]

____

[7] The smooth dependence of a map in domains which are changing, should be understood in the sense that there is smooth family of maps from a fixed domain to the domains so that the composed map is smooth.

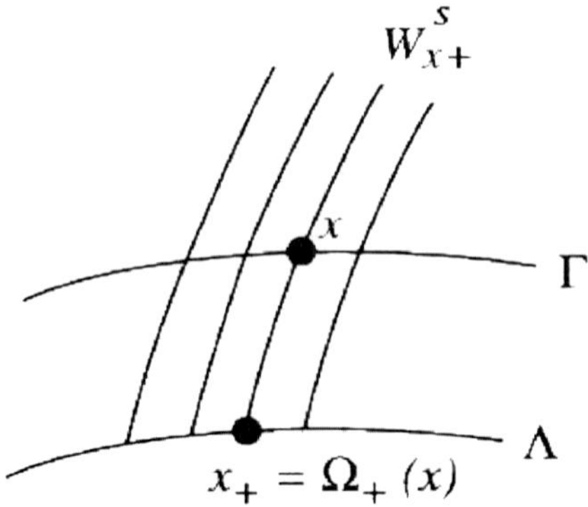

Fig. 5 Illustration of the conditions in (5).

The properties in (5) are very different. Even if the formulation of (5) does not require that the foliations $\mathscr{F}_{s,u}$ are smooth, they become more interesting when these foliations are C^1 foliations. In this case, the implicit function theorem tells that, when we move along Γ, we have to move across the foliation.

The implicit function theorem shows that, if the foliations $\mathscr{F}_{s,u}$ are C^1 – this is implied by properties of the hyperbolicity constants, so that it holds true in some C^1 open sets of examples – and (5) hold, then, $\Omega_\pm$ are locally invertible. Again, because this is just an application of the implicit function theorem and there is a good dependence on parameters, we obtain if (4), (5) are satisfied for a map, they will be satisfied – with a similar Λ, Γ – for all the small C^1 perturbations. Furthermore, if we consider smooth families of maps, there will be smooth dependence on parameters.

Remark 3.1. One could argue heuritically that (5) could fail in a codimension 1 set of Γ – transversality is a codimension 1 phenomenon. Of course, this heuristic argument, could be false. Notably, the heuristic argument is false for the models considered in [DLS00, DLS06c]. It however, applies to some examples considered in [DLS06b].

Nevertheless, as shown in [DLS06b], the existence of an open set is enough for the construction of orbits that move appreciable amounts. One can also note that one expects to have infinitely many Γ, each of which with a different scattering map. The argument does not require that all the excursions go through the same Γ, so that the set of points which cannot be moved by this argument should be empty in manu examples.

3.3 The scattering map and homoclinic intersections of submanifolds

One important application of the scattering map is that it allows us to discuss transversal intersections of $W^u_{\Sigma_1}$, $W^s_{\Sigma_2}$ where $\Sigma_1, \Sigma_2 \subset \Lambda$ are invariant manifolds under the map f. One example is, of course, the whiskered tori inside the manifold Λ that were discussed in Section 2. In Section 4 we will see other examples that are more challenging.

It was shown in [DLS00, DLS06b, DLS06c] that if, for some manifold Γ, satisfying (4) (5), we have[8]

$$s^\Gamma(\Sigma_1) \pitchfork^\Lambda \Sigma_2. \tag{6}$$

Then,

$$W^u_{\Sigma_1} \pitchfork W^s_{\Sigma_2}. \tag{7}$$

This result is useful because the hypothesis (6) can be verified by calculations on the invariant manifold Λ. The conclusions is that the (un)stable manifolds of σ_1, σ_2 are transverse in the full manifold M.

In the case that Σ_1, Σ_2 are invariant circles which are close together, the transversality of intersections is usually discussed using Melnikov theory. Notice, however that Melnikov theory – since it is based on first order calculations often done in a concrete coordinate system – requires that the manifolds $\Sigma_{1,2}$ are expressed in the same system of coordinates, in particular, they are homotopically equivalent. The above result, however, is coordinate independent. This is crucial for the applications in [DLS06b], discussed in Section 4, where $\Sigma_{1,2}$ are not topologically equivalent.

As we will see in Section 3.7 there are rather explicit – rapidly convergent – formulas for the perturbative computation of the scattering map. Therefore, the theory outlined above can give rather efficient ways of establishing intersections.

3.4 *Monodromy of the scattering map*

Even if $\Omega^\Gamma_\pm$ are locally invertible, they could fail to be invertible in a domain which is large enough to include non-contractible closed loops. One interesting example was discussed already in [DLS00] and, in more detail in [DLS06c, DLS06a]. For example, when considering stable manifolds of a periodic orbit λ, the intersection manifold Γ looks like a *helix*. That is, if we increase the phase of the intersection, then, eventually we go into a different homoclinic intersection of the time-1 map. This is a geometric counterpart of the fact that, in some calculation in first order perturbation theory of intersections of invariant manifolds – often called Melnikov theory – one has to add real variables to angle variables. See Figure 6.

3.5 *Smoothness and smooth dependence on parameters*

Note that the sufficient conditions (4), (5) that ensure the existence of the scattering map in a neighborhood are transversality conditions that are robust under

---

[8] We use the notation $\pitchfork^\Lambda$ to indicate that the manifolds intersect transversally as manifolds in Λ. In particular, when we use this symbol, we assume that the intersection is not empty.

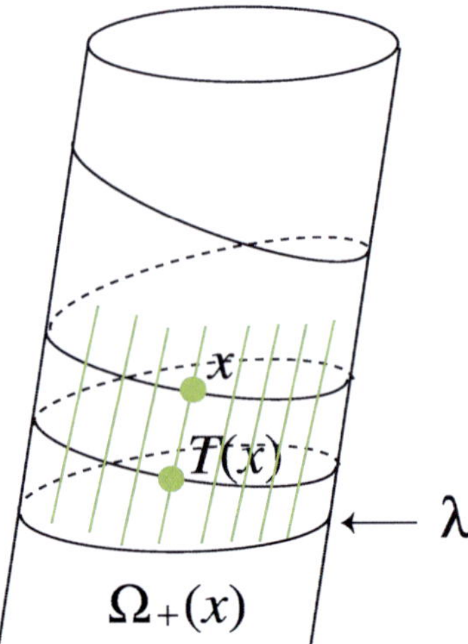

Fig. 6 Illustration of the monodromy of the scattering map for the stable manifolds of periodic orbits.

perturbations. Hence, given a concrete system, they can be established with a finite precision calculation. Later, in Section 4.1 we will see how they can be verified by perturbative calculations from an integrable system. See [DLS06b, GL06a]. The conditions can also be verified numerically if one controls the precision of the calculations [CDMR06].

It follows from the general theory of dependence on parameters that, under the conditions (4), (5), and smoothness of the foliations $\mathscr{F}_{s,u}$ then, the scattering map is smooth jointly on the manifold and on parameters.[9]

3.6 Geometric properties of the scattering map

So far, the discussion of the scattering map has only used normal hyperbolicity and regularity of the maps considered.

If the maps f_ε have some geometric structure, the scattering map also inherits some geometric properties. Notably, if f_ε is symplectic (resp. exact symplectic) and Λ_0 is a symplectic manifold (hence, exact symplectic if f_ε is exact symplectic) then s_ε is a symplectic (resp. exact symplectic) family of maps. This was proved in [DLS06a]. In the context of center manifolds it was proved in [Gar00].

---

[9] The discussion of smoothness with respect to parameters of the scattering map presents some technical annoyances such as that the domain of s_ε is Λ_ε, which changes as ε changes. An easy solution is to consider smooth (jointly with respect to the coordinates and the parameters) parameterizations k_ε of the invariant manifold Λ_ε. That is $k_\varepsilon(\Lambda_0) = \Lambda_\varepsilon$. See Appendix A.

There are two important consequences of the symplectic character.

- There are many techniques to discuss intersections of Lagrangian manifolds under symplectic mappings, see [Wei73, Wei79].
- There are very efficient perturbation theories for symplectic mappings. Historically this one of the reasons why Hamiltonian formalism was invented. We will discuss several versions of Hamiltonian perturbation theory here.

Taking advantage of both features at the same time, one gets a very efficient perturbative theory for the intersections of manifolds under the scattering map. In view of the results mentioned in Section 3.3, this is very useful to obtain transition tori.

In [DLS06a] it was proved that there is a natural smooth parameterization $k_\varepsilon(\Lambda_0) = \Lambda_\varepsilon$ such that k_0 is the immersion and that $k_\varepsilon^* \omega$ – the pull–back by k_ε of the symplectic form ω – is independent of ε. This later condition is a natural normalization and it is shown in [DLS06a] that this natural normalization determines uniquely the deformation.

Then, denoting by s_ε the scattering maps generated by a smooth family of manifolds Γ_ε satisfying (4), (5), and invertibility of Ω_-^Γ, we have that

$$\tilde{s}_\varepsilon \equiv k_\varepsilon^{-1} \circ s_\varepsilon \circ k_\varepsilon \tag{8}$$

is symplectic under $k_\varepsilon^* \omega \equiv k_0^* \omega$. Note that $\tilde{s}_\varepsilon : \Lambda_0 \to \Lambda_0$ can be thought of as the expression of s_ε in the coordinates k_ε mentioned above.

Furthermore, in [DLS06a], one can find explicit perturbative formulas for the canonical perturbation theory of $\tilde{s}_\varepsilon$. We will summarize them in Section 3.7.

3.7 Calculation of the scattering map

Given families of exact symplectic mappings there are very efficient ways of computing perturbation theories using the deformation method of singularity theory [LMM86].

If g_ε is a family of exact symplectic mappings, it is natural to study instead the vector field $\mathscr{G}_\varepsilon$ generating the family.

$$\frac{d}{d\varepsilon} g_\varepsilon = \mathscr{G}_\varepsilon \circ g_\varepsilon. \tag{9}$$

The fact that g_ε is exact symplectic for all ε is equivalent to g_0 being exact symplectic and $\iota_{\mathscr{G}_\varepsilon} \omega = dG_\varepsilon$ (here $\iota_{\mathscr{G}_\varepsilon} \omega$ is the contraction of vectors and forms). Under enough regularity conditions, (9) admits a unique solution.

Hence, it is the same to work with $\mathscr{G}_\varepsilon$ or G_ε. The interesting thing is that the family of functions G_ε satisfies much simpler equations. The reason is that the $\mathscr{G}_\varepsilon$ – and hence G_ε can be thought as infinitesimal deformations and the only equations that one can form with infinitesimal quantities are linear.

In the following, we will apply this idea to g_ε being several of the families appearing in the problem. We will keep the convention of keeping the same letter for the objects corresponding to a family. We will use caligraphic for the vector field and capitals for the Hamiltonian.

In [DLS06a], it is shown that there are remarkably simple formulas for $\tilde{S}_\varepsilon$, the generator of the the map $\tilde{s}_\varepsilon$ – the expression of s_ε in coordinates.

$$
\begin{aligned}
\tilde{S}_\varepsilon = {} & \lim_{N_\pm \to +\infty} \sum_{j=0}^{N_- -1} F_\varepsilon \circ f_\varepsilon^{-j} \circ (\Omega_{\varepsilon-}^{\Gamma_\varepsilon})^{-1} \circ s_\varepsilon^{-1} \circ k_\varepsilon - F_\varepsilon \circ f_\varepsilon^{-j} \circ s_\varepsilon^{-1} \circ k_\varepsilon \\
& + \sum_{j=1}^{N_+} F_\varepsilon \circ f_\varepsilon^{j} \circ (\Omega_{\varepsilon+}^{\Gamma_\varepsilon})^{-1} \circ k_\varepsilon - F_\varepsilon \circ f_\varepsilon^{j} \circ k_\varepsilon \\
= {} & \lim_{N_\pm \to +\infty} \sum_{j=0}^{N_- -1} F_\varepsilon \circ f_\varepsilon^{-j} \circ (\Omega_{\varepsilon-}^{\Gamma_\varepsilon})^{-1} \circ k_\varepsilon \circ s_\varepsilon^{-1} - F_\varepsilon \circ k_\varepsilon \circ r_\varepsilon^{-j} \circ s_\varepsilon^{-1} \\
& + \sum_{j=1}^{N_+} F_\varepsilon \circ f_\varepsilon^{j} \circ (\Omega_{\varepsilon+}^{\Gamma_\varepsilon})^{-1} \circ k_\varepsilon - F_\varepsilon \circ k_\varepsilon \circ r_\varepsilon^{j}
\end{aligned}
\tag{10}
$$

Similarly, for Hamiltonian flows, we have

$$
\begin{aligned}
S_\varepsilon = {} & \lim_{T_\pm \to \infty} \int_{-T_-}^{0} \frac{dH_\varepsilon}{d\varepsilon} \circ \Phi_{u,\varepsilon} \circ (\Omega_{\varepsilon-}^{\Gamma_\varepsilon})^{-1} \circ (s_\varepsilon)^{-1} \circ k_\varepsilon \\
& - \frac{dH_\varepsilon}{d\varepsilon} \circ \Phi_{u,\varepsilon} \circ (s_\varepsilon)^{-1} \circ k_\varepsilon \\
& + \int_{0}^{T_+} \frac{dH_\varepsilon}{d\varepsilon} \circ \Phi_{u,\varepsilon} \circ (\Omega_{\varepsilon+}^{\Gamma_\varepsilon})^{-1} \circ k_\varepsilon - \frac{dH_\varepsilon}{d\varepsilon} \circ \Phi_{u,\varepsilon} \circ k_\varepsilon
\end{aligned}
\tag{11}
$$

It is not difficult to see that the sums or the integrals converge uniformly.

The formulas (10) and (11) give the hamiltonian of the deformation as the integral of the generator of the perturbation over the homoclinic orbit minus the generator of the perturbation evaluated on the asymptotic orbits.

Note that, because of the exponential convergence of the homoclinic orbits and their asymptotic orbits, it is not difficult to see that the integrals in (10) and (11) converge exponentially fast. In [DLS06a] one can also find that derivatives up to an order (which is given by ratios of convergence exponents) also converge exponentially fast.

The effect of the homoclinic excursions on slowly changing variables can be computed using more conventional methods – we will present some of these computations in Section 4.1.

One novelty of the geometric theory presented in this section is that it allows computation of the effect of the homoclinic excursions not only on the slow variables, but also on the fast variables.

Notice also that, we can compute the intersection between objects of different topologies very simply. This extends many calculations usually done using

Melnikov theory. It suffices to apply (6). Note that the present theory only involves convergent integrals. This was somewhat controversial in the so-called Melnikov theory. See [Rob88].[10]

The Hamiltonian theory is particularly effective when the manifolds Σ are level sets of a function. We will see some examples in Section 4.6.

4 The large gap model

The model is basically a rotor coupled to one or several penduli and subject to a periodic perturbation.

This model was introduced in [HM82], but it appears naturally as a model of the motion near a multiplicity 1 resonance. A fuller treatment of multiplicity 1 resonances appears in [DLS07].

One could consider that it is a version of the example (1) when we set $\varepsilon = 1$ (hence rename as ε the parameter μ in (1)) but we allow the perturbing term to be a general one. In the paper [GL06b] it was remarked that the fact that the pendulum variables have only 1 degree of freedom can be easily removed and one could consider many penduli. Hence, the geometric treatment can be easily generalized to the case that the hyperbolic variables have several components.

Hence, we consider the model

$$H_\varepsilon(p_1,\dots,p_n,q_1,\dots,q_n,I,\phi,t) = \sum_{i=1}^{n} \pm \left(\frac{1}{2}p_i^2 + V_i(q_i) \right) + h_0(I)$$
$$+ \varepsilon h(p_1,\dots,p_n,q_1,\dots,q_n,I,\phi,t;\varepsilon), \tag{1}$$

where (p_i,q_i), (I,ϕ) are symplectically conjugate. We will assume that $V_i'(0) = 0$, $V_i''(0) > 0$. This means that V_i has a non-degenerate local minimum – that we set at 0. We will also assume that the pendulum P_i has a homoclinic orbit to 0. This is implied by the fact that there is no other critical point p with $V_i(p) = 0$. Both conditions are implied by V_i being a Morse function.

The version of (1) considered in [HM82, DLS06b, GL06a] consider only the case $n = 1$, but, as we will see, the complications introduced by several variables is not too important. A full treatment of (1) for general n appears in [GL06b]. We will explain it in Section 4.1.

One extra assumption in [DLS06b] – which we will maintain in the discussion in this section – is that the perturbation term h was a trigonometric polynomial in the angle variables. This assumption simplifies the calculations since there is only a

---

[10] Unfortunately, many references in Melnikov theory still invoke the use of Melnikov functions given by integrals of quasi-periodic functions. The textbook explanation is that these integrals converge along subsequences. Unfortunately, the resulting limit – and hence the predictions of these theories – depend on the sequence taken, so that the textbook explanation cannot be true. The real explanation is that these references forgot to take into account some important effect. In many cases, it is the change of the target manifold.

finite number of resonances to be studied. It allows us to emphasize the geometric objects appearing at each resonance. When h is not a polynomial, for each value of $\varepsilon > 0$ it suffices to study a finite number of resonances, but the number of resonances to be considered is $\varepsilon^{-\alpha}$. One needs to do some rather explicit quantitative estimates on the resonances. The assumption that the perturbation is a polynomial has been removed by very different methods. The paper [DH06] contains a very deep study of resonances taking into account the effect of the size of the Fourier coefficients on the size of the resonant region. The paper [GL06b] considers very large windows, much larger than the resonance zones and uses the method of correctly aligned windows to conclude existence of diffusion **without having to analyze** what happens in the region of resonance. This leads to less conditions than the analysis in [DLS06b, GL06a]. Also, the method in [GL06b] leads to optimal estimates on the time.

The analysis of (1) we will present starts by noting that $\Lambda_0 = \{p_i = 0, q_i = 0\}$ is a normally hyperbolic invariant manifold for the time-1 map. Applying the theory of normally hyperbolic manifolds, we conclude that, for ε small enough, it persists. In contrast with the example (1), the motion on the invariant manifold will not remain integrable. Indeed, the foliation of KAM tori will present gaps of size $\approx \varepsilon^{1/2}$. In the rest of the section, we will describe how to construct orbits that indeed jump over the resonance zone.[11]

4.1 Generation of intersections. Melnikov theory for normally hyperbolic manifolds

In the model (1), even if the manifold Λ_0 is normally hyperbolic, its stable and unstable manifolds coincide.

In this section, we want to argue that, under some non-degeneracy conditions on h which we will make explicit, for $0 < |\varepsilon|$, there is a manifold Γ_ε satisfying the conditions (4), (5). Furthermore, one can define the scattering map in a patch which is rather large and uniform with respect to ε.

The fact that there is a Γ_ε which depends smoothly on parameters and, in particular, can be continued through $\varepsilon = 0$ is well known to experts and we present the ideas of a simple proof later. See also [GL06b]. These are sometimes called *primary intersections* of the stable and unstable manifolds, to distinguish them from other intersections which do not have a limit as $\varepsilon \to 0$. See [Mos73, p. 99 ff.]. Subsequent steps of the construction of diffusing orbits could use any of these intersections for which the next non-degeneracy assumptions can be verified. The calculations we will develop here will work just as well for any of the primary intersections. The use of the secondary intersections deserves more study.

---

[11] The paper [HM82] showed only that there were heteroclinic intersections between some whiskered tori. The length of the heteroclinic chains constructed in [HM82] goes to 0 as $\varepsilon \to 0$. This was the meaning of *Arnol'd diffusion* adopted in that paper. It is very interesting to compare the Melnikov theory developed there with the based on the scattering map.

Very elegant geometric theories of intersections of stable and unstable manifolds can be found in [LMS03]. In these lectures, we will follow [GL06b] and present a very simpleminded calculation for the model using coordinates. The paper [GL06b] contains significantly more details than those presented here.

We call attention that the calculation here does not assume that the variables I, ϕ in (1) are one-dimensional. This will play a role in Section 7.

A key observation is that, by the theory of normally hyperbolic manifolds, we already know that Λ_ε, $W^{s,u}_{\Lambda_\varepsilon}$ depend smoothly on parameters. We just need to compute explicitly what are the derivatives of these objects. The non-degeneracy conditions alluded above are just that the first order in ε calculation predicts an intersection satisfying (4), (5). If the first order perturbation predicts a transversal intersection, the implicit function theorem allows us to conclude that indeed there is an intersection, and that the formal calculation gives the leading order.

For this calculation, the fundamental theorem of calculus will play an important role, hence it is better to consider flows rather than time-1 maps. To make it autonomous, we will just add a variable t. We will use the notation $\tilde{\Lambda}$ to refer to the invariant manifold in these coordinates.

For each of the penduli, we choose a homoclinic orbit x_i and consider the unperturbed homoclinic manifold $\{(x_1(\tau_1), x_2(\tau_2), \ldots, x_n(\tau_n))\}$.[12] The variables τ_i are variables parameterizing the separatrix of the i pendulum.

We note that in a neighborhood of the homoclinic manifold – excluding a neighborhood of the critical points – we can extend the variables τ_i. The variables τ_i and P_i constitute a good system of coordinates in this neighborhood. See Figure 7 for an illustration of this system of coordinates.

Again, appealing to the smoothness of the dependence of the stable manifolds on parameters, we know that the perturbed manifolds can be written as the graph of a function that gives the P_i as a function of τ, I, ϕ, t. Furthermore, this function will depend smoothly on parameters. Our only goal, then, is to compute the first order expansion of this function, knowing already that such an expansion exists.

We will denote the time evolution of a point by Ψ^s_ε. Remember that, to make the system autonomous, we consider t as a variable, which takes values on a circle. We will denote the invariant manifolds in the extended phase space as $\tilde{\Lambda}$.

Let x be a point in $W^s_{\tilde{\Lambda}_\varepsilon}$, by the fundamental theorem of calculus, we have, for any T,

$$P_i(x) - P_i(\Omega^\varepsilon_+ x) = P_i(\Psi^T_\varepsilon(x)) - P_i(\Psi^T_\varepsilon(\Omega^\varepsilon_+ x))$$

$$- \int_0^T \frac{d}{ds}\left[P_i(\Psi^s_\varepsilon(x)) - P_i(\Psi^s_\varepsilon(\Omega^\varepsilon_+ x))\right] ds$$

---

[12] Note that, in general, each of the penduli will have two homoclinic orbits to the critical point (one going in one direction and the other going in the opposite direction). So that, there will be 2^n homoclinic manifolds with parameterizations similar to the ones considered in the text. Since the conditions we will considering be are sufficient conditions for existence of unstable orbits, having many orbits at our disposal makes it more likely that we have instability.

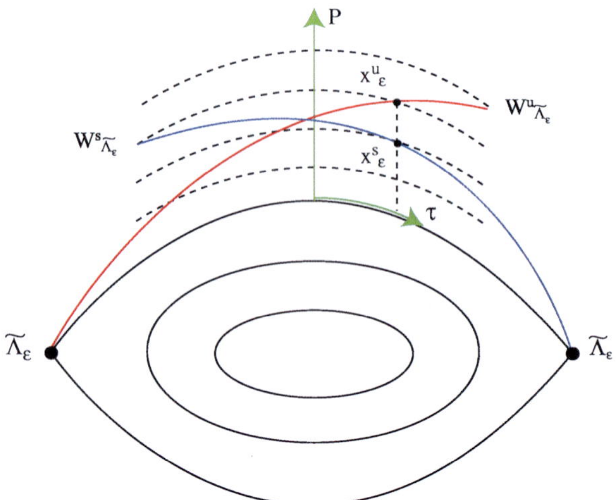

Fig. 7 Illustration of the system of coordinates in a neighborhood of the homoclinic manifold.

and, taking limits $T \to \infty$, we obtain

$$P_i(x) - P_i(\Omega^\varepsilon_+ x) = -\int_0^\infty \frac{d}{ds}\left[P_i(\Psi^s_\varepsilon(x)) - P_i(\Psi^s_\varepsilon(\Omega^\varepsilon_+ x))\right] ds \tag{2}$$

Now, recalling that we are only computing up to order ε, we can simplify significantly the formula.

We note that because P_i has a critical point at 0, we have $P_i(\Omega^\varepsilon_+ x) = O(\varepsilon^2)$, We also note that

$$\frac{d}{ds}\left[P_i(\Psi^s_\varepsilon(x)) - P_i(\Psi^s_\varepsilon(\Omega^\varepsilon_+ x))\right] = \varepsilon\left(\{P_i, h\} \circ \Psi^s_\varepsilon(x)) - \{P_i, h\} \circ \Psi^s_\varepsilon(\Omega^\varepsilon_+(x))\right)$$

$$= O(\varepsilon)$$

where $\{\cdot, \cdot\}$ is the Poisson bracket.

Notice also that the integrand in (2) is converging exponentially fast to zero. Hence, we have:

$$P_i(x) = -\varepsilon \int_0^{c|\log(\varepsilon)|} ds, \left[\{P_i, h\}(\Psi^s_\varepsilon(x)) - \{P_i, h\}(\Psi^s_\varepsilon(\Omega^\varepsilon_+ x))\right] + O(\varepsilon^2)$$

Since the integral is over a finite interval, we observe that, if $|s| \leqslant c|\ln(\varepsilon)|$, then

$$|\Psi^s_\varepsilon(x) - \Psi^s_0(x)| \leqslant c|\ln(\varepsilon)|\varepsilon$$

Also, using the smooth dependence of the stable and unstable foliations, we obtain that

$$|\Psi^s_\varepsilon(\Omega^x_+) - \Psi^s_0(\Omega^0_+ x)| \leqslant c|\ln(\varepsilon)|\varepsilon$$

Hence, we can transform the integral into

$$P_i(x) = -\varepsilon \int_0^{c|\log(\varepsilon)|} ds, \left[\{P_i, h\}(\Psi_0^s(x)) - \{P_i, h\}(\Psi_0^s(\Omega_+^0 x)) \right] + O(\varepsilon^2 |\ln(\varepsilon)|)$$

Remark 4.1. The above calculation identifies the derivative of the manifold with respect to ε when we consider the C^0 topology of functions.

In the case that we know that the derivative in C^r sense exists, the previous expression has to be the derivative in the C^r sense too.

In [GL06b], one can find justification of the slightly stronger result that the integrals above converge uniformly in C^r – provided that the Hamiltonians are uniformly C^{r+2}.

A very similar formula – reversing the time – can be obtained for an expression of the unstable manifold as a graph. Subtracting them, we obtain an expression for the first order expansion of the separation Δ of the P_i coordinates of the manifolds as a function of the τ_i, I, ϕ, t

$$\Delta_i(\underline{\tau}, I, \phi, t; \varepsilon) = \varepsilon \Delta_i^0(\underline{\tau}, I, \phi, t) + O(\varepsilon^2)$$

where the $O(\varepsilon^2)$ can be understood in the sense that the C^1 norm is bounded by $C\varepsilon^2$.

The implicit function theorem shows that if we find a zero of $\Delta_i^0 = 0$ which is non-degenerate (i.e., rank $D_\tau \Delta^0 = n$) then we can find $\tau^*(\varepsilon, I, \phi, t)$ such that $\Delta(\tau^*(\varepsilon, I, \phi, t), I, \phi, t; \varepsilon) = 0$. Hence, substituting in the variables P we can onbtain a parameterization of the intersection.

A more detailed analysis shows that the expressions of Δ_i are derivaties of a potential function with some periodicities [DR97]. Hence they have to have zeros. The assumption that these zeros are non-degenerate is a mild non-degeneracy assumption that can be verified in practical problems. It also holds generically. The case $n = 1$ is studied in great detail in [DLS06b]. In [GL06b] one can find an study of how to produce several of these solutions for $n > 1$.

4.2 Computation of the scattering map

The calculation of the scattering map in this case can be done as a particular case of the general theory of Section 3.2.

Notice that the formulas (10) are given in terms of limit of the intersection as $\varepsilon \to 0$, which we computed in the previous section using the easy part of the Melnikov theory.

The calculation in [DLS06b], was done by a different method since at the time that [DLS06b] was written, the authors were not aware of the symplectic theory of the scattering map.

The method of [DLS06b] was more elementary. Only the effect of the scattering map in one of the coordinates was computed. This was done using the fact that one of the coordinates in the invariant manifold – namely the energy – has a slow variation, so that in the calculation of the change of energy along a homoclinic excursion, one can use – up to the accuracy needed – just the fundamental theorem of calculus integrating over the unperturbed trajectory. The calculation can be done in very similar way to the calculation done in Section 4.1.[13] The fact that in [DLS06b] one only got control on one of the variables made the calculation of subsequent properties more complicated than what is nowadays possible using geometric theory. See [DLS07]. On the other hand, the calculation based on estimating the change of energy is natural for the purposes of the study of the intersection with KAM tori – which are given as level sets of the averaged energy.

For the purpose of this exposition, we will just mention that, for the model considered, once we settle on one primary homoclinic intersection, the scattering map can be computed as an explicit perturbation series with well controlled remainders. As in all the steps of this strategy, the calculations required can be done by very different methods. The more modern methods, taking more advantage of geometric cancellations seem more efficient even if the older methods can compute some features faster.

The conclusions is that – under conditions which can be checked explicitly and which, in particular, hold generically – the domain of definition of the scattering map contains a set which is independent of ε as $\varepsilon \to 0$. We call attention to the fact that the formulas for the scattering map depend heavily on the behavior of the perturbation along the whole homoclinic excursion.

4.3 The averaging method. Resonant averaging

The averaging method for nearly integrable systems goes back at least to [LP66]. Modern expositions are [LM88, AKN88, DG96]. An introduction for practitioners is [Car81]. See also [Mey91].

The basic idea is very simple. Given a quasi-integrable system, one tries to make changes of variables that reduce the perturbed system to another integrable system up to high powers in the perturbation parameter. This is accomplished by solving recursively cohomology equations.

There are many contexts and variations which make the literature extensive, even if there is only one guiding principle. For example one can consider autonomous perturbations or periodic perturbations, maps, flows etc. There are different possible meanings of "as simple as possible". One difference that leads to several variants is the fact that one can parameterize perturbations in different ways (generating functions, several types of Lie Series, deformation method, etc.) A systematic comparison of differences between these perturbation theories was undertaken in [LMM86].

---

[13] The actual calculation done in [DLS06b] uses not the energy – which is easily seen to be an slow variable – but rather a linear approximation to the energy. This makes only higher order differences. This linear approximation had been used customarily in the literature. At the time that [DLS06b] was written, it was important to make contact with the previous literature.

In the present problem, we consider periodic perturbations of integrable flows with one degree of freedom. To make comparisons with the literature easier, it will be convenient to make the system autonomous and symplectic by adding an extra variable A symplectically conjugated to t

$$H_\varepsilon(I,\phi,t,A) = H^0(I) + A + \varepsilon H^1(I,\phi,t) + \varepsilon^2 H^2(I,\phi,t) + \cdots \qquad (3)$$

where, of course, $H(I,\phi,t+1,A) = H(I,\phi,t,A)$, so that t can be considered as an angle variable. The A is added to keep the symplectic structure. Notice that it does not enter into the evolution of the other variables.

Again, for the sake of expediency in this presentation, we will omit considerations of issues of differentiability, estimates of reminders, etc. We refer to [DLS06b, Section 8], but the averaging method is covered in many other references, including some of the lectures in this volume.

For simplicity also, we will assume that all the terms in the expansion in ε are trigonometric polynomials with the same set of indices. That is,

$$H^i(I,\phi,t) = \sum_{k,l \in \mathcal{N}_i \subset \mathbb{Z}^2} H^i_{k,l}(I) \exp(k\phi + lt). \qquad (4)$$

Note that in the Appendix A, we show that this assumption for the case that we are interested in, follows from the assumption that the h in (1) is a trigonometric polynomial. The general theory of averaging does not require this assumption, but it involves several analysis consideration, which we prefer to avoid in an exposition.

We try to find a time periodic family of symplectic changes of variables $k_\varepsilon(I,\phi,t) = (I,\phi) + O(\varepsilon)$ in such a way that $H_\varepsilon(k_\varepsilon(I,\phi,t),t)$ is as simple as possible.

One possible way to try to generate the k_ε's is to write them as the time-1 solutions of a differential equation

$$\frac{d}{ds}k^s_\varepsilon = \varepsilon J \nabla K_\varepsilon \circ k^s_\varepsilon, \quad k^0_\varepsilon = \mathrm{Id},$$

where J is the symplectic matrix. In this case, we consider the evolution in the p,q,A,I,ϕ,t variables and the ε is just a parameter (this is not what we did in the section on deformation method). The gradient ∇ refers to the p,q,A,I,ϕ,t variables. The function K_ε is called the Hamiltonian. This way of parameterizing changes of variables is one of the variants of Lie transforms, [Car81, Mey91]. We will assume that $K_\varepsilon = \varepsilon K^1 + \varepsilon^2 K^2 + \cdots$,

It is well known from Hamiltonian mechanics [Arn89, AM78, Car81, Mey91] that

$$H_\varepsilon \circ k_\varepsilon = H^0 + \varepsilon(H^1 + \{H^0 + A, K^1\}) + O(\varepsilon^2)$$

where $\{\cdot,\cdot\}$ denotes the Poisson bracket in the variables I,ϕ,A,t.

Therefore, our goal is to find K^1 in such a way that

$$R^1 \equiv H^1 + \{H^0 + A, K^1\} \qquad (5)$$

is somewhat simple (we will make precise what "simple" means in our case). Since R^1 is the dominant term in $H_\varepsilon \circ k_\varepsilon$, one can hope that the dynamics expressed in the new coordinates is simple.

In terms of Fourier coefficients, (5) is equivalent to

$$R^1_{k,l}(I) = H^1_{k,l}(I) + i(k\omega(I) + l)K^1_{k,l}(I), \tag{6}$$

where $\omega(I) = \frac{\partial}{\partial I}H_0(I)$. The assumptions include that H_0 is twist. That is that $\omega(I)$ is monotonic, so that for each k,l there is one and only one $p_{k,l}$ such that $k\omega(I_{k,l}) + l = 0$. Of course, $I_{nk,nl} = I_{n,k}$. The points $I_{k,l}$ are called resonances.

Because of the assumption that the perturbation is a polynomial, we have to consider k,l ranging only over the finite set $\mathcal{N} \subset \mathbb{Z}^2$.

We see that (6) has very different character depending on whether $(k\omega(I)+l) = 0$ or not. If $(k\omega(I)+l) = 0$, we have to set $R^1_{j,k}(I) = H^1_{j,k}(I)$ but we can choose $K^i_{k,l}(I)$ arbitrary. Since we want that our solutions are differentiable, we have to make sure that the choices are made in a differentiable way. A particularly simple way – used in [DLS06b] to make these choices is to take a fixed C^∞ cut-off function Ψ and a fixed number L so that denoting $\Psi_L(t) = \Psi(t/L)$, we take the choice

$$R^1(I,\phi,t) = \sum_{k,l\in\mathcal{N}} \Psi_L(I - I_{k,l})H^1_{k,l}(I_{k,l})\exp(i(k\phi + lt)),$$

$$K^1(I,\phi,t) = \sum_{k,l\in\mathcal{N}} (1 - \Psi_L(I - I_{k,l}))/i(\omega(I)k + l)H^1_{k,l}(I)\exp(i(k\phi + lt). \tag{7}$$

If we choose conveniently L – we are considering only a finite number of resonances – we can ensure that the intervals $[-2L + I_{k,l}, 2L + I_{k,l}]$ do not intersect for different resonances.

So, we can divide the phase space into two regions:

● One "*non-resonant region*" where – in the appropriate coordinates – the system is integrable up to an error of order ε^2.

● A finite number of "*resonant regions*". Each of the resonant regions can be labeled by a frequency l/k expressed in an irreducible fraction. In one of these resonant regions, in the appropriate coordinates, the Hamiltonian is[14]:

$$H_0(I)+A + \varepsilon \sum_{n\in\natural} H^1_{nk,nl}(I_{k,l})\exp(in(k\phi + lt)) + O(\varepsilon^2)$$

$$= \quad H_0(I) + A + \varepsilon V(k\phi + lt) + O(\varepsilon^2). \tag{8}$$

The dynamics of the Hamiltonian (8) are easy to understand. If we introduce the variables $\tilde{\phi} = k\phi + lt$, $\tilde{I} = I - I_{k,l}$ – this change of variables is not symplectic, but it just multiplies the symplectic structure by a constant, so that the equations of motion – up to a constant change in time are also given by a Hamiltonian. Note also

---

[14] Again, we ignore regularity issues. It is not hard to show that if we assume that the function H^1 is C^r, then, K^1, R^1 are C^{r-2} so that the error term in (8) can be considered in the C^{r-2} norm. Again, we refer to [DLS06b].

that in this change of variables, the period of the angle variables is changed. Hence, in the new variables, the Hamiltonian is:

$$\alpha H_{k,l}(\tilde{I}) + A + \varepsilon V_{k,l}(\tilde{\phi}) + O(\varepsilon^2) \tag{9}$$

Since at the resonance the variable $\tilde{\phi}$ has frequency 0, we have that

$$H_{k,l}(\tilde{I}) = \alpha_{k,l} I^2 + O(\tilde{I}^3)$$

Furthermore, α will not be zero since it will be close to the second derivative of the unperturbed Hamiltonian, which we assumed is strictly positive (twist condition).

Note that the dynamics of (9) is very similar to the dynamics of a pendulum with a potential of size ε. In this case, the variable A does not play any role at all. There will be homoclinic orbits to the maximum of the potential. These orbits will be given by the conservation of energy and the form of the kinetic energy as

$$\tilde{I} = \pm \varepsilon^{1/2} \sqrt{\alpha^{-1}(\max V - V(\tilde{\phi}))} + O(\varepsilon). \tag{10}$$

Inside these curves, the system does a rotation.

If the maximum is non-degenerate – another hypothesis which is easy to verify in practice and which holds for generic V – we see that the orbits described in (10) are orbits that start and end in a critical point, which is hyperbolic. They are at the same time the stable and the unstable manifolds of this hyperbolic fixed point.

Note that these orbits are very different from the KAM tori. This is the reason why the KAM foliation gets interrupted by gaps of order $\varepsilon^{1/2}$.

It is important to remark that the stable and unstable manifolds of these periodic points have Lyapunov exponents $O(\varepsilon^{1/2})$. This is much smaller than the Lyapunov exponents in the transverse directions, which are independent of ε. Hence, when we talk about the stable manifolds restricted to Λ this is not the same as the W^s in the sense of the theory of normally hyperbolic invariant manifolds , which requires convergence at an exponential rate of order 1.

The dynamics of the averaged system – we will see that many of these features are preserved in the full system – consists of the foliation of – more or less horizontal – curves given by the orbits of the integrable system interrupted by a group of *eyes* or *islands*. At a resonance of type k, l we obtain k eyes. The amplitude of these eyes is $O(\varepsilon^{1/2})$.

Remark 4.2. The above classification ignores some stripes of width $O(\varepsilon)$ near the separation of the regions. The conclusions remain valid if we realize that the separation between the zones – the choice of L – was a choice we made. We can repeat the same analysis with an slightly different L and see that the ambiguous zones are different in the two procedures. So that by doing the analysis twice with slightly different L one establishes the conclusions above for all the phase space.

Remark 4.3. The choice of separation between the resonances zones is rather wasteful (even if it makes the estimates and the concepts easier). We assign the same width to all the resonances even if it is clear that the real width will decrease with ε. (In particular, we expect that the optimal size would be close to $\varepsilon^{1/2}$). Furthermore, if the original Hamiltonian is several times differentiable, then, its Fourier coefficients will decrease at least like a power of k, l. Hence the $V_{k,l}$ will become smaller with k, l. Hence, if for a fixed ε we decide to consider only resonant regions of size ε^B, we only need to consider a finite number of resonances – which will grow as $\varepsilon \to 0$ if $B > 1/2$.

Considerations of these type were known heuristically since at least [Chi79]. A rigorous implementation appears in [DH06]. The paper [DH06] includes also considerations of repeated averaging – discussed in the next section – and a very detailed analysis of the motion in each resonance with error terms.

4.4 Repeated averaging

The method of averaging can be applied several times. Indeed, in celestial mechanics it has been common for centuries to do at least two steps of averaging.

In the region that was marked as integrable in the first step, after we perform the change of variables, we are left with a quasi-integrable system. The perturbation parameter is ε^2. We can restart the procedure and get again some regions where the system can be made integrable up to $O(\varepsilon^2)$ and new resonant regions in which the dynamics has *eyes*, which will now be of size ε rather than $\varepsilon^{1/2}$.

In the resonant regions, nothing much happens except that the resonant potential $V_{k,l}$ gets deformed.

In the case that the perturbation is a trigonometric polynomial, the number of resonances we get at each step is finite and given a number of steps, we can get an L which works for all cases.

The result of applying averaging twice is depicted crudely in Figure 8.[15] For future analysis, the only important thing is that near resonances, we encounter separatrices well approximated by other tori and that, outside the resonances the system is very approximately integrable.

4.5 Invariant objects generated by resonances: secondary tori, lower dimensional tori

The resonant averaging described above, gives very accurate predictions of the dynamics.

The difference between the perturbed system expressed in a system of coordinates and the true system – in a smooth norm – is smaller than $C_N \varepsilon^N$. The constants C_N grow very fast.

---

[15] We have ignored, for example, the fact that inside the big islands of size $\varepsilon^{1/2}$ there are other *baby islands* of size ε going around.

Fig. 8 Schematic description of the predictions for the dynamics by the averaging method.

This can be taken advantage off in two different ways:

A) If some perturbation theories apply, we can conclude that some of the invariant objects for the integrable system, persist for the true system.
B) We have good control of some long orbits that, using some conditions can be glued together or shadowed.

This can be applied to the two types of geometric programs mentioned in Section 1.1.

In this section, we will be concerned mainly with point A) and will produce invariant objects. We will come to point B) in Section 6.

If we consider the averaged system, we see that near resonances of order j, we obtain hyperbolic orbits, whose Lyapunov exponents are $C\varepsilon^{j/2} + O(\varepsilon^{(j+1)/2})$ and such that the angle between the stable and unstable directions are $C\varepsilon^{j/2} + O(\varepsilon^{(j+1)/2})$. Then, applying the implicit function theorem if $N > j$, we get that there are periodic orbits that persist.[16] More importantly for our later applications we obtain that the stable and unstable manifolds are very similar to those of the integrable system. The results are depicted in Figure 9.

We also can show that some of the quasi-periodic orbits with sufficiently large Diophantine constant persist. It is important to note that, one can get invariant tori of two types. One is tori which *"go across"*. These are the *"primary tori"* which are continuous deformations of the tori that were present in the unperturbed system. The tori inside the eyes of the resonance are of a completely different type. These are the *"secondary tori"* which were not present in the unperturbed system, but rather were created by the resonances. Note that as $\varepsilon \to 0$, the eyes become flatter and the limit

---

[16] There are many versions of this argument on persistence of periodic orbits. The basic idea goes back at least to Poincaré and Birkhoff.

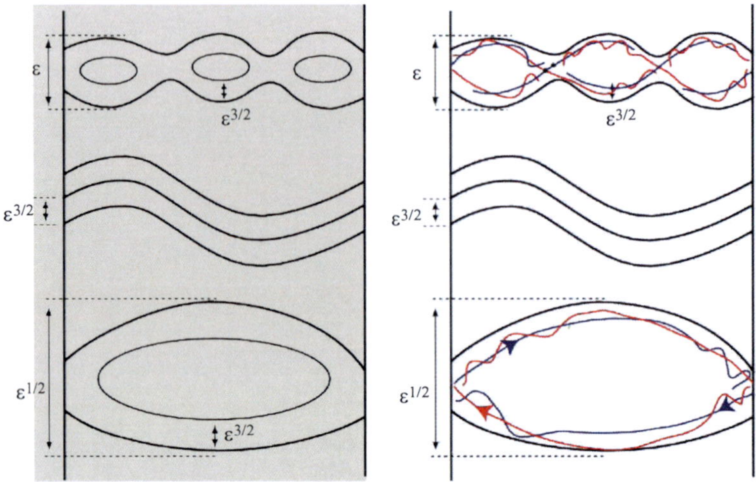

Fig. 9 Illustration of an scaffolding of invariant objects in Λ. These invariant objects are $\varepsilon^{3/2}$ dense in the manifold.

of the tori is just a segment of periodic points. The tori merge with the stable and unstable manifolds. So that at the limit $\varepsilon = 0$ there is change of the topology.

One point which is important is that there are invariant tori very close to the resonances both from the inside and from the outside. These problems had been considered in [Ne184, Her83] under slightly different hypothesis. The method used in [DLS06b] was, mainly, to study in detail the expansion of the action-angle coordinates in a neighborhood of the separatrix. Using the – more or less explicit – formulas one can find in textbooks, it is possible to show that the C^r norm of the change to action angle variables can be bounded by d^{-rA} where A is an explicit number. As it turns out the twist constant does not degenerate – the frequency is singular, but in the good direction that the twist becomes infinite. On the other hand, remember that the error of the averaging method was less than $C_N \varepsilon^N$. It follows that one can apply the KAM theorem at a distance $\varepsilon^{N/B}$. So that, one can get KAM tori – both rotating or librating – faster that a power of ε. The power is arbitrarily large assuming that the system is differentiable enough.

The paper [DLS06b, Section 8] contained other considerations on properties of the KAM tori as graphs and how the set of KAM tori close to the invariant circle can be interpolated with the others where the averaging method is different. The problem is somewhat difficult because depending on how does one relate the ε to the distance to the separatrix, and to the fixed point, the expression of the KAM tori has different leading expressions.

It seems possible that using more the geometric methods developed after [DLS06b] was written, many of these technical calculations can be eliminated or improved in many ways. A significant extension of the results can be found in [DH06]. Another line of argument that seems promising is the use of KAM theory without action angle coordinates – the singularity of the action angle variables

and the different expressions in different regions is one of the source of problems –
[dlLGJV05, FS07] so that one can prove directly the persistence of the orbits in the
level sets of the averaged energy. We hope to come back to this.

In summary, it is possible to show that one can get persistence of many of the
orbits predicted by the averaging method. For our purposes, it is enough to claim
that we get an scaffolding of orbits which are much closer that ε – the size of the
effect of the scattering map.

4.6 Heteroclinic intersections between the invariant objects generated by resonances

Now we want to argue that the objects discussed in the previous sections possess
heteroclinic intersections. Since these objects have different topologies and very
different characteristics, it is useful to use the scattering map and the argument discussed in Section 3.3.

To establish this intersection, we just compute the image of these invariant objects under the scattering map and check whether one can verify (6).

Given that we have computed rather explicitly the leading expansions of the scattering map and the leading expansions of the invariant objects, it is possible to compute the angles of intersections of manifolds. If these angles are not zero in the
leading approximation, then, the implicit function theorem will establish that the
true invariant manifolds satisfy (6).

The effect of the scattering map on the invariant objects is depicted schematically
in Figure 10.

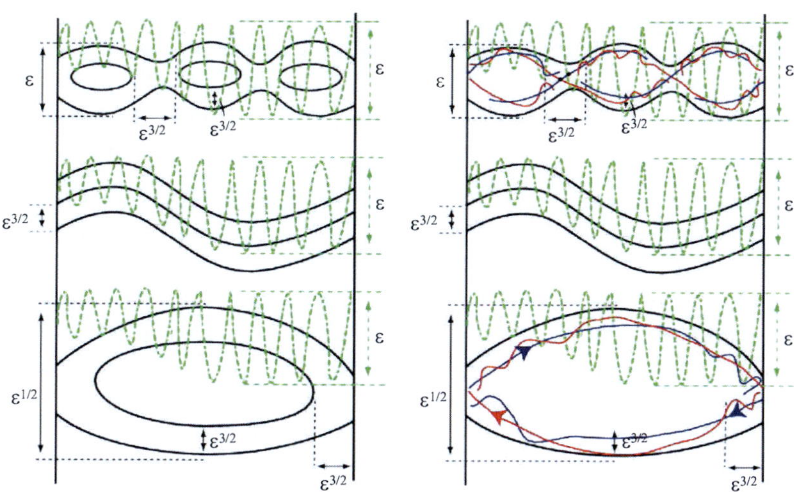

Fig. 10 Effect of the scattering map on the invariant objects found in Figure 9.

Therefore, the above calculation gives – rather explicit – expressions so that, if they do not vanish, then indeed we can obtain heteroclinic excursions between a primary torus below the resonance, to a secondary torus inside the resonance, and then to another torus above the resonance.

The non-vanishing of these explicit expressions giving the angles is a non-degeneracy assumption on the perturbation.

It is intuitively clear that the conditions hold rather generically. Basically, they are a comparison of two effects: the deformation of the invariant objects in Λ and the effect of the scattering map. We note that the first effect, is very much affected by the behavior of the perturbation near Λ, but not by the behavior of the perturbation near Γ. The scattering map has the opposite properties. Hence, if by some miracle, the angles happened to be zero, some perturbation near Γ could destroy this coincidence.

Remark 4.4. The calculation of the scattering map in [DLS06b] was based on traditional methods of perturbations of slow variables. This had the consequence that only the energy component of the scattering map could be computed.

The use of the symplectic properties, which was developed in [DLS06a] and explained in Section 3.7, simplifies and extends the calculation. Note also that we mentioned that the invariant objects are very close to the level sets of a function Ψ_ε. Since the scattering map is a symplectic map close to the identity, the images of the level sets of Ψ_ε will be level sets of the function $\Psi_\varepsilon + \varepsilon\{\Psi_\varepsilon, S_0\} + $ h.o.t. See [DH06].

5 The method of correctly aligned windows

The method of correctly aligned windows is a way of proving that given segments of orbits – with some extra conditions – one can get an orbit that tracks them. Since we never have to consider more than finite orbits, in principle, we do not need the existence of invariant objects. On the other hand, considerations about times become relevant. This is the reason why one gets explicit estimates on diffusion time.

The method has its origins in [Eas78, EM79, Eas89]. The version we will discuss comes from [ZG04, GZ04].

One can think of a window, as a topological version of a rectangle with some marked sides. Windows are correctly aligned when the image of one *stretches across the other*.

A window in a n-dimensional manifold M is a compact subset W of M together with a C^0-coordinate system $(x, y) : U \to \mathbb{R}^u \times \mathbb{R}^s$ defined in neighborhood U of W, where $u + s = n$, such that the homeomorphic image of W through this coordinate system is the rectangle $[0, 1]^u \times [0, 1]^s$. The subset W^- of W that corresponds through the coordinates (x, y) to $\partial[0, 1]^u \times [0, 1]^s$ is called the 'exit set' and the subset W^+ of W that corresponds through the local coordinates (x, y) to $[0, 1]^u \times \partial[0, 1]^s$ is called the 'entry set' of W. Here ∂ denotes the topological boundary of a set. If we want to specify the dimension u of the unstable-like direction and the dimension s of the stable-like direction of a window W, we refer to W as an (u,s)-window. We will assume that $u > 0$.

Let W_1, W_2 be two (u,s)-windows in M, and let $(x_1,y_1) : U_1 \to \mathbb{R}^n$ and $(x_2,y_2) : U_2 \to \mathbb{R}^n$ be the corresponding coordinates systems. Let f be a continuous map on M; we will denote its expression $(x_2,y_2) = f(x_1,y_1)$ in local coordinates also by f. Assume $f(U_1) \subseteq U_2$. We say that W_1 is correctly aligned with W_2 under f provided that the following conditions are satisfied:

(i) $f(\partial[0,1]^u \times [0,1]^s) \cap [0,1]^u \times [0,1]^s = \emptyset,\quad f([0,1]^u \times [0,1]^s) \cap ([0,1]^u \times \partial[0,1]^s) = \emptyset.$

(ii) there exists a point $y_0 \in [0,1]^s$ such that

(a)
$$f([0,1]^u \times \{y_0\}) \subseteq \mathrm{int}\left([0,1]^u \times [0,1]^s \cup (\mathbb{R}^u \setminus (0,1)^u) \times \mathbb{R}^s\right),$$

(b)   The map $A_{y_0} : \mathbb{R}^u \to \mathbb{R}^u$ defined by $A_{y_0}(x) = \pi_1\left(f(x,y_0)\right)$ satisfies

$$A_{y_0}\left(\partial[0,1]^u\right) \subseteq \mathbb{R}^u \setminus [0,1]^u,$$
$$\deg(A_{y_0},0) \neq 0.$$

The main result is that "One can see through correctly aligned windows". See [ZG04, GZ04].

Let W_i be a collection of (u,s)-windows in M, where $i \in \mathbb{Z}$ or $i \in \{0,\ldots,d-1\}$, with $d > 0$ (in the latter case, for convenience, we let $W_i := W_{(i \bmod d)}$ for all $i \in \mathbb{Z}$). Let f_i be a collection of continuous maps on M. If W_i is correctly aligned with W_{i+1}, for all i, then there exists a point $p \in W_0$ such that

$$f_i \circ \ldots \circ f_0(p) \in W_{i+1},$$

Moreover, if $W_{i+k} = W_i$ for some $k > 0$ and all i, then the point p can be chosen so that

$$f_{k-1} \circ \ldots \circ f_0(p) = p.$$

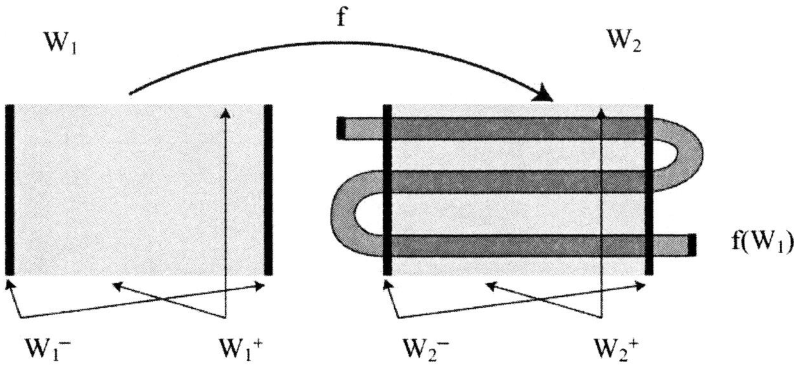

If one takes very small windows, the behavior of the windows is determined by the derivative of the orbit. If the orbit is hyperbolic, by choosing the rectangles as products of balls along the stable direction and the unstable direction with the unstable being the exit direction, we can get the correct alignment. Then the result that one can see through chains of correctly aligned windows becomes the standard shadowing result. On the other hand, the method is more flexible since we can choose the sizes of the windows and the time we take to put them along the orbits. This has some advantages for non-uniformly hyperbolic systems. See the proof of the non-uniformly hyperbolic closing lemma in [Pol93].

On the other hand, the windows do not need to be small. As we will see in the next section, one can take advantage of large scale effects to get the alignment of windows. Notably, when one has some twist – shear – that causes some stretching, this can be used in place of the stretching caused by the hyperbolicity. It is also important to notice that, to check whether windows are well aligned or not, one can just study what happens on the boundary.

In our applications the time of diffusion can be computed by the time that it takes the windows to stretch.

An important technical tool [GL06a] is that, for systems that are close to product of systems, one can construct product windows and verify the alignment checking conditions on each of the factors.

6 The large gap model: the method of correctly aligned windows

The method of correctly aligned windows has been applied to the large gap model in [GL06a, GL06b].

The construction of windows adapted to the problem of diffusion basically requires to choose the parameters of a sequence of windows (the length of the sides, the center in a good coordinate system) and choose the times taken to go from one to the next. Then, one has to verify that all the steps match. In practice this amounts to choosing two dozen of parameters and verifying about a dozen of trivial inequalities.

Even if verifying the validity of the choices is not very hard, coming up with the good choices requires a good understanding of the behavior of the model. We now discuss some of the reasons behind the choices.

We have already discussed the pseudo-orbits that appear. We go from the intersection to the manifold, rotate around and then escape back again.

It is important to note that even the unperturbed system is not hyperbolic. The vector along the separatrices of the pendulum contracts both in the future and in the past. So that, these vectors in the intersection of the stable and unstable subspace and the forward Lyapunov exponent is different from the backward Lyapunov exponent.[17]

---

[17] The equality of these two exponents was called *regularity* by Lyapunov and plays a very important role. See [BP01].

The construction of windows, however, can take advantage of the fact that there are some direction with good hyperbolicity for a long stretch ($O(|\ln(\varepsilon)|)$) of time while the orbit moves from Γ to Λ or back. The fact that one can control the behavior in the hyperbolic directions is possible because of the transversal intersection. (On the other hand, the windowing method, being a topological method could work with much weaker assumptions [GR04].)

The treatment of the center directions is much more interesting. Of course, the windows that start close to Λ, go to Γ and come back to Λ are very well described by the scattering map. One does not have any hyperbolicity in these directions, but on the other hand, the twist does distort the windows and one can use this distortion to construct windows that are correctly aligned. This is very similar to the *torsion-hyperbolicity* mechanism.

In the paper, [GL06a] the windows were taken very thin in the action variables, but they were taken of order 1 in the angle. This allowed to avoid discussions of *ergodization times* and produced rather concrete estimates on the time. In [GL06b] the windows are chosen in a scale $O(1/|\ln(\varepsilon)|)$. This, of course, goes to zero, but it is much larger than the scales of the resonance. The orbits also do not come too close to the manifold. This has the effect that the method does not need to analyze what happens in the resonances. This method also leads to times of order $O(\varepsilon|\ln(\varepsilon)|)$ that – up to, perhaps, a constant – match the upper bounds obtained in [BBB03]. Similar results appear in [Tre04].

7 The large gap model in higher dimensions

Some of the analysis in Section 4 can be adapted to higher dimensional models. See [DLS07].

We consider the same model as in (1), but now I, ϕ are higher dimensional variables. Again, for simplicity, for the moment, we assume that the perturbation h is a trigonometric polynomial.

The averaging method described in Section 4.3 can be carried out pretty much the same way. The only difference is that now, that the resonances $\omega(I) \cdot k = n$ are codimension 1 manifolds. If the number of degrees of freedom is more than 1, there will be intersections of these resonant surfaces. The intersection of two independent resonances are called *multiple resonances*. The multiplicity of the resonance – not to be confused with the order – is the dimension of the module of vectors k, n for which there is resonance relation. The order is the power of ε of the terms that cannot be eliminated.

The mathematical analysis of multiple resonances and their role in diffusion remains a very interesting problem. Very important progress has been done in [Hal97, Hal99].

Nevertheless in [DLS07] it is argued that there exist diffusing orbits – under the assumption that h is polynomial – plus some non-degeneracy assumptions.

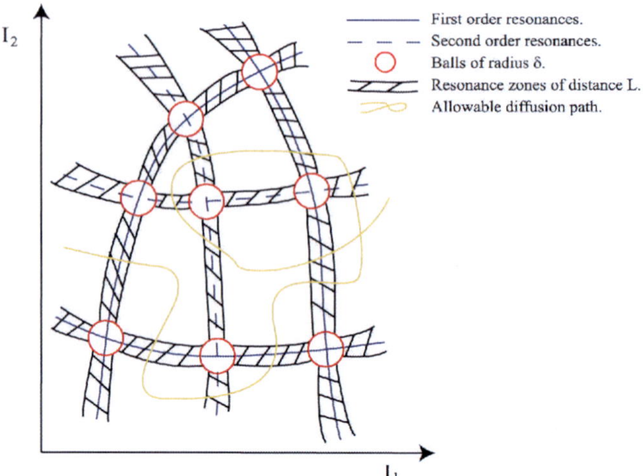

Fig. 11 Illustration of the paths of diffusion avoiding higher order resonances.

The key observation is that, under a twist condition, the multiple resonances can be contoured. (Since they happen on sets of codimension 2 or higher, there are paths that go around them.)

The analysis of resonances of order 1 in higher dimensional systems is very similar to the analysis carried in Section 4.[18]

The upshot is that, under explicit non-degeneracy conditions, for any path in the space of actions that crosses only multiplicity one resonances, for $0 < \varepsilon$ small enough there orbits whose actions evolve along the path – up to errors that go to zero with ε.

8 Instability caused by normally hyperbolic laminations

One of the standard heuristics in the numerical studies is that of *modulational diffusion* [Chi79, TLL80]. It is often described as saying that *A degree of freedom becomes chaotic and drives another one.*

Mathematically, one can formulate this as perturbing a system which is the product of a system with some hyperbolic behavior, and another system which is integrable: $F_\varepsilon = F_h \times F_i + O(\varepsilon)$, where $F_h(\Lambda) = \Lambda$ and Λ is a hyperbolic set, and $F_i : M \mapsto M$ is an integrable map.

In the mathematical literature, some rigorous results have been obtained. The paper [MS02] constructed a specific system of this type. The paper [Moe02] used topological methods in two dimensions. Closely related to this paper is [EMR01].

---

[18] The scattering map does not require any change, but the persistence of tori of lower dimension becomes more complicated (one has to use KAM theory rather than the implicit function theorem). Also the secondary tori require some extra considerations.

One systematic way to make sense of the above [Lla04, dlL06] is to observe that the set $\cup_{x \in \Lambda} \{x\} \times M$ is a normally hyperbolic lamination for F_0. See Appendix A.3. The laminae are $\{x\} \times M$ are permuted by the map and the normal directions are hyperbolic.

It was shown in [HPS77, Ch. 15] that these structures persist under perturbations in the sense that one can get slightly deformed collections of laminae which are also permuted under the map F_ε. Of course, the dynamics on these laminae is not integrable anymore. The dynamics on the integrable parts is a random composition of maps, which one can consider as uncoupled as in [MS02].

8.1 Models with two time scales: geodesic flows, billiards with moving boundaries, Littlewood problems

The above mechanism is particularly effective in systems that have two time scales.

One important system is the model of a geodesic flow perturbed by a periodic or quasi-periodic potential considered by other methods in [Mat96, BT99, DLS00, DLS06c].

This dynamical system is defined on the cotangent bundle T^*M of a compact manifold M. It has the form:

$$\dot{p} = -\nabla V(q, \omega t), \quad \dot{q} = p, \tag{1}$$

where the potential $V : M \times \mathbb{T}^d$ and $\omega \in \mathbb{R}^d$ is a non-resonant vector. When $d = 1$, the potential depends periodically on time.

We note that the system satisfies some scaling properties. Setting $p = \varepsilon \tilde{p}$, $q = \tilde{q}$, $t = \varepsilon^{-1} \tilde{t}$ and denoting by $'$ the derivative with respect to $\tilde{t}$, the system, above becomes

$$\tilde{p}' = -\varepsilon^2 \nabla V(\tilde{q}, \varepsilon \omega t) \quad \tilde{q}' = \tilde{q} \tag{2}$$

So that, for high energy, the potential can be considered as a slow and weak perturbation.

We will assume that the unperturbed geodesic flow has a horseshoe in the unit energy surface. Using the above scaling, we obtain that, considering the system for all the energies it possesses an invariant lamination. By the theory of persistence of normally hyperbolic invariant laminations, we obtain that this structure just gets deformed.

If $\gamma_1, \ldots, \gamma_N$ are periodic orbits in the horseshoe, we denote $|\gamma_i|$ the period and define:

$$G_i(t) = \frac{1}{|\gamma_i|} \int_0^{|\gamma_i|} \frac{\partial}{\partial t} V(\gamma_i(s), t) \, ds$$

This has the meaning of the gain of energy per unit time for orbits that stay in a close proximity to the periodic orbit. Note that $\int_0^1 G_i(t) \, dt = 0$.

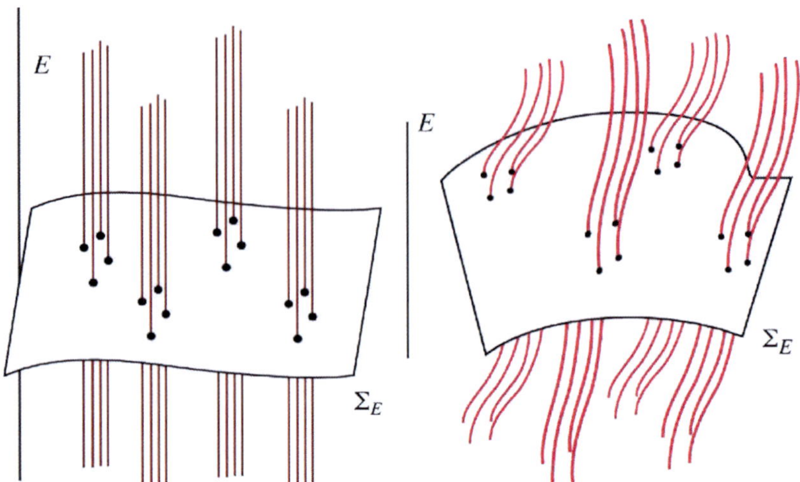

Fig. 12 Invariant normally hyperbolic laminations associated to the geodesic flow and the periodic geodesic flow.

Recall that, in the horseshoe, we have a symbolic dynamics for the hyperbolic orbits. That is, if we fix neighborhoods of these orbits, we can move from one to the other in arbitrary order. Each of the steps can be accomplished in a fixed time.

By the persistence of the normally hyperbolic laminations, the same property persists when we consider the perturbation by the potential. So, we can switch from a neighborhood of an orbit to another one in a fixed time for the geodesic flow. For the potential, this is a slow time.

In the periodic case, $d = 1$, we assume without loss of generality that $V(q,t + 1) = V(q,t)$, $\omega = 1$. If we assume that there exist $0 = a_0 < a_1 < \cdots < a_N = 1$ in such a way that

$$A \equiv \sum_{i=1}^{N} \int_{a_{i-1}}^{a_i} G_i(t)\, dt > 0 \tag{3}$$

then, we can construct orbits whose energy as function of time is larger than $At - B$. The idea is very simple. We stay close to γ_1 during the macroscopic times $[a_0, a_1]$. Using the symbolic dynamics, we can move to γ_2, etc. Hence, during a cycle, we have gained roughly A.

In the quasi-periodic case, we just need to assume that it is possible to write $\mathbb{T}^d = \cup_{i=1}^{N} O_i$ where O_i are sets with smooth boundary transversal to the rotation, which only overlap in the boundary, and such that $A \equiv \sum_{i=1}^{N} \int_{O_i} G_i(\tau)\, d\tau > 0$.

If we look at the symbolic dynamics, we see that the space of sequences that lead to linear gain in energy has positive Hausdorff dimension. Then, using that the conjugacy given by the stability, we obtain that, when the (3) are satisfied, the orbits with energy growing linearly are of positive Hausdorff dimension.

It is shown in [Lla04] that if the metric is of negative curvature and, in case that it has dimension $\geqslant 3$, that it satisfies some pinching conditions, then, the only C^3

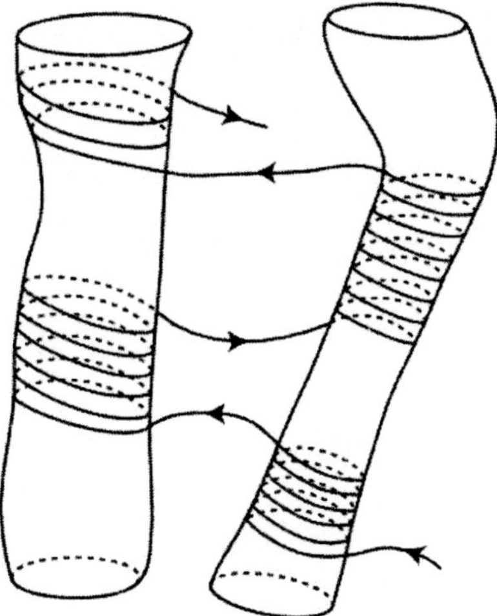

Fig. 13 Illustration of the mechanisms of gain of energy based in locally hyperbolic manifolds.

potentials for which it is impossible to find orbits satisfying the hypothesis of the above result are the potentials of the form $V(q,t) = V_1(q) + V_2(t)$.

Very similar analysis applies to other systems which have two scales.

One example is what we call the *Littlewood models* in higher dimensions.

$$H(p,q,t) = \frac{1}{2}p^2 + V_n(q) + V_m(q,t) \tag{4}$$

where $p,q \in \mathbb{R}^d$, $d \geqslant 2$, V_n, V_m are homogeneous of degree n, m respectively, $n > m$, $n > 2, V_n > 0$, V_m periodic or quasi-periodic in t. The fact that different terms have different homogeneities makes the geometric analysis similar to that of the geodesic flows.

In the case $d = 1$, [Lit66a, Lit66b] constructed examples of potentials – which are not polynomials and with not very smooth dependence on time – with orbits with unbounded energy. Unfortunately, the papers contain a serious error. The papers [LL91, LZ95] showed that for terms which are like polynomials, and with smooth quasiperiodic perturbations the orbits stay bounded. An excellent survey of the history of these models and simplification of the results is [Lev92].

When the number of degrees of freedom is greater or equal than 2, a very similar analysis to the one carried out above for geodesic flows applies. We note that if we scale, $p = \varepsilon^{m/2}\tilde{p}, q = \varepsilon\tilde{q}, t = \varepsilon^{-m}$ we get that the system (4) can be rewritten as:

$$H(\tilde{p},\tilde{q},\tilde{t}) = \frac{1}{2}(\tilde{p})^2 + V_m(\tilde{q}) + \varepsilon^{2m-n}V_n(q,\varepsilon^m t)$$

so that the low degree polynomial can be considered as a small and slow perturbation and an analysis very similar to the one carried above for the geodesic flow applies. The only difference is that one gets that the orbits grow like a power. This is optimal due to a calculation in [LZ95].

One interesting example, which does not fit in the above theory proposed as a challenge by M. Levi is the system defined by a Hamiltonian

$$\frac{1}{2}p^2 + q_1^6 + q_1^4 + \eta q_1^2 q_2^2 + q_1 f(t)$$

This is a challenging model because for large energy, the dominant term is the one degree of freedom system for which the theorem of [LZ95] applies.

Another model which has scaling behavior is the billiard with moving boundaries. A higher dimensional model of the Fermi acceleration.

For all these systems, when they are sufficiently chaotic, it seems possible to derive – heuristically – stochastic models for the growth of energy. These stochastic models can be analyzed rigorously and the final results compared satisfactorily with numerical simulations [DdlL06]. Even if parts of a stochastic theory of diffusion can be made rigorous, deriving a fully rigorous stochastic theory of diffusion remains a very challenging problem.

Acknowledgements The work or M. G. and R. L has been supported by NSF grants. Visits of R. L to Barcelona were supported by ICREA. Some of the work reported here was supported ny MCyT-FEDER Grants BFM2003-9504 and MTM2006-00478.

We thank the organizers of the Institute for a very warm hospitality (besides efficient organization). The participants in the institute made very valuable comments and suggestions. Particularly, R. Calleja, J. Mireles-James, P. Roldán and R. Treviño, helped in the preparation of the notes. H. Lomelí made many suggestions.

References

[AA67] V.I. Arnold and A. Avez. *Ergodic problems of classical mechanics.* Benjamin, New York, 1967.

[AKN88] V.I. Arnold, V.V. Kozlov, and A.I. Neishtadt. *Dynamical Systems III*, volume 3 of *Encyclopaedia Math. Sci.* Springer, Berlin, 1988.

[Ale68a] V. M. Alekseev. Quasirandom dynamical systems. I. Quasirandom diffeomorphisms. *Mat. Sb. (N.S.)*, 76 (118):72–134, 1968.

[Ale68b] V. M. Alekseev. Quasirandom dynamical systems. II. One-dimensional nonlinear vibrations in a periodically perturbed field. *Mat. Sb. (N.S.)*, 77 (119):545–601, 1968.

[Ale69] V. M. Alekseev. Quasirandom dynamical systems. III. Quasirandom vibrations of one-dimensional oscillators. *Mat. Sb. (N.S.)*, 78 (120):3–50, 1969.

[Ale81] V.M. Alekseev. Quasirandom oscillations and qualitative questions in celestial mechanics. *Transl., Ser. 2, Am. Math. Soc.*, 116:97–169, 1981.

[AM78] R. Abraham and J. E. Marsden. *Foundations of mechanics.* Benjamin/Cummings Publishing Co. Inc. Advanced Book Program, Reading, MA, 1978. Second edition, revised and enlarged, With the assistance of Tudor Raţiu and Richard Cushman.

[Ang87] Sigurd Angenent. The shadowing lemma for elliptic PDE. In *Dynamics of infinite-dimensional systems (Lisbon, 1986)*, pages 7–22. Springer, Berlin, 1987.

[AO98] J. Miguel Alonso and R. Ortega. Roots of unity and unbounded motions of an asymmetric oscillator. *J. Differential Equations*, 143(1):201–220, 1998.

[Arn63] V. I. Arnol'd. Small denominators and problems of stability of motion in classical and celestial mechanics. *Russian Math. Surveys*, 18(6):85–191, 1963.

[Arn64]   V.I. Arnold. Instability of dynamical systems with several degrees of freedom. *Sov. Math. Doklady*, 5:581–585, 1964.

[Arn89]   V. I. Arnol'd. *Mathematical methods of classical mechanics*. Springer, New York, second edition, 1989. Translated from the Russian by K. Vogtmann and A. Weinstein.

[Ban88]   V. Bangert. Mather sets for twist maps and geodesics on tori. In *Dynamics reported, Vol. 1*, pages 1–56. Teubner, Stuttgart, 1988.

[BBB03]   M. Berti, L. Biasco, and P. Bolle. Drift in phase space: a new variational mechanism with optimal diffusion time. *J. Math. Pures Appl. (9)*, 82(6):613–664, 2003.

[BCV01]   U. Bessi, L. Chierchia, and E. Valdinoci. Upper bounds on Arnold diffusion times via Mather theory. *J. Math. Pures Appl. (9)*, 80(1):105–129, 2001.

[Bes96]   U. Bessi. An approach to Arnol'd's diffusion through the calculus of variations. *Nonlinear Anal.*, 26(6):1115–1135, 1996.

[BK94]   I. U. Bronstein and A. Ya. Kopanskiĭ. *Smooth invariant manifolds and normal forms*, volume 7 of *World Scientific Series on Nonlinear Science. Series A: Monographs and Treatises*. World Sci. Publ., River Edge, NJ, 1994.

[BK05]   J. Bourgain and V. Kaloshin. On diffusion in high-dimensional Hamiltonian systems. *J. Funct. Anal.*, 229(1):1–61, 2005.

[BLZ98]   P. W. Bates, K. Lu, and C. Zeng. Existence and persistence of invariant manifolds for semiflows in Banach space. *Mem. Amer. Math. Soc.*, 135(645):viii+129, 1998.

[BP01]   L. Barreira and Ya. Pesin. Lectures on Lyapunov exponents and smooth ergodic theory. In *Smooth ergodic theory and its applications (Seattle, WA, 1999)*, pages 3–106. Amer. Math. Soc., Providence, RI, 2001. Appendix A by M. Brin and Appendix B by D. Dolgopyat, H. Hu and Pesin.

[BT79]   M. Breitenecker and W. Thirring. Scattering theory in classical dynamics. *Riv. Nuovo Cimento (3)*, 2(4):21, 1979.

[BT99]   S. Bolotin and D. Treschev. Unbounded growth of energy in nonautonomous Hamiltonian systems. *Nonlinearity*, 12(2):365–388, 1999.

[Car81]   J. R. Cary. Lie transform perturbation theory for Hamiltonian systems. *Phys. Rep.*, 79(2):129–159, 1981.

[CDMR06]   E. Canalias, A. Delshams, J. Masdemont, and P. Roldán. The scattering map in the planar restricted three body problem. *Celestial Mech. Dynam. Astronom.*, 95(1–4):155–171, 2006.

[CG98]   L. Chierchia and G. Gallavotti. Drift and diffusion in phase space. *Ann. Inst. H. Poincaré Phys. Théor.*, 60(1):1–144, 1994. Erratum, *Ann. Inst. H. Poincaré, Phys. Théor.* 68 (1):135, 1998.

[CG03]   Jacky Cresson and Christophe Guillet. Periodic orbits and Arnold diffusion. *Discrete Contin. Dyn. Syst.*, 9(2):451–470, 2003.

[Chi79]   B.V. Chirikov. A universal instability of many-dimensional oscillator systems. *Phys. Rep.*, 52(5):264–379, 1979.

[CI99]   G. Contreras and R. Iturriaga. *Global minimizers of autonomous Lagrangians*. 22° Colóquio Brasileiro de Matemática. [22nd Brazilian Mathematics Colloquium]. Instituto de Matemática Pura e Aplicada (IMPA), Rio de Janeiro, 1999.

[Con68]   C. C. Conley. Low energy transit orbits in the restricted three-body problem. *SIAM J. Appl. Math.*, 16:732–746, 1968.

[Con78]   C. Conley. *Isolated invariant sets and the Morse index*, volume 38 of *CBMS Regional Conference Series in Mathematics*. Amer. Math. Soc., Providence, RI, 1978.

[Cre97]   J. Cresson. A λ-lemma for partially hyperbolic tori and the obstruction property. *Lett. Math. Phys.*, 42(4):363–377, 1997.

[CY04a]   C.-Q. Cheng and J. Yan. Arnold diffusion in Hamiltonian systems: 1 a priori unstable case. Preprint 04–265, mp_arc@math.utexas.edu, 2004.

[CY04b]   C.-Q. Cheng and J. Yan. Existence of diffusion orbits in a priori unstable Hamiltonian systems. *J. Differential Geom.*, 67(3):457–517, 2004.

[DdlL06]   D. Dolgopyat and R. de la Llave. Stochastic acceleration. Manusrcript, 2006.

[DDLLS00] A. Delshams, R. de la Llave, and T. M. Seara. Unbounded growth of energy in periodic perturbations of geodesic flows of the torus. In *Hamiltonian systems and celestial mechanics (Pátzcuaro, 1998)*, volume 6 of *World Sci. Monogr. Ser. Math.*, pages 90–110. World Sci. Publ., River Edge, NJ, 2000.

[DG96] A. Delshams and P. Gutiérrez. Effective stability and KAM theory. *J. Differential Equations*, 128(2):415–490, 1996.

[DH06] A. Delshams and G. Huguet. The large gap problem in arnold diffusion for non polynomial perturbations of an a-priori unstable hamiltonian system. Manuscript, 2006.

[DLC83] R. Douady and P. Le Calvez. Exemple de point fixe elliptique non topologiquement stable en dimension 4. *C. R. Acad. Sci. Paris Sér. I Math.*, 296(21):895–898, 1983.

[dlL06] R. de la Llave. Some recent progress in geometric methods in the instability problem in Hamiltonian mechanics. In *International Congress of Mathematicians. Vol. II*, pages 1705–1729. Eur. Math. Soc., Zürich, 2006.

[dlLGJV05] R. de la Llave, A. González, À. Jorba, and J. Villanueva. KAM theory without action-angle variables. *Nonlinearity*, 18(2):855–895, 2005.

[dlLRR07] R. de la Llave and R. Ramirez-Ros. Instability in billiards with moving boundaries. Manuscript, 2007.

[DLS00] A. Delshams, R. de la Llave, and T.M. Seara. A geometric approach to the existence of orbits with unbounded energy in generic periodic perturbations by a potential of generic geodesic flows of $\mathbf{T}^2$. *Comm. Math. Phys.*, 209(2):353–392, 2000.

[DLS03] Amadeu Delshams, Rafael de la Llave, and Tere M. Seara. A geometric mechanism for diffusion in Hamiltonian systems overcoming the large gap problem: announcement of results. *Electron. Res. Announc. Amer. Math. Soc.*, 9:125–134 (electronic), 2003.

[DLS06a] A. Delshams, R. de la Llave, and T. M. Seara. Geometric properties of the scattering map to a normally hyperbolic manifold. *Adv. Math.*, 2006. To appear.

[DLS06b] A. Delshams, R. de la Llave, and T. M. Seara. A geometric mechanism for diffusion in Hamiltonian systems overcoming the large gap problem: heuristics and rigorous verification on a model. *Mem. Amer. Math. Soc.*, 179(844):viii+141, 2006.

[DLS06c] A. Delshams, R. de la Llave, and T. M. Seara. Orbits of unbounded energy in quasi-periodic perturbations of geodesic flows. *Adv. Math.*, 202(1):64–188, 2006.

[DLS07] A. Delshams, R. de la Llave, and T. M. Seara. Instability of high dimensional hamiltonian systems: multiple resonances do not impede diffusion. 2007.

[Dou88] R. Douady. Regular dependence of invariant curves and Aubry-Mather sets of twist maps of an annulus. *Ergodic Theory Dynam. Systems*, 8(4):555–584, 1988.

[Dou89] R. Douady. Systèmes dynamiques non autonomes: démonstration d'un théorème de Pustyl'nikov. *J. Math. Pures Appl. (9)*, 68(3):297–317, 1989.

[DR97] A. Delshams and R. Ramírez-Ros. Melnikov potential for exact symplectic maps. *Comm. Math. Phys.*, 190:213–245, 1997.

[Eas78] R. W. Easton. Homoclinic phenomena in Hamiltonian systems with several degrees of freedom. *J. Differential Equations*, 29(2):241–252, 1978.

[Eas89] R. W. Easton. Isolating blocks and epsilon chains for maps. *Phys. D*, 39(1):95–110, 1989.

[EM79] R. W. Easton and R. McGehee. Homoclinic phenomena for orbits doubly asymptotic to an invariant three-sphere. *Indiana Univ. Math. J.*, 28(2):211–240, 1979.

[EMR01] R. W. Easton, J. D. Meiss, and G. Roberts. Drift by coupling to an anti-integrable limit. *Phys. D*, 156(3-4):201–218, 2001.

[Fen72] N. Fenichel. Persistence and smoothness of invariant manifolds for flows. *Indiana Univ. Math. J.*, 21:193–226, 1971/1972.

[Fen77] N. Fenichel. Asymptotic stability with rate conditions. II. *Indiana Univ. Math. J.*, 26(1):81–93, 1977.

[Fen79] N. Fenichel. Geometric singular perturbation theory for ordinary differential equations. *J. Differential Equations*, 31(1):53–98, 1979.

[Fen74]   N. Fenichel. Asymptotic stability with rate conditions. *Indiana Univ. Math. J.*, 23:1109–1137, 1973/74.

[FGL05]   C. Froeschlé, M. Guzzo, and E. Lega. Local and global diffusion along resonant lines in discrete quasi-integrable dynamical systems. *Celestial Mech. Dynam. Astronom.*, 92(1–3):243–255, 2005.

[FLG06]   C. Froeschlé, E. Lega, and M. Guzzo. Analysis of the chaotic behaviour of orbits diffusing along the Arnold web. *Celestial Mech. Dynam. Astronom.*, 95(1–4):141–153, 2006.

[FM00]   E. Fontich and P. Martín. Differentiable invariant manifolds for partially hyperbolic tori and a lambda lemma. *Nonlinearity*, 13(5):1561–1593, 2000.

[FM01]   E. Fontich and P. Martín. Arnold diffusion in perturbations of analytic integrable Hamiltonian systems. *Discrete Contin. Dynam. Systems*, 7(1):61–84, 2001.

[FM03]   E. Fontich and P. Martín. Hamiltonian systems with orbits covering densely submanifolds of small codimension. *Nonlinear Anal.*, 52(1):315–327, 2003.

[FS07]   R. Fontich, E. de la Llave and Y. Sire. Construction of invariant whiskered tori by a parameterization method. 2007. Manuscript.

[Gar00]   Antonio García. Transition tori near an elliptic fixed point. *Discrete Contin. Dynam. Systems*, 6(2):381–392, 2000.

[GL06a]   M. Gidea and R. de la Llave. Arnold diffusion with optimal time in the large gap problem. Preprint, 2006.

[GL06b]   M. Gidea and R. de la Llave. Topological methods in the instability problem of Hamiltonian systems. *Discrete Contin. Dyn. Syst.*, 14(2):295–328, 2006.

[GLF05]   M. Guzzo, E. Lega, and C. Froeschlé. First numerical evidence of global Arnold diffusion in quasi-integrable systems. *Discrete Contin. Dyn. Syst. Ser. B*, 5(3):687–698, 2005.

[GR04]   M. Gidea and C. Robinson. Symbolic dynamics for transition tori II. In *New advances in celestial mechanics and Hamiltonian systems*, pages 95–109. Kluwer, Dordrecht, The Netherlands, 2004.

[GZ04]   M. Gidea and P. Zgliczyński. Covering relations for multidimensional dynamical systems. II. *J. Differential Equations*, 202(1):59–80, 2004.

[Hal97]   G. Haller. Universal homoclinic bifurcations and chaos near double resonances. *J. Statist. Phys.*, 86(5-6):1011–1051, 1997.

[Hal99]   G. Haller. *Chaos near resonance.* Springer, New York, 1999.

[HdlL06a]   À. Haro and R. de la Llave. Manifolds on the verge of a hyperbolicity breakdown. *Chaos*, 16(1):013120, 8, 2006.

[HdlL06b]   À. Haro and R. de la Llave. A parameterization method for the computation of invariant tori and their whiskers in quasi-periodic maps: numerical algorithms. *Discrete Contin. Dyn. Syst. Ser. B*, 6(6):1261–1300 (electronic), 2006.

[HdlL06c]   A. Haro and R. de la Llave. A parameterization method for the computation of invariant tori and their whiskers in quasi-periodic maps: rigorous results. *J. Differential Equations*, 228(2):530–579, 2006.

[HdlL07]   A. Haro and R. de la Llave. A parameterization method for the computation of invariant tori and their whiskers in quasi-periodic maps: explorations and mechanisms for the breakdown of hyperbolicity. *SIAM J. Appl. Dyn. Syst.*, 6(1):142–207 (electronic), 2007.

[Hed32]   G. A. Hedlund. Geodesics on a two-dimensional Riemannian manifold with periodic coefficients. *Ann. of Math.*, 33:719–739, 1932.

[Her83]   M. R. Herman. *Sur les courbes invariantes par les difféomorphismes de l'anneau. Vol. 1*, volume 103 of *Astérisque*. Société Mathématique de France, Paris, 1983.

[HM82]   P. J. Holmes and J. E. Marsden. Melnikov's method and Arnold diffusion for perturbations of integrable Hamiltonian systems. *J. Math. Phys.*, 23(4):669–675, 1982.

[HP70]   M. W. Hirsch and C. C. Pugh. Stable manifolds and hyperbolic sets. In S. Chern and S. Smale, editors, *Global Analysis (Proc. Sympos. Pure Math., Vol. XIV, Berkeley, Calif., 1968)*, pages 133–163, Amer. Math. Soc. Providence, RI, 1970.

[HPS77] M. W. Hirsch, C. C. Pugh, and M. Shub. *Invariant manifolds*, volume 583 of *Lecture Notes in Math.* Springer, Berlin, 1977.

[Hun68] W. Hunziker. The s-matrix in classical mechanics. *Com. Math. Phys.*, 8(4):282–299, 1968.

[HW99] B. Hasselblatt and A. Wilkinson. Prevalence of non-Lipschitz Anosov foliations. *Ergodic Theory Dynam. Systems*, 19(3):643–656, 1999.

[KP90] U. Kirchgraber and K. J. Palmer. *Geometry in the neighborhood of invariant manifolds of maps and flows and linearization.* Longman Scientific & Technical, Harlow, 1990.

[KPT95] T. Krüger, L. D. Pustyl′nikov, and S. E. Troubetzkoy. Acceleration of bouncing balls in external fields. *Nonlinearity*, 8(3):397–410, 1995.

[Lev92] M. Levi. On Littlewood's counterexample of unbounded motions in superquadratic potentials. In *Dynamics reported: expositions in dynamical systems*, pages 113–124. Springer, Berlin, 1992.

[Lit66a] J. E. Littlewood. Unbounded solutions of an equation $\ddot{y} + g(y) = p(t)$, with $p(t)$ periodic and bounded, and $g(y)/y \to \infty$ as $y \to \pm\infty$. *J. London Math. Soc.*, 41:497–507, 1966.

[Lit66b] J. E. Littlewood. Unbounded solutions of $\ddot{y} + g(y) = p(t)$. *J. London Math. Soc.*, 41:491–496, 1966.

[LL91] S. Laederich and M. Levi. Invariant curves and time-dependent potentials. *Ergodic Theory Dynam. Systems*, 11(2):365–378, 1991.

[Lla01] R. de la Llave. A tutorial on KAM theory. In *Smooth ergodic theory and its applications (Seattle, WA, 1999)*, pages 175–292. Amer. Math. Soc., Providence, RI, 2001.

[Lla02] R. de la Llave. Orbits of unbounded energy in perturbations of geodesic flows by periodic potentials. a simple construction. Preprint, 2002.

[Lla04] R. de la Llave. Orbits of unbounded energy in perturbation of geodesic flows: a simple mechanism. Preprint, 2004.

[LM88] P. Lochak and C. Meunier. *Multiphase Averaging for Classical Systems*, volume 72 of *Appl. Math. Sci.* Springer, New York, 1988.

[LMM86] R. de la Llave, J. M. Marco, and R. Moriyón. Canonical perturbation theory of Anosov systems and regularity results for the Livšic cohomology equation. *Ann. of Math. (2)*, 123(3):537–611, 1986.

[LMS03] P. Lochak, J.-P. Marco, and D. Sauzin. On the splitting of invariant manifolds in multidimensional near-integrable Hamiltonian systems. *Mem. Amer. Math. Soc.*, 163(775):viii+145, 2003.

[LP66] M. de La Place. *Celestial mechanics. Vols. I–IV.* Translated from the French, with a commentary, by Nathaniel Bowditch. Chelsea Publishing, Bronx, NY, 1966.

[LT83] M. A. Lieberman and Jeffrey L. Tennyson. Chaotic motion along resonance layers in near-integrable Hamiltonian systems with three or more degrees of freedom. In C. Wendell Horton, Jr. and L. E. Reichl, editors, *Long-time prediction in dynamics (Lakeway, Tex., 1981)*, pages 179–211. Wiley, New York, 1983.

[LY97] M. Levi and J. You. Oscillatory escape in a Duffing equation with a polynomial potential. *J. Differential Equations*, 140(2):415–426, 1997.

[LZ95] M. Levi and E. Zehnder. Boundedness of solutions for quasiperiodic potentials. *SIAM J. Math. Anal.*, 26(5):1233–1256, 1995.

[Mañ97] R. Mañé. Lagrangian flows: the dynamics of globally minimizing orbits. *Bol. Soc. Brasil. Mat. (N.S.)*, 28(2):141–153, 1997.

[Mat93] J. N. Mather. Variational construction of connecting orbits. *Ann. Inst. Fourier (Grenoble)*, 43(5):1349–1386, 1993.

[Mat96] J. N. Mather. Graduate course at Princeton, 95–96, and Lectures at Penn State, Spring 96, Paris, Summer 96, Austin, Fall 96.

[Mey91] K. R. Meyer. Lie transform tutorial. II. In Kenneth R. Meyer and Dieter S. Schmidt, editors, *Computer aided proofs in analysis (Cincinnati, OH, 1989)*, volume 28 of *IMA Vol. Math. Appl.*, pages 190–210. Springer, New York, 1991.

[Moe96]   R. Moeckel. Transition tori in the five-body problem. *J. Differential Equations*, 129(2):290–314, 1996.

[Moe02]   R. Moeckel. Generic drift on Cantor sets of annuli. In *Celestial mechanics (Evanston, IL, 1999)*, volume 292 of *Contemp. Math.*, pages 163–171. Amer. Math. Soc., Providence, RI, 2002.

[Moe05]   R. Moeckel. A variational proof of existence of transit orbits in the restricted three-body problem. *Dyn. Syst.*, 20(1):45–58, 2005.

[Mor24]   M. Morse. A fundamental class of geodesics on any closed surface of genus greater than one. *Trans. Amer. Math. Soc.*, 26:26–60, 1924.

[Mos69]   J. Moser. On a theorem of Anosov. *J. Differential Equations*, 5:411–440, 1969.

[Mos73]   J. Moser. *Stable and random motions in dynamical systems*. Princeton University Press, Princeton, NJ, 1973. With special emphasis on celestial mechanics, Hermann Weyl Lectures, the Institute for Advanced Study, Princeton, NJ, *Annals Math. Studies*, No. 77.

[MS02]   J.-P. Marco and D. Sauzin. Stability and instability for Gevrey quasi-convex near-integrable Hamiltonian systems. *Publ. Math. Inst. Hautes Études Sci.*, (96):199–275 (2003), 2002.

[Nei84]   A. I. Neishtadt. The separation of motions in systems with rapidly rotating phase. *J. Appl. Math. Mech.*, 48(2):133–139, 1984.

[Nel69]   E. Nelson. *Topics in dynamics. I: Flows*. Mathematical Notes. Princeton University Press, Princeton, NJ, 1969.

[Nie07]   L. Niederman. Prevalence of exponential stability among nearly integrable Hamiltonian systems. *Ergodic Theory Dynam. Systems*, 27(3):905–928, 2007.

[Ort97]   R. Ortega. Nonexistence of invariant curves of mappings with small twist. *Nonlinearity*, 10(1):195–197, 1997.

[Ort04]   R. Ortega. Unbounded motions in forced newtonian equations. Preprint, 2004.

[Pal00]   K. Palmer. *Shadowing in dynamical systems*, volume 501 of *Mathematics and its Applications*. Kluwer, Dordrecht, The Netherlands, 2000. Theory and applications.

[Pes04]   Y. B. Pesin. *Lectures on partial hyperbolicity and stable ergodicity*. Zurich Lectures in Advanced Mathematics. Eur. Math. Soc. (EMS), Zürich, 2004.

[Pil99]   S. Yu. Pilyugin. *Shadowing in dynamical systems*, volume 1706 of *Lecture Notes in Mathematics*. Springer, Berlin, 1999.

[Poi99]   H. Poincaré. *Les méthodes nouvelles de la mécanique céleste*, volume 1, 2, 3. Gauthier-Villars, Paris, 1892–1899.

[Pol93]   M. Pollicott. *Lectures on ergodic theory and Pesin theory on compact manifolds*, volume 180 of *London Mathematical Society Lecture Note Series*. Cambridge University Press, Cambridge, 1993.

[PS70]   C. Pugh and M. Shub. Linearization of normally hyperbolic diffeomorphisms and flows. *Invent. Math.*, 10:187–198, 1970.

[Pus77a]   L. D. Pustyl'nikov. Stable and oscillating motions in nonautonomous dynamical systems. II. *Trudy Moskov. Mat. Obšč.*, 34:3–103, 1977.

[Pus77b]   L. D. Pustyl'nikov. Stable and oscillating motions in nonautonomous dynamical systems. II. *Trudy Moskov. Mat. Obšč.*, 34:3–103, 1977. English translation: *Trans. Moscow Math. Soc.*, 1978, *Issue 2*, pages 1–101. Amer. Math. Soc., Providence, RI, 1978,

[Pus95]   L. D. Pustyl'nikov. Poincaré models, rigorous justification of the second law of thermodynamics from mechanics, and the Fermi acceleration mechanism. *Uspekhi Mat. Nauk*, 50(1(301)):143–186, 1995.

[Rob71]   C. Robinson. Differentiable conjugacy near compact invariant manifolds. *Bol. Soc. Brasil. Mat.*, 2(1):33–44, 1971.

[Rob88]   C. Robinson. Horseshoes for autonomous Hamiltonian systems using the Melnikov integral. *Ergodic Theory Dynam. Systems*, 8:395–409, 1988.

[Rob02]   C. Robinson. Symbolic dynamics for transition tori. In *Celestial mechanics (Evanston, IL, 1999)*, volume 292 of *Contemp. Math.*, pages 199–208. Amer. Math. Soc., Providence, RI, 2002.

[RS02] P. H. Rabinowitz and E. W. Stredulinsky. A variational shadowing method. In *Celestial mechanics (Evanston, IL, 1999)*, volume 292 of *Contemp. Math.*, pages 185–197. Amer. Math. Soc., Providence, RI, 2002.

[Sac65] R. J. Sacker. A new approach to the perturbation theory of invariant surfaces. *Comm. Pure Appl. Math.*, 18:717–732, 1965.

[Shu78] M. Shub. *Stabilité globale des systèmes dynamiques*. Société Mathématique de France, Paris, 1978. With an English preface and summary.

[Sim99] C. Simó, editor. *Hamiltonian systems with three or more degrees of freedom*, Kluwer, Dordrecht, The Netherlands, 1999.

[Sit53] K. A. Sitnikov. On the possibility of capture in the problem of three bodies. *Mat. Sbornik N.S.*, 32(74):693–705, 1953.

[Ten82] J. Tennyson. Resonance transport in near-integrable systems with many degrees of freedom. *Phys. D*, 5(1):123–135, 1982.

[Thi83] W. Thirring. Classical scattering theory. In *Conference on differential geometric methods in theoretical physics (Trieste, 1981)*, pages 41–64. World Sci. Publ., Singapore, 1983.

[TLL80] J. L. Tennyson, M. A. Lieberman, and A. J. Lichtenberg. Diffusion in near-integrable Hamiltonian systems with three degrees of freedom. In Melvin Month and John C. Herrera, editors, *Nonlinear dynamics and the beam-beam interaction (Sympos., Brookhaven Nat. Lab., New York, 1979)*, pages 272–301. Amer. Inst. Physics, New York, 1980.

[Tre02a] D. Treschev. Multidimensional symplectic separatrix maps. *J. Nonlinear Sci.*, 12(1):27–58, 2002.

[Tre02b] D. Treschev. Trajectories in a neighbourhood of asymptotic surfaces of a priori unstable Hamiltonian systems. *Nonlinearity*, 15(6):2033–2052, 2002.

[Tre04] D. Treschev. Evolution of slow variables in a priori unstable Hamiltonian systems. *Nonlinearity*, 17(5):1803–1841, 2004.

[Wei73] A. Weinstein. Lagrangian submanifolds and Hamiltonian systems. *Ann. of Math. (2)*, 98:377–410, 1973.

[Wei79] A. Weinstein. *Lectures on symplectic manifolds*, volume 29 of *CBMS Regional Conference Series in Mathematics*. Amer. Math. Soc. Providence, RI, 1979. Corrected reprint.

[Zas02] G. M. Zaslavsky. Chaos, fractional kinetics, and anomalous transport. *Phys. Rep.*, 371(6):461–580, 2002.

[ZG04] P. Zgliczyński and M. Gidea. Covering relations for multidimensional dynamical systems. *J. Differential Equations*, 202(1):32–58, 2004.

[ZZN+89] G. M. Zaslavskiı, M. Yu. Zakharov, A. I. Neıshtadt, R. Z. Sagdeev, D. A. Usikov, and A. A. Chernikov. Multidimensional Hamiltonian chaos. *Zh. Èksper. Teoret. Fiz.*, 96(11):1563–1586, 1989.

Appendix

A: Normally hyperbolic manifolds

In this section, we recall some results in the literature on normally hyperbolic manifolds. Good references are [Fen72, Fen74, Fen77, HPS77, Pes04].

For simplicity, we will discuss only the case of diffeomorphisms. The case of flows is very similar. For many of the applications (persistence of invariant manifolds, regularity) the case of flows follows from the case of diffeomorphism by taking time-1 maps.

Let M be a smooth d-dimensional manifold, $f : M \to M$ a C^r diffeomorphism, $r \geqslant 1$.

Definition 9.1. Let $\Lambda \subset M$ be a C^1 submanifold invariant under f, $f(\Lambda) = \Lambda$. We say that Λ is a normally hyperbolic invariant manifold if there exist a constant $C > 0$, rates $0 < \lambda < \mu^{-1} < 1$ and a splitting for every $x \in \Lambda$

$$T_x M = E_x^s \oplus E_x^u \oplus T_x \Lambda$$

in such a way that

$$
\begin{aligned}
v \in E_x^s &\Leftrightarrow |Df^n(x)v| \leqslant C\lambda^n |v| & n \geqslant 0 \\
v \in E_x^u &\Leftrightarrow |Df^n(x)v| \leqslant C\lambda^{|n|} |v| & n \leqslant 0 \\
v \in T_x \Lambda &\Leftrightarrow |Df^n(x)v| \leqslant C\mu^{|n|} |v| & n \in \mathbb{Z}
\end{aligned}
\tag{1}
$$

In this exposition, we will assume that Λ is compact and, without loss of generality, connected.

Remark 9.1. The set up can be weakened in several directions which appear in applications.

For example, as remarked in [HPS77], instead of assuming that Λ is compact, it suffices to assume that f is C^r in a neighborhood of Λ with all the derivatives of order up to r uniformly bounded. The non-compact case involves some complications such as study of extension operators. These considerations become much more important in the extension of the theory to infinite dimensional Banach spaces, which we will also not consider [BLZ98]. In these infinite dimensional cases, the standard arguments often give one or two derivatives less in the conclusions than the finite dimensional compact arguments.

We also note that some parts of the theory are also true for manifolds with boundary such that $f(\Lambda) \subset \Lambda$, $d(f(\partial \Lambda), \partial \Lambda) > 0$ (inflowing) or $f(\Lambda) \subset \Lambda$, $d(f(\partial \Lambda), \partial \Lambda) > 0$ (outflowing). Note that the definition of stable (resp. unstable) directions in (1) requires serious changes in the outflowing (resp. inflowing) cases. An adaptation of the theory to the inflowing and outflowing cases is done in [Fen72]. Note that, even if these definitions become possible, the resulting objects may lack some of the properties of the more standard definitions. For example, the stable spaces are not unique in the inflowing case, so that issues of regularity are more delicate, even if well understood in the literature.

In some applications to instability, one often gets systems with two time scales, so that the hyperbolicity degenerates. Therefore it is useful to keep explicit track of how C, λ, μ, the parameters affecting the quality of the hyperbolicity in (1) enter in the hypothesis of the theorems. See [Fen79].

A self-contained detailed treatment of a case that involves several of these complications can be found in Appendix A of [DLS06c].

It follows from (1) that E_x^s, E_x^u depend continuously on x. In particular, the dimension of E_x^s, E_x^u are independent of x. In fact, using the invariant section theorem [HP70] or some direct arguments [Fen74, Fen77] they are $C^{\ell-1}$,

$$\ell < \min \left(r, \frac{|\log \lambda|}{\log \mu} \right). \tag{2}$$

Indeed, using some variants of these arguments, it is possible to show that the invariant manifold Λ is C^ℓ – even if the hypothesis of the definition only require it is C^1. In general, one cannot improve on these regularities. [Mos69] contains explicit examples – even trigonometric polynomials – where the regularity claimed above is sharp, and [HW99] shows that this regularity is indeed sharp for generic examples. Hence, in general, one cannot expect that the normally hyperbolic invariant manifolds are C^∞ even if f is a polynomial. One can however have uniform lower bounds for all the C^r maps which are in a C^1 neighborhood. The regularity of overflowing (resp. inflowing) manifolds is even more problematic since the stable (resp. unstable) bundles are not uniquely defined, hence the hyperbolicity constants do not have a unique value.

Given a normally hyperbolic invariant manifold Λ we define

$$W_\Lambda^s = \{ y \in M \mid d(f^n(y), \Lambda) \leqslant C_y \lambda^n, \ n \geqslant 0 \}$$
$$W_\Lambda^u = \{ y \in M \mid d(f^n(y), \Lambda) \leqslant C_y \lambda^{|n|}, \ n \leqslant 0 \}$$

Furthermore, for each $x \in \Lambda$, we define

$$W_x^s = \{ y \in M \mid d(f^n(x), f^n(y)) \leqslant C_{x,y} \lambda^n, \ n \geqslant 0 \}$$
$$W_x^u = \{ y \in M \mid d(f^n(x), f^n(y)) \leqslant C_{x,y} \lambda^{|n|}, \ n \leqslant 0 \}$$

and we note that $E_x^s = T_x W_x^s$ and $E_x^u = T_x W_x^u$. It is a fact that

$$W_\Lambda^s = \bigcup_{x \in \Lambda} W_x^s$$
$$W_\Lambda^u = \bigcup_{x \in \Lambda} W_x^u \tag{3}$$

Moreover, $x \neq \tilde{x} \Rightarrow W_x^s \cap W_{\tilde{x}}^s = \emptyset, \ W_x^u \cap W_{\tilde{x}}^u = \emptyset$.

The decomposition (3) can expressed geometrically saying that $\{W_x^s\}_{x \in \Lambda}$, $\{W_x^u\}_{x \in \Lambda}$ are a foliation of W_Λ^s, W_Λ^u, respectively. We will refer to these foliations as $\mathscr{F}_s, \mathscr{F}_u$.

Dynamically, the above statement means that, when the orbit of a point is approaching Λ, it approaches the orbit of a single point. This, as well as the uniqueness can be established easily by noting that, for two points in Λ, we have $d(f^n(x), f^n(x)) \geqslant C\mu^{-n}$. Since $\lambda \mu < 1$, we can see that if we fix y there can only be one x such that $d(f^n(x), f^n(y)) \leqslant C\lambda^n$.

We recall that in these circumstances we have that

1. Λ is a C^ℓ manifold with ℓ given in (2).
2. W_Λ^s, W_Λ^u are $C^{\ell-1}$ manifolds
3. W_x^s, W_x^u are C^r manifolds

4. The maps $x \mapsto W_x^s$, W_x^u are $C^{\ell-1-j}$, when W_x^s, W_x^u are given the C^j topologies in compact sets.
5. When $x \in \Lambda$, we have

$$T_x W_\Lambda^{s,u} = E_x^{s,u} \qquad T_x W_x^{s,u} = E_x^{s,u}$$

6. As a consequence of the above, using the implicit function theorem, we have:
   Denote by $W_\Lambda^{s,\delta}$ a δ-neighborhood of Λ in W_Λ^s and by $W_x^{s,\delta}$ a δ neighborhood of x in W_x^s.

   Then, for sufficiently small δ, there is a $C^{\ell-1}$ diffeomorphism h^s from $W_\Lambda^{s,\delta}$ to a neighborhood of the zero section in E^s. Furthermore, $h^s(W_x^s) \subset E_x^s$.

Note, that, even if W_x^s are as smooth as the map, the dependence of the point on the base point has only some finite regularity that depends on the regularity exponents entering in (1).

The manifold W_Λ^s is invariant. That is $f(W_\Lambda^s) = W_\Lambda^s$. Analogously, of course, the unstable manifolds.

On the other hand, the manifolds W_x^s are not invariant. They, however satisfy a covariance property

$$f(W_x^s) = W_{f(x)}^s \tag{4}$$

The local behavior in a neighborhood of a normally hyperbolic invariant manifold is described very precisely by the following theorem in [HPS77, PS70], who show who show that if Λ is a normally hyperbolic invariant manifold , then there is a homeomorphism h from a neighborhood of the zero section in $T\Lambda$ to a neighborhood in Λ in such a way that if $x \in \Lambda$, $\eta \in T_x M$ and $|\eta|$ is sufficiently small, we have

$$f \circ h(x, \eta) = h(f(x), Df(x)\eta) \tag{5}$$

The homeomorphism h is, of course, highly non-unique. Note that, in the case that Λ is just a point, the theorem reduces to the celebrated Hartman–Grobman theorem. Indeed the proof of the references above, after some clever reductions, becomes the Hartman–Grobman theorem in infinite dimensions.

An important consequence of the linearization theorem is that if Λ is a normally hyperbolic invariant manifold , then, for any sufficiently small open neighborhood U of Λ we have

$$\Lambda = \bigcup_{n \in \mathbb{Z}} f^n(U)$$

Of course, if $\Lambda \subset V \subset U$, then $\Lambda = \bigcup_{n \in \mathbb{Z}} f^n(V)$.

The homeomorphism h solving (5) is not unique and there are really terrible choices.[19] Nevertheless, there are choices which are continuous and indeed Hölder in some of the variables. We also have that, $W_x^{s,uloc} = h(x, E_x^{s,u} \cup B_\delta)$.

The linearization (5) is a generalization of Hartman–Grobman theorem. Under appropriate non-resonance conditions on the possible rates of growth of the vectors

---

[19] The lovers of pathologies can amuse themselves using the axiom of choice – Argh!! – to produce h solving (5) which are not measurable.

on $T_x M|_{x \in \Lambda}$ it is possible to obtain more precise linearizations [Rob71,KP90,BK94]. In contrast with the Sternberg Linearization theorem, the non-resonance conditions can fail in C^1 open sets of diffeomorphisms. When the conditions for the linearization apply, then one can obtain very good estimates for the orbits that "fly by" the invariant manifold. In particular, one can get very detailed information about the separatrix map. Note that the time that one can spend in a "fly by" is unbounded, so that linearization gives information over trajectories that go over a long time.

A.1 Persistence and dependence on parameters

One of the most important results of the theory of normally hyperbolic invariant manifolds is that they persist under perturbations and that they depend smoothly under parameters.

Persistence means, roughly, that if a map f has an invariant manifold Λ_f and g is sufficiently C^1 close to f, then g also has an invariant manifold Λ_g.

In these cases, the results on dependence on parameters and can be obtained very economically from the results on persistence by considering an extended system.

Let $f(x, \varepsilon) : M \times \Sigma \to M$ is a family of maps (ε is the parameter). <– We will also use $f_\varepsilon = f(\cdot, \varepsilon)$. We consider $\tilde{f} = f \times \mathrm{Id}$ and $\tilde{f}_0 = f_0 \times \mathrm{Id}$.

We note that if Λ_0 is a normally hyperbolic invariant manifold for f_0, then $\Lambda_0 \times \Sigma$ is a normally hyperbolic invariant manifold for $\tilde{f}_0$. Furthermore, the hyperbolicity for $\tilde{f}_0$ admits the same constants in (1) than f_0. Hence, if $\tilde{f}$ is C^1 close to $\tilde{f}_0$, the persistence result implies that we can find a manifold $\tilde{\Lambda}$ that is invariant for $\tilde{f}$. Because $\tilde{f}$ is the identity in the ε variable, we have that $\tilde{\Lambda}$ has to have the form $\bigcup_\varepsilon \Lambda_\varepsilon \times \{\varepsilon\}$, where Λ_ε is invariant under f_ε.

Another important result in the theory of persistence of invariant manifolds is that the change in the hyperbolicity constants can be controlled by the C^1 distance of the maps. This is important since the regularity of the foliations $\mathscr{F}_{s,u}$ can be bounded uniformly in sets which are the intersection of C^1 open sets and C^r. For example in Section 3.2 it was convenient to assume that the foliations $\mathscr{F}_{s,u}$ are C^1. The previous remark implies that this assumption holds in some open sets, characterized by ratios in the contractions exponents.

A very efficient way of describing the results of persistence and smooth dependence on parameters is to use a parameterization method.

We write Λ_ε as $k_\varepsilon(\Lambda_0)$ where $k : \Lambda_0 \times \Sigma \to M$ The fact that Λ_ε is invariant is equivalent to

$$f_\varepsilon \circ k_\varepsilon = k_\varepsilon \circ r_\varepsilon \tag{6}$$

where $r_\varepsilon : \Lambda_0 \to \Lambda_0$ is a representation of dynamics of f_ε restricted to the invariant manifold.

The result that Λ_ε is C^ℓ, means that k_ε can be chosen to be C^ℓ in $\Lambda_0 \times \Sigma$. Hence, $\frac{\partial^j}{\partial \varepsilon^j} k_\varepsilon(x)$ is $C^{\ell-j}$. So that the map $\varepsilon - \Lambda_\varepsilon$ is $C^{\ell-j}$ when the manifolds are given the C^j topology.

Even if in this presentations we have argued that the standard theory of normally hyperbolic invariant manifolds implies the existence of solutions of (6), it is possible consider (6) as an equation for $k_\varepsilon, r_\varepsilon$ and show that there are solutions. This is an alternative approach to the theory of existence of normally hyperbolic invariant manifolds developed in [HdlL07]. This has several advantages from the point of view of numerical computation. See [HdlL06c, HdlL06b, HdlL06a] for some simpler cases.

Notice that (6) is a geometrically natural equation. We also note that – since all geometrically natural equations are invariant under the choice of a system of coordinates in Λ_0 – if $k_\varepsilon, r_\varepsilon$ is a solution of (6) and $h_\varepsilon : \Lambda_0 \to \Lambda_0$ is a diffeomorphism we have that $\tilde{k}_\varepsilon = k_\varepsilon \circ h_\varepsilon$, $\tilde{r}_\varepsilon = h_\varepsilon^{-1} \circ r_\varepsilon \circ h_\varepsilon$ is also a solution of (6). This lack of uniqueness can be chosen to impose some supplementary conditions. For example, in [DLS06a] it is shown that if f_ε preserve a symplectic form ω, there is one and only one k_ε such that $k_\varepsilon^* \omega = k_0^* \omega \equiv \omega|_{\Lambda_0}$. (This choice also has other geometric properties, we refer to the [DLS06a].

Remark 9.2. There is a large literature on formal perturbation theories based on "*expanding to first order*" and solving the resulting equations. This, in general, is not a correct procedure, but in the case that we know that there is a derivative, it is easy to show that this derivative satisfies a functional equation (which is the equation considered by the formal expansion). If the solution of this equation is unique, then, the solution of this equation will be the derivative.

A.2 The λ-lemma and the exchange lemma

The simplest version of the λ lemma states that if there is manifold Σ which intersects transversally W_x^s, then, for large n, $f^n(\Sigma)$ will have a patch which is exponentially close – in a smooth topology – to $W^u(U_n)$ where $U_n \subset \Lambda$ is an open set around $f^n(x)$.

The sizes of the U_n may decrease exponentially– but the rate is bounded by μ^{-n} –

A.3 Normally hyperbolic laminations

This is a very interesting concept developed in [HPS77, Ch. 15]. See the results in Section 8.

In the simplest formulation, a lamination is a closed set of manifolds which do not intersect. $\{\Lambda_\sigma\}_{\sigma \in \Sigma}$.

A lamination is invariant if $f(\Lambda_\sigma) \subset \Lambda_{\Phi(\sigma)}$. A lamination is normally hyperbolic if, for $x \in \Lambda_\sigma$ we can find decompositions $T_x M = T_x \Lambda_\sigma \oplus E_x^s \oplus E_x^u$ satisfying estimates similar to those in (1).

The result of [HPS77, Ch. 15] is that this situation is stable under perturbations. Some improvements were developed in [Lla02]. Namely, that we can find another lamination Λ_σ^g and a map $h_\sigma : \Lambda_\sigma^f \to \Lambda_\sigma^g$ in such a way that $g \circ h_\sigma = h_\sigma \circ f$.

A heuristic point of view which is useful is that one can consider the laminae as points, so that the above result is just the structural stability. As shown in [HPS77], there are also shadowing theorems and many other results analogue to the results for hyperbolic sets.

In a way similar to the stability of normally hyperbolic invariant manifolds , it is convenient to describe the stability of invariant laminations using a parameterization method.

If $F_0(L_\sigma) = L_{f(\sigma)}$, satisfying the hypothesis of normal hyperbolicity, we can try to find $h_\sigma^\varepsilon : L_\sigma \to M$ and $r_\sigma^\varepsilon : L_\sigma \to L_{f(\sigma)}$ that $F_\varepsilon \circ h_\sigma^\varepsilon = h_\sigma^\varepsilon \circ r_\sigma^\varepsilon$. Clearly $L_\sigma^\varepsilon = h^\varepsilon(L_\sigma)$ satisfy the invariance properties of laminations, $F_\varepsilon(L_\sigma^\varepsilon) = L_{f(\sigma)}^\varepsilon$.

The h_σ^ε r_σ^ε are parameterizations of the new laminae in terms of the old and the r_σ^ε are expressions of the dynamics.

It follows from the results in [HPS77] that, for fixed σ the $h_\sigma^\varepsilon(x)$, $r_\sigma^\varepsilon(x)$, are C^ℓ on (ε, x), where ℓ depends on the exponents.

One small improvement from the results of [HPS77] that is found in [Lla04] is the observation that, the mappings $\sigma \mapsto h_\sigma^\varepsilon, r_\sigma^\varepsilon$ are Hölder when the h, r are given a C^ℓ topology.

Variational methods for the problem of Arnold diffusion

Chong-Qing Cheng[1]

Abstract The problem of Arnold diffusion is raised for nearly integrable Hamiltonian systems. It concerns whether there exists an orbit along which the action undergoes substantial variation. Variational method has been shown a powerful tool for the study of Arnold diffusion of Hamiltonian systems convex in actions. In variational language, it amounts to construct an orbit connecting two different Aubry sets. This is the main content of the lecture notes.

1 Introduction to Mather theory

Let M be a closed and connected C^∞-manifold. We denote by TM the tangent bundle of the manifold M. Usually, $M = \mathbb{T}^n$, for instance, when we study the problem of Arnold diffusion. Let $L \in C^r(TM \times \mathbb{T}, \mathbb{R})$ be a Lagrangian, we assume it satisfies the Tonelli's condition:

POSITIVE DEFINITENESS. For each $(x,t) \in M \times \mathbb{T}$, the Lagrangian function is strictly convex in velocity: the Hessian $L_{\dot{x}\dot{x}}$ is positive definite.

SUPER-LINEAR GROWTH. We assume that L has fiber-wise superlinear growth: for each $(x,t) \in M \times \mathbb{T}$, we have $L/\|\dot{x}\| \to \infty$ as $\|\dot{x}\| \to \infty$.

COMPLETENESS. All solutions of the Lagrangian equations are well defined for all $t \in \mathbb{R}$.

This lagrangian is uniquely related to a Hamiltonian $H(x,y,t)$ via Legendre transformation

$$L(x,\dot{x},t) = \max \langle \dot{x}, y \rangle - H(x,y,t),$$

and the Hamiltonian equation

$$\dot{x} = \frac{\partial H}{\partial y}, \qquad \dot{y} = -\frac{\partial H}{\partial x} \tag{1}$$

[1] Department of Mathematics, Nanjing University, Nanjing 210093, China
e-mail: chengcq@nju.edu.cn

W. Craig (ed.), *Hamiltonian Dynamical Systems and Applications*, 337–365.
© 2008 *Springer Science + Business Media B.V.*

are equivalent to the Lagrangian equation

$$\frac{d}{dt}\left(\frac{\partial L}{\partial \dot{x}}\right) - \frac{\partial L}{\partial x} = 0. \tag{2}$$

This Lagrange equation corresponds to the critical point of the functional

$$A(\gamma) = \int L(\gamma, \dot{\gamma}, t)dt.$$

We now introduce the concept of minimal measure. The method we use here basically follows Mañé. Let $\bar{M}$ be a covering space of M, $\bar{\gamma}\colon [t_0, t_1] \to \bar{M}$ be an absolutely continuous curve, we define its action as

$$A(\bar{\gamma}) = \int_{t_0}^{t_1} L(\pi\dot{\gamma}(t), \pi\gamma(t), t)dt = \int_{t_0}^{t_1} L(\pi d\gamma(t), t)dt.$$

Lemma 6 (TONELLI) *Let $t_0 < t_1 \in \mathbb{R}$, $x_0, x_1 \in \bar{M}$. The conditions of positive definiteness and completeness guarantee that the action takes a finite minimum value over the set of absolutely continuous curves $\bar{\gamma}\colon [t_0, t_1] \to \bar{M}$ such that $\bar{\gamma}(t_0) = x_0$, $\bar{\gamma}(t_1) = x_1$. Moreover, if the completeness condition is assumed, then the minimizer is C^1 and satisfies the Lagrangian equation.*

With this lemma, it is easy to show the existence of k-periodic orbit for each $k \in \mathbb{Z}^+$. Let

$$-\alpha_k(0) = \min_{x\in M} \inf_{\gamma(0)=\gamma(k)=x} \frac{1}{k}\int_0^k L(d\gamma(t), t)dt,$$

and denote the minimizer γ_k. Since M is compact, the minimum can be reached. Clearly, $\gamma_k \in C^1[0, k]$. We claim that

$$\lim_{t\searrow 0} \dot{\gamma}_k(0) = \lim_{t\nearrow k} \dot{\gamma}_k(k).$$

Indeed, if this is not true, it would imply that the loop has a corner at $\gamma(0)$. We join the point $\gamma(k-\delta)$ to the point $\gamma(\delta)$ by a minimal curve $\xi\colon [-\delta, \delta] \to M$ such that $\xi(-\delta) = \gamma(k-\delta)$ and $\xi(\delta) = \gamma(\delta)$. Since the minimizer is C^1, it implies that the action along $\xi|_{[-\delta,\delta]}$ is smaller than the action along $\gamma|_{[k-\delta,k]\cup[0,\delta]}$. But this contradicts to the fact that the minimum is reached at $\gamma|_{[0,k]}$ for all k-closed curves. Therefore, we obtain an k-periodic orbit $(\gamma(t), \dot{\gamma}(t))$.

With this periodic orbit, we define a probability measure μ_k on $TM \times \mathbb{T}$ such that

$$\int_{TM\times\mathbb{T}} f d\mu_k = \frac{1}{k}\int_0^k f(d\gamma_k(s), s)ds$$

for all $f \in C^0(TM \times \mathbb{T})$. This measure is clearly invariant to the Euler–Lagrange flow ϕ_L^t determined by L.

$$\int_{TM\times\mathbb{T}} f d\phi'^{*}_{L}\mu_k = \frac{1}{k}\int_0^k f(d\gamma_k(s+t),s+t)ds.$$

Let $-\alpha(0) = \liminf_{k\to\infty}\int Ld\mu_k$, let k_i be the subsequence reaching the limit infimum. Obviously, $\{\mu_{k_i}\}$ is compact in the weak-* topology, consequently, there exists at least one accumulation point μ which is of course invariant to ϕ^t_L. Let $\mathfrak{M}_L$ be the set of those probability measures which is invariant to ϕ^t_L, clearly, we have

$$\int Ld\mu = \inf_{v\in\mathfrak{M}_L}\int Ldv.$$

We call μ the invariant minimal measure, call $-\alpha(0) = \liminf_{k\to\infty}-\alpha_k(0)$ the average action.

Let us now consider

$$A_c(\gamma) = \int (L-\eta_c)(d\gamma(t),t)dt,$$

where $\eta_c = \langle\eta(x),\dot{x}\rangle$, $\eta(x)$ denotes a closed 1-form $\langle\eta(x),dx\rangle$ evaluated at x, and its de-Rham cohomology $[\langle\eta(x),dx\rangle] = c \in H^1(M,\mathbb{R})$. For convenience and without danger of confusion, we call η_c closed 1-form also. As η is closed, it has no contribution to the Euler–Lagrange equation, the variational derivative of the functional $A_c(\gamma)$ is the same as the derivative of $A(\gamma)$. However, the minimal measure for L may different from the minimal measure for $L-\eta_c$. The minimal measure only depends on the cohomology class, it is independent of which closed form it takes. We usually denote by μ_c the invariant minimal measure for the class c, by $-\alpha(c)$ the minimal average action.

Theorem 6 [Ma1] *For each $c \in H^1(M,\mathbb{R})$, there exists at least one probability measure $\mu_c \in \mathfrak{M}_L$ which minimizes $L-\eta_c$:*

$$\infty > -\alpha(c) = \int (L-\eta_c)d\mu = \inf_{v\in\mathfrak{M}_L}\int (L-\eta_c)dv.$$

$\alpha(c)$ is convex in c.

We usually call $\alpha(c)$ the α-function. Since it is convex, we can consider its dual through Legendre transformation

$$\beta(\rho) = \max_c\{\langle c,\rho\rangle - \alpha(c)\}$$

it is called β-function, which is convex in the first homology class, corresponding to the rotation vector. It is worthwhile to point out that the rotation vector of some minimal measure may different from the rotation vector of those orbits in the support of the measure. A typical example is the Hedlund geodesic flow in $\mathbb{T}^3$ equipped with nontrivial Riemannian metric.

According to Mather's result, both α-function and β-function are convex and conjugate by the Legendre transformation. By definition,

$$\beta_L(\rho) + \alpha_L(c) \geqslant \langle \rho, c \rangle, \qquad \forall \rho \in H_1(\mathbb{T}^n, \mathbb{R}),\ c \in H^1(\mathbb{T}^n, \mathbb{R}). \tag{3}$$

The map

$$\mathscr{L}_\beta : H_1(\mathbb{T}^n, \mathbb{R}) \Rightarrow H^1(\mathbb{T}^n, \mathbb{R})$$

defined by letting $\mathscr{L}_\beta(\rho)$ be the set of $c \in H^1(\mathbb{T}^n, \mathbb{R})$ for which the inequality in (3) becomes equality, is called *Fenchel–Legendre transformation*. Obviously, $\mathscr{L}_\beta(\rho)$ is a compact, convex, non-empty subset of $H^1(\mathbb{T}^n, \mathbb{R})$.

With well defined minimal measure, we can define Mather set as follows:

$$\tilde{\mathscr{M}}(c) = \overline{\bigcup \mathrm{supp}\mu_c}.$$

In the variational construction of diffusion orbits, two objects play important role, they are so-called Aubry set as well as Mañé set. To introduce the definitions, we introduce the idea of semi-static and static curves, due to Mañé. Let $\gamma : \mathbb{R} \to M$ be an absolutely continuous curve. We call it the minimizer of L_c if for any $a < b$ and any absolutely continuous curve $\zeta : [a,b] \to M$ with $\zeta(a) = \gamma(a)$, $\zeta(b) = \gamma(b)$ we have

$$\int_a^b (L - \eta_c)(d\gamma(t), t)dt \leqslant \int_a^b (L - \eta_c)(d\zeta(t), t)dt,$$

where we use $d\gamma$ to denote $(\gamma, \dot{\gamma})$ for abbreviation. We define

$$h_c((m,t),(m',t')) = \min_{\substack{\gamma \in C^1([t,t'],M) \\ \gamma(t)=m,\gamma(t')=m'}} \int_t^{t'} (L - \eta_c)(d\gamma(s), s)ds + (t' - t)\alpha(c),$$

$$F_c((m,s),(m',s')) = \inf_{\substack{t=s \bmod 1 \\ t'=s' \bmod 1, \\ t'-t \geqslant 1}} h_c((m,t),(m',t'))$$

$$h_c^\infty(m,m') = \liminf_{\substack{t,t' \in \mathbb{Z} \\ t'-t \to \infty}} h_c((m,t),(m',t')), \tag{4}$$

$$h_c^\infty((m,t),(m',t')) = \liminf_{t'-t \to \infty} h_c((m,t),(m',t')), \tag{5}$$

Given an absolutely continuous curve $\gamma : [a,b] \to M$, we use the notation

$$[A_c(\gamma)] = \int_a^b (L - \eta_c)(d\gamma(s), s)ds + \alpha(c))|b - a|.$$

A curve $\gamma \in C^1(\mathbb{R}, M)$ is called c-semi-static if

$$[A_c(\gamma|_{[a,b]})] = F_c(\gamma(a), \gamma(b), a \bmod 1, b \bmod 1)$$

for each $[a,b] \subset \mathbb{R}$. A curve $\gamma \in C^1(\mathbb{R},M)$ is called c-static if, in addition

$$[A_c(\gamma|_{[a,b]})] = -F_c(\gamma(b),\gamma(a),b \bmod 1, a \bmod 1)$$

for each $[a,b] \subset \mathbb{R}$. An orbit $X(t) = (d\gamma(t), t \bmod 1)$ is called c-static (semi-static) if γ is c-static (semi-static). We call the Mañé set $\tilde{\mathcal{N}}(c)$ the union of global c-semi-static orbits, and call the Aubry set $\tilde{\mathcal{A}}(c)$ the union of c-static orbits. It is proved in [Be3] that these Mather sets, Aubry sets and Mañé sets are symplectic invariants.

We use $\mathcal{M}(c)$, $\mathcal{A}(c)$ and $\mathcal{N}(c)$ to denote the standard projection of $\tilde{\mathcal{M}}(c)$, $\tilde{\mathcal{A}}(c)$ and $\tilde{\mathcal{N}}(c)$ from $TM \times \mathbb{T}$ to $M \times \mathbb{T}$ respectively. The inverse of the projection is Lipschitz when it is restricted to $\mathcal{A}(c)$ and $\mathcal{M}(c)$. We use the symbol $\tilde{\mathcal{N}}_s(c) = \tilde{\mathcal{N}}(c)|_{t=s}$ to denote the s-time section of a Mañé set, and so on. The following inclusions are shown in [Be2]

$$\tilde{\mathcal{M}}(c) \subseteq \tilde{\mathcal{A}}(c) \subseteq \tilde{\mathcal{N}}(c).$$

Aubry set has Lipschitz property (see [Ma1], [Ma2]):

Theorem 7 $\pi\colon \tilde{\mathcal{A}}(c) \to M \times \mathbb{T}$ *is injective. Its inverse from $\mathcal{A}(c)$ to $\tilde{\mathcal{A}}(c)$ is Lipschitz, i.e. there exists a constant C such that for any $(x,t),(x',t') \in \mathcal{A}(c)$ we have*

$$\mathrm{dist}(\pi^{-1}(x,t),\pi^{-1}(x',t')) \leqslant C\mathrm{dist}((x,t),(x',t')).$$

If the minimal measure is uniquely ergodic, then the Mañé set is same as the Aubry set. We mention that the definition of Mañé set depends on what configuration manifold we choose. If we choose some finite covering manifold, the Mañé set could be larger.

The structure of Mather set can be very complicated, however, in several cases its structure is well known. KAM torus, some lower dimensional torus with hyperbolic type, Aubry–Mather set and minimal periodic orbit are those examples.

2 Existence of Homoclinic Orbits

Recall the example of Arnold in [Ar1], the diffusion orbits are in a small neighborhood of a chain of heteroclinic orbits, these heteroclinic orbits appear when the loop of homoclinic orbits break. So, as the first step, let us consider the existence of homoclinic orbits to some Aubry sets. A sufficient condition is

$$H_1(M,\mathcal{A}_0(c),\mathbb{Z}) \neq 0.$$

Let d be the dimension of the group $H_1(M,\mathcal{A}_0,\mathbb{Z})$, we have (see [Be1])

Theorem 8 *There are at least $d+1$ minimal homoclinic orbits to $\tilde{\mathcal{A}}(c)$.*

Proof. Consider an absolutely continuous curve $\xi_k\colon [-k,k] \to M$ such that $\xi_k(-k) = \xi_k(k) \in \mathcal{A}_0(c)$. In this case it makes sense to define $[\xi_k] \in H_1(M,\mathcal{A}_0(c),\mathbb{Z})$. Given e_i, a given generator of the group, we choose those ξ_k ($k = 1, 2, \cdots$) such that $[\gamma_k] = e_i$ for each k. Consider the action along each of these curves

$$[A(\xi_k)] = \frac{1}{2k} \int_{-k}^{k} (L - \eta_c)(d\xi_k(t), t)dt + 2k\alpha(c).$$

As the Aubry set is compact, there exists a curve denoted by γ_k such that

$$[A(\gamma_k^*)] = \min[A(\xi_k)]$$

where the minimum is taken among all those closed curves ξ_k such that $[\xi_k] = e_i$. Obviously, we have

$$\sup_k [A(\gamma_k)] < \infty.$$

Consider a subsequence k_i such that

$$\lim_{i \to \infty} [A(\gamma_{k_i})] = \liminf_{k \to \infty} [A(\gamma_k)]$$

Given a δ-neighborhood of the Aubry set $\mathscr{A}(c) + \delta \subset M \times \mathbb{T}$, there exists $K \in \mathbb{Z}$ such that $(\gamma_{k_i}(t), t) \in \mathscr{A}(c) + \delta$ for each $t \in [-k_i, \ell_i] \cup [\ell_i + K, k_i]$ if k_i is sufficiently large. If not, we would obtain the property that $[A(\gamma_k)] \to \infty$.

Let us consider the set of curves $\{\gamma_i(t - j_i)\}$. Two facts are obvious: γ_i is the solution of the Euler–Lagrange equation, and there exist some points $\gamma_i(t - j_i) \notin \mathscr{A}(c) + \delta$. As the matrix $\partial_{\dot{x}^2} L$ is positive definite, this set is compact in C^1-topology if t is restricted in $[-j, j]$ for each finite j. By diagonal argument, we see that some $\gamma \colon \mathbb{R} \to M$ exists such that $\gamma_{k_{i_m}}(t - j_{i_m}) \to \gamma(t)$. Thus, we have that both the α- and the ω-limit set of $d\gamma$ are contained in the Aubry set and $[\gamma] \neq 0$. This completes the proof of the existence of homoclinic orbit. To see the existence of the second homoclinic orbit, we only need to consider some closed curve whose first relative homology is different $[\gamma]$, such procedure can be repeated for d-times.

M has several kinds of finite lift, we can choose $d + 1$ covering $\bar{M}$ such that the lift of each of these $d + 1$ homoclinic orbits is in the Mañé set with respect to one lift of M.

The existence of homoclinic orbit does not implies the existence of some heteroclinic orbit immediately, it is not naturally granted. It depends on the condition whether the homoclinic orbits are discrete or not. Before we show the existence of heteroclinic orbit, we have the following section.

3 Pseudo-connecting orbit set $\tilde{\mathscr{C}}_{\eta,\mu,\psi}$

The construction of connecting orbits are based on two properties, one is the upper semi-continuity of $\tilde{\mathscr{C}}_{\eta,\mu,\psi}$ and $\tilde{\mathscr{N}}(c)$, another one is some kind of topological triviality of Mañé sets.

Lemma 7 (see [Be2], [CY1]) *We assume $L \in C^r(TM \times \mathbb{R}, \mathbb{R})$ $(r \geqslant 2)$ satisfies the positive definite, superlinear-growth and completeness conditions, where M is a compact, connected Riemanian manifold. Considered as the function of t, L is assumed periodic for $t \in (-\infty, 0]$ and for $t \in [1, \infty)$. Let $\tilde{\mathscr{G}}_L \subset TM \times \mathbb{R}$ be the set of minimal orbits for L. Then the map $L \to \tilde{\mathscr{G}}_L$ is upper semi-continuous.*

To construct some connecting orbits between two different Mañé sets, we consider a modified Lagrangian

$$L_{\eta,\mu,\psi} = L - \eta - \mu - \psi$$

where η is a closed 1-form such that $[\eta] = c$, μ is a 1-form depending on t in the way that the restriction of μ on $\{t \leqslant 0\}$ is 0, the restriction on $\{t \leqslant 1\}$ is a closed 1-form $\bar{\mu}$ such that $[\bar{\mu}] = c' - c$, we call this μ U-step 1-form. ψ is a function on $TM \times \mathbb{R}$, $\psi(\cdot, t) = 0$ for all $t \in (-\infty, 0] \cup [1, \infty)$ and the support of ψ is compact in TM for all t. Let $m, m' \in M$, we define

$$h^{T_0, T_1}_{\eta,\mu,\psi}(m, m') = \inf_{\substack{\gamma(-T_0)=m \\ \gamma(T_1)=m'}} \int_{-T_0}^{T_1} L_{\eta,\mu,\psi}(d\gamma(t), t)dt + T_0\alpha(c) + T_1\alpha(c').$$

We take the limit infimum which is clearly is bounded

$$h^{\infty}_{\eta,\mu,\psi}(m, m') = \liminf_{T_0, T_1 \to \infty} h^{T_0, T_1}_{\eta,\mu,\psi}(m, m') \leqslant C_{\eta,\mu,\psi}.$$

Let $\{T_0^i\}_{i \in \mathbb{Z}_+}$ and $\{T_1^i\}_{i \in \mathbb{Z}_+}$ be the sequence of positive integers such that $T_j^i \to \infty$ $(j = 0, 1)$ as $i \to \infty$ and the following limit exists

$$\lim_{i \to \infty} h^{T_0^i, T_1^i}_{\eta,\mu,\psi}(m, m') = h^{\infty}_{\eta,\mu,\psi}(m, m').$$

Let $\gamma_i(t, m, m') \colon [-T_0^i, T_1^i] \to M$ be a minimizer connecting m and m'

$$h^{T_0^i, T_1^i}_{\eta,\mu,\psi}(m, m') = \int_{-T_0^i}^{T_1^i} L_{\eta,\mu,\psi}(d\gamma_i(t), t)dt + T_0^i\alpha(c) + T_1^i\alpha(c').$$

It is not difficult to see that for any compact interval $[a, b]$, the set $\{\gamma_i\}$ is pre-compact in $C^1([a, b], M)$.

Lemma 8 *Let $\gamma \colon \mathbb{R} \to M$ be an accumulation point of $\{\gamma_i\}$. If $s \geqslant 1$ then*

$$A_{L_{\eta,\mu,\psi}}(\gamma|[s, \tau]) = \inf_{\substack{\tau_1 - \tau \in \mathbb{Z}, \tau_1 > s \\ \gamma^*(s) = \gamma(s) \\ \gamma^*(\tau_1) = \gamma(\tau)}} \int_s^{\tau_1} L_{\eta,\mu,\psi}(d\gamma^*(t), t)dt \tag{1}$$

$$+ (\tau_1 - \tau)\alpha(c');$$

if $\tau \leqslant 0$ then

$$A_{L_{\eta,\mu,\psi}}(\gamma|[s,\tau]) = \inf_{\substack{s_1-s\in\mathbb{Z},s_1<\tau \\ \gamma^*(s_1)=\gamma(s) \\ \gamma^*(\tau)=\gamma(\tau)}} \int_{s_1}^{\tau} L_{\eta,\mu,\psi}(d\gamma^*(t),t)dt \qquad (2)$$

$$- (s_1-s)\alpha(c);$$

if $s \leqslant 0$ and $\tau \geqslant 1$ then

$$A_{L_{\eta,\mu,\psi}}(\gamma|[s,\tau]) = \inf_{\substack{s_1-s\in\mathbb{Z},\tau_1-\tau\in\mathbb{Z} \\ s_1\leqslant 0,\tau_1\geqslant 1 \\ \gamma^*(s_1)=\gamma(s) \\ \gamma^*(\tau_1)=\gamma(\tau)}} \int_{s_1}^{\tau_1} L_{\eta,\mu,\psi}(d\gamma^*(t),t)dt \qquad (3)$$

$$- (s_1-s)\alpha(c) + (\tau_1-\tau)\alpha(c').$$

With this lemma it is natural to define

$$\tilde{\mathscr{C}}_{\eta,\mu,\psi} = \{d\gamma \in \tilde{\mathscr{G}}_{L_{\eta,\mu,\psi}} : (1),(2) \text{ and } (3) \text{ hold}\}.$$

Although the elements in this set are not necessarily the orbits of the Lagrangian flow determined by L, the α-limit set of each element is contained in $\tilde{\mathscr{A}}(c)$, and the ω-limit set of each element is contained in $\tilde{\mathscr{A}}(c')$. Due to this reason, we call it pseudo connecting orbit set. Clearly, $\tilde{\mathscr{C}}_{\eta,0,0} = \tilde{\mathscr{N}}(c)$. For convenience we may drop the subscript ψ in the symbol when it is equal to zero, $\tilde{\mathscr{C}}_{\eta,\mu} := \tilde{\mathscr{C}}_{\eta,\mu,0}$.

Lemma 9 *The map $(\eta,\mu) \to \tilde{\mathscr{C}}_{\eta,\mu,\psi}$ is upper semi-continuous. Consequently, the map $c \to \tilde{\mathscr{N}}(c)$ is upper semi-continuous.*

Proof. Let $\eta_i \to \eta$, $\mu_i \to \mu$ and $\psi_i \to \psi$, let $\gamma_i \in \tilde{\mathscr{C}}_{\eta_i,\mu_i,\psi_i}$ and let γ be an accumulation point of the set $\{\gamma_i \in \tilde{\mathscr{C}}_{\eta_i,\mu_i,\psi_i}\}_{i\in\mathbb{Z}^+}$. Clearly, $\gamma \in \tilde{\mathscr{C}}_{\eta,\mu,\psi}$. If $\gamma \notin \tilde{\mathscr{C}}_{\eta,\mu,\psi}$ there would be two point $\gamma(s),\gamma(\tau) \in M$ such that one of the following three possible cases takes place. Either $\gamma(s)$ and $\gamma(\tau) \in M$ can be connected by another curve γ^*: $[s+n,\tau] \to M$ with smaller action

$$A_{\eta,\mu,\psi}(\gamma|[s,\tau]) < A_{\eta,\mu,\psi}(\gamma^*|[s+n,\tau]) - n\alpha(c)$$

in the case $\tau < 0$; or there would a curve γ^*: $[s,\tau+n] \to M$ such that

$$A_{\eta,\mu,\psi}(\gamma|[s,\tau]) < A_{\eta,\mu,\psi}(\gamma^*|[s,\tau+n]) - n\alpha(c')$$

in the case $s \geqslant 1$, or when $s \leqslant 0$ and $\tau \geqslant 1$ there would be a curve γ^*: $[s+n_1,\tau+n_2] \to M$ such that

$$A_{\eta,\mu,\psi}(\gamma|[s,\tau]) < A_{\eta,\mu,\psi}(\gamma^*|[s+n_1,\tau+n_2]) - n_1\alpha(c) - n_2\alpha(c')$$

where $s + n_1 \leqslant 0$, $\tau + n_2 \geqslant 1$. Since γ is an accumulation point of γ_i, for any small $\varepsilon > 0$, there would be sufficiently large i such that $\|\gamma - \gamma_i\|_{C^1[s,t]} < \varepsilon$, it follows that $\gamma_i \notin \widetilde{\mathscr{C}}_{\eta_i,\mu_i,\psi_i}$ but that is absurd.

Let us consider the case that $\mu = 0$ and $\psi = 0$. In this case, $L - \eta$ is periodic in t. If some orbit $\gamma \in \widetilde{\mathscr{C}}_{\eta,0,0} : \mathbb{R} \to M$ is not semi-static, then there exist $s < \tau \in \mathbb{R}$, $n \in \mathbb{Z}$, $\Delta > 0$ and a curve $\gamma^* : [s, \tau + n] \to M$ such that $\gamma^*(s) = \gamma(s)$, $\gamma^*(\tau + n) = \gamma(\tau)$ and

$$A_{\eta,0,0}(\gamma|[s,\tau]) \geqslant A_{\eta,0,0}(\gamma^*|[s,\tau + n]) - n\alpha(c) + \Delta.$$

We can extend γ^* to $[s_1, \tau_1 + n] \to M$ such that $s_1 \leqslant \min\{s, 0\}$, $\min\{\tau_1, \tau_1 + n\} \geqslant 1$, $\tau_1 \geqslant \tau$ and

$$\gamma^* = \begin{cases} \gamma(t), & s_1 \leqslant t \leqslant s, \\ \gamma^*(t), & s \leqslant t \leqslant \tau + n, \\ \gamma(t - n), & \tau + n \leqslant t \leqslant \tau_1 + n. \end{cases}$$

Since $L - \eta$ is periodic in t, we would have

$$A_{\eta,0,0}(\gamma|[s_1,\tau_1]) \geqslant A_{\eta,0,0}(\tau^*\gamma|[s_1, \tau_1 + n]) - n\alpha(c) + \Delta.$$

but this contradicts to (3).

4 Existence of heteroclinic orbits

Let μ_c be such a minimal measure that its rotation vector satisfies some resonant condition $[\rho(\mu_c)]_i = 0$. We assume its support is restricted in a neighborhood of a lower dimensional torus $\{x_i = 0\}$. It is generic that the α-function has a flat at $\rho(\mu_c)$. Indeed, under small perturbation of potential, for instance, $L(x, \dot{x}, t) \to L(x, \dot{x}, t) + \varepsilon P(x)$ where $P = 1$ when $x_i = \frac{1}{2}$ and $\mathrm{supp}P \subset \{|x_i - \frac{1}{2}| \leqslant \frac{1}{4}\}$ the Aubry set is also restricted in a neighborhood of the lower dimensional torus. According to the study in the second section (see Theorem 8), there exists a homoclinic orbit $d\gamma$ such that the ith component of its relative homology is not zero: $[\gamma]_i \neq 0$. Such homoclinic orbit is not in the Mañé set $\mathscr{N}(c)$ if c is not at the boundary of the flat $\mathscr{L}_\beta(\rho(\mu_c))$, (the interior of a convex set with m-dimension is defined in the way we treat it as a set in m-dimensional space) it is in the Mañé set $\mathscr{N}(c, \tilde{M})$ for a finite covering space $\tilde{M}$.

Let $\tilde{M} = \mathbb{T}^{i-1} \times (2\mathbb{T}) \times \mathbb{T}^{n-i}$ be the covering space of $M = \mathbb{T}^n$, let π_1 be the covering map $\tilde{M} \to M$: $\pi_1(x_1, \cdots, x_i \cdots, x_n) = (x_1, \cdots, [x_i] \cdots, x_n)$ where $[x_i] = x_i$ if $x_i \leqslant 1$, $[x_i] = x_i - 1$ if $1 \leqslant x_i \leqslant 2$. We also use $\pi_1 : T\tilde{M} \to TM$ to denote the standard projection, $\pi_1(x, \dot{x}) = (\pi_1 x, \dot{x})$.

Lemma 10 *Let* $c \in \mathring{I}_c$, $\tilde{M} = \mathbb{T}^{i-1} \times (2\mathbb{T}) \times \mathbb{T}^{n-i}$. *If* $d\gamma : \mathbb{R} \to T M$ *is a homoclinic orbit such that*

$$\liminf_{\substack{T_0 \to \infty \\ T_1 \to \infty}} \left\{ \int_{-T_0}^{T_1} (L - \eta_c)(d\gamma(t),t)dt + (T_0 + T_1)\alpha(c) \right\}$$

$$= \liminf_{\substack{T_0 \to \infty \\ T_1 \to \infty}} \min_{\substack{\xi(-T_0) \in V_c \\ \xi(T_1) \in V_c \\ [\xi]_i \neq 0}} \left\{ \int_{-T_0}^{T_1} (L - \eta_c)(d\xi(t),t)dt + (T_0 + T_1)\alpha(c) \right\},$$

then $\{d\gamma(t),t\} \subset \pi_1 \tilde{\mathcal{N}}(c,\tilde{M})$.

Proof. If we think $\tilde{M}$ as the configuration manifold, V_c has two lifts denoted by V_c' and V_c^*. In this case, the minimal measure has two ergodic components, the support of one component is in V_c', another one is in V_c^*. The lift of the homoclinic orbit is just an orbit joining the lift of the support of the minimal measure in V_c' with another lift in V_c^*. Recall the definition of the barrier function (cf. [Ma2])

$$B_c^*(m) = \min\{h_c^\infty(\xi,m) + h_c^\infty(m,\zeta) - h_c^\infty(\xi,\zeta) : \forall\, \xi, \zeta \in \mathcal{M}_0(c)\},$$

we obtain the result immediately.

We can also define the Mañé set $\tilde{\mathcal{N}}(c,\tilde{M})$ from another point of view.

Let $c \in \dot{I}_c$, the interior of $\mathcal{L}_\beta(\rho(\mu_c))$, $m_0 \in V_c$, $m_1 \in V_c$, we define

$$h_{c,e_i}^k(m_0,m_1) = \inf_{\substack{\gamma(0)=m_0 \\ \gamma(k)=m_1 \\ [\gamma]_i \neq 0}} \int_0^k (L - \eta_c)(d\gamma(t),t)dt + k\alpha(c),$$

$$h_{c,e_i}^{k_1,k_2}(m_0,\xi,m_1) = \inf_{\substack{\gamma(-k_1)=m_0 \\ \gamma(0)=\xi \\ \gamma(k_2)=m_1 \\ [\gamma]_i \neq 0}} \int_{-k_1}^{k_2} (L - \eta_c)(d\gamma(t),t)dt + (k_1 + k_2)\alpha(c),$$

$$h_{c,e_i}^\infty(m_0,m_1) = \liminf_{k \to \infty} h_{c,e_i}^k(\xi,\zeta), \tag{1}$$

$$h_{c,e_i}^\infty(m_0,\xi,m_1) = \liminf_{\substack{k_1 \to \infty \\ k_2 \to \infty}} h_{c,e_i}^{k_1,k_2}(m_0,\xi,m_1), \tag{2}$$

$$B_{c,e_1}^*(\xi) = \inf\{h_{c,e_1}^\infty(m_0,\xi,m_1) - h_{c,e_1}^\infty(m_0,m_1) : m_0,m_1 \in \mathcal{M}_0(c)\}. \tag{3}$$

Recall we have introduced a modified Lagrangian $L_{\eta,\mu,\psi} = L - \eta - \mu - \psi$. Let $T_0 \in \mathbb{Z}_+$, $T_1 \in \mathbb{Z}_+$, we define

$$h_{\eta,\mu,\psi,e_i}^{T_0,T_1}(m_0,m_1) = \inf_{\substack{\xi(-T_0)=m_0 \\ \xi(T_1)=m_1 \\ [\xi]_i \neq 0}} \int_{-T_0}^{T_1} L_{\eta,\mu,\psi}(d\gamma(t),t) + T_0\alpha(c) + T_1\alpha(c').$$

$$h^{T_0,T_1}_{\eta,\mu,\psi,e_i}(m_0,\xi,m_1) = \inf_{\substack{\xi(-T_0)=m_0 \\ \xi(T_1)=m_1 \\ \xi(0)=\xi \\ [\xi]_1 \neq 0}} \int_{-T_0}^{T_1} L_{\eta,\mu,\psi}(d\gamma(t),t) + T_0\alpha(c) + T_1\alpha(c').$$

$$h^{\infty}_{\eta,\mu,\psi,e_i}(m_0,m_1) = \liminf_{\substack{k_1\to\infty \\ k_2\to\infty}} h^{T_0,T_1}_{\eta,\mu,\psi,e_i}(m_0,m_1), \tag{4}$$

$$h^{\infty}_{\eta,\mu,\psi,e_i}(m_0,\xi,m_1) = \liminf_{\substack{k_1\to\infty \\ k_2\to\infty}} h^{T_0,T_1}_{\eta,\mu,\psi,e_i}(m_0,\xi,m_1). \tag{5}$$

Clearly, we have

Lemma 11 *Let $\tilde{M} = \mathbb{T}^{i-1} \times (2\mathbb{T}) \times \mathbb{T}^{n-i}$. For each $c \in \dot{I}_c$, we have*

$$\pi_1 \mathcal{N}_0(c,\tilde{M}) = \{B^*_{c,e_i} = 0\} \cup \{B^*_c = 0\},$$

$$\pi_1 \mathcal{N}_0(c,\tilde{M}) \backslash \mathcal{N}_0(c,M) \neq \varnothing.$$

For $\tilde{\mathscr{C}}_{\eta,\mu,\psi}(\tilde{M})$, we have the similar result if we consider another minimal measure $\mu_{c'}$ with c' being close to c. In this case, $\mathcal{N}(c') \subset V_{c'}$, which is guaranteed by the upper semi-continuity of Mañé set on cohomology.

Lemma 12 *Let $c \in \dot{I}_c$, $[\eta] = c$ and μ is a U-step 1-form with $[\bar{\mu}] = c' - c$. If $\|\psi\|_{C^0}$ is suitably small and $\mathrm{supp}(\psi) \cap V_c = \varnothing$, then*

$$\pi_1 \mathscr{C}_{\eta,\mu,\psi}(\tilde{M}) \backslash \mathscr{C}_{\eta,\mu,\psi}(M) \neq \varnothing.$$

Proof. For $m_0,m_1 \in \mathcal{M}_0(c)$, positive integers $T_0^i, T_1^i \in \mathbb{Z}_+$, we choose $\gamma_i(t,m_0,m_1,e_i)$: $[-T_0^i, T_1^i] \to M$ be a minimal curve joining m_0 and m_1 such that $[\gamma_i]_1 \neq 0$ and

$$h^{T_0^i,T_1^i}_{\eta,\mu,\psi,e_i}(m_0,m_1) = \int_{-T_0^i}^{T_1^i} L_{\eta,\mu,\psi}(d\gamma_i(t),t)dt + T_0^i\alpha(c) + T_1^i\alpha(c').$$

Let $\{T_0^i\}_{i\in\mathbb{Z}_+}$ and $\{T_1^i\}_{i\in\mathbb{Z}_+}$ be the sequence of positive integers such that $T_j^i \to \infty$ ($j = 0, 1$) as $i \to \infty$ and the following limit exists

$$\lim_{i\to\infty} h^{T_0^i,T_1^i}_{\eta,\mu,\psi,e_i}(m_0,m_1) = \liminf_{T_0,T_1\to\infty} h^{T_0,T_1}_{\eta,\mu,\psi,e_i}(m_0,m_1) = h^{\infty}_{\eta,\mu,\psi,e_i}(m_0,m_1).$$

Let $\tilde{\gamma}_i$ be the lift of γ_i in the covering space $\tilde{M}$, it is a $\tilde{M}$-minimal curve. Clearly, the set of accumulation points of the set $\{\gamma_i\}$ contains a curve $\gamma \colon \mathbb{R} \to M$ with $[\gamma]_i \neq 0$.

On the other hand, if $\|\psi\|_{C^0}$ is suitably small and $m_0, m_1 \in V_{c'}$, the fact that c is in the interior guarantees that

$$h^{\infty}_{\eta,\mu,\psi}(m_0,m_1) < h^{\infty}_{\eta,\mu,\psi,e_i}(m_0,m_1).$$

In other words, these $\tilde{M}$-minimal curves $\{\gamma_i\}$ are not M-minimal curve. Consequently, γ is not a M-minimal curve. This completes the proof.

We can now investigate the existence of heteroclinic orbits, local connecting orbits, in other words. As heuristical example, let us consider Arnold's work in [Ar1] from variational point of view. There, each Mañé set under consideration properly contains the corresponding Mather set if we study the problem in a covering manifold $\tilde{M} = \mathbb{T} \times 2\mathbb{T}$. Under the small perturbation the stable manifold of each invariant circle transversally intersects the unstable manifold of the same invariant circle. In terms of variational language, the set $\pi_1 \mathcal{N}_0(c, \tilde{M}) \backslash (\mathcal{M}_0(c, M) + \delta)$ is non-empty but its first homology is trivial for each c under consideration. In this case, there exist local connecting orbits between $\mathcal{A}_{c'}$ and $\mathcal{A}_c$ if c' is close to c. Let us describe it in general:

Definition 1 *Let $c \in \mathring{I}_c$, $[\eta] = c$ and μ is a U-step 1-form with $[\bar{\mu}] = c' - c$. We assume that $\tilde{\mathcal{N}}(c) \subset V_c, \tilde{\mathcal{N}}(c') \subset V_c$. Let $\gamma \colon \mathbb{R} \to M$ be an absolutely continuous curve such that $\gamma(t) \in V_{c'}$ when $|t| \geqslant T$, and $[\gamma]_i \neq 0$. We say $d\gamma$ is a local minimal orbit of L of the first type that connects $\tilde{\mathcal{A}}(c)$ to $\tilde{\mathcal{A}}(\bar{c})$ if*

1. $d\gamma(t)$ is the solution of the Euler–Lagrange equation, the α- and ω-limit sets of $d\gamma$ are in $\tilde{\mathcal{A}}(c)$ and $\tilde{\mathcal{A}}(\bar{c})$ respectively;

2. There exist a closed 1-form η with $[\eta] = c$, a U-step 1-form μ with $[\bar{\mu}] = c' - c$ and a bump function ψ such that $d\gamma(t) \in \tilde{\mathscr{C}}_{\eta,\mu,\psi}(t)$ is a local minimal curve of the Lagrangian $L_{\eta,\mu,\psi}$ in the following sense: there exist two open balls O_0, O_1 and two positive integers T_0, T_1 such that $\bar{O}_0 \subset V_c \backslash \mathcal{M}_0(c)$, $\bar{O}_1 \subset V_c \backslash \mathcal{M}_0(c')$, $\gamma(-T_0) \in O_0$, $\gamma(T_1) \in O_1$ and

$$\min \left\{ h_{\eta,\mu,\psi,e_i}^{T_0,T_1}(m_0,m_1) + h_c^\infty(\xi,m_0) + h_{c'}^\infty(m_1,\zeta) : \right. \tag{6}$$

$$\left. \xi \in \mathcal{M}_0(c) \cap \pi(\alpha(d\gamma)|_{t=0}), \zeta \in \mathcal{M}_0(c') \cap \pi(\omega(d\gamma)|_{t=0}) \right\}$$

$$- \liminf_{\substack{T_0' \to \infty \\ T_1' \to \infty}} \int_{-T_0'}^{T_1'} L_{\eta,\mu,\psi}(d\gamma(t),t)dt - T_0'\alpha(c) - T_1'\alpha(c')$$

$$> 0$$

holds for any $(m_0,m_1) \in \partial(O_0 \times O_1)$.

Since $\pi(\omega(d\gamma)) \subset \mathcal{A}(c') \subset V_c$ and $\pi(\alpha(d\gamma)) \subset \mathcal{A}(c) \subset V_c$, $[\gamma|_{T_1 \leqslant t < \infty}]$ and $[\gamma|_{-\infty < t \leqslant -T_0}]$ are well defined. Indeed, $[\gamma|_{T_1 \leqslant t < \infty}] = 0$ and $[\gamma|_{-\infty < t \leqslant -T_0}] = 0$. That is why we can use $h_c^\infty(\xi,m_0)$ and $h_{c'}^\infty(m_1,\zeta)$ in this definition.

Obviously, (6) is equivalent to

$$h_{\eta,\mu,\psi,e_i}^{T_0,T_1}(m_0,m_1) + h_c^\infty(\xi,m_0) + h_{c'}^\infty(m_1,\zeta) \tag{7}$$

$$- \int_{-T_0}^{T_1} L_{\eta,\mu,\psi}(d\gamma(t),t)dt - T_0'\alpha(c) - T_1'\alpha(c')$$

$$- h_c^\infty(\xi,\gamma(-T_0)) - h_{c'}^\infty(\gamma(T_1),\zeta)$$

$$> 0$$

for each $\xi \in \mathcal{M}_0(c) \cap \pi(\alpha(d\gamma)|_{t=0})$ and each $\zeta \in \mathcal{M}_0(c') \cap \pi(\omega(d\gamma)|_{t=0})$.

Lemma 13 *If $\mathcal{N}_0(c,M) \subset V_c$ and $\pi_1\mathcal{N}_0(c,\tilde{M})\backslash V_c$ is totally disconnected, then there exist $\varepsilon_1 > 0$, a U-step form μ, a bump function ψ, a small number $t_0 > 0$ and an open disk O such that if $[\mu] = c'$, $\|c' - c\| \leqslant \varepsilon_1$, then*

$$\varnothing \neq \left\{ \pi_1\mathscr{C}_{\eta,\mu,\psi}(\tilde{M})\backslash\mathscr{C}_{\eta,\mu,\psi}(M) \right\}_{0\leqslant t\leqslant t_0} \subset O \qquad (8)$$

and each $d\gamma(t) \in \tilde{\mathscr{C}}_{\eta,\mu,\psi}(\tilde{M})\backslash\tilde{\mathscr{C}}_{\eta,\mu,\psi}(M)|_t$ determines a minimal orbit of L of first type which connecting $\mathscr{A}_c$ with $\mathscr{A}_{c'}$.

Proof. Since $\pi_1\mathcal{N}_0(c,\tilde{M})\backslash V_c$ is totally disconnected, there exist an open, connected set O which can shrink to one point by continuous deformation, and a small positive number $t_0 > 0$ such that

$$O \cap \pi_1\mathcal{N}(c,\tilde{M})|_{0\leqslant t\leqslant t_0}\backslash V_c \neq \varnothing,$$

$$O \cap V_c = \varnothing, \qquad \partial O \cap \pi_1\mathcal{N}(c,\tilde{M})|_{0\leqslant t\leqslant t_0} = \varnothing.$$

Clearly, we can find a small $\delta_1 > 0$ and define a non-negative function $f \in C^r(M,\mathbb{R})$ such that

$$f(x) \begin{cases} = 0 & x \in V_c \cup \left(\pi_1\mathcal{N}(c,\tilde{M})|_{0\leqslant t\leqslant t_0}\backslash(O+\delta_1) \right), \\ = 1 & x \in O, \\ < 1 & \text{elsewhere.} \end{cases}$$

We choose a C^r-function $\rho : \mathbb{R} \to [0,1]$ such that $\rho = 0$ on $t \in (-\infty,0] \cup [t_0,\infty)$, $0 < \rho \leqslant 1$ on $t \in (0,t_0)$. Let $\lambda \geqslant 0$ be a positive number,

$$\psi(x,t) = \lambda\rho(t)f(x),$$

By the upper semi-continuity of the set function $(\eta,\mu,\psi) \to \mathscr{C}_{\eta,\mu,\psi}(\tilde{M})$ we see that $\mathscr{C}_{\eta,0,\psi}(\tilde{M})|_{0\leqslant t\leqslant t_0} \cap \partial O = \varnothing$ if $\lambda > 0$ is suitably small. By the choice of ψ, we have $\mathscr{C}_{\eta,0,\psi}(M) = \mathcal{N}(c,M)$. Consequently, by using the similar argument to prove the lemma 12 we find

$$\varnothing \neq \left\{ \pi_1\mathscr{C}_{\eta,0,\psi}(\tilde{M})\backslash\mathscr{C}_{\eta,0,0}(M) \right\}_{0\leqslant t\leqslant t_0} \subset O.$$

Since O is homotopically trivial, for any cohomology class c', there exists a closed 1-form $\bar{\mu}$ such that $[\bar{\mu}] = c' - c$ and $\mathrm{supp}(\bar{\mu}) \cap O = \varnothing$. Let $\rho_1 \in C^r(\mathbb{R},[0,1])$ such that $\rho_1 = 0$ on $(-\infty,0]$, $0 < \rho_1 < 1$ on $(0,t_0)$ and $\rho_1 = 1$ on $[t_0,\infty)$, let $\mu = \rho_1(t)\bar{\mu}$ and set $L_{\eta,\mu,\psi} = L - \eta - \mu - \psi$. By using the upper semi-continuity and the similar argument to prove the lemma 12 again we obtain (7) if $\|\mu\|$ is suitably small. Let $d\gamma \in \pi_1\tilde{\mathscr{C}}_{\eta,\mu,\psi}(\tilde{M})\backslash\tilde{\mathscr{C}}_{\eta,\mu,\psi}(M)$. Note that $f \equiv 1$ in O, $\mathrm{supp}(\bar{\mu}) \cap O = \varnothing$, $d\gamma$: $TM \to \mathbb{R}$ is obviously a solution of the Euler–Lagrange equation, $\alpha(d\gamma) \subset \tilde{\mathcal{N}}(c)$ and $\omega(d\gamma) \subset \tilde{\mathcal{N}}(c')$.

Since $\pi_1 \mathscr{C}_0(c,\tilde{M}) \backslash N_\delta$ is assumed totally disconnected in O, by the upper semi-continuity, there obviously are two open and connected sets O_0 and O_1 such that $\bar{O}_0 \subset V_c \backslash \mathscr{M}_0(c)$, $\bar{O}_1 \subset V_c \backslash \mathscr{M}_0(c')$ and (6) holds.

Let us compare $\pi_1 \mathscr{C}_{\eta,0,\psi}(\tilde{M}) \backslash \mathscr{C}_{\eta,0,\psi}(M)$ with $\pi_1 \mathscr{N}(c,\tilde{M}) \backslash \mathscr{N}(c,M)$. If $\gamma(t)$ is a minimal curve in $\pi_1 \mathscr{N}(c,\tilde{M}) \backslash \mathscr{N}(c,M)$, then its time k translation $\gamma(t+k)$ is also a minimal curve for each $k \in \mathbb{Z}$. By the choice of the open set O and the function ψ, we see that each orbit $d\gamma$ in $\pi_1 \mathscr{N}(c,\tilde{M}) \backslash \mathscr{N}(c,M)$ might be an orbit of the Euler–Lagrange equation determined by $L - \psi$ still, but only those curves remain to be minimal if they pass through O when $t \in [0,t_0]$.

Next, let us consider a resonant minimal measure μ_c which consists of more than one ergodic component. We study the case that the minimal measure consists of finitely many ergodic components which has been proved generic in [CP]. Let $\tilde{\mathscr{A}}_c^i$ $(i \in \Lambda = \{1,2,\cdots m\})$ be the Aubry class such that $\tilde{\mathscr{A}}(c) = \cup \tilde{\mathscr{A}}_c^i$. A reflexive partial order $\preccurlyeq$ in the set of static classes $\{\tilde{\mathscr{A}}_c^i\}_{i \in \Lambda}$ is defined:

(a) $\preccurlyeq$ is reflexive;

(b) $\preccurlyeq$ is transitive;

(c) If there is $z \in \tilde{\mathscr{N}}_0(c)$ such that the α-limit set $\alpha(d\phi_L^t(z,0)) \subseteq \tilde{\mathscr{A}}_c^i$, and the ω-limit set $\omega(d\phi_L^t(z,0)) \subseteq \tilde{\mathscr{A}}_c^j$, then $\tilde{\mathscr{A}}_c^i \preccurlyeq \tilde{\mathscr{A}}_c^j$.

Theorem 9 (see [CP]) *Suppose that the number of static classes is finite. Then given $\tilde{\mathscr{A}}_c^i$ and $\tilde{\mathscr{A}}_c^j$ in $\{\mathscr{A}_c^i\}_{i \in \Lambda}$, we have that $\tilde{\mathscr{A}}_c^i \preccurlyeq \tilde{\mathscr{A}}_c^j$.*

This theorem can be also proved by variational method which is useful for the arguments in the following.

Proof. Let $x_i \in \mathscr{A}_c^i$, $x_j \in \mathscr{A}_c^j$ and consider the sequence $\{k_i \in \mathbb{Z}\}$ with $k_i \to \infty$ such that $\lim_{k_i \to \infty} h_c^{k_i}(x,x') = h_c^\infty(x,x')$, let $\gamma_c^{k_i}(t)\colon [0,k_i] \to M$ be an absolutely continuous curve with $\gamma_c^{k_i}(0) = x$, $\gamma_c^{k_i}(k_i) = x'$ which realizes the quantity $h_c^{k_i}(x,x')$. By the C^1-compactness of these curves we see that there exists at least one forward c-semi static curve $\gamma_c^+(t)\colon [0,\infty) \to M$ as well as at least one backward c-semi static curve $\gamma_c^-(t)\colon (-\infty,0] \to M$ such that $d\gamma_c^{k_i'}(0) \to d\gamma_c^+(0)$ and $d\gamma_c^{k_i''}(k_i'') \to d\gamma_c^-(0)$, where $\{k_i'\}$ and $\{k_i''\}$ are two subsequences of $\{k_i\}$. Indeed, the existence of such forward (backward) c-semi static curve in this case is unique because of the following lemma:

Lemma 14 (see [Me2] and [CDI]) *For each $(x,t) \in \mathscr{A}(c)$ there exists a unique $v \in T_x M$ such that $(x,v,t) \in \tilde{\mathscr{N}}^\pm(c)$. In fact, $(x,v,t) \in \tilde{\mathscr{A}}(c)$.*

Therefore, $\{d\gamma_c^+(t)\}_{[0,+\infty)} \subset \tilde{\mathscr{A}}_c^i$, $\{d\gamma_c^-(t)\}_{(-\infty,0]} \subset \tilde{\mathscr{A}}_c^j$, $d\gamma_c^{k_i}(0) \to d\gamma_c^+(0)$ and $d\gamma_c^{k_i}(k_i) \to d\gamma_c^-(0)$ as $k_i \to \infty$. Let $\delta > 0$ be small number so that there is no other static class in $\mathscr{A}_c^i + \delta$ and in $\mathscr{A}_c^j + \delta$. Define

$$t_\delta^{i-} = \min\{t \in \mathbb{Z} : \gamma_c^{k_i}(t) \notin \mathscr{A}_c^i + \delta\},$$
$$t_\delta^{i+} = \max\{t \in \mathbb{Z} : \gamma_c^{k_i}(t) \notin \mathscr{A}_c^j + \delta\},$$

we have $t_\delta^{i-} \leqslant t_\delta^{i+}$, $t_\delta^{i-} \to \infty$ and $k_i - t_\delta^{i+} \to \infty$ as $k_i \to \infty$. Let us consider the sequence of minimizers $\{\gamma_c^{k_i}(t - t_\delta^{i-})\}$. Because it has C^1-compactness, there is a subsequence $\{k_i^{(1-)}\}$ of $\{k_i\}$ and a curve $\gamma_c^{(1-)}: \mathbb{R} \to M$ such that $\gamma_c^{k_i^{(1-)}}(t - t_\delta^{i-}) \to \gamma_c^{(1-)}(t)|_{[-N,N]}$. Clearly, the curve $\gamma_c^{(1-)}$ is c-semi static and its α-limit set $\alpha(d\gamma_c^{(1-)}) \subseteq \tilde{\mathscr{A}}_c^i$. In the same way, there is a c-semi static curve $\gamma_c^{(1+)}: \mathbb{R} \to M$ which is, restricted on any closed interval $[-N,N] \subset \mathbb{R}$, is a limit of a subsequence of curves $\{\gamma_c^{k_i^{(1+)}}(t - t_\delta^{i+})\}$ and of which the ω-limit set is contained in $\tilde{\mathscr{A}}_c^j: \omega(d\gamma_c^{(1+)}) \subseteq \tilde{\mathscr{A}}_c^j$.

There are three possibilities for the relations between these two curves $\gamma_c^{(1-)}$ and $\gamma_c^{(1+)}$:

1. $\omega(d\gamma_c^{(1-)}) \subseteq \tilde{\mathscr{A}}_c^j$, or $\alpha(d\gamma_c^{(1+)}) \subseteq \tilde{\mathscr{A}}_c^i$, or $\gamma_c^{(1-)}(t) = \gamma_c^{(1+)}(t + t_0)$ for some $t_0 \in \mathbb{Z}$

2. there is another Aubry class $\tilde{\mathscr{A}}_c^m$ such that $\omega(d\gamma_c^{(1-)}) \subseteq \tilde{\mathscr{A}}_c^m$ and $\alpha(d\gamma_c^{(1+)}) \subseteq \tilde{\mathscr{A}}_c^m$

3. there are two different Aubry classes $\tilde{\mathscr{A}}_c^m$ and $\tilde{\mathscr{A}}_c^{m'}$ such that $\omega(d\gamma_c^{(1-)}) \subseteq \tilde{\mathscr{A}}_c^m$ and $\alpha(d\gamma_c^{(1+)}) \subseteq \tilde{\mathscr{A}}_c^{m'}$

The first two cases imply that theorem has been proved. In the third case, by the same argument we can find two c-semi static curves $\gamma_c^{(2\pm)}$ which are the limit of the subsequence of $\{\gamma_c^{k_i^{(1\pm)}}(t - t_\delta^{m\pm})\}$ respectively. The α-limit set $\alpha(d\gamma_c^{(2-)}) \subseteq \tilde{\mathscr{A}}_c^m$ and the ω-limit set $\omega(d\gamma_c^{(2+)}) \subseteq \tilde{\mathscr{A}}_c^{m'}$. As $d_c(\mathscr{A}_c^i, \mathscr{A}_c^m) > 0$, the ω-limit set of $d\gamma_c^{(2-)}$ can not be contained in $\tilde{\mathscr{A}}_c^i$ because both $\gamma_c^{(1-)}$ and $\gamma_c^{(2-)}$ can be approximated by a sequence of minimal curves which would induce the property that $d_c(\mathscr{A}_c^i, \mathscr{A}_c^m) = 0$. By the same argument, the α-limit set of $d\gamma_c^{(2+)}$ can not be contained in $\tilde{\mathscr{A}}_c^j$ as $d_c(\mathscr{A}_c^j, \mathscr{A}_c^{m'}) > 0$. Since there are only finitely many Aubry classes, we can see that there are some c-semi static curves $\gamma_c^k: \mathbb{R} \to M$ $(k = 1, 2, \cdots, k_0)$ such that $d_c(\mathscr{A}_c^i, \pi(\alpha(d\gamma_c^1))) = 0$, $d_c(\pi(\omega(d\gamma_c^k)), \pi(\alpha(d\gamma_c^{k+1}))) = 0$ for $k = 1, 2, \cdots, k_0 - 1$, and $d_c(\pi(\omega(d\gamma_c^{k_0})), \mathscr{A}_c^j) = 0$.

When a minimal measure has more than one ergodic component, the rotation vector of an ergodic component is not necessarily the same as of the whole measure. When there are finite ergodic components, we may assume that the Aubry set $\tilde{\mathscr{A}}(c)$ contains more than one class, it is obviously generic. Denote the ergodic component of c-minimal measure and of c'-minimal measure by μ_c^i and $\mu_{c'}^i$ respectively. As $c' \to c$, each ergoic component of $\mu_{c'}$ converges to some ergodic component of μ_c. If $\mu_{c'}^1 \rightharpoonup \mu_c^1$ as $c' \to c$.

Lemma 15 *Assume μ_c has m ergodic components, $\mathscr{N}_0(c)\backslash(\mathscr{A}_0(c) + \delta)$ is totally disconnected. Then,*

(1) There exists small $\varepsilon > 0$ such that if $|c' - c| < \varepsilon$ there is a closed 1-forms η with $[\eta] = c$, a U-step 1-form v with $[v] = c' - c$, a bump function ψ and an open

disk O such that each orbits $d\gamma(t) \in \tilde{\mathscr{C}}_{\eta,v,\psi}(t)$ is an orbit of the Lagrange flow ϕ_L^t, the support of $\bar{v}$ is not contained in O and

$$\varnothing \neq \mathscr{C}_{\eta,v,\psi}(M)|_{0 \leqslant t \leqslant t_0} \subset O. \tag{9}$$

(II) If $\mu_{c'}^i \rightharpoonup \mu_c^i$, and if there exists an orbit $d\xi \colon \mathbb{R} \to TM$ such that $\cup_{t \in \mathbb{R}} d\xi(t) \in \tilde{\mathscr{N}}(c)$, $\alpha(d\xi) \subseteq \tilde{\mathscr{A}}_c^i$ and $\omega(d\xi) \subseteq \tilde{\mathscr{A}}_c^j$, then we can choose the open disc O such that $\tilde{\mathscr{C}}_{\eta,v,\psi}(t)$ contains an orbit $d\gamma \colon \mathbb{R} \to TM$ with $\alpha(d\gamma) \subseteq \tilde{\mathscr{A}}_{c'}^i$ and $\omega(d\gamma) \subseteq \tilde{\mathscr{A}}_c^j$.

Proof. According to the Theorem 9, $\mathscr{N}(c) \backslash (\mathscr{A}(c) + \delta)$ is not empty. In this case we do not need to lift M to its finite covering. Since $\mathscr{N}_0(c) \backslash (\mathscr{A}_0(c) + \delta)$ is totally disconnected, there is a shrinkable open set $O \subset M$ and a small positive number $t_0 > 0$ such that

$$O \cap (\mathscr{N}(c) \backslash (\mathscr{A}(c) + \delta))|_{0 \leqslant t \leqslant t_0} \neq \varnothing,$$

$$\partial O \cap \mathscr{N}(c)|_{0 \leqslant t \leqslant t_0} = \varnothing, \qquad O \cap (\mathscr{A}_0(c) + \delta) = \varnothing.$$

Remaining argument for the proof of the first part is similar to the proof of the lemma 13.

To prove the second part, let us observe a fact. Let $\tilde{\mathscr{N}}_c^{i,j} \subset \tilde{\mathscr{N}}(c)$ be the set consisting of those orbits whose α-limit set is contained in $\tilde{\mathscr{A}}_c^i$ and the ω-limit set is contained in $\tilde{\mathscr{A}}_c^j$. It is easy to see that this set is a Lipschitz graph of the map π^{-1}, let $\mathscr{N}_c^{i,j}$ denote its projection. Thus, we can find an open neighborhood $O_{i,j}$ of $\cup_{k \neq i,j} \text{supp} \mu_c^k$ and some c-semi static curve $\gamma \colon \mathbb{R} \to M$ with $\cup_{t \in \mathbb{R}} \gamma(t) \subset \mathscr{N}_c^{i,j}$ such that $\gamma(t) \notin O_{i,j}$. We introduce non-negative function $\phi \colon M \to \mathbb{R}$ with $\text{supp} \phi = \bar{O}_{i,j}$. Thus, the minimal measure of the perturbed Lagrangian $L - \varepsilon \phi$ contains only two ergodic components μ_c^i and μ_c^j, and the Mañé set contains some c-semi static orbits for L which connects μ_c^i with μ_c^j. To construct the pseudo-connecting orbit set we choose a small open disc O such that $\gamma(t)|_{k \leqslant t \leqslant k+t_0} \in O$. Thus $d\gamma(t-k) \in \tilde{\mathscr{C}}_{\eta,v,\psi}$. Let $\varepsilon \to 0$, we obtain the conclusion in the second part from the upper semi-continuity on Lagrangian.

The orbits in $\tilde{\mathscr{C}}_{\eta,v,\psi}$ has some local minimal property as the orbits of the first type have. To describe it, let us observe a fact first. When the minimal measure has finite ergodic components, as a set-valued function, the Aubry set is upper semi-continuous on the cohomology. It is a consequence of fact that the Mañé is same as the Aubry set when the minimal measure is uniquely ergodic.

For each Aubry class $\mathscr{A}_c^i$ we choose a small neighborhood U_i. When $|c' - c|$ is sufficiently small, each Aubry class $\mathscr{A}_{c'}^i$ is contained in some U_i. For each $d\gamma \in \tilde{\mathscr{C}}_{\eta,v,\psi}$, if $\alpha(d\gamma) \subset \tilde{\mathscr{A}}_{c'}^i$ and $\omega(d\gamma) \subset \tilde{\mathscr{A}}_{c'}^j$, there exist two open balls O_0 and O_1 such that $\bar{O}_0 \subset U_i$, $\bar{O}_1 \subset U_j$, $\gamma(-T_0) \in O_0$, $\gamma(T_1) \in O_1$ and

$$h_{\eta,v,\psi}^{T_0,T_1}(m_0,m_1) + h_c^\infty(\xi,m_0) + h_{c'}^\infty(m_1,\zeta) \tag{10}$$

$$- \int_{-T_0}^{T_1} L_{\eta,v,\psi}(d\gamma(t),t)dt - T_0'\alpha(c) - T_1'\alpha(c')$$

$$> 0.$$

holds for any $(m_0, m_1) \in \partial(O_0 \times O_1)$ and for each $\xi \in \mathcal{M}_0(c) \cap \pi(\alpha(d\gamma)|_{t=0})$ and each $\zeta \in \mathcal{M}_0(c') \cap \pi(\omega(d\gamma)|_{t=0})$. In this case, we call the element of $\tilde{\mathscr{C}}_{\eta,v,\psi}$ local minimal orbits of the second type.

We consider another type of local minimal orbits which connects $\tilde{\mathscr{N}}(c)$ with $\tilde{\mathscr{N}}(c')$.

Lemma 16 *We assume that there is an open neighborhood V of $\mathcal{N}_0(c)$ such that $H_1(V, \mathbb{R}) = 0$, then there exists small $\varepsilon > 0$, for each c' with $\|c' - c\| \leqslant \varepsilon$ there exist a closed 1-form η and a U-step 1-form μ such that $[\eta] = c$, $\bar{\mu} = c' - c$ and each orbit in $\tilde{\mathscr{C}}_{\eta,\mu}$ is an orbit of the Lagrange flow ϕ_L^t.*

We call $d\gamma$ in such $\tilde{\mathscr{C}}_{\eta,\mu}$ local minimal orbit of the third type. Earlier version of this lemma was formulated by Mather in [Ma2].

Proof. Since V is topologically trivial, for any $c' \in H^1(M, \mathbb{R})$ there exists a closed 1-form $\bar{\mu}$ such that $\mathrm{supp}\,\bar{\mu} \cap V = \varnothing$. We take the U-step 1-form in the way such that $\mu = 0$ when $t \leqslant 0$ and $\mu = \bar{\mu}$ when $t \geqslant t_0$ where $t_0 > 0$ is suitably small. By the upper-semi continuity of the map $(\eta, \mu) \to \tilde{\mathscr{C}}_{\eta,\mu}$, we find that $d\gamma(t)$ $(0 \leqslant t \leqslant t_0)$ is in V if $d\gamma \in \tilde{\mathscr{C}}_{\eta,\mu}$ and if $\|c' - c\|$ is sufficiently small. Therefore, $d\gamma$ is a solution of the Euler–Lagrangian equation determined by L.

5 Construction of global connecting orbits

We call $d\gamma\colon \mathbb{R} \to \mathbb{R}$ is a global connecting orbit if its ω-limit set $\omega(d\gamma) \subset \tilde{\mathscr{A}}(c_0)$, its α-limit set $\alpha(d\gamma) \subset \tilde{\mathscr{A}}(c_1)$ and the variation from c_0 to c_1 is large. A sufficient condition for the existence of orbit connecting $\tilde{\mathscr{A}}(c_0)$ to $\tilde{\mathscr{A}}(c_1)$ is the existence of a generalized transition chain in $H^1(M, \mathbb{R})$ that connects c_0 to c_1.

Definition 2 *Let $\tilde{M}$ be a finite covering of a compact manifold M and let c_0, c_1 be two cohomolgy classes in $H^1(M, \mathbb{R})$. We say that c_0 is joined with c_1 by a generalized transition chain if there is a continuous curve $\Gamma\colon [0, 1] \to H^1(M, \mathbb{R})$ such that $\Gamma(0) = c_0$, $\Gamma(1) = c_1$, for each $\tau \in [0, 1]$ at least one of the following cases takes place:*

(I) There is small $\delta_\tau > 0$ such that $\pi_1 \mathcal{N}_0(\Gamma(\tau), \tilde{M}) \backslash (\mathscr{A}_0(\Gamma(\tau), M) + \delta_\tau)$ is non-empty and totally disconnected

(II) The Aubry set consists of m classes $(m > 1)$, $\mathscr{A}(\Gamma(\tau)) = \cup_{i=1}^m \mathscr{A}_{\Gamma(\tau)}^i$

(III) $\mathcal{N}_0(\Gamma(\tau), M)$ is homologically trivial, i.e. it has a neighborhood U_τ such that the inclusion map $H_1(U_\tau, \mathbb{R}) \to H_1(M, \mathbb{R})$ is a zero map

Theorem 10 *Let $M = \mathbb{T}^n$, $\tilde{M} = \mathbb{T}^{i-1} \times (2\mathbb{T}) \times \mathbb{T}^{n-i}$. Let $\Gamma : [0, 1] \to H^1(M, \mathbb{R})$ be a generalized transition chain connects c_0 to c_1. Then there exists an orbit of the Euler–Lagrange flow ϕ_L^t $d\gamma\colon \mathbb{R} \to TM$ that connects $\tilde{\mathscr{A}}(c_0)$ to $\tilde{\mathscr{A}}(c_1)$: $\alpha(d\gamma) \subset \tilde{\mathscr{A}}(c_0)$ and $\omega(d\gamma) \subset \tilde{\mathscr{A}}(c_1)$.*

Proof. Since the map $c \to \tilde{\mathcal{N}}(c, M)$ is upper semi-continuous, once the Mañé set $\tilde{\mathcal{N}}(\Gamma(\tau))$ is in the case I (or III), then for those τ' closed to τ, the $\tilde{\mathcal{N}}(\Gamma(\tau))$ is also in the case I (or III). The case II does not have such property.

For each $\tau \in [0, 1]$, according to the study in the last section, each Aubry set $\tilde{\mathcal{A}}(c_\tau)$ can be connected to some $\tilde{\mathcal{A}}(c_{\tau'})$ by either the first type, or the second, or the third type of local minimal orbits, if τ' is close to τ. Thus, there is a sequence $0 = \tau_0 < \tau_1 < \cdots < \tau_k = 1$ such that for each $0 \leqslant j < k$ $\tilde{\mathcal{A}}(\Gamma(\tau_j))$ is connected to $\tilde{\mathcal{A}}(\Gamma(\tau_{j+1}))$ by some local minimal orbits.

For the convenience of notation, we divide the subindex set into m groups $\{0, 1, \cdots k\} = \{0, 1, \cdots, i_1, i_1 + 1, \cdots, i_2, \cdots, i_m - 1, i_m = k\}$. The rule to make such division is that for all $i = i_j, i_j + 1, \cdots, i_{j+1} - 1$, $\tilde{\mathcal{A}}(\tau_i)$ is connected to $\tilde{\mathcal{A}}(\tau_{i+1})$ by a local minimal orbit of the same type. Let Λ_1, Λ_2 and Λ_3 be the subset of $\{i_1, i_2, \cdots, i_m\}$, $\Lambda_1 \cup \Lambda_2 \cup \Lambda_3 = \{i_1, i_2, \cdots i_m\}$, $\Lambda_i \cap \Lambda_j = \varnothing$ for $i \neq j$. If $i_j \in \Lambda_\iota$, then for all $i = i_\iota, i_\iota + 1, \cdots, i_{\iota+1} - 1$, $\tilde{\mathcal{A}}(\tau_i)$ is connected to $\tilde{\mathcal{A}}(\tau_{i+1})$ by a local minimal orbit of the ι-th type (ι-th = first, second, or third). In the following we write $c_i = \Gamma(\tau_i)$.

More precisely, for each integer $i \in \bigcup_{i_j \in \Lambda_1} \{i_j, i_j + 1, \cdots, i_{j+1} - 1\}$:

1. There exists a local minimal orbit of the first type $d\gamma_i \colon \mathbb{R} \to TM$ such that it solves the the Euler–Lagrange equation determined by L, $\alpha(d\gamma_i) \subset \tilde{\mathcal{A}}(c_i)$ and $\omega(d\gamma_i) \subset \tilde{\mathcal{A}}(c_{i+1})$.

When $\tilde{\mathcal{A}}(c_i)$ is connected to $\tilde{\mathcal{A}}(c_{i-1})$ by $d\gamma_{i-1}$, and connected to $\tilde{\mathcal{A}}(c_{i+1})$ by $d\gamma_i$, we write $\tilde{\mathcal{A}}^-(c_i)$ the c_i-static class which contains the α-limit set $\alpha(d\gamma_i)$, write $\tilde{\mathcal{A}}^+(c_i)$ the c_i-static class which contains the ω-limit set $\omega(d\gamma_i)$. $\mathcal{A}^-(c_i)$ is not necessarily the same as $\mathcal{A}^+(c_i)$. According to Theorem 9, there exist some Aubry classes $\tilde{\mathcal{A}}^-(c_i) = \tilde{\mathcal{A}}^1(c_i), \tilde{\mathcal{A}}^2(c_i), \cdots, \tilde{\mathcal{A}}^{i_k}(c_i) = \tilde{\mathcal{A}}^+(c_i)$ and semi-static curves γ_i^j ($j = 1, 2, \cdots, i_k - 1$) such that $\alpha(d\gamma_i^j) \subseteq \tilde{\mathcal{A}}^j(c_i)$ and $\omega(d\gamma_i^j) \subseteq \tilde{\mathcal{A}}^{j+1}(c_i)$;

2. Given a small number λ_i there is a non-negative function $\psi_i(x, t) \leqslant \lambda_i$ such that $\psi_i = 0$ when $t \in (-\infty, 0] \cup [1, \infty)$. For each fixed t, the support of ψ_i is contained in a small neighborhood of the open disk O_i and $\psi_i = $ constant when it is restricted in O_i.

$$O_i \cap (\mathcal{N}(c_i, \tilde{M})|_{0 \leqslant t \leqslant t_0} \setminus (\mathcal{A}(c_i) + \delta)) \neq \varnothing,$$

$$\partial O_i \cap \mathcal{N}(c_i, \tilde{M})|_{0 \leqslant t \leqslant t_0} = \varnothing,$$

$$O_i \cap (\mathcal{A}(c_i) + \delta) = \varnothing;$$

3. There exist a closed 1-forms η_i with $[\eta_i] = c_i$ and a U step 1-form μ_i such that the restriction on $\{t \geqslant t_0\}$ is a closed 1-form $\bar{\mu}_i$ on M with $[\bar{\mu}_i] = c_{i+1} - c_i$. The support of μ_i is disjoint with O_i. According to the lemma 4.2, we can see that the set $\tilde{\mathcal{C}}_{\eta_i, \mu_i, \psi_i}(\tilde{M})$ has the property:

$$\varnothing \neq \pi_1 \tilde{\mathcal{C}}_{\eta_i, \mu_i, \psi_i}(\tilde{M}) \setminus \mathcal{C}_{\eta_i, \mu_i, \psi_i}(M)|_{0 \leqslant t \leqslant t_0} \subset O_i, \tag{1}$$

each orbit $d\gamma(t) \in \tilde{\mathcal{C}}_{\eta_i, \mu_i, \psi_i}(\tilde{M}) \setminus \tilde{\mathcal{C}}_{\eta_i, \mu_i, \psi_i}(M)|_t$ determines a local minimal orbit of L of the first type, which connects $\tilde{\mathcal{A}}(c_i)$ to $\tilde{\mathcal{A}}(c_{i+1})$. Consequently, there exist two open disks V_i^- and V_{i+1}^+ with $\bar{V}_i^- \subset (\mathcal{A}_0^-(c_i) + \delta) \setminus \mathcal{A}_0(c_i)$, $\bar{V}_{i+1}^+ \subset (\mathcal{A}_0^+(c_{i+1}) + \delta) \setminus \mathcal{A}_0(c_{i+1})$, two positive integers $\tilde{T}_i^0$, $\tilde{T}_i^1$ and a positive small number $\varepsilon_i^* > 0$ such that

$$\min \left\{ h_{c_i}^{\infty}(\xi,m_0) + h_{\eta_i,\mu_i,\psi_i,e_1}^{\tilde{T}_i^0,\tilde{T}_i^1}(m_0,m_1) + h_{c_{i+1}}^{\infty}(m_1,\zeta) : \right.$$

$$\left. (m_0,m_1) \in \partial(V_i^- \times V_{i+1}^+) \right\}$$

$$\geqslant \min \left\{ h_{c_i}^{\infty}(\xi,m_0) + h_{\eta_i,\mu_i,\psi_i,e_1}^{\tilde{T}_i^0,\tilde{T}_i^1}(m_0,m_1) + h_{c_{i+1}}^{\infty}(m_1,\zeta) : \right.$$

$$\left. (m_0,m_1) \in V_i^- \times V_{i+1}^+ \right\} + 5\varepsilon_i^* \tag{2}$$

where $\xi \in \mathcal{M}_0(c_i), \zeta \in \mathcal{M}_0(c_{i+1})$.

For each integer $i \in \bigcup_{i_j \in \Lambda_2} \{i_j, i_j+1, \cdots, i_{j+1}-1\}$, the situation is similar, the difference is that we do not need to lift M to its covering $\tilde{M}$.

1. There exists a local minimal orbit of the second type $d\gamma_i : \mathbb{R} \to TM$ such that it solves the the Euler–Lagrange equation determined by L, $\alpha(d\gamma_i) \subset \tilde{\mathcal{A}}(c_i)$ and $\omega(d\gamma_i) \subset \tilde{\mathcal{A}}(c_{i+1})$:

2. Given a small number λ_i there is a non-negative function $\psi_i(x,t) \leqslant \lambda_i$ such that $\psi_i = 0$ when $t \in (-\infty,0] \cup [1,\infty)$. For each fixed t, the support of ψ_i is contained in a small neighborhood of the open disk O_i and $\psi_i = $ constant when it is restricted in O_i.

$$O_i \cap (\mathcal{N}(c_i)|_{0 \leqslant t \leqslant t_0} \setminus (\mathcal{A}(c_i)+\delta)) \neq \varnothing,$$

$$\partial O_i \cap \mathcal{N}(c_i)|_{0 \leqslant t \leqslant t_0} = \varnothing, \qquad O_i \cap (\mathcal{A}(c_i)+\delta) = \varnothing;$$

3. There exist a closed 1-forms η_i with $[\eta_i] = c_i$ and a U step 1-form μ_i such that the restriction on $\{t \geqslant t_0\}$ is a closed 1-form $\bar{\mu}_i$ on M with $[\bar{\mu}_i] = c_{i+1} - c_i$. The support of μ_i is disjoint with O_i. According to lemma 15, we can see that the set $\tilde{\mathcal{C}}_{\eta_i,\mu_i,\psi_i}(M)$ has the property:

$$\varnothing \neq \tilde{\mathcal{C}}_{\eta_i,\mu_i,\psi_i}(M) \subset O_i, \tag{3}$$

each orbit $d\gamma(t) \in \tilde{\mathcal{C}}_{\eta_i,\mu_i,\psi_i}(M)$ determines a local minimal orbit of L of the second type, which connects $\tilde{\mathcal{A}}(c_i)$ to $\tilde{\mathcal{A}}(c_{i+1})$. There are $\tilde{\mathcal{A}}_{c_i}^j \subset \tilde{\mathcal{A}}(c_i)$, $\tilde{\mathcal{A}}_{c_{i+1}}^k \subset \tilde{\mathcal{A}}(c_{i+1})$ and an orbit $d\gamma \in \tilde{\mathcal{C}}_{\eta_i,\mu_i,\psi_i}(M)$ such that $\alpha(d\gamma) \subset \tilde{\mathcal{A}}_{c_i}^j$ and $\omega(d\gamma) \subset \tilde{\mathcal{A}}_{c_{i+1}}^k$. Consequently, there exist two open disks V_i^- and V_{i+1}^+ with $\bar{V}_i^- \subset (\tilde{\mathcal{A}}_{c_i}^j|_{t=0}+\delta) \setminus \mathcal{A}_0(c_i)$, $\bar{V}_{i+1}^+ \subset (\tilde{\mathcal{A}}_{c_{i+1}}^k|_{t=0}+\delta) \setminus \mathcal{A}_0(c_{i+1})$, two positive integers $\tilde{T}_i^0, \tilde{T}_i^1$ and a positive small number $\varepsilon_i^* > 0$ such that

$$\min \left\{ h_{c_i}^{\infty}(\xi,m_0) + h_{\eta_i,\mu_i,\psi_i}^{\tilde{T}_i^0,\tilde{T}_i^1}(m_0,m_1) + h_{c_{i+1}}^{\infty}(m_1,\zeta) : \right.$$

$$\left. (m_0,m_1) \in \partial(V_i^- \times V_{i+1}^+) \right\}$$

$$\geqslant \min \left\{ h_{c_i}^{\infty}(\xi,m_0) + h_{\eta_i,\mu_i,\psi_i}^{\tilde{T}_i^0,\tilde{T}_i^1}(m_0,m_1) + h_{c_{i+1}}^{\infty}(m_1,\zeta) : \right.$$

$$\left. (m_0,m_1) \in V_i^- \times V_{i+1}^+ \right\} + 5\varepsilon_i^* \tag{4}$$

where $\xi \in \tilde{\mathcal{A}}_{c_i}^j|_{t=0}, \zeta \in \tilde{\mathcal{A}}_{c_{i+1}}^k|_{t=0}$.

For each integer $i \in \bigcup_{i_j \in \Lambda_3} \{i_j, i_j + 1, \cdots, i_{j+1} - 1\}$, there exist two closed 1-forms η_i, $\bar{\mu}_i$ defined on M, a U-step 1-form μ_i defined on $(u,t) \in M \times \mathbb{R}$ and an open set $U_i \subset M$ such that $[\eta_i] = c_i$, μ_i is closed on $U_i \times [0,t_0]$, $\mu_i = 0$ when $t \leqslant 0$, $\mu_i = \bar{\mu}_i$ when $t \geqslant t_0 > 0$, $[\bar{\mu}_i] = c_{i+1} - c_i$ and there is a small number $\delta_i > 0$ such that

$$\mathscr{C}_{\eta_i,\mu_i}(t) + \delta_i \subset U_i, \qquad \text{when } t \in [0,t_0]. \tag{5}$$

All orbits in $\tilde{\mathscr{C}}_{\eta_i,\mu_i}$ are the local minimal orbits of the second type of L, they connect $\tilde{\mathscr{N}}(c_i)$ to $\tilde{\mathscr{N}}(c_{i+1})$.

By the compactness of the manifold M, for a small $\varepsilon_i^* > 0$ there exists $(\check{T}_i^0, \check{T}_i^1) = (\check{T}_i^0, \check{T}_i^1)(\varepsilon_i^*) \in (\mathbb{Z}^+, \mathbb{Z}^+)$ such that

$$h_{\eta_i,\mu_i}^{T_0,T_1}(m_0,m_1) \geqslant h_{\eta_i,\mu_i}^{\infty}(m_0,m_1) - \varepsilon_i^* \tag{6}$$

holds for all $T_0 \geqslant T_i^0$, $T_1 \geqslant T_i^1$ and for all $(m_0,m_1) \in M \times M$. Obviously, given (m_0,m_1) there are infinitely many $T_0 \geqslant T_i^0$ and $T_1 \geqslant T_i^1$ such that

$$|h_{\eta_i,\mu_i}^{T_0,T_1}(m_0,m_1) - h_{\eta_i,\mu_i}^{\infty}(m_0,m_1)| \leqslant \varepsilon_i^*. \tag{7}$$

Let $\gamma_i(t, m_0, m_1, T_0, T_1): [-T_0, T_1] \to M$ be the minimizer of $h_{\eta_i,\mu_i}^{T_0,T_1}(m_0,m_1)$, it follows from lemma 8 that if $\varepsilon_i^* > 0$ is sufficiently small, $T_0 > \check{T}_i^0$ and $T_1 > \check{T}_i^1$ are chosen sufficiently large so that (7) holds, then

$$d\gamma_i(t, m_0, m_1, T_0, T_1) \in \tilde{\mathscr{C}}_{\eta_i,\mu_i}(t) + \delta_i, \qquad \forall \, 0 \leqslant t \leqslant 1. \tag{8}$$

By the Lipschitz property of $h_{\eta_i,\mu_i}^{T_0,T_1}(m_0,m_1)$ in (m_0,m_1) there exist $\hat{T}_i^0(\varepsilon_i^*) > \check{T}_i^0(\varepsilon_i^*)$ and $\hat{T}_i^1(\varepsilon_i^*) > \check{T}_i^1(\varepsilon_i^*)$ so that for each (m_0,m_1) there are $T_j = T_j(m_0,m_1)$ with $\check{T}_i^j(\varepsilon_i^*) \leqslant T_j \leqslant \hat{T}_i^j(\varepsilon_i^*)$ $(j = 0, 1)$ such that both (7) and (8) hold. Note that for different (m_0,m_1) we may need different $T_j \geqslant \check{T}_i^j$ $(j = 0, 1)$.

Before we formulate the variational principle, let us observe some fact.

Let us consider these two orbits of the Lagrangian flow: one orbit of ϕ_L^t, $d\gamma$: $\mathbb{R} \to TM$, has the property that the α-limit set $\alpha(d\gamma) \subseteq \tilde{\mathscr{A}}(c)$ and the ω-limit set $\omega(d\gamma) \subseteq \tilde{\mathscr{A}}(c')$, another orbit of ϕ_L^t, $d\gamma'$: $\mathbb{R} \to TM$, has the property that the α-limit set $\alpha(d\gamma') \subseteq \tilde{\mathscr{A}}(c')$ and the ω-limit set $\omega(d\gamma') \subseteq \tilde{\mathscr{A}}(c'')$. It is not necessary that $d_{c'}(\omega(d\gamma), \alpha(d\gamma')) = 0$. However, under the condition II, we can use some c'-semi static orbits to connect them in the sense of pseudo-metric $d_{c'}$ (cf Theorem 9).

Proposition 17 *Assume the Aubry distance from an Aubry class $\mathscr{A}_0^i(c)$ to other Aubry classes has a positive lower bound, $d_c(\mathscr{A}_0^i(c), \mathscr{A}_0^j(c)) \geqslant d > 0$ for all $j \neq i$, then there is an open neighborhood $N_c^i \supset \mathscr{A}_0^i(c)$, for all $m_0, m_1 \in N_c^i$ and for all $x \in \mathscr{A}_0^i(c)$, we have*

$$h_c^{\infty}(m_0,x) + h_c^{\infty}(x,m_1) = h_c^{\infty}(m_0,m_1); \tag{9}$$

for all $m_0, m_1 \in N_c^i$ and any $x \in \mathscr{A}_0(c) \backslash \mathscr{A}_0^i(c)$ we have

$$h_c^\infty(m_0, x) + h_c^\infty(x, m_1) \geqslant h_c^\infty(m_0, m_1) + \frac{d}{2}. \tag{10}$$

Proof. For each pair of points $(m_0, m_1) \in M \times M$, we claim that there exists some Aubry class $\mathscr{A}_0^j(c)$ such that

$$h_c^\infty(m_0, m_1) = h_c^\infty(m_0, \xi) + h_c^\infty(\xi, m_1)$$

holds for each $\xi \in \mathscr{A}_0^j(c)$. Indeed, let $k_i \to \infty$ be a subsequence of integers such that

$$\lim_{i \to \infty} h_c^{k_i}(m_0, m_1) = h_c^\infty(m_0, m_1)$$

let $\gamma_c^{k_i}\colon [0, k_i] \to M$ be the minimizer for $h_c^{k_i}(m_0, m_1)$. For any large but finite number $N > 0$, the set $\{\gamma_c^{k_i}|_{[0,N]}\}$ is compact in C^1-topology, thus we obtain forward semi static curve $\gamma_c^s\colon [0, \infty) \to M$. The ω-limit set of $d\gamma_c^s$ must be some Aubry class, let's say $\omega(d\gamma_c^s) \subseteq \mathscr{A}^i(c)$. Obviously, for any $\xi \in \mathscr{A}^i(c)$ the equality (9) holds.

Choose a neighborhood N_c^i of $\mathscr{A}_0^i(c)$ such that

$$N_c^i = \left\{ m \in M : d_c(m, x) \leqslant \frac{d}{6\max\{1, C_L\}}, \ \forall x \in \mathscr{A}_0^i(c) \right\}$$

where C_L is the Lipschitz constant of the barrier function. Given $m \in N_c^i$, we claim that (9) and (10) hold if we let $m_0 = m_1 = m$. In fact, let $k_i \to \infty$ be a sequence such that $\lim_{k_i \to \infty} h_c^{k_i}(m, m) = h_c^\infty(m, m)$ and let $\gamma_m^{k_i}(t)\colon [0, k_i] \to M$ be the minimizer of $h_c^{k_i}(m, m)$, the ordinary distance $d(\gamma_m^{k_i}(t), \mathscr{A}_0^j(c)) \geqslant d' > 0$ for all integer $t \in [0, k_i]$ and $j \neq i$. Otherwise we would obtain from the property that $d_c(\mathscr{A}_0^i(c), \mathscr{A}_0^j(c)) \geqslant d > 0$ for all $j \neq i$ and the Lipschitz property of $h_c(x, y)$ on x and y that

$$h_c^\infty(m, m) \geqslant d - 2C_L \frac{d}{6\max\{1, C_L\}} \geqslant \frac{3}{5} d.$$

On the other hand, the Lipschitz property of the Barrier function $B_c(x)$ in x induces that

$$h_c^\infty(m, m) \leqslant \frac{2}{5} d.$$

This contradiction verifies our claim.

Now we consider any two points $m_0, m_1 \in N_c^i$. For any $x \in \mathscr{A}_0^j(c)$ with $j \neq i$, we let $\zeta_u^i(t, m_0, x)\colon [0, k_i] \to M$ be the curve which minimizes the quantity $h_c^{k_i}(m_0, x)$ and $\lim_{k_i \to \infty} h_c^{k_i}(m_0, x) = h_c^\infty(m_0, x)$, let $\zeta_s^i(t, x, m_1)\colon [0, k_i'] \to M$ be the curve which minimizes the quantity $h_c^{k_i'}(x, m_1)$ and $\lim_{k_i' \to \infty} h_c^{k_i'}(x, m_1) = h_c^\infty(x, m_1)$. Let $\ell = 0, 1$, $\gamma_{m_\ell}^i\colon [0, k_\ell^i] \to M$ be the minimizer of $h_c^{k_\ell^i}(m_\ell, m_\ell)$ with $\lim_{k_\ell^i \to \infty} h_c^{k_\ell^i}(m_\ell, m_\ell) = h_c^\infty(m_\ell, m_\ell)$, clearly $\exists x_\ell \in \mathscr{A}_0^i(c)$ and integer $t_{m_\ell}^i \in [0, k_\ell^i]$ such that $\gamma_{m_\ell}^i(t_\ell^i) \to x_\ell$

358   C.-Q. Cheng

and $t^i_\ell \to \infty$ as $k^i_\ell \to \infty$. Let $\xi^i_{01}\colon [0,k^i_{01}] \to M$ be the minimizer of $h^{k^i_{01}}_c(x_0,x_1)$ with $\lim_{k^i_{01}\to\infty} h^{k^i_{01}}_c(x_0,x_1) = h^\infty_c(x_0,x_1)$, let $\xi^i_{10}\colon [0,k^i_{10}] \to M$ be the minimizer of $h^{k^i_{10}}_c(x_1,x_0)$ with $\lim_{k^i_{10}\to\infty} h^{k^i_{10}}_c(x_1,x_0) = h^\infty_c(x_1,x_0)$. Given arbitrarily small $\delta > 0$, we have sufficiently large k_i, k'_i, k^i_0, k^i_1, k^i_{01} and k^i_{10} such that

$$|h^\infty_c(m_0,x) - h^{k_i}_c(m_0,x)| < \delta,$$

$$|h^\infty_c(x,m_1) - h^{k'_i}_c(x,m_1)| < \delta,$$

$$|h^\infty_c(m_\ell,m_\ell) - h^{k^i_\ell}_c(m_\ell,m_\ell)| < \delta, \qquad \ell = 0,1$$

$$|h^\infty_c(x_0,x_1) - h^{k^i_{01}}_c(x_0,x_1)| < \delta,$$

$$|h^\infty_c(x_1,x_0) - h^{k^i_{10}}_c(x_1,x_0)| < \delta,$$

Since $x_0,x_1 \in \mathscr{A}^i_0(c)$, we have $d_c(x_1,x_0) = 0$. Consequently,

$$h^{t^i_0}_c(m_0,x_0) + h^{k^i_{01}}_c(x_0,x_1) + h^{k^i_1 - t^i_1}_c(x_1,m_1) \tag{11}$$
$$+ h^{t^i_1}_c(m_1,x_1) + h^{k^i_{10}}_c(x_1,x_0) + h^{k^i_0 - t^i_0}_c(x_0,m_0)$$
$$\leqslant \frac{2}{6}d + 6\delta.$$

Since x is in another Aubry class, we have

$$h^{k^i}_c(m_0,x) + h^{k'^i}_c(x,m_1) \tag{12}$$
$$+ h^{t^i_1}_c(m_1,x_1) + h^{k^i_{10}}_c(x_1,x_0) + h^{k^i_0 - t^i_0}_c(x_0,m_0)$$
$$\geqslant d - \frac{1}{6}d - 5\delta.$$

Because δ is arbitrarily small, we obtain from (11) and (12) that

$$h^\infty_c(m_0,x) + h^\infty_c(x,m_1) - \frac{1}{2}d$$
$$\geqslant h^\infty_c(m_0,x_0) + h^\infty_c(x_0,x_1) + h^\infty_c(x_1,m_1)$$
$$\geqslant h^\infty_c(m_0,m_1)$$

it verifies (10). Since (10) holds for any point in any other Aubry class, (9) must hold for some, thus for all $x \in \mathscr{A}^i_0(c)$.

Consider each integer $i \in \bigcup_{i_j \in \Lambda_1 \cup \Lambda_2} \{i_j, i_j + 1, \cdots, i_{j+1} - 1\}$. In view of the proposition (17), we see that for any two points $m_0, m_1 \in N^j_{c_i}$ there exists $\check{T}_i(\varepsilon^*_i) > 0$, independent of m_0 and m_1, such that for all $T \geqslant \check{T}_i(\varepsilon^*_i)$

$$h^T_{c_i}(m_0,m_1) \geqslant h^\infty_{c_i}(m_0,\zeta) + h^\infty_{c_i}(\zeta,m_1) - \varepsilon^*_i, \quad \forall\, \zeta \in \mathscr{A}^j_0(c_i), \tag{13}$$

and there exists $\hat{T}_i(\varepsilon_i^*) > \check{T}_i(\varepsilon_i^*)$ such that for each $(m_0, m_1) \in N_{c_i}^j \times N_{c_i}^j$ we have some integers T between $\check{T}_i(\varepsilon_i^*)$ and $\hat{T}_i(\varepsilon_i^*)$ so that

$$|h_{c_i}^T(m_0, m_1) - h_{c_i}^\infty(m_0, \zeta) - h_{c_i}^\infty(\zeta, m_1)| \leqslant \varepsilon_i^*, \quad \forall \, \zeta \in \mathscr{A}_0^j(c_i). \tag{14}$$

Since there are finitely many Aubry class for each c, we can choose those $\check{T}_i(\varepsilon_i^*)$ and $\hat{T}_i(\varepsilon_i^*)$ which apply to each Aubry class.

Let $d\gamma_i$ be a local minimal orbit of the first or second type, connecting μ_{c_i} to $\mu_{c_{i+1}}$. The subindex for each Aubry class $\mathscr{A}_{c_i}^j$ is chosen so that $\mathscr{A}_{c_i}^1 \supseteq \omega(d\gamma_{i-1})$, $\mathscr{A}_{c_i}^{k_i} \supseteq \alpha(d\gamma_i)$ and there is a c_i-semi static orbit connecting $\mathscr{A}_{c_i}^j$ to $\mathscr{A}_{c_i}^{j+1}$ ($j = 1, 2, \cdots, k_i - 1$). The condition II guarantees these orbits possess the property of local minimality: there exist two open disks $V_{i,j}^-$ and $V_{i,j+1}^+$ with $\bar{V}_{i,j}^- \subset (\mathscr{A}_{c_i}^j|_{t=0} + \delta) \backslash \mathscr{A}_0(c_i)$, $\bar{V}_{i,j+1}^+ \subset (\mathscr{A}_{c_i}^{j+1}|_{t=0} + \delta) \backslash \mathscr{A}_0(c_i)$, a positive integer $T_{i,j}$ and a small number $\varepsilon_i^* > 0$ such that

$$\min \left\{ h_{c_i}^\infty(\xi, m_0) + h_{c_i}^{T_{i,j}}(m_0, m_1) + h_{c_i}^\infty(m_1, \zeta) : \right.$$
$$\left. (m_0, m_1) \in \partial(V_{i,j}^- \times V_{i,j+1}^+) \right\}$$
$$\geqslant \min \left\{ h_{c_i}^\infty(\xi, m_0) + h_{c_i}^{T_{i,j}}(m_0, m_1) + h_{c_i}^\infty(m_1, \zeta) : \right.$$
$$\left. (m_0, m_1) \in V_{i,j}^- \times V_{i,j+1}^+ \right\} + 5\varepsilon_i^* \tag{15}$$

where $\xi \in \mathscr{A}_{c_i}^j|_{t=0}$, $\zeta \in \mathscr{A}_{c_i}^{j+1}|_{t=0}$.

We define τ_i inductively for $0 \leqslant i \leqslant i_m$. Let $\tau_0 = 0$. For each $i \in \bigcup_{i_j \in \Lambda_1 \cup \Lambda_2} \{i_j, i_j + 1, \cdots, i_{j+1} - 1\}$ with $i - 1 \in \bigcup_{i_j \in \Lambda_1 \cup \Lambda_2} \{i_j, i_j + 1, \cdots, i_{j+1} - 1\}$ also, we define $\tau_{i,0}, \tau_{i,1}, \cdots, \tau_{i,i_k}$ and let $\tau_i = \tau_{i,0}$, $\tau_{i+1} = \tau_{i,i_k}$ such that

$$\tilde{T}_{i-1}^1 + \check{T}_i \leqslant \tau_{i,1} - \tau_i \leqslant \tilde{T}_{i-1}^1 + \hat{T}_i, \tag{16}$$

$$T_{i,j} + \check{T}_i \leqslant \tau_{i,j} - \tau_{i,j-1} \leqslant T_{i,j} + \hat{T}_i, \qquad \forall \, j = 2, 3, \cdots, i_k - 1, \tag{17}$$

$$\tilde{T}_i^0 + \check{T}_i + T_{i,i_k-1} \leqslant \tau_{i+1} - \tau_{i,i_k-1} \leqslant \tilde{T}_i^0 + \hat{T}_i + T_{i,i_k-1}, \tag{18}$$

see (2) for the definition of $\tilde{T}_{i-1}^1, \tilde{T}_i^0$, see (13), (14) for the definition of $\check{T}_i, \hat{T}_i$ and see (15) for the definition of $T_{i,j}$. For $i \in \bigcup_{i_j \in \Lambda_3} \{i_j, i_j + 1, \cdots, i_{j+1} - 1\}$ with $i - 1 \in \bigcup_{i_j \in \Lambda_3} \{i_j, i_j + 1, \cdots, i_{j+1}\}$ also, we choose those τ_i such that

$$\max\{\check{T}_i^0, \check{T}_{i-1}^1 + 1\} \leqslant \tau_i - \tau_{i-1} \leqslant \max\{\hat{T}_i^0, \hat{T}_{i-1}^1 + 1\}. \tag{19}$$

If there is a local minimal orbit of the first or the second type $d\gamma_i$ such that $\alpha(d\gamma_i) \subseteq \tilde{\mathscr{A}}(c_i)$ and $\omega(d\gamma_i) \subseteq \tilde{\mathscr{A}}(c_i)$, there is a local minimal orbit of the third type $d\gamma_{i+1}$ such that $\alpha(d\gamma_{i+1}) \subseteq \tilde{\mathscr{A}}(c_{i+1})$ and $\omega(d\gamma_{i+1}) \subseteq \tilde{\mathscr{A}}(c_{i+2})$. In this case we note that both $\hat{T}_{i+1}$ and $\hat{T}_{i+1}^0$ can be taken large enough such that for any $m_0, m_1 \in M$ there exist $T(m_0, m_1), T_0(m_0, m_1)$ with $\max\{\check{T}_{i+1}, \check{T}_{i+1}^0\} \leqslant T(m_0, m_1), T_0(m_0, m_1) \leqslant \max\{\hat{T}_{i+1}, \hat{T}_{i+1}^0\}$ such that (7) holds if we set $T_0 = T_0(m_0, m_1)$ there; (14) holds if we set $T = T(m_0, m_1)$ there; (6) and (13) hold for each $T_0, T \geqslant \max\{\check{T}_{i+1}, \check{T}_{i+1}^0\}$. Thus, we choose $\tau_{i,0}, \tau_{i,1}, \cdots, \tau_{i,i_k-1}$ in the way given by (16) and (17) and choose τ_{i+1} so that

$$T_{i,i_k-1} + \max\{\check{T}_i, \check{T}_{i+1}^0\} \leqslant \tau_{i+1} - \tau_{i,i_k-1} \leqslant T_{i,i_k-1} + \max\{\hat{T}_i, \hat{T}_{i+1}^0\}. \tag{20}$$

If both $\mathcal{N}_0(c_{i-1})$ and $\mathcal{N}_0(c_i)$ are homologically trivial and μ_i can be connected to μ_{i+1} by local connecting orbits of the first or second type, we can choose suitably large $\hat{T}_i$ and $\hat{T}_i^1$ and set the range for $\tau_{i,1}, \tau_{i,2}, \cdots, \tau_{i,i_k}$ in the way given by (17) and (18)

$$\tilde{T}_i^0 + \max\{\check{T}_{i-1}^1, \check{T}_i\} \leqslant \tau_{i,1} - \tau_{i-1} \leqslant \tilde{T}_i^0 + \max\{\hat{T}_{i-1}^1, \hat{T}_i\}. \tag{21}$$

Consider τ as the time translation $\tau^*\phi(x,t) = \phi(x, t+\tau)$ on $M \times \mathbb{R}$, let $\psi_i \equiv 0$ for $i \in \bigcup_{i_j \in \Lambda_3}\{i_j, i_j + 1, \cdots, i_{j+1} - 1\}$, we define a modified Lagrangian

$$\tilde{L} = L - \eta_0 - \sum_{i=0}^{i_m-1} (-\tau_i)^*(\mu_i + \psi_i). \tag{22}$$

For $i \in \bigcup_{i_j \in \Lambda_1 \cup \Lambda_2}\{i_j, i_j + 1, \cdots, i_{j+1} - 1\}$, we let $\tau_i = (\tau_{i,0}, \tau_{i,1}, \cdots, \tau_{i,i_k-1})$, for $i \in \bigcup_{i_j \in \Lambda_3}\{i_j, i_j + 1, \cdots, i_{j+1} - 1\}$, we let $\tau_i = \tau_i$. Define

$$\tau = (\tau_0, \tau_1, \cdots, \tau_{i_m-1}). \tag{23}$$

We define an index set for τ:

$$\Lambda = \left\{ \tau \in \mathbb{Z}^{i_m-1} : (16 \sim 21) \text{ hold} \right\}. \tag{24}$$

For $i \in \bigcup_{i_j \in \Lambda_1 \cup \Lambda_2}\{i_j, i_j + 1, \cdots, i_{j+1}\}$, we let $\mathbf{z}_i^\pm = (z_{i,1}^\pm, z_{i,2}^\pm, \cdots, z_{i,i_k}^\pm)$ and $\mathbf{V}_i^\pm = (V_{i,1}^\pm, V_{i,2}^\pm, \cdots, V_{i,i_k}^\pm)$, and define

$$\mathbf{z} = \left(z_0^-, \mathbf{z}_i^+, \mathbf{z}_i^-, z_{i_m}^+, i \in \bigcup_{i_j \in \Lambda_1 \cup \Lambda_2}\{i_j, i_j + 1, \cdots, i_{j+1}\}\right), \tag{25}$$

its domain is restricted in

$$\mathbf{V} = \left(V_0^-, \mathbf{V}_i^+, \mathbf{V}_i^-, V_{i_m}^+, i \in \bigcup_{i_j \in \Lambda_1 \cup \Lambda_2}\{i_j, i_j + 1, \cdots, i_{j+1}\}\right). \tag{26}$$

For $(m, m') \in M \times M$ and $\mathbf{z} \in \mathbf{V}$, we define

$$h_{\tilde{L}}^{K,K'}(m_0, m_1, \mathbf{z}, \tau) = \inf \int_{-K}^{K' + \tilde{T}_{i_m-1}^1 + \hat{T}_{i_m} + \tau_{i_m-1}} \tilde{L}(d\gamma(t), t)\,dt$$
$$+ \sum_{i=1}^{i_m-1} (\tau_i - \tau_{i-1})\alpha(c_i) + K\alpha(c_0) + K'\alpha(c_{i_m})$$

where the infimum is taken under the conditions: $\gamma(-K) = m_0$, $\gamma(\bar{K}' + \tau_{i_m-1}) = m_1$; for $i \in \bigcup_{i_j \in \Lambda_1} \{i_j, i_j + 1, \cdots, i_{j+1} - 1\}$, $[\gamma|_{t \in [\tau_i - \tilde{T}_i^0, \tau_i + \tilde{T}_i^1]}]_i \neq 0$; for $i \in \bigcup_{i_j \in \Lambda_1 \cup \Lambda_2} \{i_j, i_j + 1, \cdots, i_{j+1} - 1\}$, $\gamma(\tau_i - \tilde{T}_i^0) = z_{i,i_k}^-$, $\gamma(\tau_i + \tilde{T}_i^1) = z_{i+1,1}^+$; $\gamma(\tau_{i,j}) = z_{i,j}^-$ and $\gamma(\tau_{i,j} + T_{i,j}) = z_{i,j+1}^+$ for each $j = 1, 2, \cdots, i_k - 1$.

Let $h_{\tilde{L}}^{K,K'}(m_0, m_1)$ be the minimizer of $h_{\tilde{L}}^{K,K'}(m_0, m_1, \mathbf{z}, \tau)$ over $\mathbf{V}$ in $\mathbf{z}$ and over Λ in τ respectively:

$$h_{\tilde{L}}^{K,K'}(m_0, m_1) = \min_{\tau \in \Lambda, \mathbf{z} \in \mathbf{V}} h_{\tilde{L}}^{K,K'}(m_0, m_1, \mathbf{z}, \tau),$$

let $K_j, K_j' \to \infty$ be the subsequence such that

$$\lim_{K_j, K_j' \to \infty} h_{\tilde{L}}^{K_j, K_j'}(m_0, m_1) = \liminf_{K \to \infty K' \to \infty} h_{\tilde{L}}^{K,K'}(m_0, m_1),$$

and denote the corresponding minimal curve by $\gamma(t; K_j, K_j', m_0, m_1)$, we claim that $d\gamma(t; K_j, K_j', m_0, m_1)$ is a solution of the Euler–Lagrange equation determined by L if K_j and K_j' are sufficiently large. Indeed,

1. For each $i \in \bigcup_{i_j \in \Lambda_3} \{i_j, i_j + 1, \cdots, i_{j+1} - 1\}$, we have

$$(-\tau_i)^* \gamma(t; K_j, K_j', m_0, m_1) \in \mathscr{C}_{\eta_i, \mu_i}(t) + \delta_i \subset U_i, \tag{27}$$

when $\tau_i \leqslant t \leqslant \tau_i + 1$. To see it, let us choose $m_i = \gamma(\tau_{i-1} + 1)$, $m_i' = \gamma(\tau_{i+1})$. Since the curve $\gamma(t; K_j, K_j', m_0, m_1)$ is the minimizer of $h_{\tilde{L}}^{K,K'}(m_0, m_1, Z, \tau)$ over Λ, thus

$$A_{\tilde{L}}((-\tau_i)^* \gamma|_{\tau_{i-1}+1}^{\tau_{i+1}}) + (\tau_i - \tau_{i-1} + 1)\alpha(c_i) + (\tau_{i+1} - \tau_i \alpha(c_{i+1})$$

$$= \inf_{\substack{\xi(-T_0)=m_i \\ \xi(T_1)=m_i' \\ \tilde{T}_i^0 \leqslant T_0 \leqslant \hat{T}_i^0 \\ \tilde{T}_i^1 \leqslant T_1 \leqslant \hat{T}_i^1}} \int_{-T_0}^{T_1} (L - \eta_i - \mu_i)(d\xi(t), t)dt + T_0\alpha(c_i) + T_1\alpha(c_{i+1}).$$

Thus we obtain (27) from this formula, (5), (8) and (19). Consequently, $\gamma(t; K_j, K_j')|_{\tau_i \leqslant t \leqslant \tau_i + 1}$ falls into the region where $(-\tau_i)^* \mu_i$ is closed. So, $d\gamma(t; K_j, K_j', m_0, m_1)$ is the solution of the Euler–Lagrange equation determined by L when $\tau_i \leqslant t \leqslant \tau_i + 1$;

2. For $i \in \bigcup_{i_j \in \Lambda_1} \{i_j, i_j + 1, \cdots, i_{j+1} - 1\}$, we claim that

$$(-\tau_i)^* \gamma(t)|_{0 \leqslant t \leqslant t_0} \in \mathrm{int}(O_i). \tag{28}$$

It is the consequence of (1). In fact, if $d\gamma \in \pi_1 \mathscr{C}_{\eta_i, \mu_i, \psi_i}(\tilde{M}) \backslash \mathscr{C}_{\eta_i, \mu_i, \psi_i}(M)$ then γ must pass through O_i during the time interval $[0, t_0]$. Note, the function $L_{\eta_i, \mu_i, \psi_i}$ is no longer time-periodic, if $d\gamma_i \in \pi_1 \mathscr{C}_{\eta_i, \mu_i, \psi_i}(\tilde{M}) \backslash \mathscr{C}_{\eta_i, \mu_i, \psi_i}(M)$, $k^* d\gamma_i$ is not a minimizer of this kind, $d\gamma_i(k) \notin \pi_1 \mathscr{C}_{\eta_i, \mu_i, \psi_i}(\tilde{M})|_{t=0}$ for each $k \in \mathbb{Z} \backslash \{0\}$.

Given a smooth curve $d\gamma \in \tilde{\mathscr{C}}_{\eta_i,\mu_i,\psi_i}(\tilde{M}) \backslash \tilde{\mathscr{C}}_{\eta_i,\mu_i,\psi_i}(M)$, we assume that $\alpha(d\gamma) \subset \tilde{\mathscr{A}}_{c_i}^{i_k}$ and $\omega(d\gamma) \subset \tilde{\mathscr{A}}_{c_{i+1}}^1$. For any $m_i \in \mathscr{M}_{c_i}^{i_k}|_{t=0}$, $m_{i+1} \in \mathscr{M}_{c_{i+1}}^1|_{t=0}$ and , if $\mathbb{Z}^+ \ni T_0^k \to \infty$ and $\mathbb{Z}^+ \ni T_1^k \to \infty$ (as $k \to \infty$) are two sequences such that $\gamma(-T_0^k) \to m_i$ and $\gamma(T_1^k) \to m_{i+1}$, then

$$\lim_{k\to\infty} \int_{-T_0^k}^{T_1^k} L_{\eta_i,\mu_i,\psi_i}(d\gamma(t),t)dt + T_0^k\alpha(c_i) + T_1^k\alpha(c_{i+1}) = h_{\eta_i,\mu_i,\psi_i,e_1}^{\infty}(m_i,m_{i+1}).$$

Let $\zeta \colon \mathbb{R} \to M$ be an absolutely continuous curve such that $[\zeta]_i \neq 0$, $\zeta(t') \notin int(O_i)$ for some $t' \in [0,t_0]$, $\zeta(-T_0^k) \to m_i$, $\zeta(T_1^k) \to m_{i+1}$ as $k \to \infty$. Since $\tilde{\mathscr{C}}_{\eta_i,\mu_i,\psi_i}|_{t=\mathrm{constant}}$ is closed, there exists a positive number $d > 0$ such that

$$\liminf_{\substack{T_0^k \to \infty \\ T_1^k \to \infty}} \int_{-T_0^k}^{T_1^k} L_{\eta_i,\mu_i,\psi_i}(d\zeta(t),t)dt + T_0^k\alpha(c_i) + T_1^k\alpha(c_{i+1})$$

$$\geqslant h_{\eta_i,\mu_i,\psi_i,e_1}^{\infty}(m_i,m_{i+1}) + d.$$

Recall the construction of the modified Lagrangian $\tilde{L}$ (see (22)) and γ is the minimizer of $h_{\tilde{L}}^{K,K'}(m_0,m_1,\mathbf{z},\tau)$ over $\mathbf{V}$ in $\mathbf{z}$ and over Λ in τ respectively. Given any small number $\varepsilon > 0$, by choosing sufficiently large $\hat{T}_i - \check{T}_i$, we can see that there are sufficiently large $K_i^-, K_i^+ \in \mathbb{Z}$ with the properties that $\tau_{i-1} + \tilde{T}_{i-1}^1 + K_i^- + K_i^+ = \tau_i - \tilde{T}_i^0$, $\check{T}_i \leqslant K_i^- + K_i^+ \leqslant \hat{T}_i$ and

$$\|\gamma(\tau_i - \tilde{T}_i^0 - K_i^+) - m_i\| < \varepsilon, \qquad \|\gamma_i(\tau_i + \tilde{T}_i^1 + K_{i+1}^-) - m_{i+1}\| \leqslant \varepsilon.$$

If there was $t' \in [0,t_0]$ such that

$$(-\tau_i)^*\gamma(t') \notin int(O_i),$$

from the Lipschitz continuity of $h_{\eta_i,\mu_i,\psi_i,e_i}^{\infty}(m,m')$ in (m,m') we would obtain

$$\int_{\tau_i - \tilde{T}_i^0 - K_i^+}^{\tau_i + \tilde{T}_i^1 + K_{i+1}^-} L_{\eta_i,\mu_i,\psi_i}(d\gamma(t),t)dt + (\tilde{T}_i^0 + K_{i-1}^+)\alpha(c_i) + (\tilde{T}_i^1 + K_{i+1}^-)\alpha(c_{i+1})$$

$$\geqslant h_{\eta_i,\mu_i,\psi_i,e_1}^{\infty}(m_i,m_{i+1}) + \frac{3}{4}d.$$

On the other hand, there is suitably large $\bar{K}_i^-, \bar{K}_i^+ \in \mathbb{Z}$ with the properties that $\check{T}_i \leqslant \bar{K}_i^- + \bar{K}_i^+ \leqslant \hat{T}_i$ and

$$\|\gamma_i(\tau_i - \tilde{T}_i^0 - \bar{K}_i^+) - m_i\| \leqslant \varepsilon, \qquad \|\gamma_i(\tau_i + \tilde{T}_i^1 + \bar{K}_{i+1}^-) - m_{i+1}\| \leqslant \varepsilon.$$

Because $d\gamma_i(t') \in \tilde{\mathscr{C}}_{\eta_i,\mu_i,\psi_i}|_{t=t'}$, we have

$$\int_{\tau_i-\tilde{T}_i^0-\bar{K}_i^+}^{\tau_i+\tilde{T}_i^1+\bar{K}_{i+1}^-} L_{\eta_i,\mu_i,\psi_i}(d\gamma_i(t),t)dt + (\tilde{T}_i^0+\bar{K}_{i-1}^+)\alpha(c_i) + (\tilde{T}_i^1+\bar{K}_{i+1}^-)\alpha(c_{i+1})$$

$$\leqslant h_{\eta_i,\mu_i,\psi_i,e_i}^{\infty}(m_i,m_{i+1}) + \frac{1}{4}d$$

if ε is sufficiently small. It implies that γ is not a minimizer. This contradiction verifies our claim.

The formula (28) implies that $d\gamma(t;K_j,K_j')$ is the solution of the Euler–Lagrange equation determined by L for $i \in \bigcup_{i_j\in\Lambda_1}\{i_j,i_j+1,\cdots,i_{j+1}-1\}$.

3. For $i \in \bigcup_{i_j\in\Lambda_2}\{i_j,i_j+1,\cdots,i_{j+1}-1\}$, we claim that (28) also holds. In this case, we have (3) instead of (1). The argument is the same if we replace the quantity $h_{\eta_i,\mu_i,\psi_i,e_i}^{\infty}(m_0,m_1)$ by $h_{\eta_i,\mu_i,\psi_i}^{\infty}(m_0,m_1)$.

4. We claim that the curve γ does not touch the boundary of V_{i,i_k}^- at the time $t = \tau_i - \tilde{T}_i^0$ and does not touch the boundary of $V_{i+1,1}^+$ at the time $t = \tau_i + \tilde{T}_i^1$ for each integer $i \in \bigcup_{i_j\in\Lambda_1}\{i_j,i_j+1,\cdots,i_{j+1}-1\}$. If $(\gamma(\tau_i-\tilde{T}_i^0),\gamma(\tau_i+\tilde{T}_i^1)) = (z_{i,i_k}^-,z_{i+1,1}^+) \in \partial(V_{i,i_k}^- \times V_{i+1,1}^+)$, let $z_{i,i_k}^+ = \gamma(\tau_{i-1,(i-1)_k-1}+T_{i-1,(i-1)_k-1})$ and $z_{i+1,1}^- = \gamma(\tau_{i,1})$, from (2) we can see that there exist $(\bar{z}_{i,i_k}^-,\bar{z}_{i+1,1}^+) \in V_{i,i_k}^- \times V_{i+1,1}^+$ such that for $\xi \in \mathscr{M}_{c_i}^{i_k}|_{t=0}$, $\zeta \in \mathscr{M}_{c_{i+1}}^1|_{t=0}$:

$$h_{c_i}^{T_i}(z_{i,i_k}^+,z_{i,i_k}^-) + h_{\eta_i,\mu_i,\psi_i,e_i}^{\tilde{T}_i^0,\tilde{T}_i^1}(z_{i,i_k}^-,z_{i+1,1}^+) + h_{c_{i+1}}^{T_{i+1}}(z_{i+1,1}^+,z_{i+1,1}^-)$$

$$\geqslant h_{c_i}^{\infty}(\xi,z_{i,i_k}^-) + h_{\eta_i,\mu_i,\psi_i,e_i}^{\tilde{T}_i^0,\tilde{T}_i^1}(z_{i,i_k}^-,z_{i+1,1}^+) + h_{c_{i+1}}^{\infty}(z_{i+1,1}^+,\zeta)$$

$$+ h_{c_i}^{\infty}(z_{i,i_k}^+,\xi) + h_{c_{i+1}}^{\infty}(\zeta,z_{i+1,1}^-) - 2\varepsilon_i^*$$

$$\geqslant h_{c_i}^{\infty}(\xi,\bar{z}_{i,i_k}^-) + h_{\eta_i,\mu_i,\psi_i,e_i}^{\tilde{T}_i^0,\tilde{T}_i^1}(\bar{z}_{i,i_k}^-,\bar{z}_{i+1,1}^+) + h_{c_{i+1}}^{\infty}(\bar{z}_{i+1,1}^+,\zeta)$$

$$+ h_{c_i}^{\infty}(z_{i,i_k}^+,\xi) + h_{c_{i+1}}^{\infty}(\zeta,z_{i+1,1}^-) + 3\varepsilon_i^*$$

$$\geqslant h_{c_i}^{T_i}(z_{i,i_k}^+,\bar{z}_{i,i_k}^-) + h_{\eta_i,\mu_i,\psi_i,e_i}^{\tilde{T}_i^0,\tilde{T}_i^1}(\bar{z}_{i,i_k}^-,\bar{z}_{i+1,1}^+) + h_{c_{i+1}}^{T_{i+1}}(\bar{z}_{i+1,1}^+,z_{i+1,1}^-) + \varepsilon_i^*$$

where $T_i,T_{i+1},T_i',T_{i+1}'$ satisfy the condition $\check{T}_j \leqslant T_j, T_j' \leqslant \hat{T}_j$ $(j=i-1,i)$. In above arguments, (13) and (14) are used to obtain the first and the third inequality, (2) is used to obtain the second inequality. But this contradicts to the fact that γ is a minimal curve of $\tilde{L}$ on $\mathbf{V}$ and Λ. The case for $i \in \bigcup_{i_j\in\Lambda_2}\{i_j,i_j+1,\cdots,i_{j+1}-1\}$ can be treated in the same way. Therefore, the minimizer γ is differentiable at the time $t = \tau_i - \tilde{T}_i^0$ and $t = \tau_i + \tilde{T}_i^1$ for each i.

5. We claim that the curve γ does not touch the boundary of $V_{i,j}^-$ at the time $t = \tau_{i,j}$ and does not touch the boundary of $V_{i,j+1}^+$ at the time $t = \tau_{i,j} + T_{i,j}$ for each integer $i \in \bigcup_{i_j\in\Lambda_1\cup\Lambda_2}\{i_j,i_j+1,\cdots,i_{j+1}-1\}$ and $j \in \{1,2,\cdots,i_k-1\}$. The demonstration is similar to the case 4, based on (13), (14) and (15).

Let $K_j,K_j' \to \infty$ and let $\gamma_\infty: \mathbb{R} \to M$ be an accumulation point of the curves $\{\gamma(t,K_j,K_j')\}$. Obviously, $\alpha(d\gamma_\infty) \subset \tilde{\mathscr{A}}(c)$ and $\omega(d\gamma_\infty) \subset \tilde{\mathscr{A}}(c')$. This completes the proof.

The Theorem 10 provides us a possible way to prove Arnold diffusion is a generic phenomenon for positive definite systems. However, the verification of the condition I is difficult, when the conditions II and III do not hold. It contains two two key points: for each c in the chain the Aubry set does not generate the first homology group $H_1(M, \mathscr{A}_0(c), \mathbb{Z}) \neq 0$ and the barrier function is not constant when it is restricted outside of the Aubry set. In general case, we are unable to show the genericity of these conditions, however, it can be down in some interesting cases.

6 Application to a priori unstable systems

Here we study a typical example of a priori unstable system:

$$H(I, \phi, x, y, t) = h_0(I) + h_1(x, y) + \varepsilon P(I, \phi, x, y, t),$$

where $(I, \phi) \in \mathbb{R} \times \mathbb{T}$, $(x, y, t) \in \mathbb{T}^n \times \mathbb{R}^n \times \mathbb{T}$. Let

$$L(\dot{\phi}, \phi, \dot{x}, x, t) = \ell_0(\dot{\phi}) + \ell_1(\dot{x}, x) + \varepsilon L_1(\dot{\phi}, \phi, \dot{x}, x, t)$$

be the Legendre transformation of the Hamiltonian. We call it priori unstable when $(\dot{x}, x) = (0, 0)$ is a hyperbolic fixed point which corresponds to the minimum of the action of ℓ_1.

Under the a priori unstable condition, the time-1-map Φ of the Hamiltonian flow has an invariant cylinder which is a small deformation of the standard cylinder $\Sigma = \{(I, \phi) \in \mathbb{R} \times \mathbb{T}\}$. The restriction of Φ on Σ is area-preserving and twist. In virtue of many works before, we have very well understanding on the dynamics on the cylinder. Under Legendre transformation, this cylinder and the dynamics have their correspondence in the space of tangent bundle. Without danger of confusion, we use the same name for the object and its correspondence under the Legendre transformation.

Let $\Gamma: [0, 1] \to H^1(\mathbb{T}^{n+1}, \mathbb{R})$ be a path such that $\Gamma(s) = (c_1(s), 0, \cdots, 0)$ where the first component represents for ϕ. As the hyperbolic property is assumed, for each $s \in [0, 1]$, the Aubry set is in on the cylinder and the support of the minimal measure is either invariant curve, or the Aubry–Mather set, or the minimal periodic orbit. Therefore, the condition $H_1(M, \mathscr{A}_0(c), \mathbb{Z}) \neq 0$ is satisfied. To verify the condition:

(I) *There is small $\delta_\tau > 0$ such that $\pi_1 \mathscr{N}_0(\Gamma(\tau), \tilde{M}) \backslash (\mathscr{A}_0(\Gamma(\tau), M) + \delta_\tau)$ is non-empty and totally disconnected*

We need to study some regularity of the barrier function on c. We find that it is Hölder continuous with exponent $\frac{1}{2}$ at the points where there is an invariant curve. Consequently, we are able to show this condition is also generic (cf. [CY2]). Therefore, Arnold diffusion is a generic phenomenon in a priori unstable systems.

References

[Ar1] Arnol'd V. I., *Instability of dynamical systems with several degrees of freedom,* (Russian, English) Sov. Math., Dokl., **5**(1964), 581–585; translation from Dokl. Akad. Nauk SSSR, **156**(1964), 9–12.

[Be1] Bernard P., *Homoclinic orbits to invariant sets of quasi-integrable exact maps,* Ergod. Theory Dynam. Syst. **20**(2000), 1583–1601.

[Be2] Bernard P., *Connecting orbits of time dependent Lagrangian systems,* Ann. Inst. Fourier, **52**(2002), 1533–1568.

[Be3] Bernard P., *The dynamics of pseudographs in convex Hamiltonian systems,* preprint (2004)

[CY1] Cheng C. -Q. & Yan J., *Existence of diffusion orbits in a priori unstable Hamiltonian systems,* J. Differential Geometry, **67**(2004), 457–517.

[CY2] Cheng C. -Q. & Yan J., *Arnold diffusion in Hamiltonian systems: a priori unstable case,* preprint (2004).

[CDI] Contreras G., Delgado J. & Iturriaga R., *Lagrangian flows: the dynamics of globally minimizing orbits II,* Bol. Soc. Bras. Mat. **28**(1997), 155–196.

[CP] Contreras G. & Paternain G. P., *Connecting orbits between static classes for generic Lagrangian systems,* Topology, **41**(2002), 645–666.

[Fa1] Fathi A., *Théorème KAM faible et théorie de Mather sue les systèmes,* C. R. Acad. Sci. Paris Sér. I Math. **324**(1997), 1043–1046.

[Fa2] Fathi A., *Weak KAM Theorem in Lagrangian Dynamics* (to appear) Cambridge University Press.

[Ma1] Mather J., *Action minimizing invariant measures for positive definite Lagrangian systems,* Math. Z., **207(2)**(1991), 169–207.

[Ma2] Mather J., *Variational construction of connecting orbits,* Ann. Inst. Fourier (Grenoble), **43(5)**(1993), 1349–1386.

[Me2] Mañé R., *Generic properties and problems of minimizing measures of Lagrangian systems,* Nonlinearity, **9**(1996), 169–207.

The calculus of variations and the forced pendulum

Paul H. Rabinowitz[1]

Abstract Consider the equation of forced pendulum type:

$$u'' + V_u(t,u) = 0 \quad (*)$$

where $' = d/dt$ and V is smooth and 1-periodic in its arguments. We will show how to use elementary minimization arguments to find a variety of solutions of $(*)$. We begin with periodic solutions of $(*)$ and then find heteroclinic solutions making one transition between a pair of periodics. Then we construct heteroclinics and homoclinics making multiple (even infinitely many) transitions between periodics. If time permits, we may also discuss the construction of related mountain pass orbits of $(*)$.

1 Introduction

The goal of these lectures is to show how elementary variational techniques, in particular minimization arguments, can be used to extract a considerable amount of information about dynamical behavior. We do this for the setting of a forced pendulum model problem. This is a favorite proving ground for many techniques. Among works that are related to ours, we mention in particular [Mor], [A], [Ma82], [Ma93], [B88], [B89], and [Mos86].

The approach taken here uses essentially nothing from the theory of dynamical systems other than the uniqueness of solutions of the initial value problem. Therefore, these techniques can also be used for certain classes of problems for partial differential equations. In part our arguments are simplifications of ones used in [RS]. A disadvantage of our approach is that it does not capture finer dynamical structure that can be obtained using stable and unstable manifolds or notions like hyperbolicity.

---

[1] University of Wisconsin-Madison, Mathematics Department, 480 Lincon Dr, Madison WI 53706-1388

e-mail: rabinowi@math.wisc.edu

W. Craig (ed.), Hamiltonian Dynamical Systems and Applications, 367–390.
© 2008 *Springer Science + Business Media B.V.*

Fig. 1 Schematic of the physical pendulum.

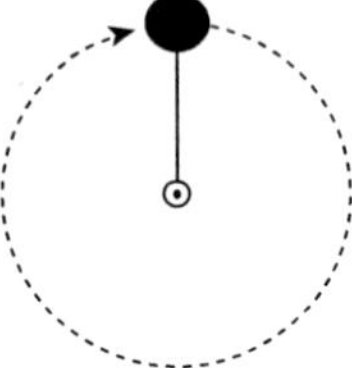

Fig. 2 Schematic of an orbit asymptotic from v to w.

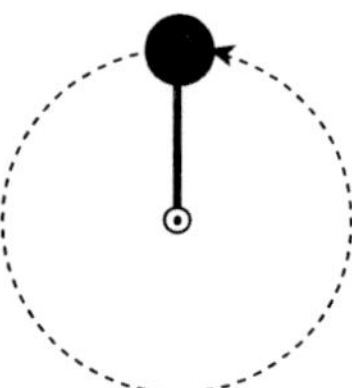

Fig. 3 Schematic of an orbit asymptotic from w to v.

The simple pendulum is modeled by $u'' + \sin u = 0$, u representing the angle made with the vertical direction. More generally we will consider a forced model

$$\text{(DE)} \quad -u'' + V_u(t,u) = 0,$$

where V satisfies

$(\mathbf{V_1})$ $V \in C^2(\mathbb{R}^2, \mathbb{R})$ *and is 1-periodic in t and in u.*
  Equivalently $V \in C^2(\mathbb{T}^2, \mathbb{R})$, where $\mathbb{T}^2$ is the 2-torus.

A caveat is in order here: V is the negative of the usual potential energy.

The simplest solutions of (DE) are periodic ones, e.g. if $V_u(t,z) = 0$ for all $t \in \mathbb{R}$ and $z \in \mathbb{Z}$, each such z is an equilibrium, and therefore periodic solution of (DE). By (V_1), if v is a solution of (DE), so is $v + k$ for all $k \in \mathbb{Z}$. Therefore we can seek solutions of (DE) that are asymptotic to a pair of periodics v and w.

We say such a solution is *heteroclinic* from v to w (Fig. 4). Such solutions undergo one 'transition'. Likewise we can try to find 2, k or infinite transition solutions. Thus a 2-transition solution is *homoclinic* to v or w (see Figs. 2–4). It turns out there are infinitely many solutions of each type, distinguished by the amount of time they spend near v or w.

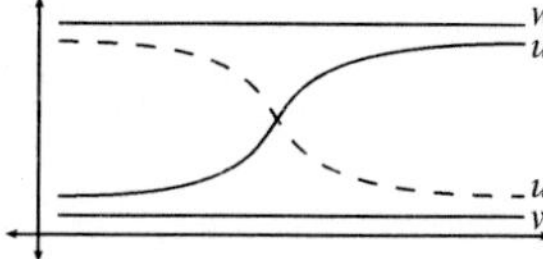

Fig. 4 Graphs of 1-transition orbits between v and w.

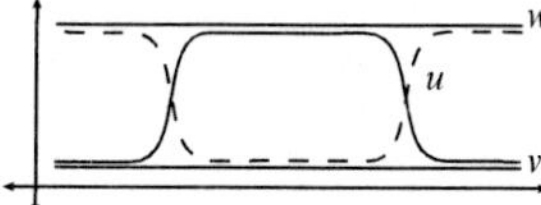

Fig. 5 Graphs of 2-transition orbits between v and w.

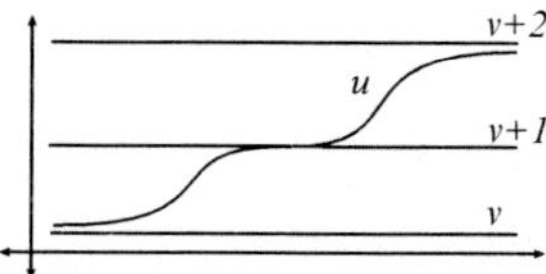

Fig. 6 A monotonic orbit asymptotic to v in the past and to $v+2$ in the future.

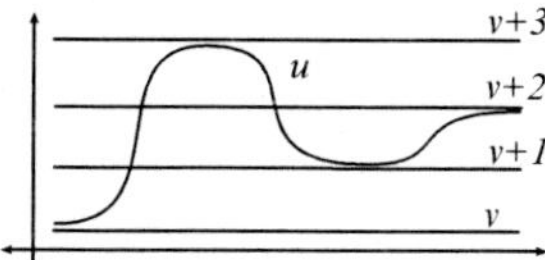

Fig. 7 An orbit which makes several transitions.

There is another kind of 2-transition solution which is monotone: $u(t+1) > u(t)$ (Fig. 1). In the simplest case, such a solution is heteroclinic from v to v + 2. Likewise, there are k and infinite-translation such solutions, and we can concatenate these two types of solutions (Fig. 7).

Within each type of solution as well as for the mixed type, one can seek a so-called symbolic dynamics of solutions that will be described later.

We will show how elementary minimization arguments can be used to find some of these solutions. Unfortunately we will not have enough time to treat the monotone and mixed cases. We begin with the simplest case of periodic solutions and then treat progressively more complex cases.

2 Periodic solutions

Periodic solutions are the easiest to find. We assume V satisfies (V_1). Set $E = W^{1,2}(\mathbb{T}^1)$, the class of 1-periodic functions having square integrable derivatives, i.e.

$$||u||_E^2 = ||u||_{W^{1,2}}^2 = \int_0^1 \left((u')^2 + u^2 \right) dt.$$

Note that $u \in E$ implies $u \in C(\mathbb{T}^1)$, in fact $u \in C^{1/2}(\mathbb{T}^1)$, i.e. u is Hölder continuous of order $1/2$. Let

$$L(u) = \frac{1}{2}|u'|^2 + V(t,u),$$

be the Lagrangian associated with (DE) with the corresponding functional

$$I(u) = \int_0^1 L(u)\,dt.$$

Then $I \in C(E,\mathbb{R})$ (even C^2) and for $u,\phi \in E$, the Frechet derivative, $I'(u)\phi$ is given by

$$\begin{aligned}
I'(u)\phi &= \lim_{h \to 0} \frac{1}{h}\left(I(u+h\phi) - I(u)\right) \\
&= \int_0^1 \left(u'\phi' + V_u(t,u)\phi\right) dt.
\end{aligned}$$

If $I'(u) = 0$, we say u is a *critical point* of I and $c = I(u)$ is called a *critical value* of I. Note also, if

$$\int_0^1 \left(u'\phi' + V_u(t,u)\phi\right) dt = 0 \tag{1}$$

for all $\phi \in E$, u is called a *weak solution* of (DE). Then we have a "regularity" theorem:

Theorem 2.1. *u is a classical solution of (DE) if and only if $u \in E$ and u is a weak solution of (DE).*

Theorem 2.1 reduces the existence of periodic solutions of (DE) to finding critical points of I in E. In the study of partial differential equations, such regularity theorems are often rather delicate. For the above special case, the proof is quite direct. Since the regularity question will also come up in more complicated settings later, we treat it here for the simplest case.

Proof of Theorem 2.1. If u is a classical solution of (DE), multiplying (DE) by $\phi \in E$ and integrating over $[0,1]$ yields (1). Conversely suppose u is a weak solution of (1). Taking $\phi = 1$ shows

$$\int_0^1 V(t,u)dt \equiv [V(t,u)] = 0,$$

i.e. the constant term in the Fourier expansion of $V(t,u)$ vanishes. It is a calculus exercise to show there is a unique $q \in C^2(\mathbb{T}^1,\mathbb{R})$ solving

$$-\ddot{q} + V_u(t,u) = 0 \ , \ [q] = 0. \tag{2}$$

Multiplying (2) by $\phi \in E$ and integrating over $[0,1]$ shows

$$\int_0^1 \left(q'\phi' + V_q(t,u)\phi\right) dt = 0. \tag{3}$$

Subtracting (3) from (1) gives

$$\int_0^1 (u' - q')\, \phi'\, dt = 0 \tag{4}$$

for all $\phi \in E$. Choosing $\phi = u - q$, (4) implies $u' - q' = 0$ and therefore $u = q +$ const $\in C^2(\mathbb{T}^1, \mathbb{R})$. $\square$

How do we find critical points of I? The simplest possibilities are minima. Thus set

$$c = \inf_{u \in E} I(u). \tag{5}$$

Note that I is bounded from below by $V_0 = \min_{\mathbb{R}^2} V$. Let (u_n) be a minimizing sequence for (5), i.e. $I(u_n) \to c$ as $n \to \infty$. Therefore there is an $M > 0$ such that

$$I(u_n) = \int_0^1 \left(\frac{1}{2}(u_n')^2 + V(t, u_n) \right) dt \leq M.$$

Hence

$$||u_n'||_{L^2}^2 \leq 2(M - V_0). \tag{6}$$

Observe that $u_n + j_n$ is also a minimizing sequence for (5) for any choice of $j_n \in \mathbb{Z}$. Therefore u_n may not be bounded. But we can choose j_n so that $[u_n + j_n] \in [0, 1]$. Thus without loss of generality, $[u_n] \in [0, 1)$. Since

$$u_n(t) - u_n(x) = \int_x^t u_n'(s)\, ds,$$

one has

$$u_n(t) = [u_n] + \int_0^1 \left(\int_x^t u_n'(s)\, ds \right) dx,$$

and therefore

$$|u_n(t)| \leq 1 + \int_0^1 ||u_n'||_{L^2}\, dx = 1 + ||u_n'||_{L^2}. \tag{7}$$

Now (6) and (7) show u_n is bounded in the Hilbert space E. Therefore there is a $v \in E$ such that along a subsequence, $u_n \rightharpoonup v$ (i.e. weakly in E). The functional I is weakly lower semicontinuous. Hence

$$c \leq I(v) \leq \lim_{n \to \infty} I(u_n) = \inf_E I = c. \tag{8}$$

Thus (8) shows $I(v) = c$ and v minimizes I over E. Moreover v is a critical point of I on E. Indeed take $\phi \in E$. Then $\psi(h) \equiv I(v + h\phi) \in C^1(\mathbb{R}, \mathbb{R})$ and has a minimum at $h = 0$. Hence

$$\psi'(0) = 0 = I'(v)\phi \tag{9}$$

for all $\phi \in E$. Thus v is a weak and therefore by Theorem 2.1, a classical solution of (DE).

As was noted above, the minimizing sequence $\{u_n\}$ is bounded in E and therefore in $C^{1/2}(\mathbb{T}^1)$. Hence the subsequence $\{u_n\}$ can be assumed to converge to v in $L^\infty(\mathbb{T}^1)$. Although it is not important here, for future reference, we have a stronger form of convergence:

Proposition 2.1. $u_n \to v$ in E (i.e. in $W^{1,2}(\mathbb{T}^1)$).

Proof. If not there is a $\delta > 0$ such that $\|u_n' - v'\|_{L^2} \geq \delta$. Set $\phi_n = u_n - v$. Then

$$
\begin{aligned}
I(u_n) &= I(v + \phi_n) \\
&= \int_0^1 \left[\frac{1}{2}|v'|^2 + v'\phi_n' + \frac{1}{2}|\phi_n'|^2 + V(t, v + \phi_n) - V(t, v) + V(t, v) \right] dt \\
&\geq I(v) + \frac{1}{2}\delta^2 + \int_0^1 \left[v'\phi_n' + V(t, v + \phi_n) - V(t, v) \right] dt.
\end{aligned}
\tag{10}
$$

As $n \to \infty$, $I(u_n) \to I(v)$ while the term on the right in (10) approaches zero. Thus $0 \geq 1/2\delta^2$, a contradiction. $\quad\square$

Set $\mathfrak{M}_0 = \{u \in E : I(u) = c\}$. We have shown $\mathfrak{M}_0 \neq \emptyset$.

Example 1: If $V \equiv 0$, then $\mathfrak{M}_0 = \mathbb{R}$.

Example 2: If $V = a(t)(\cos(2\pi u - 1))$, then $\mathfrak{M}_0 = \mathbb{Z}$.

Theorem 2.2. $\mathfrak{M}_0$ *is an ordered set, i.e.* $v, w \in \mathfrak{M}_0$ *implies* $v \equiv w$, $v < w$, *or* $v > w$.

Proof. If not, there are points $\xi, \eta \in [0, 1]$ such that $v(\xi) = w(\xi)$ and, e.g. $v(\eta) < w(\eta)$. Set $\phi = \max(v, w)$ and $\psi = \min(v, w)$. Then $\phi, \psi \in E$ and

$$
2c \leq I(\phi) + I(\psi) = I(v) + I(w) = 2c.
\tag{11}
$$

Hence by (11), $I(\phi) = c = I(\psi)$ and $\phi, \psi \in \mathfrak{M}_0$. Consequently by Theorem 2.1, ϕ and ψ are classical 1-periodic solutions of (DE). Set $\chi = \phi - \psi$ so $\chi \geq 0$, $\chi(\xi) = 0$ and therefore $\chi'(\xi) = 0$, and $\chi(\eta) > 0$. (DE) implies

$$
\chi'' + V_u(t, \phi) - V_u(t, \psi) = 0 = \chi'' + f(t)\chi,
\tag{12}
$$

where

$$
f(t) = \begin{cases} \dfrac{V_u(t, \phi(t)) - V_u(t, \psi(t))}{\psi(t) - \phi(t)} & \text{if } \phi(t) > \psi(t) \\ V_{uu}(t, \phi(t)) & \text{if } \phi(t) = \psi(t) \end{cases}
$$

and $f \in C(\mathbb{T}^1, \mathbb{R})$. Thus χ is a C^2 solution of the linear equation (12) with $\chi(\xi) = 0 = \chi'(\xi)$. Therefore the uniqueness of solutions to the initial value problem for (12) implies $\chi \equiv 0$, contrary to $\chi(\eta) > 0$. Hence $\mathfrak{M}_0$ is ordered. $\quad\square$

Next let $k \in \mathbb{Z}$. Note that V is k-periodic in t so we can seek k-periodic solutions of (DE). Let $u \in W^{1,2}(k\mathbb{T}^1) \equiv E_k$. Set

$$
I_k(u) = \int_0^k L(u) \, dt,
$$

and

$$\alpha_k = \inf_{u \in E_k} I_k(u).$$

By our above arguments,

$$\mathfrak{M}_k \equiv \{u \in E_k : I_k(u) = \alpha_k\} \neq \emptyset,$$

any $u \in \mathfrak{M}_k$ is a classical k-periodic solution of (DE), and $\mathfrak{M}_k$ is an ordered set.

Surprisingly we gain nothing new by varying k as the next result shows:

Proposition 2.2. $\mathfrak{M}_0 = \mathfrak{M}_k$ *and* $\alpha_k = kc$.

Proof. Let $v \in \mathfrak{M}_k$. Then $v(\cdot + 1) \in \mathfrak{M}_k$. If $v(t) = v(t+1)$ for all $v \in \mathfrak{M}_k$, then $\mathfrak{M}_k = \mathfrak{M}_0$ and $\alpha_k = kc$. Otherwise for some $v \in \mathfrak{M}_k$,

(a) $v(t+1) < v(t)$,

or

(b) $v(t+1) > v(t)$.

If (a) occurs, $v(t) = v(t+k) < \cdots < v(t+1) < v(t)$, a contradiction, and similarly for (b). $\square$

Proposition 2.2 can be used to show that the members of $\mathfrak{M}_0$ possess another minimality property.

Proposition 2.3. *Let* $v \in \mathfrak{M}_0$ *and* $a, b \in \mathbb{R}$ *with* $a < b$. *Set*

$$A = \{w \in W^{1,2}[a,b] : w(a) = v(a), w(b) = v(b)\}$$

and for $w \in A$, *let* $\mathscr{I}(w) = \int_a^b L(w)\,dt$. *Then*

$$\mathscr{I}(v) = \inf_{w \in A} \mathscr{I}(w) \equiv c_A. \tag{13}$$

Proof. $\mathscr{I}$ is weakly lower semi-continuous so as earlier, there is a $u \in A$ such that $\mathscr{I}(u) = c_A$. Choose $\alpha < a$, and $\beta > b$ with $\alpha, \beta \in \mathbb{Z}$. Extend u to $[\alpha, \beta]$ via $u = v$ in $[\alpha, a] \bigcup [b, \beta]$ and further extend u to $\mathbb{R}$ as a $\beta - \alpha$ periodic function. Hence $u \in E_{\beta - \alpha}$ so by Proposition 2.2,

$$I_{\beta - \alpha}(v) \leq I_{\beta - \alpha}(u). \tag{14}$$

But

$$\begin{aligned}
I_{\beta - \alpha}(v) &= \int_\alpha^a L(v)\,dt + \mathscr{I}(v) + \int_b^\beta L(v)\,dt \\
&\geq \int_\alpha^a L(u)\,dt + \mathscr{I}(u) + \int_b^\beta L(u)\,dt \\
&= I_{\beta - \alpha}(u).
\end{aligned} \tag{15}$$

Thus by (14)–(15), $\mathscr{I}(v) = \mathscr{I}(u) = c_A$. $\square$

Remark: The minimization problem (13) is a special case of

$$\inf_{w \in B} \mathscr{I}(w) \tag{16}$$

where

$$B = \{w \in W^{1,2}[a,b] : w(a) = r, w(b) = s\}.$$

By the argument of (5)–(9), problem (16) has a minimum which is a classical solution of (DE). In several future arguments we will use this observation to establish that the minimizers of certain variational problems are in fact classical solutions of (DE).

Returning to $\mathfrak{M}_0$, since it is ordered, either $\{(t, u(t)) \,|\, t \in \mathbb{R}, u \in \mathfrak{M}_0\} = \mathbb{R}^2$, i.e. $\mathfrak{M}_0$ foliates $\mathbb{R}^2$, or there are points $(x, z) \in \mathbb{R}^2$ such that $z \neq u(x)$ for any $u \in \mathfrak{M}_0$, i.e. $\mathfrak{M}_0$ merely laminates $\mathbb{R}^2$. In this latter case there is a smallest $w \in \mathfrak{M}_0$ and largest $v \in \mathfrak{M}_0$ such that $v(x) < z < w(x)$. Hence by Theorem 2.2, $v(t) < w(t)$ for all $t \in \mathbb{R}$. We then call v and w a *gap pair*. It is known that this latter case is generic; indeed given any $v \in \mathfrak{M}_0$, there is a $W \in C^2(\mathbb{T}^1, \mathbb{R})$ such that the $\mathfrak{M}_0$ associated with $V + \varepsilon W$ is $\{v + k \,|\, k \in \mathbb{Z}\}$ for all small $\varepsilon > 0$ [RS].

3 Heteroclinic solutions

Suppose $v, w \in \mathfrak{M}_0$ are a gap pair. We seek solutions of (DE) that are heteroclinic from v to w (or from w to v). A natural approach is to try to find them as minimizers of $\int_{\mathbb{R}} L(u) \, dt$ over a class of functions having the desired asymptotic behavior. However if $\int_0^1 L(v) \, dt = c = \int_0^1 L(w) \, dt \neq 0$, then for each admissible function u, $\int_{\mathbb{R}} L(u) \, dt$ will be infinite. Thus this approach must be modified. The above functional must be "renormalized" so that it is finite on the above class of functions.

This can be done merely assuming (V_1), but it is technically simpler to assume V is also time reversible. Hence suppose

$(\mathbf{V_2})$ $V(-t, z) = V(t, z)$ *for all* $t, z \in \mathbb{R}$

A key consequence of (V_2) is:

Proposition 3.1. *If V satisfies (V_1) and (V_2),*

$$\hat{c} \equiv \inf_{u \in W^{1,2}[0,1]} I(u) = c \tag{1}$$

and if $u \in \mathfrak{M}_0$, then $u(t) = u(-t)$.

Proof. Set $\hat{\mathfrak{M}} = \{u \in W^{1,2}[0,1] : I(u) = \hat{c}\}$. The existence argument of the previous section implies $\hat{\mathfrak{M}} \neq \emptyset$. Clearly $\hat{c} \leq c$. To get equality, let $u \in W^{1,2}[0,1]$. Then

$$I(u) = \int_0^{1/2} L(u) \, dt + \int_{1/2}^1 L(u) \, dt \equiv \alpha + \beta.$$

Say $\alpha \leq \beta$. Define $\phi(t) = u(t)$ for $0 \leq t \leq 1/2$, and $\phi(t) = u(1-t)$ for $1/2 \leq t \leq 1$. Then $\phi(0) = \phi(1)$ so ϕ extends naturally to an element of E and by (V_2), $I(\phi) = 2\alpha \leq I(u)$. Therefore

$$c = \inf_E I \leq \inf_{W^{1,2}[0,1]} = \hat{c}$$

so $c = \hat{c}$ and $\mathfrak{M}_0 \subset \hat{\mathfrak{M}}$. But if $u \in \hat{\mathfrak{M}}$, then $I(\phi) = c$ so $\phi \in \mathfrak{M}_0$. Since $\phi \equiv u$ on $[0, 1/2]$, uniqueness of solutions of the initial value problem for (DE) implies $u \equiv \phi$ on $[0,1]$, i.e. $u \in \mathfrak{M}_0$. Moreover $u(t) = u(1-t) = u(-t)$ via the 1-periodicity of u. $\square$

With the aid of Proposition 3.1, a renormalized functional can be introduced. For $p \in \mathbb{Z}$ and $u \in W^{1,2}_{loc}(\mathbb{R}, \mathbb{R})$, define

$$a_p(u) \equiv \int_p^{p+1} L(u)\,dt - c.$$

By Proposition 3.1, $a_p(u) \geq 0$ for all such p and u. Now we define the renormalized functional:

$$J(u) = \sum_{p \in \mathbb{Z}} a_p(u).$$

Thus $J(u) \geq 0$.

With v, w a gap pair, we define,

$$\Gamma_{-\infty} \equiv \Gamma_{-\infty}(v, w)$$
$$\equiv \{u \in W^{1,2}_{loc}(\mathbb{R}, \mathbb{R}) : ||u - v||_{L_2[i,i+1]} \to 0, i \to -\infty\}$$

$$\Gamma_{\infty} \equiv \Gamma_{\infty}(v, w)$$
$$\equiv \{u \in W^{1,2}_{loc}(\mathbb{R}, \mathbb{R}) : ||u - w||_{L^2[i,i+1]} \to 0, i \to \infty\}$$

and take as the associated class of admissible functions

$$\Gamma_1 \equiv \Gamma_1(v, w) \equiv \{u \in W^{1,2}_{loc}(\mathbb{R}, \mathbb{R}) : v \leq u \leq w\} \cap \Gamma_{-\infty} \cap \Gamma_{\infty}$$

Clearly $\Gamma_1 \neq \emptyset$ and there are u's in Γ_1 such that $J(u) < \infty$. Define

$$c_1 \equiv c_1(v, w) \equiv \inf_{u \in \Gamma_1} J(u). \tag{2}$$

Then we have

Theorem 3.1. *If V satisfies (V_1) - (V_2), and v, w are a gap pair, then*

1. $\mathfrak{M}_1 \equiv \mathfrak{M}_1(v, w) \equiv \{u \in \Gamma_1 : J(u) = c_1\} \neq \emptyset$.
2. Any $U \in \mathfrak{M}_1$ is also a classical solution of (DE).
3. $u < U < U(\cdot + 1) < w$.
4. $\mathfrak{M}_1$ is an ordered set.
5. Any $U \in \mathfrak{M}_1$ is minimal in the sense of Proposition 2.3.

Proof. Let $\{u_k\}$ be a minimizing sequence for (2). Since we are dealing with an unbounded domain, some extra care must be taken here to ensure that $\{u_k\}$ has a nontrivial limit. E.g. if $\mathfrak{M}_1 \neq \emptyset$ and $U \in \mathfrak{M}_1$, $u_k = U(\cdot - k) \in \Gamma_1$ and u_k converges in C^2_{loc} to $v \notin \Gamma_1$. To avoid such complications, $\{u_k\}$ will be normalized as follows. If $u \in \Gamma_1$ so is $u(\cdot - l)$ for any $l \in \mathbb{Z}$ and $J(u(\cdot - l)) = J(u)$. As $l \to -\infty$, $u|_l^{l+1} \to v$ in L^2 and as $l \to \infty$, $u|_l^{l+1} \to w$ in L^2. Therefore there is a unique $l = l(u) \in \mathbb{Z}$ such that

$$\begin{cases} \int_i^{i+1}(u(t-l) - v(t))\,dt < \frac{1}{2}\int_0^1 (w-v)\,dt, i < 0, i \in \mathbb{Z} \\ \int_0^1 (u(t-l) - v(t))\,dt \geq \frac{1}{2}\int_0^1 (w-v)\,dt. \end{cases} \tag{3}$$

Thus without loss of generality, $\{u_k\}$ can be chosen so that $l(u_k) = 0$.

Since $\{u_k\}$ is a minimizing sequence, there is an $M > 0$ such that for all $k \in \mathbb{N}$,

$$J(u_k) \leq M. \tag{4}$$

Hence for all $p \in \mathbb{N}$,

$$\sum_{-p}^{p} a_i(u_k) = \int_{-p}^{p+1} L(u_k)\,dt - (2p+1)c \leq M \tag{5}$$

and (5) implies

$$\int_{-p}^{p+1} |u_k'|^2\,dt \leq M_1, \tag{6}$$

where M_1 depends on p but not k. Since $v \leq u_k \leq w$, $\{u_k\}$ is bounded in $W^{1,2}_{loc}(\mathbb{R}, \mathbb{R})$. Consequently there is a $U \in W^{1,2}_{loc}$ such that along a subsequence $u_k \to U$ weakly in $W^{1,2}_{loc}$ and in L^∞_{loc}. (In fact in the spirit of Proposition 2.1, $u_k \to U$ in $W^{1,2}_{loc}$ along a subsequence, but we do not need this additional information). Since $\int_{-p}^{p+1} L(u)\,dt$ is weakly lower semi-continuous,

$$\sum_{i=-p}^{p} a_i(U) \leq M$$

for all $p \in \mathbb{N}$ and hence $J(U) \leq M$. Moreover by (3),

$$\begin{cases} \int_i^{i+1}(U - v)\,dt \leq \frac{1}{2}\int_0^1 (w-v)\,dt, i < 0, i \in \mathbb{Z} \\ \int_0^1 (U - v)\,dt \geq \frac{1}{2}\int_0^1 (w-v)\,dt. \end{cases} \tag{7}$$

We claim $U \in \Gamma_1$. The L^∞_{loc} convergence of $\{u_k\}$ implies $v \leq U \leq w$. Thus we need only show U satisfies the asymptotic requirements of Γ_1. To do so, note first that since $J(U) < \infty$, $a_p(U) \to 0$ as $|p| \to \infty$, i.e. $\int_p^{p+1} L(U)\,dt \to c$ as $|p| \to \infty$. Set $U_p(t) = U(t+p)$ for $t \in [0,1]$. Then $U_p \in W^{1,2}[0,1]$ and $I(U_p) \to c$ as $|p| \to \infty$. Hence as $|p| \to \infty$, $\{U_p\}$ is a minimizing sequence for (1). Consequently along a subsequence $\{U_p\}$ converges weakly in $W^{1,2}$ and strongly in L^∞ to $u^\pm \in \mathfrak{M}_0$. But $v \leq U_p \leq w$ implies either $u^\pm = v$ or $u^\pm = w$. By (7), as $p \to -\infty$,

$$\frac{1}{2}\int_0^1 (w-v)\,dt \geq \int_p^{p+1}(U-v)\,dt = \int_0^1 (U_p-v)\,dt \to \int_0^1 (u^- - v)\,dt.$$

Therefore $u^- = v$ and since v is the only possible limit of a subsequence of $\{U_p\}$ as $p \to -\infty$, the full sequence $U_p \to v$ as $p \to -\infty$.

It remains to prove that $U_p \to w$ as $p \to \infty$. For this, we no longer have (7) to help as for $p \to -\infty$, so more work is required. Following the argument of Proposition 2.1, we can assume $U_p \to u^+$ in $W^{1,2}[0,1]$ along our subsequence. In fact, $U_p \to u^+$ along the full sequence as $p \to \infty$ for otherwise there are a pair of subsequences such that $U_p \to v$ in $W^{1,2}$ along the first, and $U_p \to w$ in $W^{1,2}$ along the second as $p \to \infty$. But U_p cannot only be close (in $W^{1,2}[0,1]$ and therefore in $L^\infty[0,1]$) to both v and w. Therefore there is an $\varepsilon > 0$ and a third subsequence such that along it, $||U_p - \phi||_{W^{1,2}[0,1]} \geq \varepsilon$ as $p \to \infty$ for $\phi = v$ and $\phi = w$.

Now we have

Lemma 3.1. *For any $\varepsilon > 0$, there is a $\gamma(\varepsilon) > 0$ such that $||U_p - \phi||_{W^{1,2}[0,1]} \geq \varepsilon$ implies $I(U_p) \geq c + \gamma(\varepsilon)$*

Proof. Otherwise, there is a sequence of p's going to infinity such that $I(U_p) \to c$ while $||U_p - \phi||_{W^{1,2}[0,1]} \geq \varepsilon$. As above along a subsequence, $U_p \to v$ or w in $W^{1,2}[0,1]$, a contradiction. $\square$

Completion of the Proof of Theorem 3.1. Let $S = \{p \in \mathbb{N} : ||U_p - u^+||_{W^{1,2}}[0,1] \geq \varepsilon\}$. Then by Lemma 3.1,

$$J(U) \geq \sum_{p \in S} a_p(U) \geq \sum_{p \in S} \gamma(\varepsilon) = \infty,$$

contrary to $J(U) \leq M$. Thus $U_p \to u^+$ in $W^{1,2}[0,1]$ as $p \to \infty$.

Now finally to show that $u^+ = w$, suppose $u^+ = v$. By the reasoning just used and (7), there is an $i \in \mathbb{Z}$, $i \leq 0$, and $\varepsilon > 0$ such that $||U_i - \phi||_{W^{1,2}[0,1]} \geq \varepsilon$ with $\phi = v$ and $\phi = w$. Hence by Lemma 3.1

$$a_i(U) \geq \gamma(\varepsilon).$$

Therefore for large k,

$$a_i(u_k) \geq \frac{1}{2}\gamma(\varepsilon). \tag{8}$$

Choose $\delta > 0$ and free for the moment. Since $u^+ = v$, there is a $q > 0$ such that $||U_q - v||_{L^\infty[0,1]} \leq \delta/2$. Hence along our subsequence for all large k, $||u_k - U||_{L^\infty[q,q+1]} \leq \delta/2$. Thus $||v - u_k||_{L^\infty[q,q+1]} \leq \delta$ for large k. Define u_k^* to be equal to v for $t \leq q$, equal to ϕ_k for $q \leq t \leq q+1$, and equal to u_k for $q+1 \leq t$, where ϕ_k is a minimizer of the variational problem

$$\inf \int_q^{q+1} L(u)\,dt$$

over

$$K = \{u \in W^{1,2}[q, q+1] : u(q) = v(q), u(q+1) = u_k(q+1)\}.$$

The minimality properties of v and w imply $v \leq \phi_k \leq w$ and therefore $u_k^* \in \Gamma_1$. Set $u(t) = v(t) + (t-q)(u_k(q+1) - v(q+1))$ so $u \in K$.

Moreover

$$a_q(\phi_k) \leq a_q(u) \leq \beta(\delta) \tag{9}$$

where $\beta(\delta) \to 0$ as $\delta \to 0$. Now by (8)–(9)

$$J(u_k^*) - J(u_k) = \sum_{-\infty}^{\infty} [a_p(u_k^*) - a_p(u_k)]$$

$$= a_q(u_k^*) - \sum_{-\infty}^{q} a_p(u_k)$$

$$\leq \beta(\delta) - \frac{\gamma(\varepsilon)}{2}. \tag{10}$$

Choosing δ so small that $\beta(\delta) \leq \frac{1}{4}\gamma(\varepsilon)$, (10) contradicts that $\{u_k\}$ is a minimizing sequence for (2). Thus $U \in \Gamma_1$, and $J(U) \geq c_1$. On the other hand,

$$\sum_{-p}^{p} a_i(U) \leq \liminf_{k \to \infty} \sum_{-p}^{p} a_i(u_k) \leq \liminf_{k \to \infty} J(u_k) = c_1$$

so letting $p \to \infty$, we conclude $J(U) = c_1$. This establishes statement 1 of Theorem 3.1.

To prove statement 2 of Theorem 3.1, first we will obtain the minimality property 5. If it is not true, there are numbers $r < s$ and a function

$$\phi \in \{u \in W^{1,2}[r,s] : u(r) = U(r) \text{ and } u(s) = U(s)\}$$

such that

$$\int_r^s L(\phi)\, dt < \int_r^s L(U)\, dt.$$

Since v and w satisfy the minimality property, we can assume $v \leq \phi \leq w$. But then replacing $U|_r^s$ by $\phi|_r^s$ gives $U^* \in \Gamma_1$ with $J(U^*) < J(U)$, contrary to Theorem 3.1, part 1. Therefore U satisfies the minimality property and by the remark following Proposition 2.3, U is a solution or (DE).

Next, statement 4 of Theorem 3.1 follows by a mild variant of the proof of Theorem 2.2; Suppose $U, W \in \mathfrak{M}_1$. Thus $\phi = \max(U, W)$ and $\psi = \min(U, W) \in \Gamma_1$ so for all $p \in \mathbb{N}$,

$$\sum_{-p}^{p} [a_i(\phi) + a_i(\psi)] = \sum_{-p}^{p} [a_i(U) + a_i(W)].$$

Letting $p \to \infty$, this shows

$$2c_1 \leq J(\phi) + J(\psi) = J(U) + J(W) = 2c_1$$

Therefore $\phi, \psi \in \mathfrak{M}_1$, so by what has already been shown, ϕ and ψ are solutions of (DE) with $\phi \geq \psi$. The proof then concludes as for Theorem 2.2.

Lastly to verify statement 3 of Theorem 3.1, note that $v \leq U, U(\cdot + 1) \leq w$ with equality impossible by the argument of Theorem 2.2 again. Moreover since $U, U(\cdot + 1) \in \mathfrak{M}_1$, which is ordered, either; (i) $U(t) \equiv U(t+1)$, (ii) $U(t) > U(t+1)$, or (iii) $U(t) < U(t+1)$. If alternative (i) holds, U is 1-periodic and therefore $U \notin \Gamma_1$ while (ii) implies $U(t) > U(t+k) \rightarrow w(t)$ as $k \rightarrow \infty$. Thus $U > w$ and again $U \notin \Gamma_1$. Thus (iii) holds. $\square$

We conclude this section with a result that shows the gap condition is not only sufficient for there to exist minimizing heteroclinics from v to w, but also is necessary.

Theorem 3.2. *Let V satisfy (V_1) - (V_2), and further let $v, w \in \mathfrak{M}_0$ with $v < w$. Suppose there is a $U \in \Gamma_1(v, w)$ such that*

$$J(U) = \inf_{u \in \Gamma_1(v,w)} J(u).$$

Then v and w are a gap pair.

Proof. Otherwise there is a $\phi \in \mathfrak{M}_0$ such that $v < \phi < w$. There is a smallest $\alpha \in \mathbb{R}$ such $\phi(\alpha) = U(\alpha)$. Define $W(t) = U(t)$, for $t \leq \alpha$, $W(t) = \phi(t)$ when $\alpha \leq t \leq \alpha+1$, and $W(t) = U(t-1)$ when $\alpha+1 \leq t$. Then $W \in \Gamma_1(v, w)$ and $J(W) = J(U)$. Set

$$S = \{u \in W^{1,2}[\alpha - 1/2, \alpha + 1/2] : u(\alpha \pm 1/2) = W(\alpha \pm 1/2)\}.$$

The remark following Proposition 2.3 shows there is a $\psi \in S$ such that ψ is a solution of (DE) and

$$\int_{\alpha-1/2}^{\alpha+1/2} L(\psi)\, dt = \inf_{u \in S} \int_{\alpha-1/2}^{\alpha+1/2} L(u)\, dt.$$

We claim

$$A \equiv \int_{\alpha-1/2}^{\alpha+1/2} L(\psi)\, dt < \int_{\alpha-1/2}^{\alpha+1/2} L(W)\, dt \equiv B.$$

Indeed if $A = B$, W is a solution of (DE) in $(\alpha - 1/2, \alpha + 1/2)$. But $W = \phi$ in $[\alpha, \alpha + 1/2]$. Since ϕ is a solution of (DE) for all t, uniqueness for solutions of the initial value problem for (DE) imply $W = \phi$ in $(\alpha - 1/2, \alpha + 1/2)$. Since U minimizes J in $\Gamma_1(v, w)$, as in Theorem 3.1, U is a solution of (DE) on $\mathbb{R}$. But $U = W = \phi$ in $(\alpha - 1/2, \alpha)$. Therefore $U \equiv \phi$, contrary to $\|U - v\|_{L^2[i, i+1]} \rightarrow 0$ as $i \rightarrow -\infty$. Thus $A < B$. But then gluing $W|_{-\infty}^{\alpha-1/2}$ to ψ to $W|_{\alpha+1/2}^{\infty}$ produces $\Phi \in \Gamma_1(v, w)$ with $J(\Phi) < J(W)$ contrary to the minimality of W. $\square$

Remark: Theorem 3.2 does not exclude the possibility of there being a heteroclinic solution of (DE) in $\Gamma_1(v, w)$. If there is one, it cannot be a minimizer. In fact if v, f, and g, w are gap pairs with $f \leq g$, there is a monotone heteroclinic U from v to w.

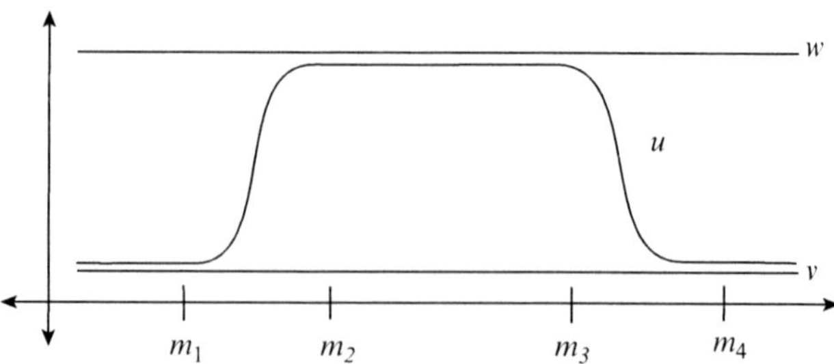

Fig. 8 An admissible u.

4 Multitransition solutions: the simplest case

Suppose v, w are a gap pair for (DE). In Section 3 we showed there are heteroclinic solutions of (DE) in $\mathfrak{M}_1(v,w)$. The same argument gives heteroclinic solutions in $\mathfrak{M}_1(w,v)$. The goal of this section is to find solutions of (DE) which lie between v and w, undergo two transitions, and are homoclinic to v or to w.

We will show there are infinitely many such solutions provided that $\mathfrak{M}_1(v,w)$ and $\mathfrak{M}_1(w,v)$ have gaps. The solutions are obtained as local minima of J on appropriate classes of functions. To introduce a suitable class of admissible functions, let $m = (m_1,\ldots,m_4) \in \mathbb{Z}^4$ and $\rho = (\rho_1,\ldots,\rho_4) \in \mathbb{R}^4$ with $m_i < m_{i+1}$ and $0 < \rho_i << 1$.

Define

$$Y_{1,2} \equiv Y_{1,2,m,\rho} \equiv \{u : u(m_1) - v(m_1) \le \rho_1,\, w(m_2) - u(m_2) \le \rho_2\}$$

$$Y_{3,4} \equiv Y_{3,4,m,\rho} \equiv \{u : w(m_3) - u(m_3) \le \rho_3,\, u(m_4) - v(m_4) \le \rho_4\}$$

$$Y \equiv Y_{m,\rho} \equiv \{u \in W^{1,2}_{loc}(\mathbb{R},\mathbb{R}) : v \le u \le w\} \cap Y_{1,2} \cap Y_{3,4}.$$

The numbers ρ_i have to be chosen in a special way which we postpone until needed.
Set

$$b \equiv b_{m,\rho} \equiv \inf_{u \in Y} J(u) \tag{1}$$

Proposition 4.1. *For all* (m,ρ), *there is a* $U = U_{m,\rho} \in Y$ *such that* $J(U) = b$.

Proof. It is straightforward to show there is a $\bar{u} \in Y$ such that $J(\bar{u}) < \infty$. Let $\{u_n\}$ be a minimizing sequence for (1). We can assume $J(u_n) \le J(\bar{u})$. As for Theorem 3.1, this implies $\{u_n\}$ is bounded in $W^{1,2}_{loc}$ and there is a $U \in W^{1,2}_{loc}$ such that along a subsequence, $u_n \to U$ weakly in both $W^{1,2}_{loc}$ and L^∞_{loc}. This latter convergence implies U satisfies the pointwise constraints of Y, so $U \in Y$. As in Section 3, $J(U) = b$. $\square$

Proposition 4.2. *U satisfies (DE) except possibly at* $t = m_i$, $1 \le i \le 4$ *(independently of ρ and m).*

Proof. This follows since U possesses a minimality property for each interval in the complement of the m_i. Eg. For $r \le s \le m_1$, U minimizes $\int_r^s L(u)\,dt$ over

$$\{u \in W^{1,2}[r,s] : u(r) = U(r), u(s) = U(s)\}$$

Hence by the remark following Proposition 2.3, U satisfies (DE) in (r,s). $\square$

Next we will show that U is asymptotic to v as $|t| \to \infty$. For this we require that ρ_1 and ρ_4 be small.

Proposition 4.3. *For ρ_1 (resp. ρ_4) sufficently small, $\|U - v\|_{W^{1,2}[i,i+1]} \to 0$ as $i \to -\infty$ (resp. $i \to \infty$).*

Proof. We treat the ρ_1 case. Since $J(U) = b < \infty$, $\|U - \phi\|_{W^{1,2}[i,i+1]} \to 0$ as $i \to -\infty$, where $\phi \in [v,w]$ via the proof of Theorem 3.1. If $\phi = w$, for any $\delta > 0$, there is an $l \in \mathbb{Z}, l < m_1$ such that $\|U - w\|_{W^{1,2}[l,l+1]} \le \delta$.
Let ψ_l be a minimizer of the problem:

$$\inf \int_{l-1}^{l} L(u)\, dt$$

over

$$\{u \in W^{1,2}[l-1,l] : u(l-1) = w(l-1), u(l) = U(l)\}.$$

As in (9)

$$a_{l-1}(\psi_l) \le \beta(\delta) \tag{2}$$

with $\beta(\delta) \to 0$ as $\delta \to 0$. Similarly let f be a minimizer of the problem

$$\inf \int_{m_1}^{m_1+1} L(u)\, dt$$

over

$$\{u \in W^{1,2}[m_1, m_1 + 1] : u(m_1) = U(m_1), u(m_1 + 1) = v(m_1 + 1)\}$$

and again as in (9),

$$a_{m_1}(f) \le \beta(\rho_1). \tag{3}$$

Set $\bar{U}$ be equal to w for $t \le l-1$, equal to ψ_l for $l-1 \le t \le l$, equal to U for $l \le t \le m_1$, equal to f for $m_1 \le t \le m_1 + 1$, and equal to v for $m_1 + 1 \le t$. Then $\bar{U} \in \Gamma_1(w,v)$ and

$$\sum_{i=l}^{m_1-1} a_i(U) = \sum_{i=l-1}^{m_1} a_i(\bar{U}) - a_{l-1}(\bar{U}) - a_{m_1}(\bar{U})$$
$$= J(\bar{U}) - a_{l-1}(\psi_l) - a_{m_1}(f)$$
$$\ge c_1(w,v) - \beta(\delta) - \beta(\rho_1) \tag{4}$$

via (2)–(3).
On the other hand, let g be a minimizer of

$$\inf \int_{m_1-1}^{m_1} L(u)\, dt$$

over

$$\{u \in W^{1,2}[m_1 - 1, m_1] : u(m_1 - 1) = v(m_1 - 1), u(m_1) = U(m_1)\}$$

Then as for (2)–(3),

$$a_{m_1-1}(g) \leq \beta(\rho_1). \tag{5}$$

Set U^* equal to v for $t \leq m_1 - 1$, and equal to g for $m_1 - 1 \leq t \leq m_1$. By the minimality property of U in $(-\infty, m_1]$, and (5)

$$\sum_{-\infty}^{m_1-1} a_i(U) \leq \sum_{-\infty}^{m_1-1} a_i(U^*) \leq \beta(\rho_1). \tag{6}$$

Since

$$\sum_{l}^{m_1-1} a_i(U) \leq \sum_{-\infty}^{m_1-1} a_i(U),$$

(4)–(6) imply

$$c_1(w, v) \leq \beta(\delta) + 2\beta(\rho_1) \tag{7}$$

which is impossible for δ and ρ_1 small. Thus U is asymptotic to v as $t \to -\infty$ and similarly as $t \to \infty$. $\square$

Next we will obtain an upper bound for $b = b_{m,\rho}$.

Proposition 4.4. *Let $\varepsilon > 0$. Then there is an $m_0(\varepsilon)$ such that if $m_2 - m_1, m_4 - m_3 \geq m_0(\varepsilon)$,*

$$b_{m,\rho} \leq c_1(v, w) + c_1(w, v) + \varepsilon$$

Proof. Let $\bar{U} \in \mathfrak{M}_1(v, w)$. Then there are $\alpha, \beta \in \mathbb{Z}$ with $\alpha \leq \beta$ such that if $\bar{f}, \bar{g}$ are respectively minimizers of

$$\int_{\alpha-1}^{\alpha} L(u)\,dt, \qquad \int_{\beta}^{\beta+1} L(u)\,dt$$

over

$$\{u \in W^{1,2}[\alpha - 1, \alpha] : u(\alpha - 1) = v(\alpha - 1), u(\alpha) = \bar{U}(\alpha)\},$$
$$\{u \in W^{1,2}[\beta, \beta + 1] : u(\beta) = \bar{U}(\beta), u(\beta + 1) = w(\beta + 1)\}.$$

Then

$$a_{\alpha-1}(\bar{f}), \quad a_\beta(\bar{g}) \leq \frac{\varepsilon}{4}. \tag{8}$$

Gluing $v|_{-\infty}^{\alpha-1}$ to $\bar{f}$ to $\bar{U}|_{\alpha}^{\beta}$ to $\bar{g}$ to $w|_{\beta+1}^{\infty}$ defines $U^* \in \Gamma_1(v, w)$ (Fig. 9). Since $J(\bar{U}) = c_1(v, w)$ by (8),

$$J(U^*) = a_{\alpha-1}(\bar{f}) + \sum_{\alpha}^{\beta-1} a_i(\bar{U}) + a_\beta(\bar{g}) \tag{9}$$

$$\leq c_1(v, w) + \frac{\varepsilon}{2}$$

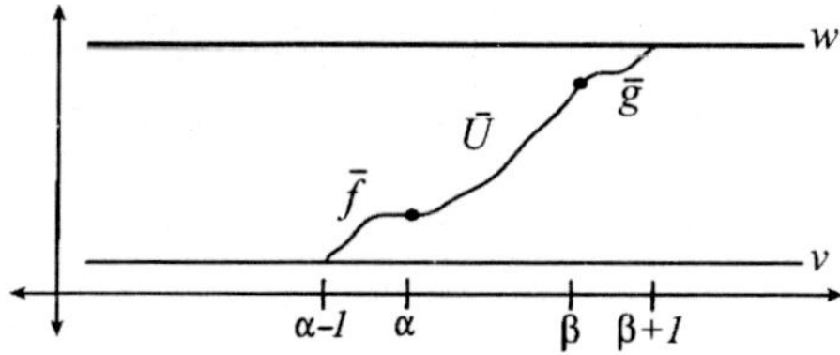

Fig. 9 The construction of U^* in Proposition 8.

Similarly let $\underline{U} \in \mathfrak{M}_1(w,v)$. As above there are $r,s \in \mathbb{Z}$ with $r < s$ such that if $\underline{f}$, respectively $\underline{g}$ are the minimizers of

$$\int_{r-1}^{r} L(u)\,dt, \qquad \int_{s}^{s+1} L(u)\,dt$$

over

$$\{u \in W^{1,2}[r-1,r] : u(r-1) = w(r-1),\, u(r) = \underline{U}(r)\},$$

$$\{u \in W^{1,2}[s,s+1] : u(s) = \underline{U}(s),\, u(s+1) = v(s+1)\}.$$

then

$$a_{r-1}(\underline{f}),\ a_s(\underline{g}) \le \frac{\varepsilon}{4}. \tag{10}$$

and gluing $w|_{-\infty}^{r-1}$ to $\underline{f}$ to $\underline{U}|_r^s$ to $\underline{g}$ to $v|_{s+1}^{\infty}$ produces $U_* \in \Gamma_1(w,v)$ with

$$J(U_*) \le c_1(w,v) + \frac{\varepsilon}{2}. \tag{11}$$

Finally set $U^{**}(t)$ equal to $U^*(t - m_2 + \beta + 1)$ for $t \le m_2$, and equal to $U_*(t - m_3 + r - 1)$ for $m_2 \le t$. By construction U^{**} satisfies the constraints of $Y_{m,\rho}$ at $t = m_2$ and m_3. For $m_2 - m_1 \ge \beta - \alpha + 2$, $U^{**}(m_1) = U^*(m_1 - m_2 + \beta + 1) = v(\alpha - 1)$ $= v(m_1)$ so U^{**} satisfies the constraint at $t = m_1$. Similarly the constraint at $t = m_4$ holds if $m_4 - m_3 \ge s - r + 2$. Therefore $U^{**} \in Y_{m,\rho}$ and by (9) and (11),

$$b_{m,\rho} \le J(U^{**}) \le c_1(v,w) + c_1(w,v) + \varepsilon \qquad \Box$$

Next we will refine our choice of ρ. Recall $\mathfrak{M}_1(v,w)$ and $\mathfrak{M}_1(w,v)$ have gaps. Define $\rho_- : \mathfrak{M}_1(v,w) \to (0, w(0) - v(0))$ via $\rho_-(u) = u(0) - v(0)$. Therefore ρ_- is a monotone function of u and $\rho_-(\mathfrak{M}_1(v,w))$ has gaps. Choose ρ_1 to lie in such a gap, i.e.

$$\rho_1 \in (0, w(0) - v(0)) \setminus \rho_-(\mathfrak{M}_1(v,w)).$$

Note that ρ_1 can be chosen as small as desired since $f \in \mathfrak{M}_1$ implies $f(\cdot - l) \in \mathfrak{M}_1(v,w)$ for any $l \in \mathbb{Z}$ so for large l, $\rho_-(f(\cdot - l))$ is near 0.

Similarly define $\rho_+ : \mathfrak{M}_1(v,w) \to (0, w(0) - v(0))$ via $\rho_+(u) = w(0) - u(0)$ so ρ_+ is also monotone and $\rho_+(\mathfrak{M}_1(v,w))$ has gaps. Choose ρ_2 in such a gap. Likewise $\rho_-, \rho_+ : \mathfrak{M}_1(w,v) \to (0, w(0) - v(0))$ as above. Choose ρ_3 and ρ_4 in associated gaps. An important consequence of this choice of the ρ_i is:

Proposition 4.5. *Let*

$$\Lambda_1(v,w) = \{u \in \Gamma_1 : u(0) - v(0) = \rho_1 \text{ or } w(0) - u(0) = \rho_2\}.$$

Set

$$d_1(v,w) = \inf_{u \in \Lambda_1(v,w)} J(u) \tag{12}$$

Then for $|\rho|$ small, $d_1(v,w) > c_1(v,w)$.

Remark Defining $\Lambda_1(w,v)$ and $d_1(w,v)$ in the obvious way, we also have $d_1(w,v) > c_1(w,v)$.

Proof of Proposition 4.5. Let $\{u_n\}$ be a minimizing sequence for (12). As in the proof of Theorem 3.1, there is a $P \in W_{loc}^{1,2}$ such that along a subsequence $u_n \to P$ weakly in $W_{loc}^{1,2}$ and also in L_{loc}^{∞}. This latter convergence implies $v \leq P \leq w$ and P satisfies one of the constraints at $t = 0$.

Also, as earlier $J(P) < \infty$ and therefore P asymptotes to v or w as $t \to -\infty$ and as $t \to \infty$. If (a) $P(0) = v(0) + \rho_1$, since ρ_1 is small, the argument of Proposition 4.3 shows $\|P - v\|_{W^{1,2}[i,i+1]} \to 0$ as $i \to -\infty$, while if (b) $w(0) = P(0) + \rho_2$, similarly $\|P - w\|_{W^{1,2}[i,i+1]} \to 0$ as $i \to \infty$.

Suppose (a) holds. Then either (c) $\|P - v\|_{W^{1,2}[j,j+1]} \to 0$ as $j \to \infty$ or (d) $\|P - w\|_{W^{1,2}[j,j+1]} \to 0$ as $j \to \infty$. If (c) occurs, $u_n(0)$ is near $v(0) + \rho_1$ along a subsequence as $n \to \infty$. Hence as in the proof of Lemma 3.1, there is a $\gamma(\rho_1) > 0$ (independent of n) such that

$$a_0(u_n) \geq \gamma(\rho_1) \tag{13}$$

for large n. Moreover for any $\delta > 0$, there is an $l = l(\delta) \in \mathbb{N}$ such that $\|u_n - v\|_{L^{\infty}[l,l+1]} \leq \delta$ for large n along the subsequence.

Now in the spirit of the proof of Proposition 4.3, set u_n^* equal to v, on $t \leq l$, equal to g_n on $l \leq t \leq l+1$, and equal to u_n on $t \geq l+1$ where g_n minimizes

$$\int_l^{l+1} L(u)\,dt$$

over

$$\{u \in W^{1,2}[l,l+1] : u(l) = v(l), u(l+1) = u_n(l+1)\}.$$

Thus as in (9) again,

$$a_l(u_n) \leq \beta(\delta). \tag{14}$$

Now by (13)–(14),

$$J(u_n) \geq a_0(u_n) + \sum_{l+1}^{\infty} a_i(u_n) \tag{15}$$

$$\geq \gamma(\rho_1) + \sum_{l}^{\infty} a_i(u_n^*) - a_l(g_n)$$

$$\geq \gamma(\rho_1) + J(u_n^*) - \beta(\delta).$$

Choosing δ so that $\beta(\delta) \leq 1/2\gamma(\rho_1)$ and noting that $u_n^* \in \Gamma_1(v,w)$, (15) yields

$$J(u_n) \geq c_1(v,w) + \frac{1}{2}\gamma(\rho_1). \tag{16}$$

Thus $d_1(v,w) \geq c_1(v,w) + \frac{1}{2}\gamma(\rho_1)$ for this case.

On the other hand, if (a) and (d) occur, $P \in \Lambda_1(v,w)$ and by earlier arguments $J(P) = d_1(v,w)$. Since $\Lambda_1(v,w) \subset \Gamma_1(v,w)$, $d_1(v,w) \geq c_1(v,w)$. If $d_1 = c_1$, then $P \in \mathfrak{M}_1(v,w)$ and by Theorem 3.1, P is a solution of (DE). Consequently $P(0) - v(0) = \rho_1 = \rho_-(P) \in \rho_-(\mathfrak{M}_1(v,w))$ contrary to the choice of ρ_1. Thus $d_1 > c_1$. The remaining cases are treated in the same fashion as above. $\square$

With the aid of Proposition 4.5, we have:

Proposition 4.6. *Set*

$$\mu = \frac{1}{2}\min(d_1(v,w) - c_1(v,w), d_1(w,v) - c_1(w,v)).$$

If $|\rho|$ is small and U satisfies an m_i constraint with equality, then for $m_3 - m_2 \gg 1$,

$$b_{m,\rho} \geq c_1(v,w) + c_1(w,v) + \mu. \tag{17}$$

Assuming Proposition 4.6 for the moment, combining Proposition 4.5 and Proposition 4.6 we have

$$\mu \leq b_{m,\rho} - c_1(v,w) - c_1(w,v) \leq \varepsilon \tag{18}$$

provided that an m_i constraint holds with equality. Here μ depends only on ρ and $m_3 - m_2$ while $m_2 - m_1$, $m_4 - m_3 \geq m_0(\varepsilon)$. Thus choosing $\varepsilon < \mu$, (18) yields a contradiction. Therefore U satisfies (DE) for all t and we have

Theorem 4.1. *If (V_1)–(V_2) hold, v and w are a gap pair, and in addition $\mathfrak{M}_1(v,w)$ and $\mathfrak{M}_1(w,v)$ have gaps, then for $|\rho|$ small and $m_{i+1} - m_i$ large, there is a $U = U_{m,\rho} \in Y_{m,\rho}$ which is a solution of (DE) with $J(U) = b_{m,\rho}$.*

Remark: That U satisfies the constraints with strict inequality implies U has a local minimization property.

Corollary 4.1. *There are infinitely many distinct 2-transition solutions of (DE).*

Proof. Simply take different sets of (m_i)'s with $m_{i+1} - m_i$ larger and larger. $\square$

To complete the proof of Theorem 4.1, we give the

Proof of Proposition 4.6. By the minimality property of $U|_{m_2}^{m_3}$,

$$\int_{m_2}^{m_3} L(U)\,dt = \inf_{u \in A} \int_{m_2}^{m_3} L(u)\,dt \tag{19}$$

where

$$A = \{u \in W^{1,2}[m_2,m_3] : u(m_2) = U(m_2), u(m_3) = U(m_3)\}.$$

Since ρ_2, ρ_3 are small and $w(m_2) - U(m_2) \leq \rho_2$, $w(m_3) - U(m_3) \leq \rho_3$, as in (9), (19) implies

$$\sum_{i=m_2}^{m_3-1} a_i(U) \leq \beta(\rho_2) + \beta(\rho_3). \tag{20}$$

We claim that given any $\sigma > 0$, there is an $\alpha(\sigma) > 0$ such that for $m_3 - m_2 \geq \alpha(\sigma)$, $\|U - w\|_{W^{1,2}[i,i+1]} \leq \sigma$ for some $q \in [m_2, m_3 - 1]$. Otherwise by Lemma 3.1,

$$\sum_{j=m_2}^{m_3-1} a_j(U) \geq (m_3 - m_2 - 1)\gamma(\sigma) \tag{21}$$

which goes to infinity as $m_3 - m_2 \to \infty$. But this is contrary to (20) which shows that the left hand side of (21) is small.

Now suppose for convenience that we have equality at an m_1 or m_2 constraint point. Set $\Phi(t)$ equal to $U(t)$ for $t \leq q$, equal to $f(t)$ for $q \leq t \leq q+1$, equal to $w(t)$ for $t \geq q+1$, where f minimizes

$$\int_q^{q+1} L(u)\,dt$$

over

$$\{u \in W^{1,2}[q,q+1] : u(q) = U(q), u(q+1) = w(q+1)\}.$$

Therefore $\Phi \in \Gamma_1(v,w)$.

Similarly set $\Psi(t)$ equal to $w(t)$ for $t \leq q$, equal to $g(t)$ for $q \leq t \leq q+1$, and equal to $U(t)$ for $t \geq q+1$, where g minimizes

$$\int_q^{q+1} L(u)\,dt$$

over

$$\{u \in W^{1,2}[q,q+1] : u(q) = w(q), u(q+1) = U(q+1)\}.$$

Therefore $\Phi \in \Gamma_1(w,v)$ and

$$d_1(v,w) + c_1(w,v) \leq J(\Phi) + J(\Psi) \leq J(U) - a_q(U) + a_q(f) + a_q(g) \tag{22}$$

Since $\|U - w\|_{W^{1,2}[q,q+1]} \leq \sigma$ it follows as in (9) again that $a_q(f) + a_q(g) \leq 2\beta(\sigma) \to 0$ as $\sigma \to 0$. Then for σ so small that

$$2\beta(\sigma) \leq \mu, \tag{23}$$

$J(U) = b$ and (22)–(23) imply (17). $\quad\square$

5 Multitransition solutions: general case

The ideas used in proving Theorem 4.1 work equally well to get k transition solutions of (DE) and then even infinite transition solutions via a limit argument, provided that the construction does not depend on k. However, given Theorem 4.1, there is a simpler geometrical argument giving the k and infinite transition cases, as well as an associated symbolic dynamics of solutions. We will illustrate with the case of $k = 3$ and then discuss the general case.

Choose $\rho, r \in \mathbb{R}^4$ and $m, n \in \mathbb{Z}^4$ such that there are associated solutions U and W of (DE) with $U \in Y_{m,\rho}(v, w)$, and $W \in Y_{n,r}(w, v)$. We seek a 3-transition solution heteroclinic from v to w. For $j \in \mathbb{Z}$, set $\tau_j u(t) = u(t - j)$. Because of their asymptotic properties for $j_1 >> 1$, $\tau_{j_1} U(t) < W(t)$ for all $t \in \mathbb{R}$.

Take $j_2 >> j_1$. Then $\tau_{j_1} U < \tau_{j_2} W$. Finally take $j_3 >> j_2$. Then $\tau_{j_3} U < \tau_{j_2} W$. For simplicity, we will take $j_1 = j, j_2 = 2j, j_3 = 3j$ for sufficiently large j. Consider $\{\tau_{-lj} W : l \in \mathbb{N}\}$ and $\{\tau_{(3+i)j} U : i \in \mathbb{N}\}$.

Delete from the region between the graphs of v and w the set of points above all of the shifted W's we have mentioned and below the shifted U's. Denote the remaining region by R and set

$$Y(R) \equiv \{u \in W_{loc}^{1,2} : (t, u(t)) \in \bar{R}\}.$$

(See Fig. 10). Define

$$c(R) = \inf_{u \in Y(R)} J(u).$$

Then we have:

Theorem 5.1. *Under the hypothesis of Theorem 4.1*

1. *$\mathfrak{M}(R) = \{u \in Y(R) : J(u) = c(R)\} \neq \emptyset$.*
2. *Any $U \in \mathfrak{M}(R)$ is a classical solution of (DE) and is interior to R.*
3. *$\|U - v\|_{L^2[i,i+1]} \to 0, i \to -\infty$, and $\|U - w\|_{L^2[i,i+1]} \to 0, i \to \infty$.*
4. *U has a local minimization property: for any $r < s$, U minimizes $\int_r^s L(u)\, dt$ over the class of $W^{1,2}[r, s]$ functions with $u(r) = U(r)$, and $u(s) = U(s)$ provided that $s - r$ sufficiently small.*

Proof. We will sketch the proof. A minimizing sequence converges as earlier to U lying in $\bar{R}$ with $J(U) = c(R)$. Since $J(\bar{U}) < \infty$, (3) of the theorem holds due

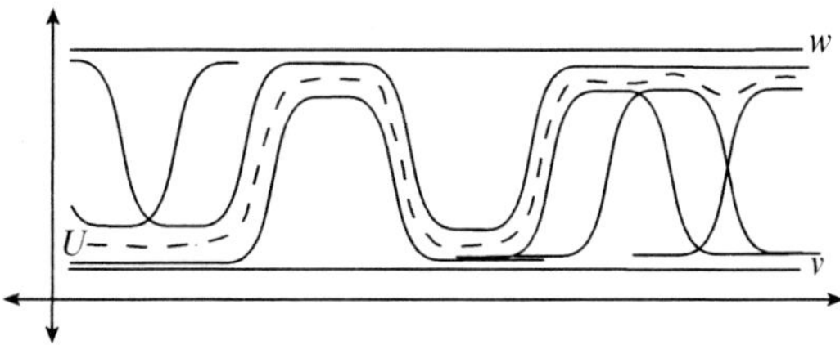

Fig. 10 A U in $Y(R)$.

to the form of R. The boundary of R consists of curves possessing local or global minimization properties and this readily implies (4), which in turn gives the first part of (2). Lastly the basic existence and uniqueness theorem for ordinary differential equations implies U cannot touch ∂R as in the proof of Theorem 2.2. $\square$

Next we will show how to generalize Theorem 5.1 and at the same time get a symbolic dynamics of solutions (Fig. 10). Choose U, W, and j as above so in particular the graphs of $\tau_{\pm j}U$ and W do not intersect. This implies the same is true of the graphs of $\tau_{ij}U$ and $\tau_{lj}W$ for all $i, l \in \mathbb{Z}$. Define

$$\Sigma \equiv \{\sigma = \{\sigma_i\}_{i \in \mathbb{Z}} : \sigma_i \in \{+, -\}\}.$$

For each $\sigma \in \Sigma$, we define a region $R(\sigma)$ lying between the graphs of v and w as follows. Set

$$S = \{(t, z) : t \in \mathbb{R}, v(t) \leq z \leq w(t)\}.$$

If $\sigma_i = +$, remove the region below $\tau_{ji}U$ from S; if $\sigma_i = -$, remove the region above $\tau_{ji}W$ from S. $R(\sigma)$ is what remains after carrying out this excision process for all $i \in \mathbb{Z}$. Then we have;

Theorem 5.2. *For each $\sigma \in \Sigma$, there is a solution $U_{R(\sigma)}$ of (DE) with the graph of $U_{R(\sigma)}$ lying in $R(\sigma)$. Moreover $U_{R(\sigma)}$ has the local minimization property of Theorem 5.2.*

Remark: If $\sigma_i = -$, $U_{R(\sigma)}$ will be L^∞ close to v on a large interval while if $\sigma_i = +$, $U_{R(\sigma)}$ will be L^∞ close to w on a large interval. In particular if $\sigma_i = -$ for all i near $-\infty$, $U_{R(\sigma)}$ asymptotes to v as $t \to -\infty$, while if $\sigma_i = +$ for all i near ∞, $U_{R(\sigma)}$ asymptotes to w as $t \to \infty$. The dynamics of the symbol σ reflect the dynamics of the solution $U_{R(\sigma)}$.

Proof of Theorem 5.2. We will sketch the proof. First we introduce four subsets of Σ:

$$\Sigma^{++} \equiv \{\sigma \in \Sigma : \sigma_i = + \text{ for all large } |i|\}$$

$$\Sigma^{--} \equiv \{\sigma \in \Sigma : \sigma_i = - \text{ for all large } |i|\}$$

$$\Sigma^{+-} \equiv \{\sigma \in \Sigma : \sigma_i = + \text{ for all large negative } i, \text{ and } \sigma_i = - \text{ for all large positive } i\}$$

$$\Sigma^{-+} \equiv \{\sigma \in \Sigma : \sigma_i = - \text{ for all large negative } i, \text{ and } \sigma_i = + \text{ for all large positive } i\}$$

Let Σ^* be the union of these four sets. Any $\sigma \in \Sigma^*$ has a finite number of changes of σ_i as i increases. For $\sigma \in \Sigma^*$, set

$$Y(\sigma) = \{u \in W^{1,2}_{loc} : (t, u(t)) \in \bar{R}(\sigma) \text{ for all } t \in \mathbb{R}\},$$

and define

$$c(\sigma) = \inf_{u \in Y(\sigma)} J(u).$$

Then $c(\sigma) < \infty$ and the proof of Theorem 5.1 shows there is a $U_{R(\sigma)} \in Y(\sigma)$ such that $U_{R(\sigma)}$ satisfies (2) and (4) of Theorem 5.1 and also possess the asymptotics associated with σ.

Next suppose $\sigma = \{\sigma_i\}_{\in \mathbb{Z}} \in \Sigma \backslash \Sigma^*$. For each $n \in \mathbb{N}$, define $f_n(\sigma) \in \Sigma^*$ via $f_n(\sigma)$ equal to σ_i, $|i| \leq n$, equal to σ_n, $i > n$, and equal to σ_{-n} when $i < -n$. Therefore by what was previously shown, there is a $U_n \in Y(f_n(\sigma))$ such that $J(U_n) = c(f_n(\sigma))$.

Since $v \leq U_n \leq w$, the functions U_n are uniformly bounded. By (DE), they are also bounded in C^2. Therefore using (DE), as $n \to \infty$, U_n converges along a subsequence in C^2 to $U(\sigma)$, a solution of (DE). Moreover for any $l \in \mathbb{N}$, if $n \geq l$, for $|t| \leq l$, the graph of U_n lies in

$$R(f_n(\sigma)) \cap \{(t,z) : |t| \leq l, v(t) < z < w(t)\}$$

$$= R(\sigma) \cap \{(t,z) : |t| \leq l, v(t) < z < w(t)\}$$

from which it follows that the graph of U lies in $R(\sigma)$. Finally the local minimality property is preserved by the L^∞_{loc} convergence of the U_n. $\square$

We conclude this section with some open questions. First, is it possible to give a variational characterization of $U(\sigma)$ for $\sigma \in \Sigma \backslash \Sigma^*$? The difficulty is that for such σ, $J(U(\sigma)) = \infty$. We suspect that a second renormalization of J can be made which allows for a direct variational characterization of $U(\sigma)$. A second question is whether it is possible to classify these multi-transition solutions. How many parameters do they really depend on?

6 The tip of the iceberg

In a sense the class of solutions of (DE) we have studied in these lectures merely represent the tip of the iceberg. All of these solutions lie between a gap pair. Even if we had had time to study the monotone solutions of (DE) mentioned in the introduction that cross a finite number of gaps, we are still only dealing with bounded solutions which therefore have rotation number 0.

For $p \in \mathbb{Z}$ and $q \in \mathbb{N}$ it is straightforward to find minimal solutions of (DE) satisfying $u(t + q) = u(t) + p$. In terms of the pendulum, they make p rotations in time q and have rotation number p/q. Thus replacing $\mathfrak{M}_0$ by such a class of minimizers, there are analogues of the results of the previous sections. There are also minimal solutions with an irrational rotation number which can be obtained as limits of the rational ones.

In addition to these minimal solutions there are nonminimal solutions that can be obtained variationally. E.g. there are mountain pass solutions lying between a gap pair v, w. In fact there is a sequence $\{u_n\}$ of such solutions with periods which go to infinity as $n \to \infty$. Likewise there are mountain pass heteroclinics between a gap pair in $\mathfrak{M}_1(v, w)$. These facts can be proven using versions of the mountain pass theorem.

References

[A]   S. Aubry, and P.Y. LeDaeron, The discrete Frenkel-Kantorova model and its extensions I-Exact results for the ground states, Physica, **8D** (1983), 381–422.

[B88]   V. Bangert, Mather sets for twist maps and geodesics on tori. Dynamics reported, Vol. 1, 1–56 (1988).

[B89]   V. Bangert, On minimal laminations of the torus, AIHP Analyses Nonlinéaire, **6** (1989), 95–138.

[Ma82]   J.N. Mather, Existence of quasi-periodic orbits for twist homeomorphisms fo the annulu, Topology, **21** (1982), 457–467.

[Ma93]   J.N. Mather, Variational construction of connecting orbits, Ann. Inst. Fourier (Grenoble) **43** (1993), 1349–1386.

[Mor]   H. Marston Morse, A fundamental class of geodesics on any closed surface of genus grater than one. Trans. Amer. Math. Soc. 26 (1924), no. 1, 25–60.

[Mos86] J. Moser, Minimal solutions of variational problems on a torus, AIHP Analyse Nonlinéaire, **3** (1986), 229–272.

[RS]   P.H. Rabinowitz, and E. Stredulinsky, Single and multitransition solutions for a class of PDE's, in progress.

Variational methods for Hamiltonian PDEs

Massimiliano Berti[1]

Abstract We present recent existence results of periodic solutions for completely resonant nonlinear wave equations in which both "small divisor" difficulties and infinite dimensional bifurcation phenomena occur. These results can be seen as generalizations of the classical finite-dimensional resonant center theorems of Weinstein–Moser and Fadell–Rabinowitz. The proofs are based on variational bifurcation theory: after a Lyapunov–Schmidt reduction, the small divisor problem in the range equation is overcome with a Nash–Moser implicit function theorem for a Cantor set of non-resonant parameters. Next, the infinite dimensional bifurcation equation, variational in nature, possesses minimax mountain-pass critical points. The big difficulty is to ensure that they are not in the "Cantor gaps". This is proved under weak non-degeneracy conditions. Finally, we also discuss the existence of forced vibrations with rational frequency. This problem requires variational methods of a completely different nature, such as constrained minimization and a priori estimates derivable from variational inequalities.

1 Finite dimensions: resonant center theorems

Consider a finite dimensional Hamiltonian system

$$\dot{x} = J\nabla H(x), \qquad x \in \mathbf{R}^{2n} \tag{1}$$

where $J = \begin{pmatrix} 0 & I \\ -I & 0 \end{pmatrix}$ is the symplectic matrix, I is the identity in $\mathbf{R}^n$, and $\nabla H(0) = 0$.

- QUESTION: Do there exist periodic solutions of (1) close to the equilibrium $x = 0$?

[1] Dipartimento di Matematica e Applicazioni, R. Caccioppoli, Università Federico II, Via Cintia, Monte S. Angelo, I-80126 Napoli, Italy
e-mail: m.berti@unina.it

W. Craig (ed.), Hamiltonian Dynamical Systems and Applications, 391–420.
© 2008 *Springer Science + Business Media B.V.*

Clearly, a necessary condition is the presence of purely imaginary eigenvalues of $J(D^2H)(0)$ so that the linearized equation

$$\dot{x} = J(D^2H)(0)x \tag{2}$$

possesses periodic solutions. In the sequel we suppose that $x = 0$ is an elliptic equilibrium with $J(D^2H)(0)$ having all eigenvalues

$$\pm i\omega_1, \dots, \pm i\omega_n \tag{3}$$

purely imaginary.

The first continuation result of periodic solutions close to an elliptic equilibrium is the celebrated Lyapunov center theorem. Assuming the non resonance condition $\omega_j - l\omega_1 \neq 0$, $\forall l \in \mathbf{Z}$, $j = 2, \dots, n$, the theorem ensures the existence of a smooth two-dimensional manifold foliated by small amplitude periodic solutions of (1) with frequencies close to ω_1, see e.g. [32].

If the above non-resonance condition is violated, no periodic solutions except the equilibrium $x = 0$ need exist. An example due to Moser [31] is provided by the Hamiltonian in $(\mathbf{R}^4, \sum_{i=1}^2 dq_i \wedge dp_i)$

$$H = \frac{q_1^2 + p_1^2}{2} - \frac{q_2^2 + p_2^2}{2} + (q_1^2 + p_1^2 + q_2^2 + p_2^2)(p_1p_2 - q_1q_2)$$

where $(q, p) = 0$ is an elliptic equilibrium with non-simple eigenvalues $\pm i$. But

$$\frac{d}{dt}(q_1p_2 + p_1q_2) = 4(p_1p_2 - q_1q_2)^2 + (q_1^2 + p_1^2 + q_2^2 + p_2^2)^2$$

so that the unique periodic solution is $q = p = 0$.

In contrast, two remarkable theorems by Weinstein [42], Moser [31] and Fadell–Rabinowitz [24] prove the existence of periodic solutions under the assumptions respectively

$$\text{(WM)} \quad (D^2H)(0) > 0, \qquad \text{(FR)} \quad \text{signature}(D^2H)(0) \neq 0.$$

Theorem 1.1. (Weinstein 1973–Moser 1976) *Let $H \in C^2(\mathbf{R}^{2n}, \mathbf{R})$ such that $(\nabla H)(0) = 0$ and $(D^2H)(0) > 0$. For all ε small enough there exist, on each energy surface $\{H(x) = H(0) + \varepsilon^2\}$, at least n geometrically distinct periodic solutions of (1).*

Note that $(D^2H)(0) > 0$ implies that the level sets of the energy $\frac{1}{2}(D^2H)(0)x \cdot x$ of (2) are ellipsoids, whence all the solutions of (2) are bounded and (3) holds.

Theorem 1.2. (Fadell–Rabinowitz 1978) *Let $H \in C^2(\mathbf{R}^{2n}, \mathbf{R})$ such that $(\nabla H)(0) = 0$ and the* signature$(D^2H)(0) = 2\nu \neq 0$. *Assume also that any non-zero solution of (2) is T-periodic (and not constant). Then, either*

(i) $x = 0$ is a non-isolated T-periodic solution of (1),
or

(ii) there exist integers $k, m \geq 0$ with $k + m \geq |\nu|$ and a left neighborhood, $\mathcal{U}_l$, resp. a right neighborhood, $\mathcal{U}_r$, of T in $\mathbf{R}$ such that $\forall \lambda \in \mathcal{U}_l \setminus \{T\}$, resp. $\mathcal{U}_r \setminus \{T\}$, there exist at least k, resp. m, distinct, non-trivial, λ-periodic solutions of (1). The L^∞-norm of the solutions tends to 0 as $\lambda \to T$.

Unlike the Lyapunov Center Theorem the periodic solutions of Theorems 1.1 and 1.2 do not, in general, vary smoothly with respect to the parameters ε (energy) and λ (period). This will cause a serious difficulty for PDEs, see Sections 5 and 6.

Both the Weinstein–Moser and the Fadell–Rabinowitz resonant center theorems follow by arguments of *variational bifurcation theory*. The Weinstein–Moser theorem, thanks to the energy constraint, reduces to find critical points of a smooth function defined on a $(2n - 1)$-dimensional sphere. The Fadell–Rabinowitz theorem is more subtle because there is no energy constraint, and one has to look for critical points of a reduced action functional defined in an open neighborhood of the origin, see e.g. [4].

For brevity, we present only a simplified version of the Fadell–Rabinowitz theorem without obtaining the optimal multiplicity results. We shall also assume that

$$(D^2 H)(0) := I \tag{4}$$

so that $\pm i$ are the multiple eigenvalues of $J(D^2 H)(0)$, and (2) possesses a $2n$-dimensional linear space of periodic solutions with the same minimal period 2π.

Theorem 1.3. (Fadell–Rabinowitz) *Under the assumptions above, either*

(i) $x = 0$ is a non-isolated 2π-periodic solution of (1),
(ii) There is a one sided neighborhood $\mathcal{U}$ of 1 such that, $\forall \lambda \in \mathcal{U} \setminus \{1\}$, equation (1) possesses at least two distinct non-trivial $2\pi\lambda$-periodic solutions,
(iii) There is a neighborhood $\mathcal{U}$ of 1 such that, $\forall \lambda \in \mathcal{U} \setminus \{1\}$, equation (1) possesses at least one non-trivial $2\pi\lambda$-periodic solution.

For the extension of these results to PDEs, due to the necessity of imposing non-resonance conditions on the frequency, it will be more convenient to give existence results with *fixed frequency* like in the Fadell–Rabinowitz theorem (see Theorems 3.2, 5.1, 6.1), and not with fixed energy as in the Weinstein–Moser theorem.

1.1 The variational Lyapunov–Schmidt reduction

With no loss of generality we suppose that $H(0) = 0$. Normalizing the period we look for 2π-periodic solutions of

$$J\dot{x} + \lambda (\nabla H)(x) = 0. \tag{5}$$

Equation (5) is the Euler–Lagrange equation of the action functional

$$\Psi(\lambda,x) := \int_{\mathbf{T}} \Big(\frac{1}{2}J\dot{x}(t)\cdot x(t) + \lambda H(x(t))\Big)\mathrm{d}t\,, \quad \mathbf{T} := (\mathbf{R}/2\pi\mathbf{Z})$$

defined and C^1, e.g. on the Sobolev space $H^1(\mathbf{T})$. To find critical points of Ψ we perform a Lyapunov–Schmidt reduction, decomposing

$$H^1(\mathbf{T}) := V \oplus V^{\perp} \qquad \text{where} \qquad V := \Big\{v \in H^1(\mathbf{T}) \mid \dot{v} = J(D^2H)(0)v\Big\}$$

is $2n$-dimensional (by (4)) and $V^{\perp} := \{w \in H^1(\mathbf{T}) \mid \int_{\mathbf{T}} w\cdot v\,\mathrm{d}t = 0,\ \forall v \in V\}$.

Projecting (5), for $x = v + w$, $v \in V$, $w \in V^{\perp}$, yields

$$\begin{cases} \Pi_V\left(J(\dot{v}+\dot{w}) + \lambda(\nabla H)(v+w)\right) = 0 & \text{bifurcation equation} \\ \Pi_{V^{\perp}}\left(J(\dot{v}+\dot{w}) + \lambda(\nabla H)(v+w)\right) = 0 & \text{range equation} \end{cases}$$

where Π_V, $\Pi_{V^{\perp}}$ denote the projectors on V, resp. $V^{\perp}$.

The range equation. We solve first the range equation with the standard implicit function theorem, finding a solution $w(\lambda,v) \in V^{\perp}$ for $v \in B_r(0)$ ($\equiv$ ball in V of radius r centered at zero), $r > 0$ small enough, and λ sufficiently close to 1. Indeed

$$\mathscr{F}(\lambda,v,w) := \Pi_{V^{\perp}}\left(J(\dot{v}+\dot{w}) + \lambda\nabla H(v+w)\right)$$

vanishes $\mathscr{F}(\lambda,0,0) = 0$, $\forall\lambda$, and its partial derivative

$$(D_w\mathscr{F})(1,0,0)[W] = J\dot{W} + (D^2H)(0)W\,, \quad \forall W \in V^{\perp}$$

is an isomorphism. By the implicit function theorem the solution $w(\lambda,v) \in V^{\perp}$ of the range equation is a C^1 function, $w(\lambda,0) = 0$ and

$$w(\lambda,v) = o(\|v\|) \qquad \text{as } v \to 0 \tag{6}$$

uniformly for λ near 1.

The bifurcation equation. It remains to solve the bifurcation equation

$$\Pi_V\left(J(\dot{v}+\dot{w}(\lambda,v)) + \lambda(\nabla H)(v+w(\lambda,v))\right) = 0$$

which is the Euler–Lagrange equation of the "reduced action functional"

$$\Phi(\lambda,\cdot) : B_r(0) \subset V \to \mathbf{R}\,, \qquad \Phi(\lambda,v) := \Psi(\lambda,v+w(\lambda,v))\,.$$

Indeed $\forall h \in V$,

$$\begin{aligned} (D_v\Phi)(\lambda,v)[h] &= (D_x\Psi)(\lambda,v+w(\lambda,v))[h + (D_vw)(\lambda,v)[h]] \\ &= (D_x\Psi)(\lambda,v+w(\lambda,v))[h] \\ &= \int_{\mathbf{T}} \Pi_V\left(J(\dot{v}+\dot{w}) + \lambda\nabla H(v+w)\right)\cdot h\,\mathrm{d}t \end{aligned} \tag{7}$$

using that

$$(D_x\Psi)(\lambda, v + w(\lambda, v))[W] = 0, \qquad \forall W \in V^\perp,$$

since $w(\lambda, v)$ solves the range equation, and $(D_v w)(\lambda, v)[h] \in V^\perp$.

Remark also that, by (7) and since $w(\lambda, v) \in C^1$, then $D_v \Phi \in C^1$ and so $\Phi \in C^2$.

To prove Theorem 1.3 we have to find non-trivial critical points of the reduced action functional $\Phi(\lambda, \cdot)$ near $v = 0$ for fixed λ near 1.

The functional $\Phi(\lambda, \cdot)$ possesses a strict local minimum or maximum at $v = 0$, according $\lambda > 1$ or $\lambda < 1$, because, by (6), and $H(0) = (\nabla H)(0) = 0$, $(D^2 H)(0) = I$,

$$\begin{aligned}
\Phi(\lambda, v) &= \int_{\mathbf{T}} \frac{1}{2} J\dot{v}(t) \cdot v(t) + \frac{\lambda}{2}(D^2 H)(0)v(t) \cdot v(t) + o(\|v\|^2) \\
&= \frac{(\lambda - 1)}{2} \int_{\mathbf{T}} |v(t)|^2 \, dt + o(\|v\|^2) = (\lambda - 1)\pi|v(0)|^2 + o(\|v\|^2). \quad (8)
\end{aligned}$$

If $v = 0$ *is not* an isolated critical point of $\Phi(1, \cdot)$ then alternative (i) of Theorem 1.3 holds. If $v = 0$ *is* an isolated critical point of $\Phi(1, \cdot)$ then, either

(a) $\Phi(1, \cdot)$ has a strict local maximum or minimum in $v = 0$

(b) $\Phi(1, \cdot)$ takes on both positive and negative values near $v = 0$.

Case (a) leads to alternative (ii) and Case (b) leads to alternative (iii) of Theorem 1.3. Figure 1 gives the idea of the existence proof.

In both cases (a) and (b), the functional $\Phi(\lambda, \cdot)$ possesses saddle critical points which can be found by a (finite dimensional) mountain pass argument [1] (see Theorem 3.1). However, the main difficulty of the minimax proof is that $\Phi(\lambda, \cdot)$ is defined only in a neighborhood of zero, see [4], [37] for details.

In the following we shall develop an analogous variational argument in infinite dimension.

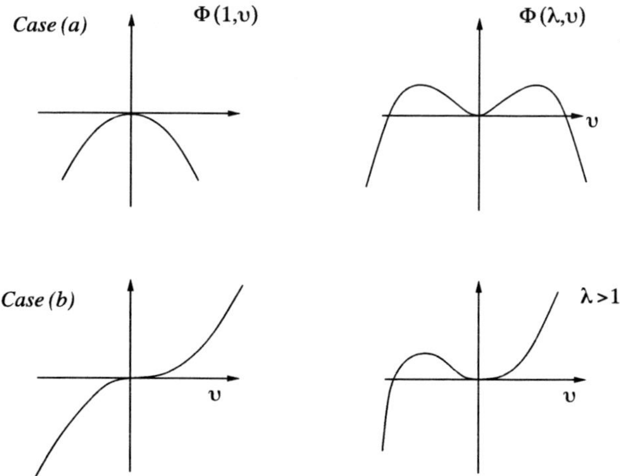

Fig. 1 In case (a), $\Phi(1, \cdot)$ has a strict local maximum at $v = 0$ and for $\lambda > 1$, $\Phi(\lambda, \cdot)$ possesses at least two non-trivial critical points. In case (b) $\Phi(1, \cdot)$ takes on both positive and negative values near $v = 0$ and for $\lambda \neq 1$, $\Phi(\lambda, \cdot)$ possesses at least one mountain pass critical point.

2 Infinite dimensions

We want to extend the local bifurcation theory of periodic solutions described in the previous section to infinite dimensional Hamiltonian PDEs, like the "completely resonant" nonlinear wave equation

$$u_{tt} - u_{xx} = f(x,u), \qquad u(t,0) = u(t,\pi) = 0 \tag{1}$$

where $f(x,0) = (\partial_u f)(x,0) = 0$. *All* the solutions of the linearized equation at $u = 0$

$$u_{tt} - u_{xx} = 0, \qquad u(t,0) = u(t,\pi) = 0 \tag{2}$$

are 2π-periodic. They can be represented like (*Fourier method*)

$$v(t,x) = \sum_{j\geq 1} a_j \cos(jt + \theta_j)\sin(jx), \qquad a_j \in \mathbf{R}$$

or like superposition of waves traveling in opposite directions (*D'Alembert method*)

$$v(t,x) = \eta(t+x) - \eta(t-x)$$

where η is any 2π-periodic function. This is the infinite dimensional analogous situation considered by the Weinstein–Moser and Fadell–Rabinowitz resonant Center theorems.

In trying to extend the Lyapunov–Schmidt reduction scheme of the previous section, we encounter two new problems:

- A "*small divisor*" problem in the range equation
- The presence of an *infinite dimensional* bifurcation equation

The "small divisor" problem is that the eigenvalues of $\partial_{tt} - \partial_{xx}$ in a space of functions $2\pi/\omega$-periodic in time and satisfying Dirichlet boundary conditions, are

$$-\omega^2 l^2 + j^2, \quad l \in \mathbf{Z}, \ j \geq 1. \tag{3}$$

Therefore, for almost every $\omega \in \mathbf{R}$, such eigenvalues accumulate to zero, the inverse $(\partial_{tt} - \partial_{xx})^{-1}$ is unbounded and the standard implicit function theorem fails.

Remark 2.1. When $\omega \in \mathbf{Q}$ the spectrum is not dense in $\mathbf{R}$. We shall discuss this case in Section 7. Existence of periodic solutions of wave equations with a rational frequency $\omega \in \mathbf{Q}$ has been proved via global minimax methods in [36] and [15]. When $\omega \notin \mathbf{Q}$ these proofs fail for a lack of compactness introduced by the small divisors. For some results in the irrational case, see [22].

The small divisors problem for Hamiltonian PDEs was first solved by Kuksin [28] and Wayne [41] using KAM theory and by Craig–Wayne [20] – who were the first to introduce the Lyapunov–Schmidt reduction method for PDEs – via a Nash–Moser implicit function technique. Other existence results of quasi-periodic solutions have been obtained by Bourgain [11, 12, 14] extending the Craig–Wayne [20] approach, and, via KAM theory, e.g. in Kuksin–Pöeschel [18], Eliasson–Kuksin [23] and Yuan [45]. See also the books [18], [29] and references therein.

The first existence result for completely resonant wave equations with Dirichlet boundary conditions was obtained in [3] for $f(u) = u^3$ (under periodic boundary conditions we quote [27]). The small divisors problem is bypassed for the zero measure set $\mathcal{W}_\gamma$ of the frequencies defined in (5) of section 3. The choice of $f(u) = u^3$ is required by a non-degeneracy condition to solve the bifurcation equation.

In [6], existence of periodic solutions, for the same zero measure set of frequencies $\mathcal{W}_\gamma$, but for a general nonlinearity, was proved. The key to remove the non-degeneracy condition is to solve the infinite dimensional bifurcation equation via variational methods. We now present these results for $f(x,u) := u^p$. This situation (where no small divisors appear) is a preparation to understand the more difficult case considered in Section 6 where we shall find solutions of the bifurcation equation on a Cantor set using variational methods, for positive measure sets of frequencies.

3 The variational Lyapunov–Schmidt reduction

Normalizing the period, and rescaling the amplitude $u \to \delta u$, we look for 2π-periodic solutions of

$$\omega^2 u_{tt} - u_{xx} = \varepsilon u^p, \qquad u(t,0) = u(t,\pi) = 0, \tag{1}$$

where $\varepsilon := \delta^{p-1}, p \geq 2$, in the Banach algebra

$$X := \left\{ u \in H^1(\Omega, \mathbf{R}) \cap L^\infty(\Omega, \mathbf{R}) \mid u(t,0) = u(t,\pi) = 0, \ u(-t,x) = u(t,x) \right\}$$

where $\Omega := \mathbf{T} \times [0, \pi]$, endowed with norm $\|u\| := \|u\|_\infty + \|u\|_{H^1}$. We consider here the easier case when p is odd.

Equation (1) is the Euler–Lagrange equation of the *Lagrangian action functional* $\Psi \in C^1(X, \mathbf{R})$ defined by

$$\Psi(u) := \int_0^{2\pi} dt \int_0^\pi \left[\frac{\omega^2}{2} u_t^2 - \frac{1}{2} u_x^2 + \varepsilon F(u) \right] dx$$

where $F(u) := u^{p+1}/p + 1$, sometimes called the "Percival variational principle". When $\omega \notin \mathbf{Q}$ this functional is highly non compact (see Remark 2.1). Therefore, to find its critical points we perform a Lyapunov–Schmidt reduction, decomposing

$$X = V \oplus W$$

where

$$V := \left\{ v = \eta(t+x) - \eta(t-x) \mid \eta(\cdot) \in H^1(\mathbf{T}), \ \eta \text{ odd} \right\} \tag{2}$$

are the solutions in X of the linear equation (2) of section 2 and

$$W := \Big\{ w \in X \,|\, (w,v)_{L^2} = 0, \, \forall v \in V \Big\} = \Big\{ \sum_{l \geq 0, j \geq 1, l \neq j} w_{l,j} \cos(lt) \sin jx \in X \Big\}.$$

Remark 3.1. Recalling (2) and the compact embedding $H^1(\mathbf{T}) \hookrightarrow L^\infty(\mathbf{T})$, also the embedding $(V, \|\cdot\|_{H^1}) \hookrightarrow (V, \|\cdot\|_\infty)$ is compact.

Projecting (1), for $u := v + w$ with $v \in V$, $w \in W$, yields

$$\omega^2 v_{tt} - v_{xx} = \varepsilon \Pi_V f(v+w) \qquad \text{bifurcation equation} \qquad (3)$$
$$\omega^2 w_{tt} - w_{xx} = \varepsilon \Pi_W f(v+w) \qquad \text{range equation} \qquad (4)$$

where $\Pi_V : X \to V$, $\Pi_W : X \to W$ are the projectors respectively on V, W.

The range equation. We first solve the range equation assuming that

$$\omega \in \mathscr{W}_\gamma := \Big\{ \omega \in \mathbf{R} \,\Big|\, |\omega l - j| \geq \frac{\gamma}{l}, \quad \forall (l,j) \in \mathbf{N} \times \mathbf{N}, \, j \neq l \Big\}. \qquad (5)$$

Lemma 3.1. *[3], [22] For $0 < \gamma \leq 1/4$ the set $\mathscr{W}_\gamma$ is uncountable, has zero measure and accumulates to $\omega = 1$ both from the left and from the right.*

Remark 3.2. By the Dirichlet Theorem, there are no real numbers ω such that $|\omega l - j| \geq (\gamma/l^\tau)$, $\forall l \neq j$, if $\tau < 1$.

For $\omega \in \mathscr{W}_\gamma$ the eigenvalues of $L_\omega := \omega^2 \partial_{tt} - \partial_{xx}$ restricted to W, satisfy

$$|-\omega^2 l^2 + j^2| = |\omega l - j||\omega l + j| \geq \frac{\gamma}{l}|\omega l + j| \geq \gamma \omega, \qquad \forall l \neq j$$

whence the inverse

$$L_\omega^{-1} w := \sum_{j \geq 1, l \geq 0, l \neq j} \frac{w_{l,j}}{-\omega^2 l^2 + j^2} \cos(lt) \sin(jx), \quad \forall w \in W$$

is a bounded operator, $\|L_\omega^{-1} w\| \leq C\gamma^{-1} \|w\|$. Fixed points of the nonlinear operator

$$\mathscr{G} : W \to W, \qquad \mathscr{G}(\varepsilon, \omega; w) := \varepsilon L_\omega^{-1} \Pi_W f(v+w)$$

are solutions of the range equation. By the contraction mapping theorem we get

Lemma 3.2. (Solution of the range equation) *Assume $\omega \in \mathscr{W}_\gamma \cap (1/2, 3/2)$. $\forall R > 0$, $\exists \, \varepsilon_0(R) > 0$, $C_0(R) > 0$ such that $\forall v \in B_{2R} := \{ v \in V \mid \|v\|_{H^1} \leq 2R \}$, $\forall 0 \leq \varepsilon \gamma^{-1} \leq \varepsilon_0(R)$ there exists a unique solution $w(\varepsilon, v) \in W$ of the range equation satisfying $\|w(\varepsilon, v)\| \leq C_0(R)\varepsilon \gamma^{-1}$. Moreover the map $v \mapsto w(\varepsilon, v)$ is in $C^1(B_{2R}, W)$.*

3.1 The bifurcation equation

It remains to solve the infinite dimensional bifurcation equation

$$\omega^2 v_{tt} - v_{xx} = \varepsilon \Pi_V f(v + w(\varepsilon, v))$$

which, arguing as in (7) of section 1.1 (see also Figure 2), is the Euler–Lagrange equation of the *reduced Lagrangian action functional* $\Phi_\varepsilon \in C^2(B_{2R}, \mathbf{R})$ defined by

$$\Phi_\varepsilon(v) := \Psi(v + w(\varepsilon, v)).$$

To find critical points of Φ_ε we expand

$$\Phi_\varepsilon(v) = \int_\Omega \frac{\omega^2}{2}(v_t + (w(\varepsilon, v))_t)^2 - \frac{1}{2}(v_x + (w(\varepsilon, v))_x)^2 + \varepsilon F(v + w(\varepsilon, v))$$

$$= \int_\Omega \frac{\omega^2}{2} v_t^2 - \frac{v_x^2}{2} + \varepsilon F(v + w(\varepsilon, v)) - \frac{\varepsilon}{2} f(v + w(\varepsilon, v)) w(\varepsilon, v)$$

because $\int_\Omega v_t w_t = \int_\Omega v_x w_x = 0$ and, since $w(\varepsilon, v)$ solves the range equation,

$$\int_\Omega \omega^2 w_t^2 - w_x^2 + \varepsilon f(v + w(\varepsilon, v)) w(\varepsilon, v) = 0.$$

Hence, using that $\|v_t\|_{L^2}^2 = \|v_x\|_{L^2}^2 = \|v\|_{H^1}^2 / 2$,

$$\Phi_\varepsilon(v) = \frac{\omega^2 - 1}{4} \|v\|_{H^1}^2 + \varepsilon \int_\Omega \left[F(v + w(\varepsilon, v)) - \frac{1}{2} f(v + w(\varepsilon, v)) w(\varepsilon, v) \right]. \quad (6)$$

Imposing the "frequency-amplitude" relation $\omega^2 - 1 = -2\varepsilon$, we get

$$\Phi_\varepsilon(v) = -\varepsilon \left[\frac{1}{2} \|v\|_{H^1}^2 - \int_\Omega \frac{v^{p+1}}{p+1} + \mathscr{R}_\varepsilon(v) \right]$$

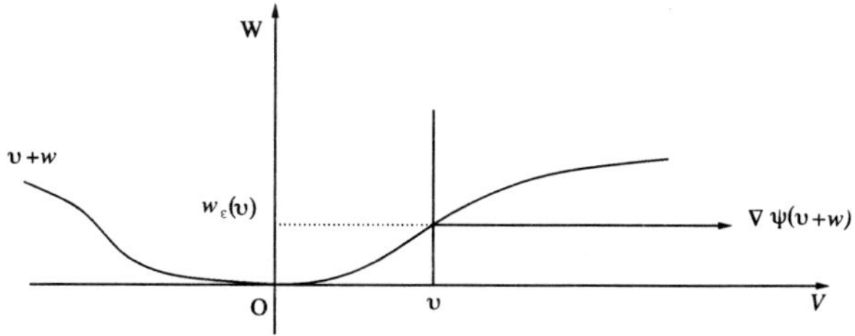

Fig. 2 Since $w(\varepsilon, v)$ solves the range equation then $v + w(\varepsilon, v)$ is a critical point of the functional $W \ni w \to \Psi(v + w)$, i.e. the gradient $(\nabla \Psi)(v + w(\varepsilon, v))$ is parallel to V.

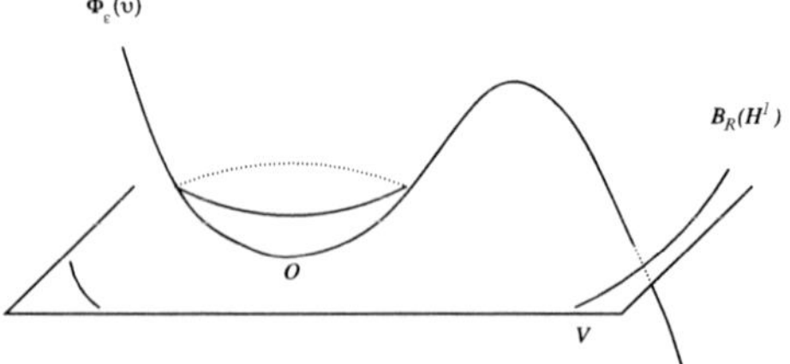

Fig. 3 The mountain Pass geometry of Φ_ε.

where, for some constant $C_1(R) > 0$,

$$|\mathscr{R}_\varepsilon(v)|, |(\nabla \mathscr{R}_\varepsilon(v), v)| \leq C_1(R)\frac{\varepsilon}{\gamma}. \tag{7}$$

The functional (that we still denote by Φ_ε)

$$\Phi_\varepsilon(v) := \frac{1}{2}\|v\|_{H^1}^2 - \int_\Omega \frac{v^{p+1}}{p+1} + \mathscr{R}_\varepsilon(v) \tag{8}$$

possesses a local minimum at the origin and one could think to prove existence of non-trivial critical points via the Mountain Pass Theorem 3.1 below, see Figure 3. Note that, since p is odd, $\int_\Omega v^{p+1} > 0, \forall v \neq 0$.

Theorem 3.1. (Mountain Pass [1]) *Let $(X, \|\cdot\|)$ be a Banach space. Suppose $\Phi \in C^1(X, \mathbf{R})$ and*

(i) $\Phi(0) = 0$
(ii) $\exists \rho, \alpha > 0$ such that $\Phi(x) \geq \alpha$ if $\|x\| = \rho$
(iii) $\exists v \in X$ with $\|v\| > \rho$ such that $\Phi(v) < 0$

Define the "Mountain Pass" value $c := \inf_{\gamma \in \Gamma} \max_{t \in [0,1]} \Phi(\gamma(t)) \geq \alpha$ where Γ is the minimax class $\Gamma := \{\gamma \in C([0, 1], X) \mid \gamma(0) = 0, \gamma(1) = v\}$.

Then there exists a Palais–Smale sequence x_n of Φ at the level c, i.e. $\Phi(x_n) \to c$, $D\Phi(x_n) \to 0$. If, up to subsequence, $x_n \to \bar{x}$ then $D\Phi(\bar{x}) = 0$ and $\Phi(\bar{x}) = c$.

Theorem 3.1 cannot be directly applied because Φ_ε is defined only close to 0.
Step 1: *Extension of Φ_ε.* We define the extended functional $\widetilde{\Phi}_\varepsilon \in C^2(V, \mathbf{R})$ as

$$\widetilde{\Phi}_\varepsilon(v) := \frac{\|v\|_{H^1}^2}{2} - \int_\Omega \frac{v^{p+1}}{p+1} + \widetilde{\mathscr{R}}_\varepsilon(v)$$

where $\widetilde{\mathscr{R}}_\varepsilon(v) := \lambda(\|v\|_{H^1}^2 R^{-2})\mathscr{R}_\varepsilon(v)$ and $\lambda : [0, +\infty) \to [0, 1]$ is a smooth cut-off function with $\lambda(x) = 1$ if $|x| \leq 1$, $\lambda(x) = 0$ if $|x| \geq 4$, and $|\lambda'(x)| < 1$. By definition

$$\widetilde{\Phi}_\varepsilon(v) = \begin{cases} \Phi_\varepsilon(v) & \text{for} \quad \|v\|_{H^1} \leq R \\ \dfrac{\|v\|_{H^1}^2}{2} - \displaystyle\int_\Omega \frac{v^{p+1}}{p+1} & \text{for} \quad \|v\|_{H^1} \geq 2R. \end{cases} \tag{9}$$

Furthermore by (7) and the definition of λ

$$|\widetilde{\mathscr{R}}_\varepsilon(v)|, |(\nabla\widetilde{\mathscr{R}}_\varepsilon(v), v)| \leq C_1(R)\frac{\varepsilon}{\gamma}. \tag{10}$$

Step 2: $\widetilde{\Phi}_\varepsilon$ *has the Mountain Pass geometry.* Let $0 < \rho < R$. For all $\|v\|_{H^1} = \rho$

$$\widetilde{\Phi}_\varepsilon(v) \overset{(9)}{=} \Phi_\varepsilon(v) \overset{(8)}{=} \frac{\|v\|_{H^1}^2}{2} - \int_\Omega \frac{v^{p+1}}{p+1} + \mathscr{R}_\varepsilon(v) \overset{(7)}{\geq} \frac{1}{2}\rho^2 - \kappa_1\rho^{p+1} - \frac{\varepsilon}{\gamma}C_1(R).$$

Fix $\rho > 0$ such that $(\rho^2/2) - \kappa_1\rho^{p+1} \geq \rho^2/4$. For $0 < \varepsilon\gamma^{-1}C_1(R) \leq \rho^2/8$

$$\widetilde{\Phi}_\varepsilon(v) \geq \frac{1}{8}\rho^2 > 0 \quad \text{if} \quad \|v\|_{H^1} = \rho$$

verifying the assumption (*ii*) of Theorem 3.1. Let us verify assumption (*iii*). By (9) for every $\|v_0\|_{H^1} = 1$, there exists $\widetilde{t}$ large enough such that

$$\widetilde{\Phi}_\varepsilon(\widetilde{t}v_0) = \frac{\widetilde{t}^2}{2} - \frac{\widetilde{t}^{p+1}}{p+1}\int_\Omega v_0^{p+1} < 0$$

because $p \geq 3$ is an odd integer. Define $\widetilde{v} := \widetilde{t}v_0$ and the Mountain Pass level

$$c_\varepsilon := \inf_{\gamma\in\Gamma} \max_{s\in[0,1]} \widetilde{\Phi}_\varepsilon(\gamma(s)) > 0 \tag{11}$$

where $\Gamma := \{\gamma \in C([0,1], V) \mid \gamma(0) = 0, \gamma(1) = \widetilde{v}\}$. By the Mountain Pass Theorem 3.1 there is a Palais–Smale sequence $v_n \in V$ for $\widetilde{\Phi}_\varepsilon$ at the level $c_\varepsilon > 0$,

$$\widetilde{\Phi}_\varepsilon(v_n) \to c_\varepsilon, \quad \nabla\widetilde{\Phi}_\varepsilon(v_n) \to 0. \tag{12}$$

Step 3: *Confinement of the Palais–Smale sequence.* By (11) and (10)

$$c_\varepsilon \leq \max_{s\in[0,1]} \widetilde{\Phi}_\varepsilon(s\widetilde{v}) \leq \max_{s\in[0,1]} \left[\frac{s^2}{2}\|\widetilde{v}\|_{H^1}^2 - \frac{s^{p+1}}{p+1}\int_\Omega \widetilde{v}^{p+1}\right] + 1 =: \kappa \tag{13}$$

for $0 < C_1(R)\gamma^{-1}\varepsilon < 1$. Then

$$\widetilde{\Phi}_\varepsilon(v_n) - \frac{(\nabla\widetilde{\Phi}_\varepsilon(v_n), v_n)}{p+1} = \left(\frac{1}{2} - \frac{1}{p+1}\right)\|v_n\|_{H^1}^2 + \widetilde{\mathscr{R}}_\varepsilon(v_n) - \frac{(\nabla\widetilde{\mathscr{R}}_\varepsilon(v_n), v_n)}{p+1}$$

$$\overset{(10)}{\geq} \left(\frac{1}{2} - \frac{1}{p+1}\right)\|v_n\|_{H^1}^2 - 1 \tag{14}$$

for $0 < 2C_1(R)\varepsilon\gamma^{-1} \leq 1$. By (12), for n large,

$$\widetilde{\Phi}_\varepsilon(v_n) - \frac{(\nabla\widetilde{\Phi}_\varepsilon(v_n), v_n)}{p+1} \leq (c_\varepsilon + 1) + \|v_n\|_{H^1} \overset{(13)}{\leq} \kappa + 1 + \|v_n\|_{H^1}$$

and, by (14), we derive $\kappa + 1 + \|v_n\|_{H^1} \geq (\frac{1}{2} - \frac{1}{p+1})\|v_n\|_{H^1}^2 - 1$. Hence, there exist $R_* > 0$ independent of ε and $C_*(R) > 0$ such that

$$\|v_n\|_{H^1} \leq R_* \qquad \text{for} \qquad 0 < \varepsilon\gamma^{-1} < C_*(R). \tag{15}$$

Step 4: *Existence of a nontrivial critical point.* Fix $\bar{R} := R_* + 1$ and take $0 < \varepsilon\gamma^{-1} \leq C_*(\bar{R})$. By (15), definitively for n large, $v_n \in B_{\bar{R}}$, and so $\widetilde{\Phi}_\varepsilon(v_n) = \Phi_\varepsilon(v_n)$. Hence

$$\nabla\widetilde{\Phi}_\varepsilon(v_n) = \nabla\Phi_\varepsilon(v_n) = v_n - \nabla G(v_n) + \nabla\mathscr{R}_\varepsilon(v_n) \to 0$$

where we have set $G(v) := \frac{1}{p+1}\int_\Omega v^{p+1}$.

By the compact embedding $(V, \|\cdot\|_{H^1}) \hookrightarrow (V, \|\cdot\|_\infty)$ (see Remark 3.1) we easily deduce that $\nabla G : V \to V$ and $\nabla\mathscr{R}_\varepsilon : B_{\bar{R}} \to V$ are compact operators. Therefore, since the Palais–Smale sequence v_n is bounded in H^1, v_n is precompact and converges to a nontrivial critical point v_ε of Φ_ε. We have finally proved

Theorem 3.2. **[6]** *Let $f(u) = u^p$ for an odd integer $p \geq 3$. $\forall\omega \in \mathscr{W}_\gamma$ with $|\omega - 1|\gamma^{-1}$ small enough, and $\omega < 1$, equation (1) of section 2 possesses at least one, non trivial, $2\pi/\omega$-periodic, small amplitude periodic solution u_ω.*

Multiplicity of solutions can also be proved. To deal with nonlinearities $f(u) = u^p$ with p even integer is more difficult because

$$\int_\Omega v^{p+1} \equiv 0, \qquad \forall v \in V \tag{16}$$

(see Lemma 7.3) and so the development (8) no longer implies the Mountain Pass geometry of Φ_ε. To find critical points of Φ_ε we have to develop at higher orders in ε the non-quadratic term, see [6].

This greater difficulty for finding periodic solutions when the nonlinearity f is non-monotonic (under Dirichlet boundary conditions) is a common feature for nonlinear wave equations, see e.g. [34, 36], [15, 16], and Section 7. In physical terms there is no "confinement effect" due to the potential. Actually, in [10] a non-existence result is proved for even power nonlinearities in the case of spatial periodic boundary conditions, highlighting that the existence result in case of Dirichlet conditions, is due to a "boundary effect".

We also remark that, since V is infinite dimensional, a Fadell–Rabinowitz type argument, working for any nonlinearity, does not apply.

4 The small divisor problem

To prove existence of periodic solutions of

$$\begin{cases} \omega^2 u_{tt} - u_{xx} = \varepsilon a(x)u^p \\ u(t,0) = u(t,\pi) = 0 \end{cases} \tag{1}$$

for positive measure set of frequencies, we have to relax the non-resonance condition in (5) of section 3 requiring

$$|\omega l - j| \geq \frac{\gamma}{l^{\tau}}, \quad \forall (l, j) \in \mathbf{N} \times \mathbf{N}, \; j \neq l$$

for some $\tau > 1$. However, for such ω we get just the bound

$$|\omega^2 l^2 - j^2| = |\omega l - j|(\omega l + j) \geq \frac{\gamma}{l^{\tau}}(\omega l + j) \geq \frac{\gamma \omega}{l^{\tau-1}}.$$

As a consequence, $(L_{\omega})^{-1}_{|W}$ "loses $(\tau - 1)$-derivatives" and the standard implicit function theorem fails.

This small divisor problem is overcome by a Nash–Moser iteration scheme. We look for 2π-periodic solutions of (1) in the Banach algebra (for $s > 1/2$)

$$X_{\sigma,s} := \left\{ u(t,x) = \sum_{l \in \mathbf{Z}} e^{ilt} u_l(x) \mid u_l \in H^1_0((0,\pi), \mathbf{R}), u_l(x) = u_{-l}(x) \right.$$

$$\left. \text{and } \|u\|^2_{\sigma,s} := \sum_{l \in \mathbf{Z}} e^{2\sigma|l|}(l^{2s} + 1)\|u_l\|^2_{H^1} < +\infty \right\}.$$

For $\sigma > 0, s \geq 0$, $X_{\sigma,s}$ is the space of all even, 2π-periodic in time functions with values in $H^1_0((0,\pi), \mathbf{R})$, which have a bounded analytic extension in the complex strip $|\mathrm{Im}\, t| < \sigma$ with trace function on $|\mathrm{Im}\, t| = \sigma$ belonging to $H^s(\mathbf{T}, H^1_0((0,\pi), \mathbf{C}))$, see [33] (recently in [9] we proved, when the nonlinearity is just C^k w.r.t. u, also existence of solutions for $\sigma = 0$, i.e. just Sobolev in time).

Projecting (1) according to the orthogonal decomposition

$$X_{\sigma,s} = (V \cap X_{\sigma,s}) \oplus (W \cap X_{\sigma,s})$$

and imposing the "frequency-amplitude" relation $\omega^2 - 1 = 2s^*\varepsilon$ with $s^* = \pm 1$ to be chosen later, yields

$$\begin{cases} \Delta v = s^* \Pi_V f(v + w) & \text{bifurcation equation} \\ L_{\omega} w = \varepsilon \Pi_W f(v + w) & \text{range equation} \end{cases} \tag{2}$$

where $\Delta v := v_{xx} + v_{tt}$, $f(u) := a(x)u^p$ and $L_{\omega} := \omega^2 \partial_{tt} - \partial_{xx}$.

When $\varepsilon = 0$ we get the "0th-order bifurcation equation"

$$\Delta v = s^* \Pi_V (a(x)v^p) \tag{3}$$

which is the Euler–Lagrange equation of the functional $\Phi_0 : V \to \mathbf{R}$

$$\Phi_0(v) := \frac{\|v\|^2_{H^1}}{2} + s^* \int_{\Omega} a(x) \frac{v^{p+1}}{p+1}, \qquad \Omega := \mathbf{T} \times [0, \pi].$$

For definiteness we suppose there exists $\tilde{v} \in V$ such that $\int_{\Omega} a(x)\tilde{v}^{p+1} < 0$ and so we choose $s^* = 1$ (otherwise we take $s^* = -1$). Let $\tilde{t} > 0$ be large enough such that $\Phi_0(\tilde{t}\tilde{v}) < 0$. The mountain pass value

$$c := \inf \left\{ \max_{s \in [0,1]} \Phi_0(\gamma(s)) \mid \gamma \in C([0,1], V), \gamma(0) = 0, \gamma(1) = \tilde{t}\tilde{v} \right\}$$

is a critical level with a non-trivial mountain pass critical set

$$\mathscr{K}_c := \left\{ v \in V \mid \Phi_0(v) = c, \nabla \Phi_0(v) = 0 \right\} \tag{4}$$

which is compact for the H^1-topology (because, with the same arguments used in Section 3.1, any Palais–Smale sequence of Φ_0 is precompact). In particular $\mathscr{K}_c$ is bounded: there exists $R_c > 0$ such that

$$\|v\|_{H^1} \leq R_c, \quad \forall v \in \mathscr{K}_c.$$

Remark 4.1. Actually Φ_0 has an unbounded sequence of critical levels tending to plus infinity [1], giving rise to multiplicity of periodic solutions of (1) close to the corresponding critical sets of Φ_0.

Since V is infinite dimensional a serious difficulty arises. If $v \in V \cap X_{\sigma,s}$ then the solution $w(\varepsilon, v)$ of the range equation, obtained with any Nash–Moser iteration scheme will have a lower regularity, e.g. $w(\varepsilon, v) \in X_{\sigma/2,s}$. Therefore, in solving next the bifurcation equation substituting $w = w(\varepsilon, v)$, the best estimate we can obtain is $v \in V \cap X_{\sigma/2,s+2}$ which makes the scheme incoherent.

We overcome this difficulty thanks to a reduction onto a *finite dimensional* bifurcation equation on a subspace of V of dimension N independent of ω decomposing

$$V = V_1 \oplus V_2$$

where

$$\begin{cases} V_1 := \left\{ v \in V \mid v(t,x) = \sum_{l=1}^{N} \cos(lt)u_l \sin(lx) \right\} & \text{``low Fourier modes''} \\ V_2 := \left\{ v \in V \mid v(t,x) = \sum_{l>N} \cos(lt)u_l \sin(lx) \right\} & \text{``high Fourier modes''}. \end{cases} \tag{5}$$

Setting $v := v_1 + v_2$, $v_1 \in V_1$, $v_2 \in V_2$, system (2) (with $s^* = 1$) is equivalent to

$$\begin{cases} \Delta v_1 = \Pi_{V_1} f(v_1 + v_2 + w) & (Q1) \\ \Delta v_2 = \Pi_{V_2} f(v_1 + v_2 + w) & (Q2) \\ L_\omega w = \varepsilon \Pi_W f(v_1 + v_2 + w) & \text{range equation} \end{cases} \tag{6}$$

where $\Pi_{V_i} : X_{\sigma,s} \to V_i$ $(i = 1, 2)$ denote the projectors on V_i.

The (Q2)-equation. We find first the solution $v_2(v_1, w)$ of the (Q2)-equation as a fixed point of

$$v_2 = \Delta^{-1} \Pi_{V_2} f(v_1 + v_2 + w)$$

by a contraction mapping argument, thanks to the compactness of the operator Δ^{-1}.

Lemma 4.1. (Solution of the (Q2)-equation) *There exist $\bar{N}$, $\bar{\sigma} > 0$, such that $\forall \sigma \in [0, \bar{\sigma}]$, $\forall \|v_1\|_{H^1} \leq 2R_c$, $\forall \|w\|_{\sigma,s} \leq R_c$, there exists a unique solution $v_2(v_1, w) \in X_{\sigma,s+2}$ of the (Q2)-equation with $\|v_2(v_1, w)\|_{\sigma,s} \leq R_c/2$. The function $v_2(\cdot, \cdot)$ is C^∞ and $\forall v \in \mathscr{K}_c$, $\Pi_{V_2} v = v_2(\Pi_{V_1} v, 0)$.*

Intuitively, to find solutions of the complete bifurcation equation close to the solutions $\mathscr{K}_c$ of the 0th order bifurcation equation (3), $\bar{N}$ must be taken large enough so that the majority of the "H^1-mass" of the functions in $\mathscr{K}_c$ is "concentrated" on the first $\bar{N}$ Fourier modes. In the sequel we consider as *fixed* the constants $\bar{N}$ and $\bar{\sigma}$ which depend only on $a(x)u^p$ and $\mathscr{K}_c$.

The range equation. We solve next the range equation

$$L_\omega w = \varepsilon \Pi_W \Gamma(v_1, w) \tag{7}$$

where $\Gamma(v_1, w) := f(v_1 + w + v_2(v_1, w))$ via a *Nash–Moser* implicit function theorem. We set $B(2R_c; V_1) := \{v_1 \in V_1 \mid \|v_1\|_{H^1} \leq 2R_c\}$.

Theorem 4.1. (Solution of the range-equation) *For $\varepsilon_0 > 0$ small enough, there exists*

$$\widetilde{w}(\cdot, \cdot) \in C^\infty([0, \varepsilon_0] \times B(2R_c; V_1), W \cap X_{\bar{\sigma}/2, s})$$

such that $\|\widetilde{w}(\varepsilon, v_1)\|_{\bar{\sigma}/2, s} \leq C\varepsilon\gamma^{-1}$ and the "large" Cantor set $B_\infty \subset [0, \varepsilon_0] \times B(2R_c; V_1)$ defined below, such that $\forall (\varepsilon, v_1) \in B_\infty$, $\widetilde{w}(\varepsilon, v_1)$ solves the range equation (7). The Cantor set B_∞ is written explicitly as

$$B_\infty := \left\{ (\varepsilon, v_1) \in [0, \varepsilon_0] \times B(2R_c; V_1) : \left| \omega l - j \right| \geq \frac{2\gamma}{(1+j)^\tau}, \right. \tag{8}$$

$$\left. \left| \omega l - j + \varepsilon \frac{M(v_1, \widetilde{w}(\varepsilon, v_1))}{2j} \right| \geq \frac{2\gamma}{(1+j)^\tau}, \forall j \geq 1, \forall l \geq \frac{1}{3\varepsilon}, l \neq j \right\}$$

where $\omega = \sqrt{1 + 2\varepsilon}$ and

$$M(v_1, w) := \frac{1}{2\pi^2} \int_\Omega (\partial_u f)(x, v_1 + w + v_2(v_1, w))\, dx dt\,.$$

To understand how the Cantor set B_∞ arises, we recall that the core of any Nash–Moser convergence method (based on a Newton's iteration scheme) is the proof of the invertibility of the linearized operators

$$h \mapsto \mathscr{L}(\varepsilon, v_1, w)[h] := L_\omega h - \varepsilon \Pi_W D_w \Gamma(v_1, w)[h]$$

where w is the approximate solution obtained at a given stage of the Nash–Moser iteration. The operators $\mathcal{L}(\varepsilon, v_1, w)$ are self-adjoint. Their eigenvalues can be estimated by

$$\lambda_{lj}(\varepsilon, v_1) = -\omega^2 l^2 + j^2 - \varepsilon M(v_1, w) + O\left(\frac{\varepsilon}{j}\right).$$

The linear operator $\mathcal{L}(\varepsilon, v_1, w)$ shall be invertible only for the (ε, v_1) where all the $\lambda_{lj}(\varepsilon, v_1) \neq 0$. This is the phenomenon giving rise to the Cantor set of "non-resonant" parameters B_∞. Some further work has to be done to get estimates for the inverse operators in $\|\ \|_{\sigma,s}$ norms. Our approach [7, 9] is different than in [20] and works also for not odd nonlinearities f with low regularity, unlike [20] works for nonlinearities which are odd and analytic in (x, u).

We underline that $\widetilde{w}(\varepsilon, v_1)$ is defined for *all* the $(\varepsilon, v_1) \in [0, \varepsilon_0] \times B(2R_c, V_1)$ and not only on the Cantor set B_∞. The function $\widetilde{w}(\cdot, \cdot)$ is a Whitney smooth interpolation.

The "Cantor gaps" in B_∞ are the main new problem to solve the bifurcation equation.

5 The $(Q1)$-equation

The last step is to find solutions of the finite dimensional $(Q1)$-equation

$$\Delta v_1 = \Pi_{V_1} \mathcal{G}(\varepsilon, v_1) \tag{1}$$

where $\mathcal{G}(\varepsilon, v_1) := f(v_1 + \widetilde{w}(\varepsilon, v_1) + v_2(v_1, \widetilde{w}(\varepsilon, v_1)))$ such that (ε, v_1) belong to the Cantor set B_∞.

Critical points of the C^∞ "reduced Lagrangian action functional"

$$\widetilde{\Phi} : B(2R_c; V_1) \to \mathbf{R}, \quad \widetilde{\Phi}(\varepsilon, v_1) := \Psi\left(v_1 + v_2(v_1, \widetilde{w}(\varepsilon, v_1)) + \widetilde{w}(\varepsilon, v_1)\right)$$

such that $(\varepsilon, v_1) \in B_\infty$ are solutions of (1) ("Percival" reduced variational principle).

As in Section 3.1, $\widetilde{\Phi}(\varepsilon, \cdot)$ possesses, for any ε small enough, a mountain pass critical point $v_1(\varepsilon)$ close to $\Pi_{V_1}\mathcal{K}_c$, where $\mathcal{K}_c$ is defined in (4) of section 4, as the mountain pass critical set of Φ_0. However, if $\mathcal{K}_c$ does not reduce to a non-degenerate solution of (3), then $v_1(\varepsilon)$ could vary in a highly irregular way as $\varepsilon \to 0$, the only information available in general is that $v_1(\varepsilon) \to \Pi_{V_1}\mathcal{K}_c$ as $\varepsilon \to 0$. Therefore for each ε the mountain pass critical point $v_1(\varepsilon)$ could belong to the complement of the Cantor set B_∞ in which the range equation (7) has been solved.

The section $E_\varepsilon := \{v_1 \mid (\varepsilon, v_1) \in B_\infty\} \equiv B(2R_c, V_1)$ has "no gaps", if and only if the frequency $\omega(\varepsilon) = \sqrt{1 + 2\varepsilon}$ belongs to the zero-measure set $\mathcal{W}_\gamma$ defined in (5) of section 3. This is why in section 3 we have proved the existence result for any nonlinearity $f(u) = u^p$.

It can be shown that in between two sections E_{ε_1}, E_{ε_2} such that $\omega(\varepsilon_1)$, $\omega(\varepsilon_2) \in \mathcal{W}_\gamma$, the complement of B_∞ is arcwise connected. Therefore it would be not sufficient to find just a continuous path of solutions $\varepsilon \mapsto v_1(\varepsilon)$ of equation (1), to conclude the existence of solutions for a positive measure set of frequencies.

This is the common principal difficulty in applying variational methods in a problem with small divisors.

The Arnold non-degeneracy condition. The simplest situation occurs when at least one solution $\bar{v} \in V$ of (3) of section 4 is non degenerate, i.e.

$$\ker \Phi''_{0|V}(\bar{v}) = \{0\}. \tag{2}$$

This condition is somehow analogous to the "Arnold non-degeneracy condition" in KAM theory, see e.g. [39] (in [20, 21] it is called "twist condition" or condition of "genuine nonlinearity").

For $\varepsilon = 0$ the $(Q1)$-equation (1) reduces to the projection of equation (3) of section 4 on V_1, namely

$$\Delta v_1 = \Pi_{V_1}\left(a(x)(v_1 + v_2(v_1,0))^p\right). \tag{3}$$

By the Arnold condition (and Lemma 4.1), $\bar{v}_1 := \Pi_{V_1}\bar{v}$ is a non-degenerate solution of (3) and, applying the implicit function theorem, there exists a C^∞ path $\varepsilon \mapsto v_1(\varepsilon)$ of solutions of (1) with $v_1(0) = \bar{v}_1$. For all ε belonging to the Cantor-like set

$$\mathscr{C} := \left\{\varepsilon \in [0,\varepsilon_0) \mid (\varepsilon, v_1(\varepsilon)) \in B_\infty\right\} \tag{4}$$

they give rise to solutions of equation (1) of section 4 like

$$u(\varepsilon) = v_1(\varepsilon) + v_2(v_1(\varepsilon), \widetilde{w}(\varepsilon, v_1(\varepsilon))) + \widetilde{w}(\varepsilon, v_1(\varepsilon)) \in X_{\bar{\sigma}/2,s}.$$

By the smoothness of $v_1(\cdot)$, the set $\mathscr{C}$ has asymptotically full measure, namely

$$\lim_{\eta \to 0^+} \frac{|\mathscr{C} \cap [0,\eta)|}{\eta} = 1. \tag{5}$$

Geometrically this estimate exploits the structure of the Cantor set B_∞ and that the curve of solutions $\varepsilon \mapsto v_1(\varepsilon)$ crosses transversally B_∞ (it is a graph), see Figure 4. This is the classical argument used in [20].

The non-degeneracy condition (2) can be verified on examples.

Theorem 5.1. [7] *Let*

$$f(x,u) = \begin{cases} a_2 u^2 & a_2 \neq 0 \\ a_3(x)u^3 & \langle a_3 \rangle := (1/\pi)\int_0^\pi a_3(x) \neq 0 \\ a_4 u^4 & a_4 \neq 0. \end{cases} \tag{6}$$

There exist $\varepsilon_0 > 0$, $\bar{\sigma} > 0$, a C^∞-curve $[0,\varepsilon_0) \ni \varepsilon \mapsto u(\varepsilon) \in X_{\bar{\sigma}/2,s}$, a Cantor set $\mathscr{C} \subset [0,\varepsilon_0)$ of asymptotically full measure, such that, $\forall \varepsilon \in \mathscr{C}$, $u(\varepsilon)$ is a solution of equation (1) of section 4 with frequency respectively

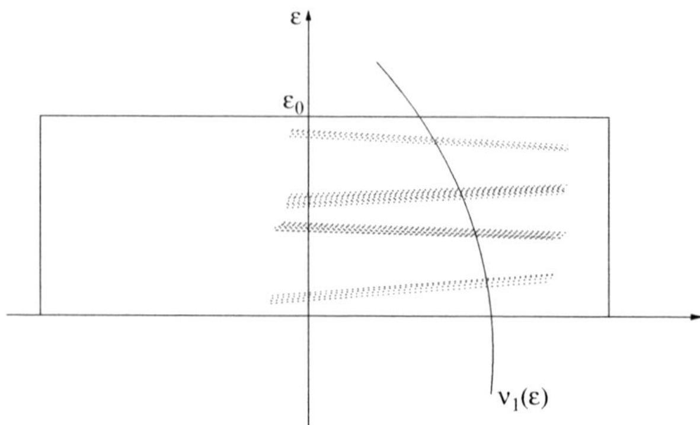

Fig. 4 The Cantor set B_∞ in which the range equation is solved and the solutions $v_1(\varepsilon)$ of the bifurcation equation (1).

$$
\omega = \begin{cases} \sqrt{1 - 2\varepsilon^2} \\ \sqrt{1 - 2\varepsilon \mathrm{sign}\langle a_3 \rangle} \\ \sqrt{1 - 2\varepsilon^2}\,. \end{cases} \tag{7}
$$

Existence of periodic solutions for completely resonant wave equations like (1) of section 2 has been proved also in [25] if $f(u) = u^3 + O(u^5)$ (to have the non-degeneracy condition) and solving the small divisors problem with the Lindsted series method.

Remark 5.1. Under periodic boundary conditions Bourgain [13] proved, when $f = u^3 + O(u^4)$, existence of periodic solutions, bifurcating from exact traveling waves $u = \delta p_0(\omega t + x)$ of $u_{tt} - u_{xx} + u^3 = 0$. More recently Yuan [44] has proved, still for periodic boundary conditions, existence of certain types of quasi-periodic solutions.

We remark that the Arnold non-degeneracy condition is generically satisfied in [20] when the bifurcation equation is two dimensional (case of the Lyapunov center theorem), but it is a difficult task yet for partially resonant PDEs like $u_{tt} - u_{xx} + a_1(x)u = f(x, u)$ where the bifurcation equation is $2m$-dimensional. In this case, considered in [21], the non-degeneracy condition is verified on examples.

6 A variational principle on a Cantor set

To relax the Arnold non-degeneracy condition, it is natural to make use of the "Percival reduced variational principle" for solving the bifurcation equation (1) of section 5. The major difficulty, explained at the beginning of Section 5, is to prove the intersection between the solutions of the bifurcation and the range equations for positive measure sets of frequencies. We present below the results and the ideas in [8] where we refer for complete proofs and details.

Remark 6.1. New ideas in variational perturbation theory of critical points can shed some light on challenging problems like the generalization of the Weinstein–Moser and Fadell–Rabinowitz theorems for quasi-periodic solutions, see e.g. [19], where the main difficulty arises exactly by a variational principle on a Cantor set. This problem is also related to degenerate KAM theory, see e.g. [38, 39] and references therein.

The weak BV-dependence on the frequency. We prove that, if there is a path of solutions $\varepsilon \mapsto v_1(\varepsilon)$ of (Q1) equation (1) of section 5 which depends just in a BV way, i.e.

$$\exists d > 0 \quad \text{such that} \quad \varepsilon_0^d \, \text{Var}_{[0,\varepsilon_0]} v_1(\varepsilon) \leq C < +\infty \tag{1}$$

then the Cantor set $\mathscr{C}$ defined in (4) of section 5 has asymptotically full measure, i.e. satisfies (5) of section 5. It's the explicit expression of the Cantor set B_∞ in (8) of section 4 to suggest a condition like this (in [8] it is stated in a slightly different way). Note that the variation of $\varepsilon \mapsto v_1(\varepsilon)$ could be very big. The idea is roughly the following. We have to bound the measure of the complementary set

$$\mathscr{C}^c = \bigcup_{(l,j)\in\mathscr{R}} S_{l,j} \quad \text{where} \quad S_{l,j} := \left\{ \varepsilon \in [0,\varepsilon_0] \mid \left| \omega(\varepsilon)l - j + \varepsilon \frac{M(\varepsilon)}{2j} \right| < \frac{2\gamma}{(l+j)^\tau} \right\},$$

$\mathscr{R} := \{(l,j) \mid l \neq j, \, l \geq 1/3\varepsilon_0, (1-4\varepsilon_0)l \leq j \leq (1+4\varepsilon_0)l\}$ (otherwise $S_{l,j} = \emptyset$) and $M(\varepsilon) := M(v_1(\varepsilon), \widetilde{w}(\varepsilon, v_1(\varepsilon)))$ satisfies a condition like (1) because $M(\cdot,\cdot)$ and $\widetilde{w}(\cdot,\cdot)$ are smooth. Calling $a_{l,j} := \inf S_{l,j}$, $b_{l,j} := \sup S_{l,j}$, the measure of each $S_{l,j}$ can be bounded like

$$|S_{l,j}| \leq C \left(\frac{\gamma}{l^{\tau+1}} + \varepsilon_0 \frac{|M(a_{l,j}) - M(b_{l,j})|}{jl} \right).$$

If all the $(a_{l,j}, b_{l,j})$ were disjoint, we have to excise all the $S_{l,j}$. The measure can be bounded like

$$\left| \bigcup_{(l,j)\in\mathscr{R}} S_{l,j} \right| \leq C \sum_{(l,j)\in\mathscr{R}} \frac{\gamma}{l^{\tau+1}} + \varepsilon_0 \sum_{\frac{1}{\varepsilon_0} \leq l \leq \frac{1}{\varepsilon_0^b}} \frac{C}{jl} + C\varepsilon_0 \sum_{l \geq \frac{1}{\varepsilon_0^b}} \frac{|M(a_{l,j}) - M(b_{l,j})|}{jl}$$

where also in the second and in the third sum $(l, j) \in \mathscr{R}$. The first and the second term are easily shown to satisfy $o(\varepsilon_0)$. The third term can be bounded as

$$C\varepsilon_0\varepsilon_0^{2b} \sum_{l \geq 1/3\varepsilon_0^b} |M(a_{l,j}) - M(b_{l,j})| \leq C\varepsilon_0^{1+2b}\text{Var}_{[0,\varepsilon_0]} M \leq \varepsilon_0^{1+2b}\varepsilon_0^{-d}C = o(\varepsilon_0)$$

taking $2b > d$. The detailed argument in Section 5.2 of [8] takes into account possible overlapping of the sets $(a_{l,j}, b_{l,j})$ to show that always $|\cup_{(l,j)\in\mathscr{R}} S_{l,j}| \leq o(\varepsilon_0) + \varepsilon_0^{1+2b}\text{Var}_{[0,\varepsilon_0]} M$.

The weak non-degeneracy condition. We are not able to ensure the BV-property (1) for any $f(x,u) = a(x)u^p$. Therefore we introduce parameter-dependent nonlinearities

$$f(\lambda, x, u) = a(x)u^p + \sum_{i=1}^{M} \lambda_i b_i(x)u^{q_i}, \qquad q_i \geq \bar{q} > p \geq 2 \tag{2}$$

where $\bar{q}$ can be arbitrarily large and $\lambda_i \in \mathbf{R}$ are the parameters. We remark that, since $q_i > p$, the nonlinearities $\lambda_i b_i(x)u^{q_i}$ do *not* change the 0th-order bifurcation equation (3) of section 4, which in particular might have only degenerate solutions. Actually, since the exponents q_i can be arbitrarily large, we are adding arbitrarily small corrections $b_i(x)u^{q_i} = o(u^p)$ for $u \to 0$.

The main idea for proving the BV-property (1) for nonlinearities like in (2) is somehow related to the Struwe "monotonicity method" [40] for parameters dependent functionals possessing the mountain pass geometry. We can infer the BV-property for the mountain pass solutions of the bifurcation equation by a BV-information on the derivatives (w.r.t λ) of the mountain pass critical levels, choosing properly the exponents q_i and the coefficients b_i.

Normalizing the period and rescaling the amplitude $u \to \delta u$, we look for solutions of

$$\begin{cases} \omega^2 u_{tt} - u_{xx} = \varepsilon g(\delta, \lambda, u) \\ u(t, 0) = u(t, \pi) = 0 \end{cases} \tag{3}$$

where $\varepsilon := \delta^{p-1}$ and $g(\delta, \lambda, u) := a(x)u^p + \sum_{i=1}^{M} \lambda_i \delta^{q_i - p} b_i(x)u^{q_i}$. Critical points of the action functional

$$\Psi(\delta, \lambda, u) := \int_\Omega \frac{\omega^2}{2} u_t^2 - \frac{u_x^2}{2} + \varepsilon a(x)\frac{u^{p+1}}{p+1} + \sum_{i=1}^{M} \lambda_i \delta^{q_i-1} \int_\Omega b_i(x)\frac{u^{q_i+1}}{q_i+1}$$

are solutions of (3). We perform the same Lyapunov–Schmidt reduction as above. Once the $(Q2)$-equation and the range equation are solved, the latter on a Cantor set B_∞, we need to find solutions $v_1(\delta, \lambda)$ of the $(Q1)$-equation. As above, we need to find a critical point v_1 of

$$\widetilde{\Phi}(\delta, \lambda, v_1) = \Psi(\delta, \lambda, v_1 + v_2(\delta, \lambda, v_1, \widetilde{w}(\delta, \lambda, v_1)) + \widetilde{w}(\delta, \lambda, v_1))$$

such that $(\delta, \lambda, v_1) \in B_\infty$. The reduced functional $\widetilde{\Phi}$ can be written $\widetilde{\Phi}(\delta, \lambda, v_1) = \varepsilon \Phi(\delta, \lambda, v_1)$ and $\Phi(\delta, \lambda, \cdot)$ has, $\forall \delta$ small, $\forall |\lambda| \leq 1$, a not empty Mountain-Pass critical set

$$\mathscr{K}(\delta, \lambda) \subset B(2R_c; V_1) \setminus \{0\}$$

at the mountain pass critical value

$$c(\lambda, \delta) = \Phi(\delta, \lambda, \mathscr{K}(\delta, \lambda)). \tag{4}$$

Furthermore $\mathscr{K}(\delta, \lambda) \to \Pi_{V_1}\mathscr{K}_c$ as $\delta \to 0$.

The key observation is that the mountain pass value $c(\delta, \lambda)$ is a *semiconcave* function, namely $c(\delta, \lambda) - K(\delta^2 + |\lambda|^2)$ is concave for some K large enough, see section 2 in [8]. Therefore $c(\delta,\lambda)$ is differentiable almost everywhere and the derivatives $\partial_{\lambda_i} c(\delta,\lambda)$ are *BV* functions. Furthermore, at the points where $c(\lambda,\delta)$ is differentiable,

$$\partial_{\lambda_i} c(\delta,\lambda) = (\partial_{\lambda_i}\Phi)(\delta,\lambda,\mathscr{K}(\delta,\lambda)), \qquad \forall i = 1,\ldots,M. \tag{5}$$

Formally (5) follows differentiating in (4) since $(\partial_u\Phi)(\delta,\lambda,\mathscr{K}(\delta,\lambda)) = 0$.

For each $|\lambda| \leq 1$ we define a path

$$\mathscr{V}_1(\cdot,\lambda) : [0,\delta_0] \longmapsto \mathscr{K}(\delta,\lambda)$$

of critical points of $\Phi(\delta,\lambda,\cdot)$. By (5) the functions $(\partial_{\lambda_i}\Phi)(\delta,\lambda,\mathscr{V}_1(\delta,\lambda))$ are BV in the variables (δ,λ). Hence, for a.e. $|\lambda| \leq 1$, the functions (of one variable)

$$\delta \longmapsto (\partial_{\lambda_i}\Phi)(\delta,\lambda,\mathscr{V}_1(\delta,\lambda)) \quad \text{are BV}. \tag{6}$$

QUESTION: How to infer the BV-property (1) for $\delta \to \mathscr{V}_1(\delta,\lambda)$ from (6)?

Differentiating the reduced action functional

$$(\partial_{\lambda_i}\Phi)(\delta,\lambda,v_1) = \delta^{q_i - p}\Big[\Phi_i(v_1) + \mathscr{R}_i(\delta,\lambda,v_1)\Big] \tag{7}$$

where $\Phi_i : B(2R_c;V_1) \to \mathbf{R}$ are

$$\Phi_i(v_1) := \frac{1}{q_i + 1} \int_\Omega b_i(x)\Big(v_1 + v_2(v_1)\Big)^{q_i+1}$$

with $v_2(v_1) := v_2(0,0,v_1,0)$, and $|\mathscr{R}_i(\delta,\lambda,v_1)| = O(\delta)$, $|\nabla_{v_1}\mathscr{R}_i(\delta,\lambda,v_1)| = O(\delta)$.

Here the choice of the nonlinearities $b_i(x)u^{q_i}$ enters into play. The required weak non-degeneracy condition is that the $\nabla\Phi_i(v_1)$ generate V_1, i.e.

$$\forall v_1 \in \Pi_{V_1}\mathscr{K}_c, \quad \text{span}\{\nabla\Phi_i(v_1), i = 1,\ldots,M\} \equiv V_1. \tag{8}$$

In this case, in any neighborhood of $\Pi_{V_1}\mathscr{K}_c$ a finite set of $\Phi_i(v_1)$ can be seen as a local chart of coordinates. Hence, by the BV dependence (6), and (7) we infer the (BV) property for the path of critical points $\delta \mapsto \mathscr{V}_1(\delta,\lambda)$. In [8], the weak non-degeneracy condition (8) is verified, proving the following theorem:

Theorem 6.1. **[8]** *For any $\bar{q} > p$ there exist $M \in \mathbf{N}$, integer exponents $\bar{q} \leq q_1 \leq \ldots \leq q_M$ and coefficients $b_1,\ldots,b_M \in H^1(0,\pi)$ depending only on $a(x)$, such that for almost every parameter $\lambda = (\lambda_1,\ldots,\lambda_M)$, $|\lambda| \leq 1$, equation (1) of section 2 with nonlinearity $f(\lambda,x,u)$ like in (2) possesses periodic solutions for an asymptotically full measure Cantor set of frequencies ω close to 1.*

Furthermore, given $a(x)u^p$, $b_i(x)u^{q_i}$, Theorem 6.1 is valid also adding *any* non-linear term $r(x,u) = \sum_{k>p} r_k(x)u^k$, with $\sum_{k>p} \|r_k\|_{H^1}\rho^k < +\infty$ for some $\rho > 0$. The term r has an influence only on the full measure set of parameters λ for which the existence result holds (in the previous argument we use just the derivatives w.r.t. λ). Theorem 6.1 can be interpreted as a genericity result in the sense of Lebesgue measure, see for details [8].

7 Forced vibrations

In this section we look for T-periodic solutions of

$$\begin{cases} u_{tt} - u_{xx} = \varepsilon f(t,x,u) \\ u(t,0) = u(t,\pi) = 0 \end{cases} \tag{1}$$

when the nonlinearity is T-periodic, namely $f(t+T,x,u) = f(t,x,u)$, $\forall t$. We suppose that the forcing frequency $\omega := 2\pi/T$ is a rational number, for simplicity

$$\omega = 1, \qquad i.e. \ T = 2\pi$$

(if $\omega \notin \mathbf{Q}$ a small divisor problem similar to the one discussed in the previous sections appears, see e.g. [2] and references therein).

The spectrum of the D'Alembert operator in a space of 2π-periodic functions is (see (3) of section 2)

$$\sigma(\partial_{tt} - \partial_{xx}) = \left\{ -l^2 + j^2 \mid l \in \mathbf{Z}, \ j \in \mathbf{N} \right\} \subset \mathbf{Z}.$$

Therefore zero is an eigenvalue of infinite multiplicity (when $|l| = j$) but the other eigenvalues are well separated from zero if $|l| \neq j$. For this reason the difficulty is not solving the range equation, but the bifurcation equation which has an intrinsic *lack of compactness*.

The first breakthrough regarding this problem was achieved by Rabinowitz [34] under the strong monotonicity assumption $(\partial_u f)(t,x,u) \geq \beta > 0$ and in [35] for weakly monotone nonlinearities like $f(t,x,u) = u^{2k+1} + G(t,x,u)$ where $G(t,x,u_2) \geq G(t,x,u_1)$ if $u_2 \geq u_1$. For several other results in the monotone case see e.g. [16] and references therein. The monotonicity of f is deeply exploited to compensate the for lack of compactness in the infinite dimensional bifurcation equation.

Little is known without the monotonicity. Willem [43], Hofer [26] and Coron [17] have proved some existence result for nonlinearities like $f(t,x,u) = g(u) + h(t,x)$, where $g(u)$ satisfies suitable linear growth conditions, $\varepsilon = 1$, and under additional symmetries or non-resonance assumptions.

We now present the recent existence results of [5] for non-monotone nonlinearities. This will highlight a completely different use of variational methods.

7.1 The variational Lyapunov–Schmidt reduction

In view of the variational argument that we shall use to solve the bifurcation equation we look for solutions u of (1) in the Banach space $E := H^1(\Omega) \cap C_0^{1/2}(\bar{\Omega})$ where $H^1(\Omega)$ is the usual Sobolev space, $\Omega := \mathbf{T} \times [0, \pi]$, and $C_0^{1/2}(\bar{\Omega})$ is the space of all the $1/2$-Hölder continuous functions satisfying $u(t,0) = u(t,\pi) = 0$ endowed with norm $\|u\|_E := \|u\|_{H^1(\Omega)} + \|u\|_{C^{1/2}(\bar{\Omega})}$.

Critical points of the Lagrangian action functional $\Psi \in C^1(E, \mathbf{R})$

$$\Psi(u) := \int_\Omega \left[\frac{u_t^2}{2} - \frac{u_x^2}{2} + \varepsilon F(t,x,u) \right] dt dx$$

where $(\partial_u F)(t,x,u) = f(t,x,u)$ are weak solutions of (1).

For $\varepsilon = 0$, the critical points of Ψ in E reduce to the space

$$V := N \cap H^1(\Omega) \subset E \tag{2}$$

where

$$N := \left\{ v = \hat{v}(t+x) - \hat{v}(t-x) =: v_+ - v_- \ : \ \hat{v} \in L^2(\mathbf{T}) \text{ and } \int_\mathbf{T} \hat{v} = 0 \right\}$$

is the L^2-closure of the classical solutions of equation (2) of section 2. We have $V \subset E$ because any $\hat{v} \in H^1(\mathbf{T})$ is $1/2$-Hölder continuous. The only difference with respect to the space V introduced in (2) of section 3 is that the functions v in (2) are not necessarily even in time.

Theorem 7.1. [5] *Let $f(t,x,u) = u^{2k} + h(t,x)$ where $h \in N^\perp$ satisfies $h(t,x) > 0$ a.e. in Ω. Then $\forall \varepsilon$ small enough, there exists at least one weak solution $u \in E$ of (1) with $\|u\|_E \leq C|\varepsilon|$.*

Theorem 7.1 is a particular case of a more general result which holds without any growth condition for f, see Theorems 1, 2 in [5]. Moreover, the solution u is proved to be more regular, when h is more regular. This is somewhat surprising: Brezis–Nirenberg [16] and Rabinowitz [36] have proved regularity of solutions if f is strictly monotone in u. For example, yet the solutions in [35] are only continuous functions. The less regular part of the solution is the component in V because of the lack of compactness.

Remark 7.1. The assumption $h \in N^\perp$ is not of technical nature: if $h \notin N^\perp$, periodic solutions of (1) do not exist in any fixed ball $\{\|u\|_{L^\infty} \leq R\}$ for ε small.

To prove Theorem 7.1 we perform a Lyapunov–Schmidt reduction, decomposing

$$E = V \oplus W \qquad \text{where} \qquad W := N^\perp \cap E$$

and $N^\perp := \{h \in L^2(\Omega) \mid \int_\Omega hv = 0, \ \forall v \in N\}$.

Projecting (1), for $u = v + w$ with $v \in V$, $w \in W$, yields

$$\begin{cases} 0 = \Pi_N f(v+w) & \text{bifurcation equation} \\ w_{tt} - w_{xx} = \varepsilon \Pi_{N^\perp} f(v+w) & \text{range equation} \end{cases} \tag{3}$$

where Π_N and $\Pi_{N^\perp}$ are the projectors from $L^2(\Omega)$ onto N and $N^\perp$.

The range equation. The inverse of the D'Alembert operator $\square^{-1} : N^\perp \to W$ defined by

$$\square^{-1} f := \sum_{j \geq 1, j \neq |l|} \frac{f_{lj}}{-l^2 + j^2} e^{ilt} \sin jx, \qquad \forall f \in N^\perp$$

is a bounded operator, i.e. $\|\square^{-1} f\|_E \leq C \|f\|_{L^2}$, see [16]. Fixed points $w \in W$ of

$$w = \varepsilon \square^{-1} \Pi_{N^\perp} f(v+w)$$

are solutions of the range equation. By the contraction mapping theorem we have:

Lemma 7.1. (Solution of the range equation) $\forall R > 0$, $\exists \; \varepsilon_0(R) > 0$, $C_0(R) > 0$, *such that* $\forall |\varepsilon| \leq \varepsilon_0(R)$ *and* $\forall v \in N$ *with* $\|v\|_{L^\infty} \leq 2R$ *there exists a unique solution* $w(\varepsilon, v) \in W$ *of the range equation satisfying*

$$\|w(\varepsilon, v)\|_E \leq C_0(R)|\varepsilon|. \tag{4}$$

Moreover the map $(\varepsilon, v) \mapsto w(\varepsilon, v)$ *is* $C^1(\{\|v\|_{L^\infty} \leq 2R\}, W)$.

7.2 The bifurcation equation

There remains the infinite dimensional bifurcation equation

$$\Pi_N f(v + w(\varepsilon, v)) = 0$$

which is the Euler-Lagrange equation of the *reduced action functional*

$$\Phi_\varepsilon : \{\|v\|_{H^1} < 2R\} \to \mathbf{R}, \qquad \Phi_\varepsilon(v) := \Psi(v + w(\varepsilon, v)).$$

Φ_ε can be written (like in (6) of section 3 with $w = 1$) as

$$\Phi_\varepsilon(v) = \varepsilon \int_\Omega \left[F(v + w(\varepsilon, v)) - \frac{1}{2} f(v + w(\varepsilon, v)) w(\varepsilon, v) \right] dt \, dx.$$

To find critical points of Φ_ε, we make a constrained minimization like in [34] (we don't have compactness to try any critical point theory). By the compact embedding $(V, \|\cdot\|_{H^1}) \hookrightarrow (V, \|\cdot\|_\infty)$ (see Remark 3.1) it easy to deduce that

Lemma 7.2. $\forall R > 0$, Φ_ε *attains a minimum* $\bar{v}$ *in* $\bar{B}_R := \{v \in V, \|v\|_{H^1} \leq R\}$.

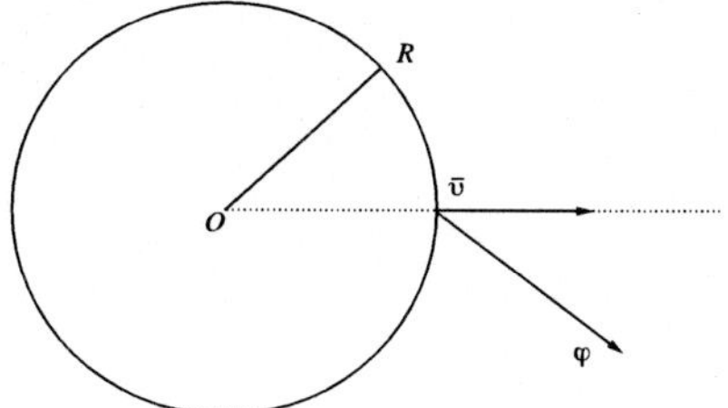

Fig. 5 The variational inequality.

If $\bar{v} \in \partial\bar{B}_R$ then we deduce only the variational inequality

$$D_v\Phi_\varepsilon(\bar{v})[\phi] = \int_\Omega f(\bar{v}+w(\varepsilon,\bar{v}))\phi \leq 0 \tag{5}$$

for any *admissible variation* $\phi \in V$ such that $\langle \bar{v}, \phi \rangle_{H^1} > 0$, see Figure 5.

The heart of the existence proof is to obtain, choosing suitable admissible variations, the *a-priori estimate* $\|\bar{v}\|_{H^1} < R_*$ for some $R_* > 0$ independent of ε.

It is here where the monotonicity of f plays its role to prove Rabinowitz's theorem [34]. On the other hand, the difficulty for dealing with non-monotone nonlinearities is well highlighted by the nonlinearities $f = u^{2k} + h(t,x)$. In this case the variational inequality (5) vanishes identically for $\varepsilon = 0$, because

$$\int_\Omega (\bar{v}^{2k} + h(t,x))\phi \not\equiv 0, \qquad \forall \phi \in V$$

since $h \in N^\perp$ and the following lemma.

Lemma 7.3. [5] *If* $v_1,\ldots,v_{2k+1} \in V$ *then* $\int_\Omega v_1\cdot\ldots\cdot v_{2k+1} = 0$. *In particular, if* $v \in V$ *then* $v^{2k} \in W$.

Therefore, for deriving, if possible, the required a priori estimates, we have to develop the variational inequality (5) at higher orders in ε. It is convenient to perform in (1) with $f = u^{2k} + h$ the change of variables $u \to \varepsilon(H + u)$ where H is a weak solution of

$$\begin{cases} H_{tt} - H_{xx} = h \\ H(t,0) = H(t,\pi) = 0 \\ H(t+2\pi,x) = H(t,x) \end{cases} \tag{6}$$

and, since $h > 0$ in Ω, we can always choose $H(t,x) > 0$, $\forall(t,x) \in \Omega$ (see the "maximum principle" theorem proved in [5]). Therefore (renaming $\varepsilon^{2k} \to \varepsilon$) we look for 2π periodic solutions of

$$\begin{cases} u_{tt} - u_{xx} = \varepsilon f(t,x,u) \\ u(t,0) = u(t,\pi) = 0 \end{cases} \qquad \text{where} \qquad f(t,x,u) := (H+u)^{2k}. \tag{7}$$

Implementing a variational Lyapunov–Schmidt reduction as above, we find the existence of a constrained minimum $\bar{v} \in \bar{B}_R$ and the variational inequality

$$\int_\Omega (H + \bar{v} + w(\varepsilon, \bar{v}))^{2k} \phi \leq 0. \tag{8}$$

The required a priori estimate for the H^1-norm of $\bar{v}$ is proved in several steps inserting into the variational inequality (8) suitable admissible variations. We consider only the more difficult case $k \geq 2$.

The following key "coercivity" estimate is heavily exploited.

Lemma 7.4. (Coercivity estimate) [5] *Let $H \in C(\bar{\Omega})$ with $H > 0$ in Ω. Then*

$$\int_\Omega Hv^{2k} \geq c_k(H) \int_\Omega v^{2k}, \qquad \forall\, v \in N \cap L^{2k}(\Omega) \tag{9}$$

where $c_k(H) := 4^{-k} \min_{\bar{\Omega}_{\alpha_k}} H > 0$, $\alpha_k := \frac{1}{4(1+2k)}$, and $\Omega_\alpha := \mathbf{T} \times (\alpha\pi, \pi - \alpha\pi)$.

The inequality (9) is not trivial because H vanishes at the boundary $\partial\Omega$ (i.e. $H(t,0) = H(t,\pi) = 0$). It holds true because $v = \hat{v}(t+x) - \hat{v}(t-x)$ is the superposition of waves which "spend a lot of time" far from $x = 0$ and $x = \pi$.

In the sequel κ_i will denote positive constants independent of ε.

Step 1: the L^{2k}-estimate. Insert $\phi := \bar{v}$ in (8). ϕ is an admissible variation since $\langle \bar{v}, \phi \rangle_{H^1} = \|\bar{v}\|_{H^1}^2 > 0$. Setting $\bar{w} := w(\varepsilon, \bar{v})$

$$\int_\Omega (\bar{v} + H)^{2k}\bar{v} = \int_\Omega (\bar{v} + H + \bar{w})^{2k}\bar{v} + \int_\Omega [(\bar{v} + H)^{2k} - (\bar{v} + \bar{w} + H)^{2k}]\bar{v}$$

$$\overset{(8)}{\leq} \int_\Omega [(\bar{v} + H)^{2k} - (\bar{v} + \bar{w} + H)^{2k}]\bar{v} \overset{(4)}{\leq} |\varepsilon|C_1(R)\|\bar{v}\|_{L^1} \leq 1 \tag{10}$$

for $|\varepsilon| \leq \varepsilon_1(R)$ where $0 < \varepsilon_1(\cdot) \leq \varepsilon_0(\cdot)$. Since $\int_\Omega \bar{v}^{2k+1} = 0$ by Lemma 7.3,

$$1 \overset{(10)}{\geq} \int_\Omega (\bar{v} + H)^{2k}\bar{v} = \int_\Omega \left[(\bar{v} + H)^{2k} - \bar{v}^{2k}\right]\bar{v} \tag{11}$$

$$= \int_\Omega 2kH\bar{v}^{2k} + \sum_{j=0}^{2k-2} \binom{2k}{j} \bar{v}^{j+1} H^{2k-j}$$

$$\overset{(9)}{\geq} 2kc_k(H)\|\bar{v}\|_{L^{2k}}^{2k} - \kappa_1\|\bar{v}\|_{L^{2k}}^{2k-1} - \kappa_2\|\bar{v}\|_{L^{2k}}$$

using Hölder inequality to estimate $\|\bar{v}\|_{L^i} \leq C_{i,k}\|\bar{v}\|_{L^{2k}}$ ($i \leq 2k - 1$). By (11)

$$\|\bar{v}\|_{L^{2k}} \leq \kappa_3 \qquad \text{for} \quad |\varepsilon| \leq \varepsilon_1(R). \tag{12}$$

Step 2: the L^∞-estimate. We now choose $\phi = q(\hat{v}(t+x)) - q(\hat{v}(t-x)) \in V$ where

$$q(\lambda) := \begin{cases} 0 & \text{if } |\lambda| \leq M \\ \lambda - M & \text{if } \lambda \geq M \\ \lambda + M & \text{if } \lambda \leq M \end{cases}$$

and

$$M := \frac{1}{2}\|\hat{\bar{v}}\|_{L^\infty(\mathbf{T})}. \tag{13}$$

We can assume $M > 0$, i.e. $\bar{v}$ is not identically zero.

Lemma 7.5. ([34]) $\phi \in V$ is an admissible variation and $\bar{v}(t,x)\phi(t,x) \geq 0$, $\forall (t,x) \in \Omega$.

Using (8), setting $\bar{w} := w(\varepsilon, \bar{v})$,

$$\int_\Omega (\bar{v}+H)^{2k}\phi = \int_\Omega (\bar{v}+H+\bar{w})^{2k}\phi + \int_\Omega [(\bar{v}+H)^{2k} - (\bar{v}+\bar{w}+H)^{2k}]\phi$$

$$\overset{(8)}{\leq} \int_\Omega [(\bar{v}+H)^{2k} - (\bar{v}+\bar{w}+H)^{2k}]\phi$$

$$\overset{(4)}{\leq} |\varepsilon|C_2(R)\|\phi\|_{L^1} \leq \|\phi\|_{L^1} \tag{14}$$

for $|\varepsilon| \leq \varepsilon_2(R)$ where $0 < \varepsilon_2(\cdot) \leq \varepsilon_1(\cdot)$. Now, using that $\int_\Omega \bar{v}^{2k}\phi = 0$ by Lemma 7.3,

$$\|\phi\|_{L^1} \overset{(14)}{\geq} \int_\Omega (\bar{v}+H)^{2k}\phi = \int_\Omega \left[(\bar{v}+H)^{2k} - \bar{v}^{2k}\right]\phi$$

$$\geq \int_\Omega 2kH\bar{v}^{2k-1}\phi - \kappa_4(\|\bar{v}\|_{L^\infty}^{2k-2} + 1)\|\phi\|_{L^1}$$

which implies, since $\int_\Omega H\bar{v}^{2k-1}\phi \geq \kappa_5 \int_{\Omega_{1/4}} \phi\bar{v}^{2k-1}$ (because $\bar{v}\phi \geq 0$), that

$$\int_{\Omega_{1/4}} \bar{v}^{2k-1}\phi \leq \kappa_6(\|\bar{v}\|_{L^\infty(\Omega)}^{2k-2} + 1)\|\phi\|_{L^1(\Omega)} \leq \kappa_7(\|\bar{v}\|_{L^\infty(\Omega)}^{2k-2} + 1)\|q(\hat{\bar{v}})\|_{L^1(\mathbf{T})}$$

since $\|\phi\|_{L^1(\Omega)} \leq \|q(\hat{\bar{v}})\|_{L^1(\mathbf{T})}$. We have to give a lower bound of

$$\int_{\Omega_{1/4}} \bar{v}^{2k-1}\phi = \int_{\Omega_{1/4}} (\bar{v}\phi)\bar{v}^{2k-2} = \int_{\Omega_{1/4}} (\bar{v}\phi)(\bar{v}_+ - \bar{v}_-)^{2k-2}.$$

By the elementary inequality $(a-b)^{2k} \geq a^{2k} + b^{2k} - 2k(a^{2k-1}b + ab^{2k-1})$, $\forall a, b \in \mathbf{R}$,

$$\int_{\Omega_{1/4}} \bar{v}^{2k-1}\phi \geq \int_{\Omega_{1/4}} \bar{v}\phi\left[\bar{v}_+^{2k-2} + \bar{v}_-^{2k-2} - (2k-2)(\bar{v}_+^{2k-3}\bar{v}_- + \bar{v}_+\bar{v}_-^{2k-3})\right]$$

$$= 2\int_{\Omega_{1/4}} \bar{v}_+^{2k-1}q_+ - \bar{v}_+^{2k-1}q_- + \bar{v}_+^{2k-2}\bar{v}_-q_- - \bar{v}_+^{2k-2}\bar{v}_-q_+$$

$$+ (2k-2)[-\bar{v}_+^{2k-2}\bar{v}_-q_+ + \bar{v}_+^{2k-2}\bar{v}_-q_- - \bar{v}_+^{2k-3}\bar{v}_-^2q_- + \bar{v}_+^{2k-3}\bar{v}_-^2q_+]$$

$$\geq 2\int_{\Omega_{1/4}} \bar{v}_+^{2k-1}q_+ \tag{15}$$

$$- 2\int_{\Omega_{1/4}} \bar{v}_+^{2k-1}q_- + (2k-1)\bar{v}_+^{2k-2}\bar{v}_-q_+ + (2k-2)\bar{v}_+^{2k-3}\bar{v}_-^2q_- \tag{16}$$

where in the last inequality we used that $\bar{v}_+ q_+$, $\bar{v}_- q_- \geq 0$ (since $\lambda q(\lambda) \geq 0$) and so $\bar{v}_+^{2k-2} \bar{v}_- q_-$, $\bar{v}_+^{2k-3} \bar{v}_-^2 q_+ \geq 0$. The dominant term (15) is estimated using the identity $\int_{\Omega_\alpha} p(t+x)\,dt\,dx = \int_{\Omega_\alpha} p(t-x)\,dt\,dx = \pi(1-2\alpha)\int_{\mathbf{T}} p(s)ds$, $\forall p \in L^1(\mathbf{T})$,

$$2\int_{\Omega_{1/4}} \bar{v}_+^{2k-1} q_+ = 2\pi\left(1 - \frac{2}{4}\right)\int_{\mathbf{T}} \hat{\bar{v}}^{2k-1}(s)q(\hat{\bar{v}}(s))ds \geq \pi M^{2k-1}\|q(\hat{\bar{v}})\|_{L^1(\mathbf{T})} \qquad (17)$$

because $\lambda^{2k-1}q(\lambda) \geq M^{2k-1}|q(\lambda)|$. The terms in (16) are estimated by

$$\left|2\int_{\Omega_{1/4}} \bar{v}_+^{2k-1} q_-\right| \leq 2\int_{\Omega}\left|\bar{v}_+^{2k-1}\right||q_-| \leq \|\hat{\bar{v}}\|_{L^{2k-1}(\mathbf{T})}^{2k-1}\|q(\hat{\bar{v}})\|_{L^1(\mathbf{T})}$$

$$\left|2\int_{\Omega_{1/4}}\left(\bar{v}_+^{2k-2} q_+\right)\bar{v}_-\right| \leq \|\hat{\bar{v}}^{2k-2}q(\hat{\bar{v}})\|_{L^1(\mathbf{T})}\|\hat{\bar{v}}\|_{L^1(\mathbf{T})} \leq (2M)^{2k-2}\|q(\hat{\bar{v}})\|_{L^1(\mathbf{T})}\|\hat{\bar{v}}\|_{L^1(\mathbf{T})}$$

$$\left|2\int_{\Omega_{1/4}} \bar{v}_+^{2k-3}\left(\bar{v}_-^2 q_-\right)\right| \leq \|\hat{\bar{v}}^{2k-3}\|_{L^1(\mathbf{T})}\|\hat{\bar{v}}^2 q(\hat{\bar{v}})\|_{L^1(\mathbf{T})} \leq (2M)^2\|\hat{\bar{v}}\|_{L^{2k-3}(\mathbf{T})}^{2k-3}\|q(\hat{\bar{v}})\|_{L^1(\mathbf{T})}$$

By the previous inequalities, (17), Hölder inequality and (12)

$$\int_{\Omega_{1/4}} \bar{v}^{2k-1}\phi \geq \pi M^{2k-1}\|q(\hat{\bar{v}})\|_{L^1(\mathbf{T})} - \kappa_7(M^{2k-2} + 1)\|q(\hat{\bar{v}})\|_{L^1(\mathbf{T})}. \qquad (18)$$

By (15) and (18)

$$M^{2k-1}\|q(\hat{\bar{v}})\|_{L^1(\mathbf{T})} \leq \kappa_8\left(\|\bar{v}\|_{L^\infty(\Omega)}^{2k-2} + M^{2k-2} + 1\right)\|q(\hat{\bar{v}})\|_{L^1(\mathbf{T})}$$

$$\leq \kappa_9\left(M^{2k-2} + 1\right)\|q(\hat{\bar{v}})\|_{L^1(\mathbf{T})}$$

and finally $M^{2k-1} \leq \kappa_9(M^{2k-2} + 1)$. Since $M := \frac{1}{2}\|\hat{\bar{v}}\|_{L^\infty(\mathbf{T})}$ (see (13))

$$\|\bar{v}\|_{L^\infty} \leq \kappa_{10} \qquad \text{for} \quad |\varepsilon| \leq \varepsilon_2(R).$$

Step 3: The H^1-estimate. The H^1-estimate is carried out inserting in (8) the variation $\phi := -D_{-h}D_h\bar{v}$ where

$$(D_h f)(t,x) := \frac{f(t+h,x) - f(t,x)}{h}$$

is the difference quotient with respect to t. Note that ϕ is *admissible* because $\langle -D_{-h}D_h\bar{v}, \bar{v}\rangle_{H^1} = \langle D_h\bar{v}, D_h\bar{v}\rangle_{H^1} > 0$. With arguments similar to the previous ones, we can deduce that, for some $0 < \varepsilon_3(R) \leq \varepsilon_2(R)$,

$$\|\bar{v}\|_{H^1} < \kappa_{11}, \qquad \forall |\varepsilon| \leq \varepsilon_3(R).$$

PROOF OF THEOREM 7.1 COMPLETED. Take $R_* := \kappa_{11}$ and $\varepsilon_* := \varepsilon_3(R_*)$. Therefore, $\forall |\varepsilon| \leq \varepsilon_*$, $\|\bar{v}(\varepsilon)\|_{H^1} < R_*$ is an interior minimum of Φ_ε in B_{R_*}.

Acknowledgements I wish to thank P. Baldi, L. Biasco, P. Bolle and M. Procesi for sharing much work together, as well as D. Bambusi, W. Craig and P. Rabinowitz for many useful discussions.

References

1. Ambrosetti A., Rabinowitz P., *Dual Variational Methods in Critical Point Theory and Applications*, Journ. Func. Anal, 14, 349–381, 1973.
2. Baldi P., Berti M., *Forced vibrations of a nonhomogeneous string*, to appear in SIAM Journal on Mathematical Analysis. preprint 2006.
3. Bambusi D., Paleari S., *Families of periodic solutions of resonant PDEs*, J. Non. Sci., 11, 69–87, 2001.
4. Berti M., Topics on *Nonlinear oscillations of Hamiltonian PDEs*, Progress in Nonlinear Differential Equations and Its Applications, Vol. 74, Birkhauser, Boston, 2008, Brezis. Ed.
5. Berti M., Biasco L., *Forced vibrations of wave equations with non-monotone nonlinearities*, Annales de l'Institute H. Poincare', Analyse nonlinaire, 23, 4, 437–474, 2006.
6. Berti M., Bolle P., *Periodic solutions of nonlinear wave equations with general nonlinearities*, Comm. Math. Phys. Vol. 243, 2, 315–328, 2003.
7. Berti M., Bolle P., *Cantor families of periodic solutions for completely resonant nonlinear wave equations*, Duke Mathematical Journal, 134, 2, 359–419, 2006.
8. Berti M., Bolle P., *Cantor families of periodic solutions for wave equations via a variational principle*, to appear in Advances in Mathematics. preprint 2006.
9. Berti M., Bolle P., *Cantor families of periodic solutions of wave equations with C^k nonlinearities*, to appear in Nonlinear Differential Equations and Applications. preprint 2007.
10. Berti M., Procesi M., *Quasi-periodic solutions of completely resonant forced wave equations*, Communications in Partial Differential Equations, 31, 959–985, 2006.
11. Bourgain J., *Construction of quasi-periodic solutions for Hamiltonian perturbations of linear equations and applications to nonlinear PDE*, Internat. Math. Res. Notices, no. 11, 1994.
12. Bourgain J., *Quasi-periodic solutions of Hamiltonian perturbations of 2D linear Schrödinger equations*, Ann. of Math., 148, 363–439, 1998.
13. Bourgain J., *Periodic solutions of nonlinear wave equations*, Harmonic analysis and partial differential equations (Chicago, IL, 1996), 69–97, Chicago Lectures in Math., Univ. Chicago Press, Chicago, IL, 1999.
14. Bourgain J., *Green's function estimates for lattice Schrödinger operators and applications*, Annals of Mathematics Studies, 158. Princeton University Press, Princeton, NJ, 2005.
15. Brezis, H., Coron, J.-M., Nirenberg, L., *Free vibrations for a nonlinear wave equation and a Theorem of P. Rabinowitz*, Comm. Pure Appl. Math. 33, no. 5, 667–684, 1980.
16. Brezis H., Nirenberg L., *Forced vibrations for a nonlinear wave equation*, Comm. Pure Appl. Math. 31, no. 1, 1–30, 1978.
17. Coron J.-M., *Periodic solutions of a nonlinear wave equation without assumption of monotonicity*, Math. Ann. 262, no. 2, 273–285, 1983.
18. Craig W., *Problèmes de petits diviseurs dans les équations aux dérivées partielles*, Panoramas et Synthèses, 9, Société Mathématique de France, Paris, 2000.
19. Craig W., *Invariant tori for Hamiltonian PDE*, Nonlinear dynamics and evolution equations, 53-66, Fields Inst., 48, 2006, AMS-Providence.
20. Craig W., Wayne E., *Newton's method and periodic solutions of nonlinear wave equation*, Comm. Pure and Appl. Math, vol. XLVI, 1409–1498, 1993.
21. Craig W., Wayne E., *Nonlinear waves and the* 1 : 1 : 2 *resonance*, Singular limits of dispersive waves, 297–313, NATO Adv. Sci. Inst. Ser. B Phys., 320, Plenum, New York, 1994.
22. De La Llave R., *Variational Methods for quasi-periodic solutions of partial differential equations*, World Sci. Monogr. Ser. Math. 6, 214–228, 2000.
23. Eliasson H., Kuksin S., *KAM for nonlinear Schrödinger equation*, to appear in Annals of Mathematics. preprint 2006.

24. Fadell E., Rabinowitz P., *Generalized cohomological index theories for the group actions with an application to bifurcation questions for Hamiltonian systems*, Inv. Math. 45, 139–174, 1978.

25. Gentile G., Mastropietro V., Procesi M., *Periodic solutions for completely resonant nonlinear wave equations*, Comm. Math. Phys. 256, no. 2, 437–490, 2005.

26. Hofer H., *On the range of a wave operator with non-monotone nonlinearity*, Math. Nachr. 106, 327–340, 1982.

27. Lidskij B. V., Shulman E. I., *Periodic solutions of the equation $u_{tt} - u_{xx} + u^3 = 0$*, Funct. Anal. Appl. 22, 332–333, 1988.

28. Kuksin S., *Hamiltonian perturbations of infinite-dimensional linear systems with imaginary spectrum*, Funktsional. Anal. i Prilozhen. 21, no. 3, 22–37, 95, 1987.

29. Kuksin S., *Analysis of Hamiltonian PDEs*, Oxford Lecture Series in Mathematics and its Applications, 19. Oxford University Press, New York, NY 2000.

30. Kuksin S., Pöschel J., *Invariant Cantor manifolds of quasi-periodic oscillations for a nonlinear Schrödinger equation*, Ann. of Math, 2, 143, no. 1, 149–179, 1996.

31. Moser J., *Periodic orbits near an Equilibrium and a Theorem by Alan Weinstein*, Comm. Pure Appl. Math., vol. XXIX, 1976.

32. Moser J., Zehnder E., *Notes on Dynamical Systems*, Courant Lecture Notes 2005.

33. Pöschel J., *Quasi-periodic solutions for a nonlinear wave equation*, Comment. Math. Helv., 71, no. 2, 269–296, 1996.

34. Rabinowitz P., *Periodic solutions of nonlinear hyperbolic partial differential equations*, Comm. Pure Appl. Math., 20, 145–205, 1967.

35. Rabinowitz P., *Time periodic solutions of nonlinear wave equations*, Manusc. Math. 5, 165–194, 1971.

36. Rabinowitz P., *Free vibrations for a semilinear wave equation*, Comm. Pure Appl. Math. 31, no. 1, 31–68, 1978.

37. Rabinowitz P., *Minimax methods in critical point theory with applications to differential equations*, CBMS Regional Conference Series in Mathematics, 65. Published for the Conference Board of the Mathematical Sciences, Washington, DC; by the American Mathematical Society, Providence, RI, 1986.

38. Sevryuk M.B., *The lack-of-parameters problem in the KAM theory revisited*, in Hamiltonian systems with three or more degrees of freedom, ed. C. Simó, NATO ASI Series C, Math Phys. Sci, vol. 533, Kluwer, Dordrecht, The Netherlands, 1999, 568–572.

39. Sevryuk M.B., *The classical KAM theory at the dawn of the twenty-first century*, Mosc. Math. J. 4 (3), 1113–1144.

40. M. Struwe, *The existence of surfaces of constant mean curvature with free boundaries*, Acta Math., 1988, 19–64.

41. Wayne E., *Periodic and quasi-periodic solutions of nonlinear wave equations via KAM theory*, Commun. Math. Phys. 127, no. 3, 479–528, 1990.

42. Weinstein A., *Normal modes for Nonlinear Hamiltonian Systems*, Inv. Math, 20, 47–57, 1973.

43. Willem M., *Density of the range of potential operators*, Proc. Amer. Math. Soc. 83, 2, 341–344, 1981.

44. Yuan X., *Quasi-periodic solutions of completely resonant nonlinear wave equations*, J. Diff. Equations, 230, no. 1, 213–274, 2006.

45. Yuan X., *A KAM theorem with applications to partial differential equations of higher dimensions*, Comm. Math. Phys. 275 (2007), no. 1, 97–137.

Spectral gaps of potentials in weighted Sobolev spaces

Jürgen Pöschel[1]

Abstract We consider the Schrödinger operator $L = -\mathrm{d}^2/\mathrm{d}x^2 + q$ on the interval $[0, 1]$ depending on an L^2-potential q and endowed with periodic or anti-periodic boundary conditions. We prove results about correspondencies between the asymptotic behaviour of the spectral gaps of L and the regularity of q in the Gevrey case, among others. The proofs are based on a Fourier block decomposition due to Kappeler & Mityagin, and a novel application of the implicit function theorem.

1 Results

We consider the Schrödinger operator

$$L = -\frac{\mathrm{d}^2}{\mathrm{d}x^2} + q$$

on the interval $[0, 1]$ depending on an L^2-potential q and endowed with periodic or anti-periodic boundary conditions. In this case, L is also known as *Hill's operator*. Its spectrum is pure point, and for real q consists of an unbounded sequence of real *periodic eigenvalues*

$$\lambda_0^+(q) < \lambda_1^-(q) \leqslant \lambda_1^+(q) < \cdots < \lambda_n^-(q) \leqslant \lambda_n^+(q) < \cdots \ .$$

Their asymptotic behaviour is

$$\lambda_n^{\pm} = n^2 \pi^2 + [q] + \ell^2(n),$$

[1] Institut für Analysis, Dynamik und Optimierung, Universität Stuttgart, Pfaffenwaldring 57
D-70569 Stuttgart
e-mail: poschel@mathematik.uni-stuttgart.de

W. Craig (ed.), Hamiltonian Dynamical Systems and Applications, 421–430.

where $[q]$ denotes the mean value of q and $\ell^2(n)$ a generic square sumable term. Equality may occur in every place with a '$\leqslant$'-sign, and one speaks of the *gap lengths*

$$\gamma_n(q) = \lambda_n^+(q) - \lambda_n^-(q), \qquad n \geqslant 1,$$

of the potential q.

For complex q, the periodic eigenvalues are still well defined, but in general not real, since L is no longer self-adjoint. Their asymptotic behaviour is the same, however, and we may order them lexicographically – first by their real part, then by their imaginary part – so that

$$\lambda_0(q) \prec \lambda_1^-(q) \preccurlyeq \lambda_1^+(q) \prec \cdots \prec \lambda_n^-(q) \preccurlyeq \lambda_n^+(q) \prec \cdots .$$

The gap lengths are then defined as before, but are now complex valued in general.

Classical Results

We are interested in the relationship between the regularity of a potential and the sequence of its gap lengths. Hochstadt [5] observed that

$$q \in C^\infty(S^1, \mathbb{R}) \quad \Leftrightarrow \quad \gamma_n(q) = O(n^{-k}) \ \text{ for all } k \geqslant 0,$$

and Marčenko & Ostrowskĭ [27] subsequently showed that

$$q \in H^m(S^1, \mathbb{R}) \quad \Leftrightarrow \quad \sum_{n \geqslant 1} n^{2m} \gamma_n^2(q) < \infty$$

for all nonnegative integers m. Trubowitz [32] then proved that

$$q \in C^\omega(S^1, \mathbb{R}) \quad \Leftrightarrow \quad \gamma_n(q) = O(\varepsilon^{-an}) \ \text{ for some } a > 0.$$

Later, due to the realization of the periodic KdV flow as an isospectral deformation of Hill's operator, other regularity classes such as Gevrey functions and non-real potentials came into focus. Recent results in this direction appear example in [2, 11, 12, 16, 17]. Within certain limits, one may think of the gap lengths as another kind of Fourier coefficients of the potential.

It is the purpose of this note to describe some of these recent developments. For more detailed statements and proofs we refer to [10] and later [3].

Weighted Sobolev Spaces

As in [16, 17] a *weight* is a *normalized, symmetric* and *submultiplicative* function $w \colon \mathscr{Z} \to \mathbb{R}$. That is, for all integers n and m, we have

$$w_n \geqslant 1, \qquad w_{-n} = w_n, \qquad w_{n+m} \leqslant w_n w_m.$$

Within the standard Sobolev space

$$\mathcal{H}^0 = L^2(S^1, \mathbb{C})$$

of square-integrable functions $u = \sum_{n \in \mathscr{Z}} u_n \varepsilon^{2\pi i n x}$ we then define the *weighted Sobolev spaces*

$$\mathcal{H}^w = \{u \in \mathcal{H}^0 : \|u\|_w^2 := \sum_{n \in \mathscr{Z}} w_n^2 |u_n|^2 < \infty\}.$$

To give some examples, let $\langle n \rangle = 1 + |n|$. The *Sobolev weights* $\langle n \rangle^r$, $r \geqslant 0$, give rise to the usual Sobolev spaces $\mathcal{H}^r$ of 1-periodic, complex-valued functions. The *Abel weights*[1] $\langle n \rangle^r \varepsilon^{a|n|}$ with $a > 0$ define spaces $\mathcal{H}^{r,a}$ of functions in $\mathcal{H}^r$, which are analytic on the complex strip $|\mathrm{Im}\, z| < a/2\pi$ and have traces in $\mathcal{H}^r$ on the boundary lines. The *Gevrey weights*

$$w_n = \langle n \rangle^r \varepsilon^{a|n|^\sigma}, \qquad r \geqslant 0,\, a > 0,\, 0 < \sigma < 1,$$

lie in between and give rise to the so called Gevrey spaces $\mathcal{H}^{r,a,\sigma}$ of smooth functions. Obviously,

$$\mathcal{H}^{r,a} = \mathcal{H}^{r,a,1} \subsetneq \mathcal{H}^{r,a,\sigma} \subsetneq \mathcal{H}^{r,a,0} = \mathcal{H}^r.$$

Yet another weight is for example

$$w_n = \langle n \rangle^r \exp\left(\frac{a|n|}{1 + \log^\alpha \langle n \rangle}\right), \qquad \alpha > 0.$$

Since $\log w_n$ is subadditive and nonnegative, the limit

$$\chi(w) := \lim_{n \to \infty} \frac{\log w_n}{n}$$

exists and is nonnegative [29, no. 98]. Naturally, we call a weight w *exponential*, if $\chi(w) > 0$. We call w *subexponential*, if $\chi(w) = 0$ *with* $n^{-1} \log w_n$ converging to zero in an eventually monotone manner. This is not an exact dichotomy, but we are not aware of any interesting weight that does not belong to either class. Clearly, Abel weights are exponential, while Sobolev and Gevrey weights are subexponential.

The Theorems

Let

$$h^w = \{u = (u_n)_{n \geqslant 1} \in \ell^2 : \sum_{n \geqslant 1} w_n^2 |u_n|^2 < \infty\},$$

and $\gamma(q) = (\gamma_n(q))_{n \geqslant 1}$.

---

[1] The term *Abel* weights is chosen to go along with *Sobolev* and *Gevrey* weights and has no deeper meaning.

Theorem 1 *If $q \in \mathcal{H}^w$, then $\gamma(q) \in h^w$. In particular,*

$$\sum_{n \geqslant N} w_n^2 |\gamma_n(q)|^2 \leqslant 9\|T_N q\|_w^2 + \frac{288}{N}\|q\|_w^4$$

for all $N \geqslant 4\|q\|_w$, where $T_N q = \sum_{|n| \geqslant N} q_n \varepsilon^{2n\pi i x}$.

There is no one-to-one converse to Theorem 1 for exponential weights, as there exist real analytic finite gap potentials such as the Weierstrass $\wp$-function, which are not entire functions. In this case, fixing any $r > 0$, we have $(\gamma_n(u)) \in h^{r,a}$ for *all* $a > 0$, but *only* $u \in \mathcal{H}^{r,a}$ for $a < a_0$. There is, however, a true converse for subexponential weights. We first consider the real case.

Theorem 2 *Suppose $q \in \mathcal{H}^0$ is real and $\gamma(q) \in h^w$. If w is subexponential, then $q \in \mathcal{H}^w$. If w is exponential, then q is real analytic.*

Corollary 3 *If q is real and w is subexponential, then*

$$q \in \mathcal{H}^w \quad \Leftrightarrow \quad \gamma(q) \in h^w.$$

For complex potentials, the spectral gap lengths alone do not suffice to determine the regularity of a potential. For instance, Gasymov [13] showed that *all* gap lengths vanish for complex potentials of the form

$$q = \sum_{n \geqslant 1} q_n \varepsilon^{2n\pi i x} = \sum_{n \geqslant 1} q_n z^n \Big|_{z = \varepsilon^{2\pi i x}}.$$

But Sansuc & Tkachenko [11] noted that the situation can be remedied by taking into account additional spectral data. In particular, they considered the quantities

$$\delta_n = \mu_n - \tau_n = \mu_n - (\lambda_n^+ - \lambda_n^-)/2,$$

where μ_n denotes the n-th Dirichlet eigenvalue. The quantities

$$\Gamma_n = |\gamma_n| + |\delta_n|,$$

may be considered as a measure of the size of the *spectral triangle* formed by the points λ_n^-, δ_n and λ_n^+. Note that $\gamma_n \leqslant \Gamma_n \leqslant 2\gamma_n$ for real potentials.

We then have the following converse theorem.

Theorem 4 *Suppose $q \in \mathcal{H}^0$ is real or complex and $\Gamma(q) \in h^w$. If w is subexponential, then $q \in \mathcal{H}^w$. If w is exponential, then q is real analytic.*

Corollary 5 *If w is subexponential, then*

$$q \in \mathcal{H}^w \quad \Leftrightarrow \quad \Gamma(q) \in h^w.$$

For the sake of brevity and simplicity, we will describe the line of reasoning for the real case. For more details, complete proofs and the complex case we refer to [10] and also [3].

2 Reduction

The idea of the proof of Theorem 1 is due to Kappeler & Mityagin [17]. They employ a Lyapunov–Schmidt reduction scheme called *Fourier block decomposition*.

The aim is to determine those λ near $n^2\pi^2$ with n sufficiently large, for which the equation $-y'' + qy = \lambda y$ admits a nontrivial 2-periodic solution f. As q can be considered small for large n, one expects its dominant modes to be $\varepsilon^{\pm n\pi ix}$. So it makes sense to separate these modes from the other ones by a Lyapunov–Schmidt reduction.

To this end we consider a similarly defined space $\mathcal{H}^w_\star$ of *2-periodic* functions, and write

$$\mathcal{H}^w_\star = \mathcal{P}_n \oplus \mathcal{Q}_n$$
$$= \operatorname{span}\{e_n, e_{-n}\} \oplus \operatorname{span}\{e_k : |k| \neq n\},$$

where $e_k = \varepsilon^{k\pi ix}$. The projections onto $\mathcal{P}_n$ and $\mathcal{Q}_n$ are denoted by P_n and Q_n, respectively. With

$$f = u + v = P_n f + Q_n f,$$

Hill's equation decomposes into the so called P- and Q-equations

$$A_\lambda u = P_n V(u+v),$$
$$A_\lambda v = Q_n V(u+v),$$

where $A_\lambda f = f'' + \lambda f$ and $Vf = qf$.

The operator A_λ has a compact inverse on $\mathcal{Q}_n$, when λ is near $n^2\pi^2$. Indeed, this holds on the complex strips $U_n = \{\lambda : |\operatorname{Re}\lambda - n^2\pi^2| \leqslant 12n\}$ for $n \geqslant 1$.

Lemma 1 *For $q \in \mathcal{H}^w$ and $\lambda \in U_n$, the operator $T_n = VA_\lambda^{-1}Q_n$ exists and is bounded on $\mathcal{H}^w_\star$ with norm*
$$\|T_n\|_w \leqslant \frac{2}{n}\|q\|_w.$$

Left-multiplying the Q-equation with VA_λ^{-1} we obtain $Vv = T_n Vu + T_n Vv$. For n large enough, T_n is a contraction on $\mathcal{H}^w_\star$, and there is a unique solution

$$Vv = \hat{T}_n T_n Vu, \qquad \hat{T}_n = (I - T_n)^{-1}.$$

Inserted into the P-equation we get $A_\lambda u = P_n Vu + P_n \hat{T}_n T_n Vu = P_n \hat{T}_n Vu$. So the P- and Q-equation reduce to the S-equation

$$S_n u = 0, \qquad S_n = A_\lambda - P_n \hat{T}_n V.$$

Since $\mathcal{P}_n$ is two-dimensional with basis e_n, e_{-n}, we have the matrix representation

$$S_n = \begin{pmatrix} \lambda - n^2\pi^2 - a_n & -c_n \\ -c_{-n} & \lambda - n^2\pi^2 - a_n \end{pmatrix},$$

with

$$a_n = \langle \hat{T}_n V e_n, e_n \rangle, \qquad c_n = \langle \hat{T}_n V e_{-n}, e_n \rangle.$$

Any nontrivial solution u gives rise to a 2-periodic solution of $A_\lambda f = V f$, and vice versa. Hence the following holds.

Lemma 2 *A complex number λ near $n^2 \pi^2$ is a periodic eigenvalue of q if and only if the determinant of S_n vanishes.*

Moreover, from the representations of a_n and c_n one easily obtains the following facts, which we need later.

Lemma 3 *For $n \geqslant 4\|q\|_w$ and $\lambda \in U_n$,*

$$|a_n - q_0|_{U_n}, \, w_n |c_n - q_n|_{U_n}, \, w_n |c_{-n} - q_{-n}|_{U_n} \leqslant \frac{4}{n} \|q\|_w^2.$$

Moreover, these coefficients are analytic functions of λ and q.

3 Gap Estimates

With the preceding results the forward problem of estimating the gap lengths of a potential is fairly straightforward. The determinant of S_n is the quadratic polynomial

$$\det S_n = (\lambda - n^2 \pi^2 - a_n)^2 - |c_n|^2,$$

and the distance of its two roots has to be of the order of $|c_n|$.

Lemma 4 *For $n \geqslant 4\|q\|_w$ the determinant of S_n has exactly two roots $\xi_n^\pm$ in U_n, which are contained in the disc $|\lambda - n^2 \pi^2| \leqslant 6\|q\|_w$ and satisfy*

$$\left| \xi_n^+ - \xi_n^- \right|^2 \leqslant 9 |c_n c_{-n}|_{U_n}.$$

A counting argument then shows that these two roots have to be the two eigenvalues $\lambda_n^\pm$. Consequently, we obtain

$$|\gamma_n|^2 = \left| \xi_n^+ - \xi_n^- \right|^2 \leqslant 9 |c_n c_{-n}|_{U_n} \leqslant 9 |q_n|^2 + 9 |q_{-n}|^2 + \frac{144}{n^2 w_n^2} \|q\|_w^4$$

by Lemma 3. Multiplying by w_n^2 and summing over $n \geqslant N$ we obtain Theorem 1.

4 Coefficient Estimates

We now turn to the more subtle problem of estimating the asymptotic behaviour of the Fourier coefficients of a potential in terms of its gap lengths. The geometric aspect is rather straightforward, at least in the real case. The off diagonal elements of S_n have to be bounded in terms of the gap lengths, that is

$$|c_n| \lessdot |\gamma_n|, \qquad n \gg 1,$$

where the dot stands for some implicit constant. The identity $c_n = \langle \hat{T}_n Ve_{-n}, e_n \rangle$ then leads to an infinite dimensional system of nonlinear equations

$$c_n = q_n + O_2(\ldots, q_k, \ldots), \qquad n \neq 0,$$

which allows us to bound the q_n in terms of the c_n and hence in terms of the γ_n. See [12] for the lengthy details of this argument.

Here we present a functional analytic approach based on the fact that the coefficients of S_n – which contain all the necessary data – are analytic functions of q. Indeed, it even suffices to consider S_n at that value of λ where its diagonal vanishes.

For $m \geqslant 1$ and any weight w we introduce the centered balls

$$B_m^w = \{ q \in \mathcal{H}^w : \|q\|_w \leqslant m/4 \} \subset \mathcal{H}^0.$$

We also assume from now on that q has zero mean, that is, $\int_0^1 q \, dx = q_0 = 0$, since adding a constant to q shifts its entire spectrum, but does not affect its gap lengths.

Using a contraction argument it is easy to show that the diagonal of S_n vanishes at a unique point α_n near $n^2\pi^2$.

Lemma 5 *For $m \geqslant 1$ and $n \geqslant m$ there exists a unique real analytic function*

$$\alpha_n : \ B_m^0 \to \mathbb{C}, \qquad \left| \alpha_n - n^2\pi^2 \right|_{B_m^0} \leqslant \frac{m^2}{4n},$$

such that $\lambda - n^2\pi^2 - a_n(\lambda, \cdot)\big|_{\lambda = \alpha_n} \equiv 0$ on B_m^0.

Given $q \in \mathcal{H}^0$ we replace its Fourier coefficients q_n for $|n|$ large enough by

$$p_n = c_n(\alpha_n(q), q) = q_n + \ldots \ .$$

The point is that

$$S_n(\alpha_n, q) = \begin{pmatrix} 0 & -p_n \\ -p_{-n} & 0 \end{pmatrix},$$

so these Fourier coefficients are well adapted to the lengths of the corresponding gaps. More precisely, we define a map $\mathscr{P}_m : B_m^0 \to \mathcal{H}^0$ by

$$\mathscr{P}_m(q) = \sum_{|n| < M_m} q_n e_{2n} + \sum_{|n| \geqslant M_m} c_n(\alpha_n(q), q) e_{2n},$$

where, say, $M_m = 2^{10} m^2$. This is a near identity map with the following properties.

Proposition 6 *For each $m \geq 1$ and every weight w, the restriction of $\mathscr{P}_m$ to B_m^w is a real analytic diffeomorphism*

$$\Phi_m\colon \quad B_m^w \to \Phi_m(B_m^w) \subset \mathcal{H}^w,$$

such that $\|D\Phi_m - I\|_{B_m^w} \leq 1/8$ *and*

$$2^{-1}\|q\|_w \leq \|\Phi_m(q)\|_w \leq 2\|q\|_w, \qquad q \in B_m^w.$$

By the last token, the image of B_m^w under $\mathscr{P}_m$ covers the ball $B_{m/2}^w$. Hence we have the following "abstract regularity result".

Proposition 7 *If $q \in B_m^0$ for some $m \geq 1$ and $\Phi_m(q) \in B_{m/2}^w$ for some weight w, then $q \in B_m^w \subset \mathcal{H}^w$.*

Thus we would like to argue as follows. Given $q \in \mathcal{H}^0$ with a certain asymptotic behaviour of its gap lengths γ_n, we know that

$$\gamma_n \asymp c_{|n|} \asymp p_{|n|}, \qquad n \gg 1.$$

Choosing $m \asymp \|q\|_0$ so that $q \in B_m^0$, we thus have $p = \mathscr{P}_m(q) \in \mathcal{H}^w$. *If we also had*

$$\|p\|_w \leq m/2, \qquad\qquad\qquad\qquad (*)$$

then $\mathscr{P}_m(q) \in B_{m/2}^w$ and thus

$$q = \mathscr{P}_m^{-1}(p) \in \mathcal{H}^w,$$

by the preceding proposition. Of course, given some $p \in \mathcal{H}^w$ there is no reason to have $\|p\|_w \leq m/2$. Simply increasing m does not help, since p depends on m.

5 Modified Weights

The idea is to *modify* the weight w in such a way that its asymptotics are preserved, while the norm $\|p\|_w$ is brought close to the norm $\|p\|_0$. For this to work flawlessly, however, we have to assume that w is *subexponential*.

So let $p \in \mathcal{H}^w$. Choosing m appropriately, we may assume that

$$0 < \|p\|_0 < m/6, \qquad \|p\|_w < \infty.$$

For $\varepsilon > 0$ we define a new function w_ε by

$$(w_\varepsilon)_n = \min(w_n, \varepsilon^{\varepsilon|n|}).$$

This is indeed a normalized, symmetric and submultiplicative function on $\mathscr{Z}$, hence a *weight*. Moreover, if w is subexponential, then clearly

$$(w_\varepsilon)_n = w_n, \qquad n \gg 1,$$

for any $\varepsilon > 0$.

   If we now choose first N sufficiently large, and then ε sufficiently small, we can arrange that

$$\|T_N p\|_{w_\varepsilon} \leqslant \|T_N p\|_w \leqslant \|p\|_0,$$
$$\|p - T_N p\|_{w_\varepsilon} \leqslant 2\|p\|_0,$$

for $T_N p = \sum_{|n| \geqslant N} p_n e_{2n}$. Altogether, we have

$$\|p\|_{w_\varepsilon} \leqslant 3\|p\|_0 \leqslant m/2.$$

According to Proposition 7 we thus have $q = \mathscr{P}_m^{-1}(p) \in \mathcal{H}^{w_\varepsilon}$. But since w_ε has the same asymptotics as w we indeed have

$$q = \mathscr{P}_m^{-1}(p) \in \mathcal{H}^{w_\varepsilon} = \mathcal{H}^w,$$

as we wanted to show.

   Essentially the same reasoning applies in the *exponential* case, with one important difference. If w is exponential, then

$$(w_\varepsilon)_n = \varepsilon^{\varepsilon|n|}, \qquad n \gg 1.$$

We thus may only conclude that

$$q = \mathscr{P}_m^{-1}(p) \in \mathcal{H}^{w_\varepsilon} = \mathcal{H}^{0,\varepsilon} \supsetneq \mathcal{H}^w.$$

So we conclude that q is real analytic, but its width of analyticity may be smaller than what the weight w may suggest.

References

1. P. DJAKOV & B. MITYAGIN, Smoothness of Schrödinger operator potential in the case of Gevrey type asymptotics of the gaps. *J. Funct. Anal.* **195** (2002), 89–128.
2. P. DJAKOV & B. MITYAGIN, Spectral triangles of Schrödinger operators with complex potentials. *Selecta Math. (N.S.)* **9** (2003), 495–528.
3. P. DJAKOV & B. MITYAGIN, Instability zones of one-dimensional periodic Schrödinger and Dirac operators. (Russian) *Uspekhi Mat. Nauk* **61** (2006), 77–182; translation in *Russian Math. Surveys* **61** (2006).
4. M. G. GASYMOV, Spectral analysis of a class of second order nonselfadjoint differential operators. *Funct. Anal. Appl.* **14** (1980), 14–19.
5. H. HOCHSTADT, Estimates on the stability interval's for the Hill's equation. *Proc. AMS* **14** (1963), 930–932.

6. T. Kappeler & B. Mityagin, Gap estimates of the spectrum of Hill's equation and action variables for KdV. *Trans. Amer. Math. Soc.* **351** (1999), 619–646.

7. T. Kappeler & B. Mityagin, Estimates for periodic and Dirichlet eigenvalues of the Schrödinger operator. *SIAM J. Math. Anal.* **33** (2001), 113–152.

8. V. A. Marčenko & I. O. Ostrowskĭ, A characterization of the spectrum of Hill's operator. *Math. USSR Sbornik* **97** (1975), 493–554.

9. G. Pólya & G. Szegö, *Problems and Theorems in Analysis I.* Springer, New York, 1976.

10. J. Pöschel, *Hill's potentials in weihtes Sobolev spaces and their spectral gaps.* Preprint 2004, www.poschel.de/pbl.html.

11. J. J. Sansuc & V. Tkachenko, Spectral properties of non-selfadjoint Hill's operators with smooth potentials. *Algebraic and Geometric Methods in Mathematical Physics (Kaciveli, 1993),* 371–385, Kluwer, 1996.

12. E. Trubowitz, The inverse problem for periodic potentials. *Comm. Pure Appl. Math.* **30** (1977), 321–342.

On the well-posedness of the periodic KdV equation in high regularity classes

Thomas Kappeler[1] and Jürgen Pöschel[2]

Abstract We prove well-posedness results for the initial value problem of the periodic KdV equation in classes of high regularity solutions. More precisely, we consider the problem in weighted Sobolev spaces, which comprise classical Sobolev spaces, Gevrey spaces, and analytic spaces. We show that the initial value problem is well posed in all spaces with subexponential growth of Fourier coefficients, and 'almost well posed' in spaces with exponential growth of Fourier coefficients.

1 Results

We consider the inital value problem for the periodic KdV equation,

$$u_t = -u_{xxx} + 6uu_x, \qquad u\big|_{t=0} = u_0, \tag{1}$$

where all functions are considered to be defined on $\mathbb{T} = \mathbb{R}/\mathscr{L}$.

According to one of the first results in this direction due to Bona & Smith [5] this problem has a unique, global solution for any initial value in one of the standard Sobolev space $\mathcal{H}^m = H^m(\mathbb{T}, \mathbb{R})$ with $m \geqslant 2$. That is, for each $u_0 \in \mathcal{H}^m$ there exists a unique continuous curve

$$\varphi: \quad \mathbb{R} \to \mathcal{H}^m, \quad t \mapsto \varphi(t, u_0)$$

solving the initial value in the sense defined below. Moreover, taken together they define a continous flow

$$\mathbb{R} \times \mathcal{H}^m \to \mathcal{H}^m, \quad (t, u_0) \mapsto \varphi(t, u_0).$$

[1] Institut für Mathematik, Universität Zürich, Winterthurerstrasse 190, CH-8057 Zürich
e-mail: tk@math.unizh.ch

[2] Institut für Analysis, Dynamik und Optimierung, Universität Stuttgart, Pfaffenwaldring 57
D-70569 Stuttgart
e-mail: poschel@mathematik.uni-stuttgart.de

W. Craig (ed.), Hamiltonian Dynamical Systems and Applications, 431–441.

Thus, the initial value problem is globally well-posed on $\mathcal{H}^m$ with $m \geqslant 2$ in the sense of Hadamard: solutions exist for all time, are unique, and depend continuously on their initial values.

Well-Posedness

Before we proceed we fix some notions. Let $\mathcal{H}^r = H^r(\mathbb{T}, \mathbb{R})$ be the usual Sobolev space of 1-periodic, real valued functions for real $r \geqslant 0$. A continuous curve $\varphi : I \to \mathcal{H}^r$ is called a *solution* of the initial value problem (1), if it solves (1) is the usual sense of distributions with $\varphi(0) = u_0$. It is called *global*, if $I = \mathbb{R}$.

We then say that the initial value problem (1) is *globally well-posed in* $\mathcal{H}^r$, if it has a global solution for each initial value in $\mathcal{H}^r$, and the resulting flow

$$\mathbb{R} \times \mathcal{H}^r \to \mathcal{H}^r, \quad (t,u) \mapsto \varphi(t,u)$$

is continuous. Moreover, we call (1) *globally uniformly well-posed in* $\mathcal{H}^r$, if it is globally well-posed, and for every compact interval I the map

$$\mathcal{H}^r \to C^0(I, \mathcal{H}^r), \quad u \mapsto \varphi(\cdot, u)$$

is uniformly continuous on bounded subsets of $\mathcal{H}^r$ with respect to the usual sup-norm on the second space. Well-posedness in the spaces $\mathcal{H}^w$ introduced later is defined analogously.

Known Results

Since the first results of Temam [31], Sjöberg [30] and Bona & Smith [5], the inital value problem for KdV and its well-posedness have been studied intensively. An excellent overview with a detailed bibliography is provided by the web site created by Colliander, Keel, Staffilani, Takaoka & Tao [11].

One focus has been on *low regularity solutions* in Sobolev spaces $\mathcal{H}^r$ with $r \leqslant 0$. We mention the works [6–10, 20–22]. As a result, KdV is now known to be globally well-posed in $\mathcal{H}^r$ for every $r \geqslant -1$, and globally uniformly well-posed in $\mathcal{H}^r$ for every $r \geqslant -1/2$. Incidentally, it is an interesting phenomenon, that an equation can be globally well-posed, but not in a uniform way.

In this paper we focus on *high regularity solutions*. These are solutions in a general class of weighted Sobolev spaces within $\mathcal{H}^0$, that encompass analytic and Gevrey spaces, among others. Some results in this direction on the *real line* can be found in [4, 14]. But in general, the question of existence and well-posedness of solutions of nonlinear pdes of high regularity have not been widely considered. We this that this topic deserves to be studied in more depth, revealing important features of the nonlinear equation considered.

Weighted Sobolev Spaces

To state our results we first introduce *weighted Sobolev spaces* within the standard space

$$\mathcal{H}^0 = L^2(\mathbb{T})$$

of square-integrable functions $u = \sum_{n \in \mathscr{Z}} u_n \varepsilon^{2\pi i n x}$. As in [16, 17] a *weight* is a *normalized, symmetric* and *submultiplicative* function $w \colon \mathscr{Z} \to \mathbb{R}$. That is, for all integers n and m, we have

$$w_n \geqslant 1, \qquad w_{-n} = w_n, \qquad w_{n+m} \leqslant w_n w_m.$$

We then define the *weighted Sobolev spaces*

$$\mathcal{H}^w := \{ u \in \mathcal{H}^0 : \|u\|_w^2 := \sum_{n \in \mathscr{Z}} w_n^2 |u_n|^2 < \infty \}.$$

To give some examples, let $\langle n \rangle = 1 + |n|$. The *Sobolev weights* $\langle n \rangle^r$, $r \geqslant 0$, give rise to the usual Sobolev spaces $\mathcal{H}^r$ of 1-periodic, complex-valued functions. In particular, for nonnegative integers m we obtain the standard spaces $\mathcal{H}^m$. The *Abel weights*[1] $\langle n \rangle^r \varepsilon^{a|n|}$ with $a > 0$ define spaces $\mathcal{H}^{r,a}$ of functions in $\mathcal{H}^r$, which are analytic on the complex strip $|\operatorname{Im} z| < a/2\pi$ and have traces in $\mathcal{H}^r$ on the boundary lines. The *Gevrey weights*

$$w_n = \langle n \rangle^r \varepsilon^{a|n|^\sigma}, \qquad r \geqslant 0, \, a > 0, \, 0 < \sigma < 1,$$

lie in between and give rise to the so called Gevrey spaces $\mathcal{H}^{r,a,\sigma}$ of smooth 1-periodic functions. Obviously,

$$\mathcal{H}^{r,a} = \mathcal{H}^{r,a,1} \subsetneq \mathcal{H}^{r,a,\sigma} \subsetneq \mathcal{H}^{r,a,0} = \mathcal{H}^r.$$

Since $\log w_n$ is subadditive and nonnegative, the limit

$$\chi(w) := \lim_{n \to \infty} \frac{\log w_n}{n}$$

exists and is nonnegative [29, no. 98]. Naturally, we call a weight w *exponential*, if $\chi(w) > 0$. We call w *subexponential*, if $\chi(w) = 0$ *with* $\log w_n / n$ converging to zero in an eventually monotone manner. This is not an exact dichotomy, but we are not aware of any interesting weight that does not belong to either class.

Theorems

Theorem 1 *The periodic KdV equation is globally uniformly well-posed in every space $\mathcal{H}^w$ with a subexponential weight w. That is, for each initial value u in one of*

---

[1] The term *Abel* weights is chosen to go along with *Sobolev* and *Gevrey* weights and has no deeper meaning.

these spaces $\mathcal{H}^w$ the associated Cauchy problem has a global solution $t \mapsto \varphi^t(u)$ in $\mathcal{H}^w$, giving rise to a continuous flow

$$\mathbb{R} \times \mathcal{H}^w \to \mathcal{H}^w, \quad (t,u) \mapsto \varphi^t(u),$$

which is even uniformly continuous on bounded subsets of $\mathcal{H}^w$.

Indeed, the flow map is even analytic, see also [3]. For exponential weights the result is not as clear cut.

Theorem 2 *The periodic KdV equation is "almost" globally well-posed in every space $\mathcal{H}^w$ with an exponential weight w. That is, for each bounded subset $\mathcal{B}$ of $\mathcal{H}^w$ there exists $0 < \rho \leqslant 1$ such that the Cauchy problem for each initial value $u \in \mathcal{B}$ has a global solution $t \mapsto \varphi^t(u)$ in $\mathcal{H}^{w^\rho}$, giving rise to a continuous flow*

$$\mathbb{R} \times \mathcal{B} \to \mathcal{H}^{w^\rho}, \quad (t,u) \mapsto \varphi^t(u).$$

Here, w^ρ is the weight with $(w^\rho)_n = w_n^\rho$, which is again normalized, symmetric and submultiplicative. Thus, for initial values u in a bounded subset $\mathcal{B}$ of $\mathcal{H}^{0,a}$, say, (1) has a global solution in $\mathcal{H}^{0,\rho a}$ with a fixed $0 < \rho \leqslant 1$. It is an open question, whether ρ can be chosen to be 1. For related results, see for example [1].

These results are not restricted to the standard KdV equation, but apply simultaneously to all equations in the KdV hierarchy, as defined for instance in [18]. The second KdV equation, for example, reads

$$u_t = u_{xxxxx} - 10uu_{xxx} - 20u_x u_{xx} + 30u^2 u_x.$$

Such a hierarchy may be defined in a variety of ways, but this is immaterial here and does not affect the statement of the following theorem.

Theorem 3 *Theorems 1 and 2 also hold for every KdV equation in the KdV hierarchy, provided that in the case of Sobolev spaces $\mathcal{H}^r$, r is sufficiently large.*

Our results naturally extend the KAM theory of Hamiltonian perturbations of KdV equations developed by Kuksin [23–25] and expounded in [18, 26]. Consider the perturbed KdV equation

$$\frac{\partial u}{\partial t} = \frac{\mathrm{d}}{\mathrm{d}x}\left(\frac{\partial H}{\partial u} + \varepsilon \frac{\partial K}{\partial u}\right).$$

If K is real analytic in u with a gradient $\partial K/\partial u$ in some standard Sobolev space $\mathcal{H}^m$, $m \geqslant 1$, then KAM for KdV asserts the persistence of *quasi-periodic* solutions for sufficiently small $\varepsilon \neq 0$. Theorems 1 and 2 may now be extended as follows – for a more precise statement we refer to [19].

Theorem 4 *Under sufficiently small Hamiltonian perturbations, the majority of the quasi-periodic solutions of the KdV equation persists, their regularity being only slightly less than the regularity of the perturbing term.*

These theorems are based on two observations. First, the periodic KdV equation is well known to be an *infinite dimensional, integrable Hamiltonian system*. As such, it even admits global Birkhoff coordinates $(x_n, y_n)_{n \geq 1}$ defined as the cartesian counterpart to global action angle coordinates $(I_n, \theta_n)_{n \geq 1}$. Second, there is a precise correspondence between the decay properties of the coordinates $(x_n, y_n)_{n \geq 1}$ and the regularity properties of u. The link is provided by the spectral properties of the associated Hill operator

$$L_u = -\frac{\mathrm{d}^2}{\mathrm{d}x^2} + u$$

on the interval $[0, 2]$ with periodic boundary conditions.

In the rest of this note we describe this approach in more detail, but without lengthy proofs. These are given in [19].

2 Birkhoff Coordinates

As is well known, the KdV equation can be written as an infinite dimensional Hamiltonian system

$$\frac{\partial u}{\partial t} = \frac{\mathrm{d}}{\mathrm{d}x}\frac{\partial H}{\partial u}$$

with Hamiltonian

$$H(u) = \int_{\mathbb{T}} (\tfrac{1}{2}u_x^2 + u^3)\,\mathrm{d}x.$$

As a phase space one may take

$$\mathcal{H}_0^m = \{u \in \mathcal{H}^w : [u] := \textstyle\int_{\mathbb{T}} u\,\mathrm{d}x = 0\}$$

with $m \geq 1$, as the KdV flow preserves mean values. The Poisson bracket proposed by Gardner,

$$\{F, G\} = \int_{\mathbb{T}} \frac{\partial F}{\partial u}\frac{\mathrm{d}}{\mathrm{d}x}\frac{\partial G}{\partial u}\,\mathrm{d}x,$$

then makes $\mathcal{H}_0^m$ into a nondegenerate Poisson manifold, such that $u_t = \{u, H\}$.

Next, we introduce the weighted sequence spaces

$$h^w = \ell^w \times \ell^w$$

with elements (x, y), where

$$\ell^w = \{x = (x_n)_{n \geq 1} : \|x\|_w^2 = \textstyle\sum_{n \geq 1} w_n^2 |x_n|^2 < \infty\}.$$

We endow h^w with the standard Poisson structure, for which $\{x_n, y_m\} = \delta_{nm}$, while all other brackets vanish. To simplify notations, we further introduce

$$h_\star^w = \ell_\star^w \times \ell_\star^w, \qquad \ell_\star^w = \{x \in \ell^w : (\sqrt{n}x_n)_{n \geq 1} \in \ell^w\}.$$

The extra weight $\sqrt{n}$ reflects the effect of the derivative d/dx in the Gardner bracket.

The following theorem was first proven in [2, 3]. A quite different approach was first presented in [15], and a comprehensive exposition is given in [18]. Note that $\mathcal{H}_0^0 = \{ u \in L^2(\mathbb{T}) : [u] = 0 \}$.

Theorem 5 *There exists a diffeomorphism*

$$\Omega : \quad \mathcal{H}_0^0 \to h_\star^0$$

with the following properties.

(i) *Ω is onto, bi-analytic, and takes the standard Poisson bracket into the Gardner bracket.*

(ii) *The restriction of Ω to $\mathcal{H}_0^m$, $m \geq 1$, gives rise to a map $\Omega : \mathcal{H}_0^m \to h_\star^m$, which is again onto and bi-analytic.*

(iii) *Ω introduces global Birkhoff coordinates for the KdV Hamiltonian on $\mathcal{H}_0^1$. That is, on $h_\star^1$ the transformed KdV Hamiltonian $H \circ \Omega^{-1}$ is a real analytic function of*

$$I_n = \frac{1}{2}(x_n^2 + y_n^2), \qquad n \geq 1.$$

(iv) *The last statement also applies to every other Hamiltonian in the KdV hierarchy, if '1' is replaced by 'm' with m sufficiently large.*

Denoting the transformed KdV Hamiltonian by the same symbol we thus obtain a real analytic Hamiltonian

$$H = H(I_1, I_2, \ldots)$$

on $h_\star^1$. Its equations of motion are the classical ones,

$$\dot{x}_n = H_{y_n}, \quad \dot{y}_n = -H_{x_n}, \qquad n \geq 1,$$

since the Poisson structure on $h_\star^1$ is the standard one. It is therefore evident that every solution of the KdV equation exists for all time, and is indeed *almost periodic*. More precisely, every solution winds around some underlying invariant torus

$$T_I = \prod_{n \geq 1} S_{I_n}, \qquad S_{I_n} = \{ x_n^2 + y_n^2 = 2I_n \},$$

which is fixed by the actions of the initial position. The speed on the n-th circle S_{I_n} is determined by the n-th frequency

$$\omega_n = H_{I_n}(I_1, I_2, \ldots),$$

and the entire flow is given by

$$\psi^t(x, y) = (x_n \cos \omega_n t, y_n \sin \omega_n t)_{n \geq 1}.$$

Obviously, ψ^t preserves all weighted norms and thus all weighted spaces $h_\star^w$.

To obtains our results about the well-posedness of the KdV equation, we now formulate two extensions of Theorem 5. First we consider subexponential weights.

Theorem 6 *For each subexponential weight w, the restriction of Ω to $\mathcal{H}_0^w$ gives rise to an onto, bi-analytic diffeomorphism $\Omega : \mathcal{H}_0^w \to h_\star^w$.*

Proof (Proof of Theorem 1). Due to its symplectic nature, Ω maps solution curves $t \mapsto \varphi^t(u)$ in function space into solution curves $t \mapsto \psi^t(x,y)$ in sequence space with $(x,y) = \Omega(u)$. Since Ω is also a diffeomorphism between $\mathcal{H}_0^w$ and $h_\star^w$ and ψ^t preserves $h_\star^w$, the diagram

$$
\begin{array}{ccc}
u \in \mathcal{H}_0^w & \xrightarrow{\ \Omega\ } & (x,y) \in h_\star^w \\[4pt]
\varphi^t \Big\downarrow & & \Big\downarrow \psi^t \\[4pt]
\varphi^t(u) \in \mathcal{H}_0^w & \xleftarrow{\ \Omega^{-1}\ } & \psi^t(x,y) \in h_\star^w
\end{array}
$$

is correct and proves the theorem.

Now we consider exponential weights. Here, the result is not as elegant.

Theorem 7 *Let w be an exponential weight. Then for every bounded subset B of $h_\star^w$ there exists $0 < \rho \leqslant 1$ such that $\Omega^{-1}(B) \subset \mathcal{H}_0^{w^\rho}$.*

Proof (Proof of Theorem 2). Let w be an exponential weight and $\mathcal{B}$ a bounded subset of $\mathcal{H}_0^w$. Then $B = \Omega(\mathcal{B})$ is a bounded subset of $h_\star^w$ by Proposition 8 below. As the flow ψ^t preserves the $h_\star^w$-norm, the set

$$
B^- = \bigcup_{t \in \mathbb{R}} \psi^t(B)
$$

is contained in the same centered ball as B. Hence, by the previous theorem there exists a $0 < \rho \leqslant 1$ such that $\mathcal{B}^- = \Omega^{-1}(B^-)$ is contained in $\mathcal{H}_0^{w^\rho}$. We obtain the commutative diagramm

$$
\begin{array}{ccc}
\mathcal{B} \subset \mathcal{H}_0^w & \xrightarrow{\ \Omega\ } & B \subset h_\star^w \\[4pt]
\varphi^t \Big\downarrow & & \Big\downarrow \psi^t \\[4pt]
\mathcal{B}^- \subset \mathcal{H}_0^{w^\rho} & \xleftarrow{\ \Omega^{-1}\ } & B^- \subset h_\star^w
\end{array}
$$

which proves the theorem.

Proof (Proof of Theorem 3). The proofs of Theorem 1 and 2 are based on the fact that the map Ω triviales the KdV flow in the Birkhoff coordinates. By item (iv) of Theorem 5, however, Ω simultaneously trivializes any other KdV flow in the KdV hierarchy. The only difference is in the frequencies ω_n associated with the circles S_{I_n}, and in the minimal regularity required for the KdV hamiltonians to make sense. Hence the preceding proofs apply to higher KdV equations as well.

3 Regularity

Theorems 6 and 7 are based on two observations. First, the asymptotics of the Birkhoff coordinates of a function u in $\mathcal{H}_0^0$ are closely related to the asymptotics of its spectral gaps. Second, these asymptotics are very closely related to the regularity of u. To keep the discussion simple, we restrict ourselves to the real case.

Spectral Gaps and Actions

For a *potential* $u \in L_0^2 = \mathcal{H}_0^0$ consider Hill's operator

$$L_u = -\frac{\mathrm{d}^2}{\mathrm{d}x^2} + u$$

on the interval $[0, 2]$ with periodic boundary conditions. As is well known, its spectrum is pure point and consists of an unbounded sequence of real eigenvalues

$$\lambda_0(u) < \lambda_1^-(u) \leqslant \lambda_1^+(u) < \lambda_2^-(u) \leqslant \cdots .$$

Its so called *spectral gaps* are the – possibly empty – intervals $(\lambda_n^-(u), \lambda_n^+(u))$, and one speaks of the *gap lengths* $\gamma_n(u) = \lambda_{2n}(u) - \lambda_{2n-1}(u)$ of u.

Proposition 8 ([18, p. 67]) *There exists a complex neighbourhood W of L_0^2 such that each quotient I_n/γ_n^2 extends analytically to W and satisfies*

$$8\pi n \frac{I_n}{\gamma_n^2} = 1 + O\left(\frac{\log n}{n}\right), \qquad n \geqslant 1,$$

locally uniformly on W, as well as uniformly on bounded subsets of L_0^2.

So we have

$$n(x_n^2 + y_n^2) \sim n I_n \sim \gamma_n^2$$

locally uniformly on W. This gives us control of $x_n^2 + y_n^2$ in terms of γ_n^2 on the *real* space L_0^2, where all quantities are *real*. This is not the case on the *complex* neighbourhood W, where a gap γ_n and thus an action I_n may vanish, while the Birkhoff coordinates x_n, y_n do *not*. See [19] for the details of this case.

Spectral Gaps and Regularity

The decay properties of spectral gaps are also closely tied to the regularity of the potential. For example, a classical result due to Marčenko & Ostrowskiĭ [27] states that

$$u \in \mathcal{H}^m \quad \Leftrightarrow \quad \sum_{n \geqslant 1} n^{2m} \gamma_n^{2m}(u) < \infty$$

for any integer $m \geqslant 0$. The forward part of this result generalizes as follows.

Theorem 9 *For any subexponential or exponential weight w,*

$$u \in \mathcal{H}^w \quad \Rightarrow \quad (\gamma_n(u)) \in h^w.$$

Consequently, for any subexponential or exponential weight w, the Birkhoff map Ω maps $\mathcal{H}_0^w$ into $h_\star^w$. Indeed,

$$u \in \mathcal{H}^w \quad \Rightarrow \quad (\gamma_n) \in h^w \quad \Rightarrow \quad (nI_n) \in h^w \quad \Rightarrow \quad (x_n, y_n) \in h_\star^w.$$

by Theorem 9 and Proposition 8.

A one-to-one converse to this theorem is only true in the subexponential case.

Theorem 10 ([12, 28]) *For a subexponential weight w,*

$$(\gamma_n(u)) \in h^w \quad \Rightarrow \quad u \in \mathcal{H}^w.$$

Consequently, in the subexponential case, the Birkhoff map Ω also maps $\mathcal{H}_0^w$ onto $h_\star^w$. Indeed,

$$(x,y) \in h_\star^w \subset h_\star^0$$
$$\Rightarrow \quad u = \Omega^{-1}(x,y) \in \mathcal{H}_0^0 \quad \text{with} \quad \gamma_n^2 \sim n(x_n^2 + y_n^2)$$
$$\Rightarrow \quad (\gamma_n(u)) \in h^w$$
$$\Rightarrow \quad u \in \mathcal{H}^w.$$

Altogether, Ω is a diffeomorphism between $\mathcal{H}_0^w$ and $h_\star^w$ whenever w is a subexponential weight. Thus, Theorem 6 is proven.

The last theorem does not extend to exponential weights, however. This is exemplified by finite gap potentials such as the Weierstrass $\wp$-function, which are not entire functions. Gasymov [13] even observed that *any* complex potential of the form

$$u = \sum_{n \geqslant 1} u_n \varepsilon^{2\pi i n x} = \sum_{n \geqslant 1} u_n z^n \Big|_{z = \varepsilon^{2\pi i x}}$$

is a 0-gap-potential. So in the complex case, the gap sequence need not contain *any* information about the regularity of the potential.

In the real case, however, we have the following classical result by Trubowitz. The very last statement is proven in [28].

Theorem 11 ([32]) *For an exponential weight w,*

$$(\gamma_n(u)) \in h^w \quad \Rightarrow \quad u \in \mathcal{H}^{w^\rho},$$

where $0 < \rho \leqslant 1$ depends on $\|u\|_{L^2}$ and $\sum_n w_n^2 \gamma_n^2$.

Consequently, for any bounded subset B of $h_\star^w$ there exists $0 < \rho \leqslant 1$ so that

$$\Omega^{-1}(B) \subset \mathcal{H}_0^{w^\rho}.$$

Indeed, $A = \Omega^{-1}(B)$ is bounded in L_0^2, and by Proposition 8,

$$\sum_{n \geq 1} w_n^2 |\gamma_n^2(u)| \leq c \sum_{n \geq 1} n w_n^2 (|x_n^2(u)| + |y_n^2(u)|)$$

uniformly on A. The latter sum is uniformly bounded by assumption, so by Theorem 11 we have $A \subset \mathcal{H}_0^{w^\rho}$ for some $0 < \rho \leq 1$. This establishes Theorem 7.

References

1. H. AIRAULT, H. MCKEAN & J. MOSER, Rational and elliptic solutions of the Kortweg-de Vries equation and a related many-body problem. *Comm. Pure Appl. Math* **30** (1977), 95–148.
2. D. BÄTTIG, A. M. BLOCH, J.-C. GUILLOT & T. KAPPELER, On the symplectic structure of the phase space for periodic KdV, Toda, and defocusing NLS. *Duke Math. J.* **79** (1995), 549–604.
3. D. BÄTTIG, T. KAPPELER & B. MITYAGIN, On the Korteweg-deVries equation: Frequencies and initial value problem. *Pacific J. Math* **181** (1997), 1–55.
4. J. BONA, Z. GRUJIĆ & H. KALISCH, Algebraic lower bounds for the uniform radius of spatial analyticity for the generalized KdV equation. *Ann. Inst. H. Poincaré Anal. Non Linéaire* **22** (2005), 783–797.
5. J. L. BONA & R. SMITH, The initial-value problem for the Korteweg-de Vries equation. *Philos. Trans. Roy. Soc. London Ser. A* **278** (1975), 555-601.
6. J. BOURGAIN, Fourier transform restriction phenomena for certain lattice subsets and applications to nonlinear evolution equations, II: The KdV-equation. *Geom. Funct. Anal.* **3** (1993), 209–262.
7. J. BOURGAIN, On the Cauchy problem for periodic KdV-type equations. In: Conference in Honor of Jean-Pierre Kahan (Orsay, France, 1993). *J. Fourier Anal. Appl. Special Issue* (1995), 17–86.
8. J. BOURGAIN, Periodic Korteweg-de Vries equation with measures as initial data. *Selecta Math. (N.S.)* **3** (1997), 115–159.
9. J. BOURGAIN, *Global Solutions of Nonlinear Schrödinger Equations.* AMS Colloq. Publ., American Mathematical Society, Providence, 1999.
10. J. COLLIANDER, M. KEEL, G. STAFFILANI, H. TAKAOKA & T. TAO, Sharp global well-posedness for KdV and modified KdV on $\mathbb{R}$ and $\mathbb{T}$. *J. Amer. Math. Soc.* **16** (2003), 705–749.
11. J. COLLIANDER, M. KEEL, G. STAFFILANI, H. TAKAOKA & T. TAO, Local and global well-posedness for non-linear dispersive and wave equations. `www.math.ucla.edu/~tao/Dispersive.`
12. P. DJAKOV & B. MITYAGIN, Smoothness of Schrödinger operator potential in the case of Gevrey type asymptotics of the gaps. *J. Funct. Anal.* **195** (2002), 89–128.
13. M. G. GASYMOV, Spectral analysis of a class of second order nonselfadjoint differential operators. *Funct. Anal. Appl.* **14** (1980), 14–19.
14. Z. GRUJIĆ & H. KALISCH, Local well-posedness of the generalized Kortweg-de Vries equation in spaces of analytic functions. *Differential Integral Equations* **15** (2002), 1325–1334.
15. T. KAPPELER & M. MAKAROV, On the Birkhoff coordinates for KdV. *Ann. Henri Poincaré* **2** (2001), 807–856.
16. T. KAPPELER & B. MITYAGIN, Gap estimates of the spectrum of Hill's equation and action variables for KdV. *Trans. Amer. Math. Soc.* **351** (1999), 619–646.
17. T. KAPPELER & B. MITYAGIN, Estimates for periodic and Dirichlet eigenvalues of the Schrödinger operator. *SIAM J. Math. Anal.* **33** (2001), 113–152.
18. T. KAPPELER & J. PÖSCHEL, *KdV & KAM.* Springer, Berlin, 2003.

19. T. KAPPELER & J. PÖSCHEL, On the periodic KdV equation in weighted Sobolev spaces. Preprint 2007.

20. T. KAPPELER & P. TOPALOV, Global wellposedness of KdV in $H^{-1}(\mathbb{T}, \mathbb{R})$. *Duke Math. J.* **135** (2006), 327–360.

21. C. E. KENIG, G. PONCE & L. VEGA, On the Cauchy problem for the Korteweg-de Vries equation in Sobolev spaces of negative indices. *Duke Math. J.* **71** (1993), 1–20.

22. C. E. KENIG, G. PONCE & L. VEGA, A bilinear estimate with applications to the KdV equation. *J. Amer. Math. Soc.* **9** (1996), 573–603.

23. S. B. KUKSIN, Perturbation theory for quasiperiodic solutions of infinite-dimensional Hamiltonian systems, and its application to the Korteweg-de Vries equation. *Matem. Sbornik* **136** (1988) [Russian]. English translation in *Math. USSR Sbornik* **64** (1989), 397–413.

24. S. B. KUKSIN, *Nearly integrable infinite-dimensional Hamiltonian systems.* Lecture Notes in Mathematics 1556, Springer, 1993.

25. S. B. KUKSIN, A KAM-theorem for equations of the Korteweg-deVries type. *Rev. Math. Phys.* **10** (1998), 1–64.

26. S. B. KUKSIN, *Analysis of Hamiltonian PDEs.* Oxford University Press, Oxford, 2000.

27. V. A. MARČENKO & I. O. OSTROWSKIĬ, A characterization of the spectrum of Hill's operator. *Math. USSR Sbornik* **97** (1975), 493–554.

28. J. PÖSCHEL, Hill's potential in weighted Sobolev spaces and their spectral gaps. Preprint, `www.poschel.de/pbl`, 2004.

29. G. PÓLYA & G. SZEGÖ, *Problems and Theorems in Analysis I.* Springer, New York, 1976.

30. A. SJÓBERG, On the Korteweg-de Vries equation: Existence and uniqueness. *J. Math. Anal. Appl.* **29** (1970), 569–579.

31. R. TEMAM, Sur un probléme non linéaire. *J. Math. Pures Appl.* **48** (1969), 159–172.

32. E. TRUBOWITZ, The inverse problem for periodic potentials. *Comm. Pure Appl. Math.* **30** (1977), 321–342.